MW00574924

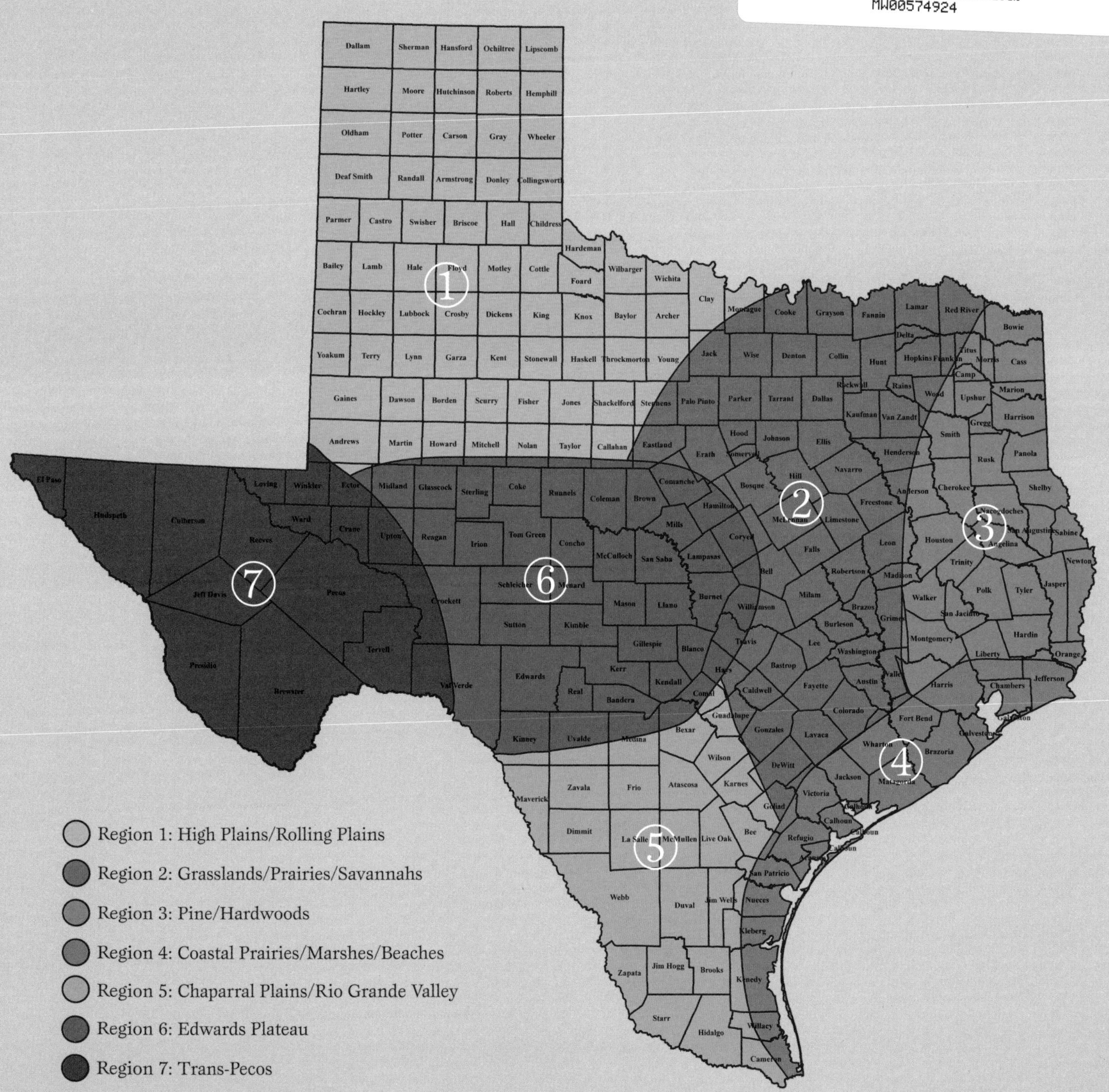
Dallam
Sherman
Hansford
Ochiltree
Lipscomb
Hartley
Moore
Hutchinson
Roberts
Hemphill
Oldham
Potter
Carson
Gray
Wheeler
Deaf Smith
Randall
Armstrong
Donley
Collingsworth
Parmer
Castro
Swisher
Briscoe
Hall
Childress
Hardeman
Bailey
Lamb
Hale
Floyd
Motley
Cottle
Foard
Wilbarger
Wichita
Clay
Montague
Cooke
Grayson
Fannin
Lamar
Red River
Bowie
Cochran
Hockley
Lubbock
Crosby
Dickens
King
Knox
Baylor
Archer
Yoakum
Terry
Lynn
Garza
Kent
Stonewall
Haskell
Throckmorton
Young
Jack
Wise
Denton
Collin
Hunt
Delta
Hopkins
Franklin
Titus
Morris
Cass
Camp
Gaines
Dawson
Borden
Scurry
Fisher
Jones
Shackelford
Stephens
Palo Pinto
Parker
Tarrant
Dallas
Rockwall
Rains
Wood
Upshur
Marion
Kaufman
Van Zandt
Gregg
Harrison
Andrews
Martin
Howard
Mitchell
Nolan
Taylor
Callahan
Eastland
Hood
Johnson
Ellis
Erath
Smith
Henderson
Rusk
Panola
El Paso
Loving
Winkler
Ector
Midland
Glasscock
Sterling
Coke
Runnels
Coleman
Brown
Comanche
Hill
Navarro
Bosque
Anderson
Cherokee
Shelby
Hudspeth
Culberson
Ward
Crane
Upton
Reagan
Irion
Tom Green
Concho
Hamilton
Mills
McLennan
Limestone
Freestone
Nacogdoches
Sabine
Reeves
McCulloch
San Saba
Lampasas
Coryell
Falls
Leon
Houston
Trinity
Angelina
Newton
Jeff Davis
Pecos
Crockett
Schleicher
Menard
Mason
Llano
Burnet
Bell
Robertson
Madison
Polk
Tyler
Jasper
Walker
San Jacinto
Milam
Williamson
Brazos
Grimes
Burleson
Sutton
Kimble
Gillespie
Blanco
Travis
Lee
Washington
Montgomery
Liberty
Hardin
Orange
Presidio
Brewster
Terrell
Val Verde
Edwards
Kerr
Real
Bandera
Kendall
Comal
Hays
Bastrop
Caldwell
Fayette
Austin
Waller
Harris
Chambers
Jefferson
Guadalupe
Colorado
Fort Bend
Galveston
Kinney
Uvalde
Medina
Bexar
Gonzales
Lavaca
Wharton
Brazoria
Wilson
DeWitt
Zavala
Frio
Atascosa
Karnes
Jackson
Matagorda
Maverick
Victoria
Goliad
Calhoun
Dimmit
La Salle
McMullen
Live Oak
Bee
Refugio
Aransas
San Patricio
Webb
Duval
Jim Wells
Nueces
Kleberg
Zapata
Jim Hogg
Brooks
Kenedy
Starr
Hidalgo
Willacy
Cameron
1
2
3
4
5
6
7
Region 1: High Plains/Rolling Plains
Region 2: Grasslands/Prairies/Savannahs
Region 3: Pine/Hardwoods
Region 4: Coastal Prairies/Marshes/Beaches
Region 5: Chaparral Plains/Rio Grande Valley
Region 6: Edwards Plateau
Region 7: Trans-Pecos

Butterfly Gardening for Texas

Palamedes Swallowtail
(Papilio palamedes)

Number Forty-six
Louise Lindsey Merrick Natural Environment Series

Butterfly Gardening for Texas

GEYATA AJILVSGI

TEXAS A&M UNIVERSITY PRESS *College Station*

(previous page)
PIPEVINE SWALLOWTAIL
(Battus philenor)

Printed in China by Everbest Printing Co.,
through FCI Print Group

First edition

This paper meets the requirements
of ANSI/NISO Z39.48-1992
(Permanence of Paper).
Binding materials have
been chosen for durability.

LIBRARY OF CONGRESS
Cataloging-in-Publication Data

Ajilvsgi, Geyata
Butterfly gardening for Texas / Geyata Ajilvsgi. —
1st ed.
p. cm. — (Louise Lindsey Merrick natural environment series ; no. 46)
Includes bibliographical references and index.
ISBN-13: 978-1-60344-806-2 (flex : alk. paper)
ISBN-10: 1-60344-806-3 (flex : alk. paper)
ISBN-13: 978-1-60344-957-1 (e-book)
ISBN-10: 1-60344-957-4 (e-book)
1. Butterfly gardening—Texas. I. Title. II. Series:
Louise Lindsey Merrick natural environment series ;
no. 46.
QL544.6.A35 2013
638'.578909764—dc23
2012031589

JANAIS PATCH
(Chlosyne janais)
LARVA

All photographs for the book were taken by the author, using 35 mm single-lens reflex (SLR) Nikon FA and Nikon 8008 camera bodies, Fujichrome 100 or Ektachrome 100 Professional film, and 35 mm, 50 mm, 55 mm, and 200 mm macro lenses with various close-up rings, and with a Canon EOS 40D digital camera and a 180 mm macro lens with various extensions. Supplementary flash was often used to stop wind motion or the flutter of a butterfly's wing.

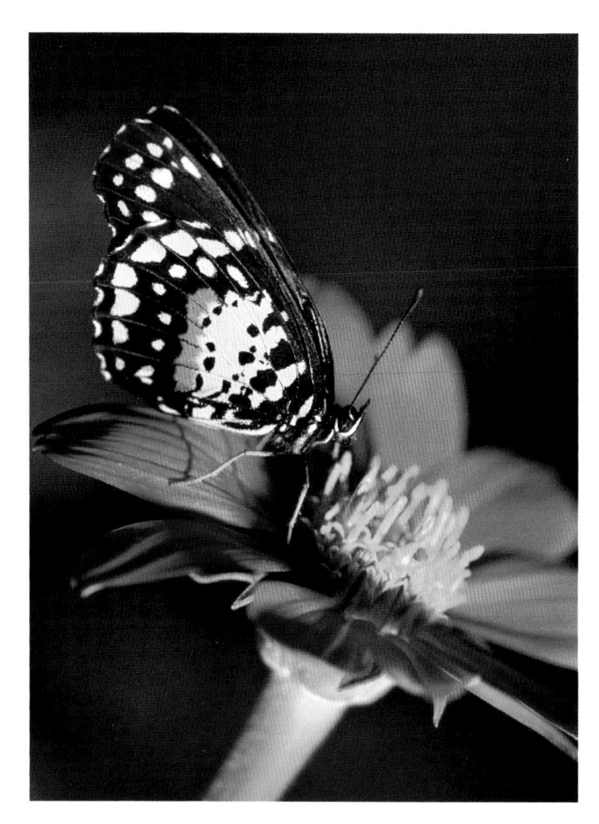

Janais Patch
(Chlosyne janais)

The purpose of butterflies . . . will not be found . . . in the few flowers that they may inadvertently pollinate . . . nor in the numbers of parasitic wasps they may support . . . and to peer beneath a microscope at their dissected fragments will in no way . . . elucidate the reason for their being. . . . their purpose is their BEAUTY. . . . and the beauty they bring into the lives of those of us who have stopped long enough from the cares of the world to LISTEN to their fascinating story . . .

—William H. Howe, *On Butterflies and Moths*

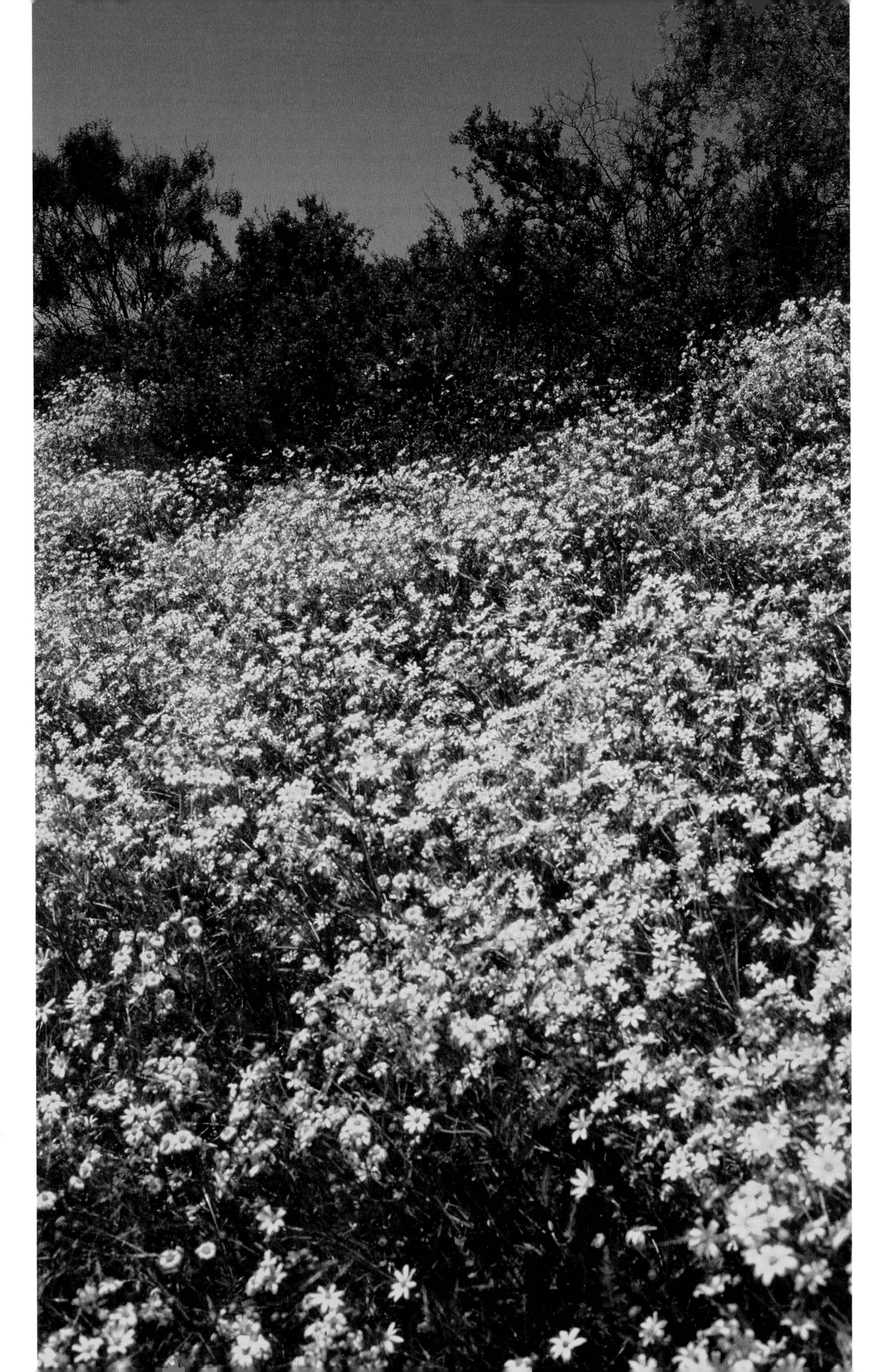

Field of fall blooming flowers attracts butterflies.

Contents

THEONA
CHECKERSPOT
(Chlosyne theona)

Acknowledgments

This book was originally written just for Texas but was expanded at the request of the publisher to include the southeastern states and titled *Butterfly Gardening for the South.* In this Texas A&M University Press edition, it has been brought back to its original focus with much updating, more information and corrections, and many new additional photographs.

Since the original publication, butterfly gardeners have taken wing with astonishing enthusiasm and with no indication of slowing down.

It has been a real joy reshaping this book—to talk to numerous gardeners who are now slanting their gardens with the purpose of attracting butterflies and to visit nurseries and see the advertised "Butterfly Plants." Some nurseries even have their own demonstration butterfly gardens—a great help to see plants growing beautifully in a garden setting and being used by butterflies in their various stages.

And as always, there are those who have helped make this book revision happen. Especially appreciated are my friends and neighbors Boyd Merworth and Wes Alexander for their frequent help with computer glitches and much-needed words of encouragement. A special thanks to my friend Elizabeth Hobson Cannedy for the generous loan of eggs and caterpillars to photograph and to Nick Grishin and Dale Clark for identification of butterfly slides. Much appreciation to artist Rose Baxter for the butterfly and landscape illustrations, and to Cynthia Lindlof, my editor, for making me look much better on the printed page.

During the preparation of the original text, much of which remains intact here, I consulted many books, magazine articles, and other sources on several subjects dealing with butterflies as well as wildflowers, flower gardens, and gardening in general. Lists of many of these sources appear in the appendix.

The original dream of this book would never have become reality without the personal help, advice, and enthusiastic support of outstanding experts in the fields of lepidopterology, botany, and gardening. Some of these were good friends from the beginning; some have become good friends in the making of this book. I have been privileged to meet and exchange information with many others. I am grateful and heartfelt thanks go to all those friends and acquaintences.

In the butterfly world I must give special heartfelt thanks to my dear friends, the late Roy and Connie Kendall, and to Timothy Friedlander for their friendship, hospitality, sharing of field trips, and inestimable sources of information. Appreciation also goes to Christopher J. Durden for much helpful advice and to Samuel A. Johnson, Kevin MacDonnel, and Gregory S. Forbes (Las Cruces, New Mexico) for sharing scrupulously compiled field notes on both the butterflies and their nectar sources.

I also wish to thank Roger Peace for sharing his firsthand experiences in the growing of various species of *Aristolochia*; Tim Friedlander for help with the regional map and for reviewing the butterfly descriptions; and Herbert K. Durand, Burr Williams, Benny Simpson, John Fairey, and Carl M. Schoenfeld for critiquing portions of the plant descriptions. Mike Rose was most helpful with the extreme close-up shots of wing scales, and Leo Mieier in "inventing" special photographic equipment for me. Stephen Myers and Paul Montgomery were of tremendous help in reviewing the photography section and generously shared suggestions, techniques, and their own experiences. I am especially grateful to Raymond W. Neck for a most thorough review of the entire manuscript as well as much help with ranges, scientific names, and answers to a multitude of questions. Thanks also to my friend Martha Bell, who provided the drawings for and reviewed the original manuscript. William F. Mahler and Barney Lipscomb at the Southern Methodist University Herbarium were most gracious and generous, as always, with help in plant identification. For help in locating gardens and needed information, I extend my appreciation to Burr Williams, Sheryl McLaughlin, Martha Henshen, and Sally Wasowski, and to Patty Leslie and Paul Cox at the San Antonio Botanical Center, Doug Williams at the Houston Arboretum, and John Koros at the Mercer Arboretum. Much appreciation goes to the Antique Rose Emporium at Brenham for opening their grounds and giving me free rein to photograph. And thanks to John Thomas of Wildseed Inc. for enough wildflower seeds to cover my entire twenty acres in Robertson County—even though it was mostly wooded. To my friend John Meeks a special thanks for the vanloads of plants he gave me to try in the butterfly garden, for gardening advice, and for the spiritual connection to keep me grounded to Mother Earth—and to my desk. I am grateful to James H. Yantis, who was a source of advice during the early writing of this book.

To those of you, too numerous to mention by name, who opened your gardens to me for photographing and took the time to answer lengthy and probing correspondence and telephone calls, I offer my deepest gratitude. With your contributions and sharing, this book is surely more worthwhile than anything I ever could have accomplished alone.

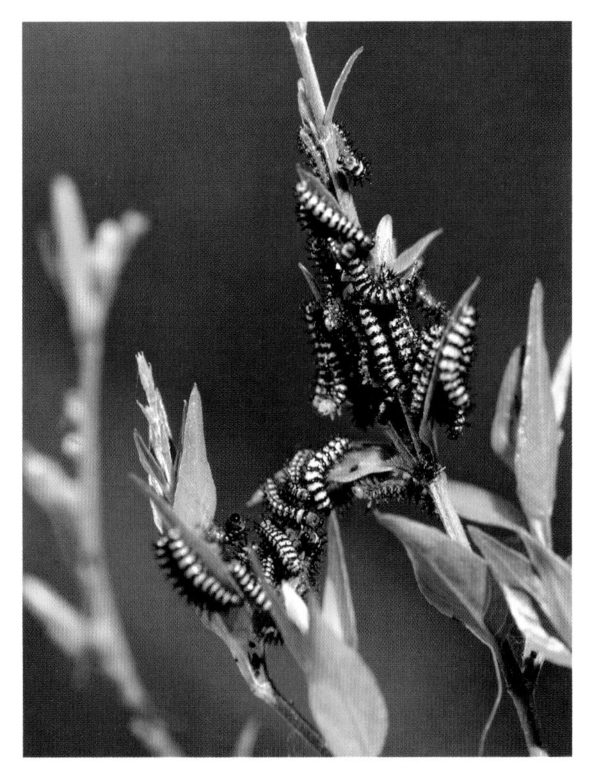

JANAIS PATCH *(Chlosyne janais)* CATERPILLARS. BUTTERFLY LARVAE CONSUME FOOD AT AN AMAZING RATE AND INCREASE THEIR INITIAL WEIGHT BY AS MUCH AS ONE THOUSAND TIMES.

Circle of Life

For Native Tribal peoples, the circle is singularly the most important symbol of our culture. For us it represents all of life with no beginning and no end, portraying the roundness of the All. Within our world, everything is of equal importance—animals, rocks, plants, the waters, the sky above, and the earth below. Within this circle of life, each being has its place and its importance. And so, as with a tree or a human, butterflies have their importance in the whole, the roundness of things, and their place to be on the Earth Mother. With this book, I share with you a small portion of the fascinating world of the butterflies, which forms one link in the roundness. I have tried to show their beauty and a few facets of their life through photographs, and I have offered suggestions on how you can bring them closer for personal enjoyment by providing their simple needs.

This book, then, is my "giveaway" to all of you who, in attracting butterflies to your gardens, come to know more of their ways and, in the knowing, recognize their place along with our own in the circle of life. And it is a gift to my people, the Cherokee, who honor the butterfly, *kamama*, in their daily lives as they honor and respect all things in the natural world.

GIANT SWALLOWTAIL
(Papilio cresphontes)

Palamedes
Swallowtail
(Papilio palamedes)

Introduction

The state of Texas encompasses numerous and varied natural land regions—some of the most diverse in North America. In its wide expanse of almost 270,000 square miles of land, many climatic zones, geologic provinces, and botanical realms come together, creating especially rich and interesting biotic communities. Forming the southern end of a grassland system extending from Canada, the high, arid plains of the Panhandle continue southward into the undulating hills of Central Texas, through the sparsely vegetated chaparral brushlands, and on into the semitropical Rio Grande Valley—a distance of more than 800 miles. Along the southern coastline, fertile swamps and marshes reach inland from the dune-bordered edges of the Gulf of Mexico. Lush, humid forest lands of the eastern region of the state slowly give way to the rich farmland and prairies to the west, the deep valleys and jagged hills of the Edwards Plateau, and on to the semideserts and tall, time-worn mountains of the far west, which are surrounded by true desert.

Such an incredible diversity of habitats provides an equally astonishing number of species of plants, animals, and insects within the state's boundaries, and in North America, Texas has no equal in regard to the number of butterflies, for it leads all other states with more than 450 species and/or subspecies.

Butterflies are complicated creatures, and it is sheer folly to make definitive, irrevocable statements about the lives of certain species. Little is known about many species, and much study must be done before any hypotheses can be ventured as to their life cycles or habitats. Even their distribution within the state is often determined by such climatic factors as unusually heavy rains, extreme droughts, uncommonly cold or mild winters, and hurricanes. Any one of these things can determine whether and where a butterfly may wander or establish a temporary colony far outside its normal breeding range. Such unpredictability only makes searching for, watching, and attracting butterflies more enjoyable.

So often a how-to work is so general it is of little help or even proves worthless for a particular purpose or area. The more specific the knowledge gleaned of the life cycles, habits, and eccentricities of these insects, the more helpful this knowledge will be. Also becoming better acquainted with the butterflies themselves can only deepen an interest and appreciation for this insect as the beautiful and complex organism it is. All butterflies and plants discussed in this text can be found within the state and within the regions specified. It is my hope that the information given here will not only be useful and make it easier in creating your personal garden for butterflies but will also bring beauty, wonder, and a sense of oneness into your life.

This book is written in a nontechnical style especially for gardeners living within the borders of Texas. In order to adequately cover the state, it has been divided into seven regions (see end sheet map): High Plains/Rolling Plains (Region 1), Grasslands/Prairies/Savannahs (Region 2), Pine/Hardwoods (Region 3), Coastal Prairies/Marshes/Beaches (Region 4), Chaparral Plains and Rio Grande Valley (Region 5), Edwards Plateau (Region 6), and Trans-Pecos (Region 7).

GIANT SWALLOWTAIL
(Papilio cresphontes)

Butterfly Gardening for Texas

AMERICAN LADY *(Vanessa virginiensis)*

1 *Understanding the Butterfly*

Butterflies, along with moths, are easily differentiated from all other insects. They belong to the order Lepidoptera, a name composed of two Greek words, *lepis* meaning "scale" and *ptera* meaning "wing," combined to mean "scale-winged," which aptly describes their most obvious feature. The wings, as well as the body, are almost always entirely covered with scales. Scales on the wings are usually flat, but the scales covering the body are often long and silky, appearing almost hairlike.

In most respects butterflies and moths are quite similar, but four characteristics usually separate them. In almost all species of butterflies, the antennae end in a club or swelling at the tip, while the antennae of moths are slender or feathery but rarely clubbed. Also, butterflies generally fly during the day, while moths fly primarily at night. However, there are numerous exceptions to the generality concerning moths, since many fly about during the day. Another defining characteristic is that butterflies usually rest with their wings closed and held vertically over the back. Exceptions to this occur during periods of basking when the wings of most species may be spread flat, and the Skippers frequently rest with their wings half spread. Conversely, most moths rest with their wings outstretched and held flat against the surface on which they are resting or drawn back tightly along the sides of the body. Last, and as a general rule, butterflies more often form an unprotected chrysalis in the open, while moths form a tough, silken cocoon in which to pupate.

Life Cycle

In their life cycle, butterflies go through four distinct stages, together known as a complete metamorphosis. These stages of growth are the embryo stage (egg or ovum), the wormlike growing stage (caterpillar or larva), the mummylike transition stage (chrysalis or pupa), and the winged reproductive stage (adult or imago).

Adult females usually lay eggs on or near the food plant that will sustain the caterpillars or larvae upon hatching. The eggs are almost always left unattended and will usually hatch in a few days. In some species eggs are left to overwinter, and in some rare cases they will not hatch for two years or more.

Eggs are very soft when first laid and are usually attached firmly to the food source with a sticky, gluelike substance. Slowly, the egg takes on its particular form and color as the shell dries and hardens. Shape is usually characteristic for the species, and under magnification each egg reveals its own beautiful markings and coloring. It may be round, domed, flattened, elongated, or shaped like a minute barrel, urn, pincushion, spindle, or sea urchin. The egg surface may be pearly smooth or elaborately sculptured with raised or sunken ribbing, horizontal furrows, pits, grooves, knots, spikes, or other ornamentations. Eggs of the Cabbage White (*Pieris rapae*) are fat yellow cones with intricate lengthwise striations. The Pipevine Swallowtail (*Battus philenor*) lays large, reddish-brown, almost perfectly spherical eggs, while eggs of the Guava Skipper (*Phocides polybius lilea*) resemble beautifully ribbed and flattened turbans. Eggs of the Blue Metalmark (*Lasaia sula*) resemble two pies, one stacked on top of the other.

Janais Patch
(Chlosyne janais)

Females of each species instinctively choose the exact food plant on which to deposit their eggs, using an intricate detection system that involves sight, feel, taste, and smell. Many butterflies lay their eggs singly, but others attach several in a single layer or in clusters of up to five hundred, usually to the underside of a leaf. Depending on the species, a female may lay between one hundred and two thousand eggs during her lifetime.

Immediately upon hatching, the young caterpillar usually eats all or a portion of the eggshell as its first meal, thus gaining vital nutrients that have been passed on from the mother. With many species the eating of the shell is extremely important, for the young larvae will not survive if they do not do so. The caterpillar then begins feeding on the leaf or flower it is on. In this nutritive stage of the larva's life, its entire purpose is to eat—and eat it does. As it grows, it ravenously consumes food at an astonishing rate. By the time the larva has finally become satiated, it may have increased its weight by as much as one thousand times.

Some caterpillars, such as those of the Clouded Sulphur (*Colias philodice*), are cannibalistic in the sense that they consume eggs or already-formed chrysalides along with foliage if they happen to be attached where the caterpillar is feeding. Caterpillars of another butterfly, the Harvester (*Feniseca tarquinius*),

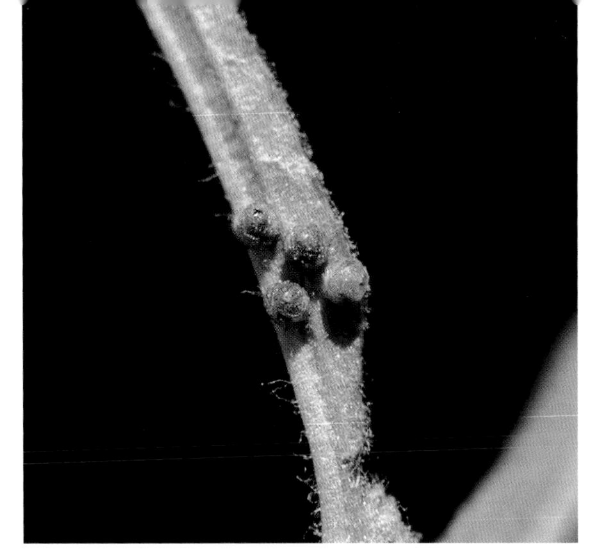

Eggs of the Pipevine Swallowtail *(Battus philenor)*

Eggs of the Guava Skipper *(Phocides polybius lilea)* resemble turbans.

eat only woolly aphids (family Eriosomatidae) for its food.

The skin, or exoskeleton, of a caterpillar is incapable of stretching, so to grow any larger, a caterpillar must shed the old, too-tight skin for a new one. Through a complicated process, the skin splits down the back, and the larva emerges with a totally new skin. At first the new skin is much too large, but as the caterpillar continues to eat, this skin eventually becomes stuffed to bursting. Again the splitting process is repeated, and the larva emerges with yet another loose skin to be filled up. This changing of skins is called molting. After each molt the old skin may be abandoned but is more commonly eaten by the caterpillar.

A caterpillar will go through this process several times, the number of molts varying with the species. Most Blues and Hairstreaks (family Lycaenidae) molt only four times, while many of the Metalmarks (family Riodinidae) change skins from four to nine times. The period of a caterpillar's life between each molt is called a stage or an instar, and the new skin of each instar is often colored and patterned differently from the previous one.

Eggs of Janais Patch *(Chlosyne janais)*. Some butterflies may lay hundreds of eggs in one cluster on the undersides of foliage.

At the end of the last instar, the insect is ready to begin the last phase of its caterpillar life and seeks the most protected place possible. Caterpillars of many butterflies will choose a stem or branch, securely attaching themselves with strands or small pads of silk. Others pupate in silk-lined leaf-nests made from their respective food plants or in ground debris at the base of the food plant. During this transformational phase, sometimes requiring several hours, the caterpillar slowly undergoes a complete change inside the skin and eventually sheds the skin to emerge in a totally different shape—as the chrysalis or pupa. In this form it will remain, mostly immobile and helpless, for several days, months, or in a few instances, years. During this period of its life, the butterfly is vulnerable and open to attack by all sorts of enemies, such as ants, parasitic wasps, birds, lizards, and environmental factors. It is only natural, then, for the caterpillar to seek a protected place when the time comes to pupate.

When the appropriate, species-specific time is completed, yet another change takes place within the pupal case. The cells have multiplied, changed, and rearranged into the various parts of the soon-to-be adult butterfly. The chrysalis changes color; its outer skin becomes almost transparent, and the wing colors of the butterfly become visible inside. After finally splitting the pupal shell and freeing its head, the butterfly uses its legs and contortions of the body to free itself. While it remains motionless, fluids are slowly pumped into the veins of the crumpled wings until completely expanded and dried into their beautiful colorings and patterns.

The Body

The body of a butterfly is separated into three main divisions: the head, thorax, and abdomen. The head bears the principal sense organs, one of the most obvious being the pair of antennae projecting from the top of the head. Each antenna is composed of short joints

Variegated Fritillary *(Euptoieta claudia)*

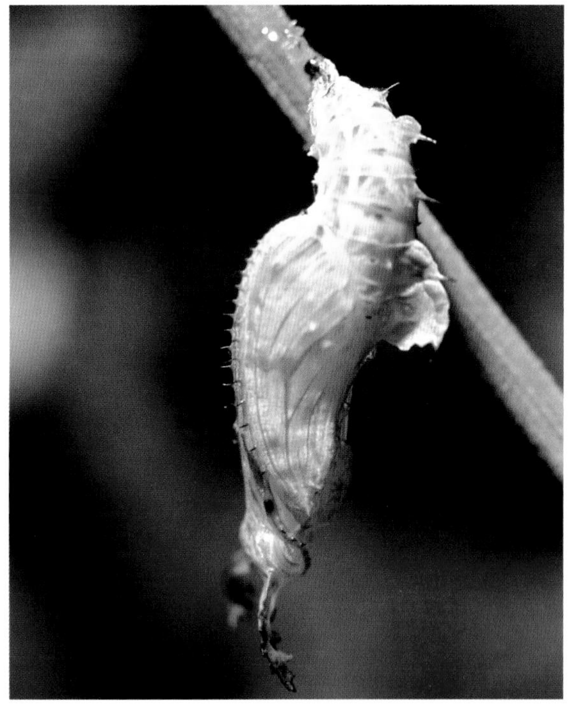

Zebra Longwing *(Heliconius charithonius vazquezae)*

or segments, with segments near the tip being larger and causing the antennae to appear knobbed or clubbed. The antennae are quite movable and are used for balancing, touching, hearing, and tasting. Also, most of the main detection sites of smell are situated in the tips. Not having long-distance vision, the butterfly depends more on this sense of smell for detecting odors in search of food. At the base of each antenna is a most essential organ used for orientation, especially while flying.

Enormous, almost hemispherical compound eyes are just below the antennae. They are among the most complex and intricately designed mechanisms used by animals for seeing. The shape, structure, and position of the butterfly's eyes enable it to see in all directions except directly beneath its body. The eyes are called compound because they are actually made up of thousands of honeycomb-like facets—between two thousand and twenty thousand facets, depending on the species. Instead of seeing a single light image or object, the butterfly sees a separate image with each tiny facet, making thousands of images. These images are then integrated by its brain into a mosaic picture.

Its eyes are well suited for detecting any type of nearby movement, including butterflies of the opposite sex as well as numerous predators. The eyes are fixed and cannot move, rendering the insect unable to keep objects approaching it in focus. Instead, as the butterfly nears an object (or the object nears it), images move toward the inner part of each eye, with the visual angle decreasing.

The spectral or color range visible to the butterfly extends from ultraviolet through yellow-orange and red, fully covering the visual spectrum of humans as well as that of other insects. This mechanism gives the butterfly the broadest spectrum of color vision known to exist in the animal kingdom.

The ability to recognize more colors gives the butterfly many advantages in communication, feeding, protection, and perpetuation of young. Besides the colors visible to us, a butterfly's wing may reflect a little or a lot of the ultraviolet spectrum to another butterfly. This ultraviolet coloring plays an important role in courtship and mating. Also, since vegetation generally absorbs ultraviolet light, reflection from flowers and foliage serves to maximize color contrast, thus aiding the female butterfly in plant identification for egg deposition.

Also on the head, situated below and extending upward to the side of the eyes, are two soft, furry or scaly palpi. The palpi are sensitive receptors that test the suitability of the food source and protect the mouth and proboscis. The proboscis is divided into two grooved parts or half tubes that are separated when the butterfly first emerges from the pupal shell. The butterfly must spend several minutes twisting and turning the parts to fasten them together securely by the interlocking spines along the edges of each half. This structure, which now forms a long, hollow, flexible tube much like a drinking straw, can be rolled up tightly or unrolled at will and is used to suck up liquid food. Some species of butterflies also have thick, tubular organs for tasting near the tip of the proboscis.

Behind the head is the thick, muscular thorax, divided into three segments and bearing six legs and four wings. Each segment of the thorax has two jointed legs, one on each side of the body. In some species of butterflies,

Lower Surface (Underside, Ventral)

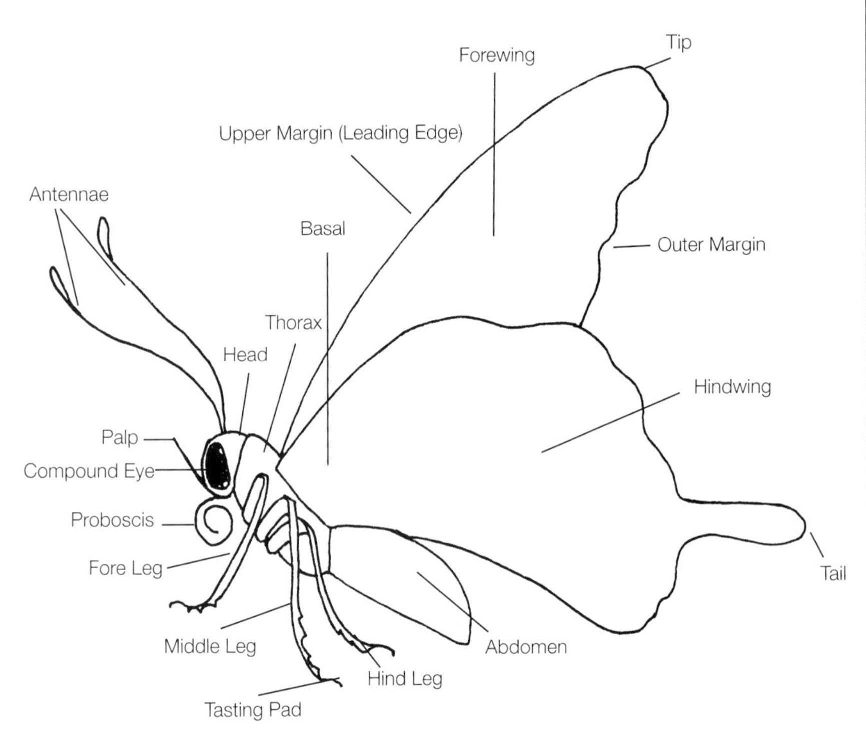

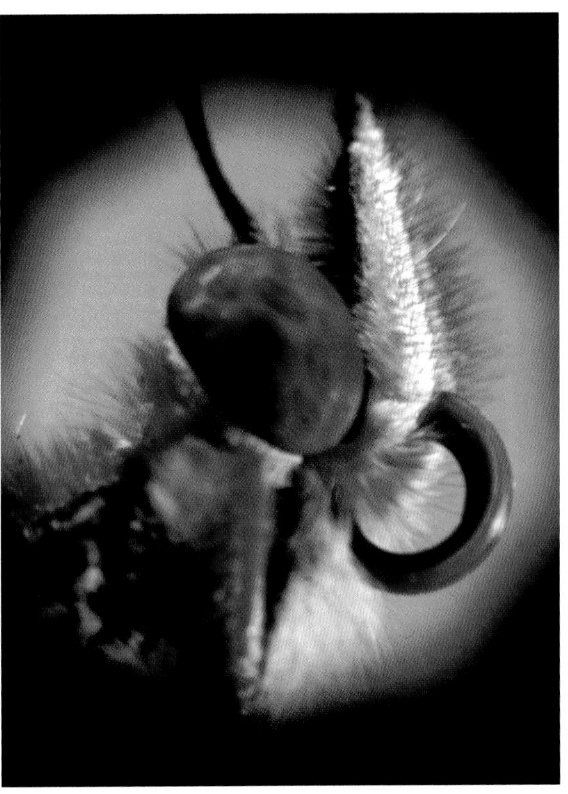

Close-up showing compound eye, palp, and rolled proboscis.

such as the Brush-footed (family Nymphalidae), legs on the first segment are undeveloped, held close to the body, and rarely used, causing the butterfly to appear to have only four legs. The last portion of each leg consists of five tiny segments that form the foot with the last segment ending in a pair of claws. These feet possess organs that enable the butterfly to taste its food. Tasting with the feet triggers an automatic, reflex action that causes the proboscis to uncoil when food is found. In some instances, the female uses the clawed segment to scratch the surface of plant leaves and stems, testing the chemical content to determine if it is the proper plant for depositing her eggs.

Above the last two pairs of legs on the thorax are the four wings. Each butterfly wing consists of two delicate membranous sheets with an inner framework of hollow, tubelike veins between the layers. This venation helps strengthen the wings and is in a distinct pattern for most families and species, becoming an important tool in identification. The membrane of the wings is usually transparent, but in most butterflies it is completely covered with thousands of tiny, flat or hairlike scales of various shapes and colors. These scales, shingled in overlapping rows, provide insulation from the cold, protection from rain or dew, and help in flying. They are extremely fragile and, if touched, will readily adhere to the fingers, appearing as colored dust.

The third portion of the butterfly's body is the abdomen, which consists of eleven segments, although only seven or eight can readily be seen. The abdomen is very soft and contains the digestive and reproductive organs. At the tip of the abdomen of the male is a pair of grasping organs used to clasp the female during mating. The abdomen of the female is usually much larger than the male's due to the large mass of unlaid eggs.

Courtship

Butterflies search for a mate primarily by either perching or patrolling, and almost always it is the male who does the courting. In perching behavior, at a specific time of day the male chooses a certain place—such as a rock, a patch of ground, a post, or a particular tree branch—that offers a visual observation point. There he waits until some moving object comes into view. Since his sight is limited to close-up vision, he must inspect everything that comes into his range of sight, be it wasp, bird, dog, cat, squirrel, human, or a butterfly of any species. If the object turns out to be an unacceptable female, the male returns to his perch to wait for the next passerby. If a male encounters another male of the same species, great exhibitions of "fighting" occur, where the two fly in close association, often spiraling upward until almost out of sight, before breaking apart and going their separate ways. The female of each species is instinctively drawn to likely spots that the male of that species ordinarily chooses, so finding a mate is not usually a problem.

The second method used in seeking a mate is patrolling. A patrolling male flies from one end of a selected site to the other almost continuously until a female of the same species

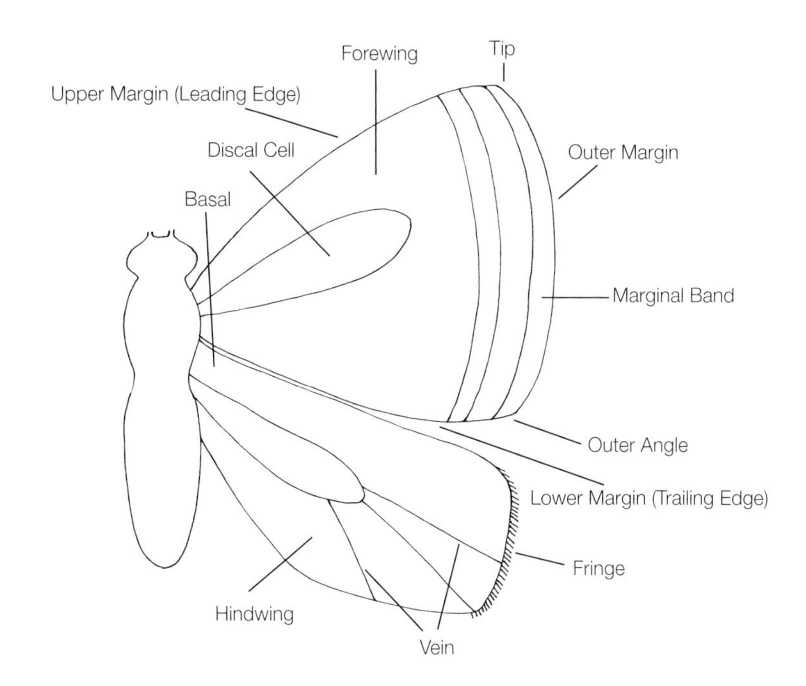

flies into the territory. Here again the female is genetically keyed to locate areas typically selected by the male.

The male butterfly uses a variety of ways to recognize and attract a female of the same species. One way is by color and its placement on the wings. Since butterflies see the colors humans see, as well as a large range within the ultraviolet spectrum, the male butterfly sees the colors of a female butterfly's wings entirely differently from how humans see them.

Scale coloration is wondrously complicated. Basically, it is of two types, pigmented and structural. Pigmented colors are actual colors, produced by a pigment within the insect itself or derived from the food plant of the caterpillar. The majority of such colors as blacks, grays, tans, browns, brownish-reds, and some yellows are forms of the chemical melanin, the same pigment that produces

THE WING SCALES OF THE PIPEVINE SWALLOWTAIL *(Battus philenor)* DISPLAY A STUNNING IRIDESCENCE.

WING SCALES OF THE PALAMEDES SWALLOWTAIL *(Papilio palamedes)*

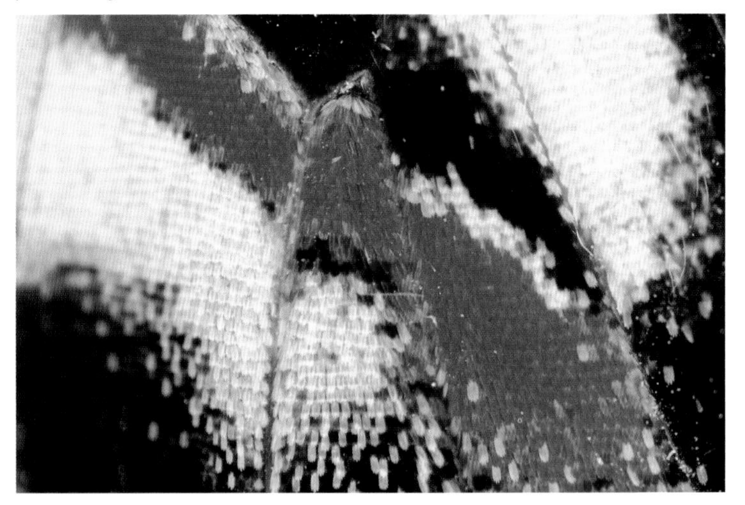

WING SCALES OF THE ZEBRA SWALLOWTAIL *(Eurytides marcellus)*

freckles and suntans in humans. Ivory to dark yellow colors are usually from organic dyes called flavones, which are retained from the larva's food plant. The yellow coloring of the Sulphurs and Yellows (family Pieridae) is produced by pterines, which are derived from an excretory uric acid. While some white coloring is made from pigment, others are the result of bubbles of air. The scattering of light by transparent particles produces a white effect in the same way that snow appears white. White hair in humans is similarly produced by tiny air bubbles within the strand of hair that replace the natural color rather than by a white coloring substance.

Some of the most striking and beautiful colors, such as most of the blues, greens, golds, and silvers, are the iridescent colors that are due to structural features of the scales or hairs rather than pigmentation. Each scale may be ridged or grooved with microscopic striations. The colors are the result of light being reflected from these physical features, much in the same manner as light refracting off a film of oil on water, which produces glittering, changing colors. Although each scale is of only one color, as a general rule there is a mixture of pigmented and structurally colored scales on almost every species of butterfly. On some, such as the Sulphurs, iridescence can hardly be noticed, but on the Common Buckeye (*Junonia coenia*) or Pipevine Swallowtail, the glow and shimmer are spectacular.

Scent or fragrance also plays an important role in the selection and seduction of a mate. Male butterflies possess special scales that act as dispensers for scents called pheromones. Placement of scent scales varies according to

Courtship display of the Common Buckeye (*Junonia coenia*)

species and may be either scattered among the regular wing scales, grouped in patches, arranged along certain veins, or located in folds along the edges of the lower wings. Some of these scents gain the interest of the female, while others calm the female prior to and during the actual mating. Such scent scales are called androconia, from two Greek words (*andr* and *konic*) meaning "male-dust." Some courtships involve the male performing an elaborate and prolonged dance in front of or above the female, bathing her in his perfume until she becomes settled and receptive for the actual mating.

Many species of freshly emerged male butterflies have a strong, noticeable fragrance, frequently comparable to that of flowers. The male Monarch (*Danaus plexippus*) has a rather musky, exotic scent somewhat resembling the fragrance of a wild rose (*Rosa* spp.). Scents of other butterflies have been likened to violets (*Viola* spp.), verbenas (*Verbena* spp.), meadowsweets (*Spiraea* spp.), heliotropes (*Heliotropium* spp.), and even chocolate. A female is readily attracted to this flower fragrance in males, since this is a scent that also entices her to nectar.

Temperature Control

Butterflies are not as cold-blooded as once believed and described. They are able, under many circumstances, to regulate their body temperature by an assortment of behavioral acts. They do, however, require some heat to begin operations. Most species do not begin

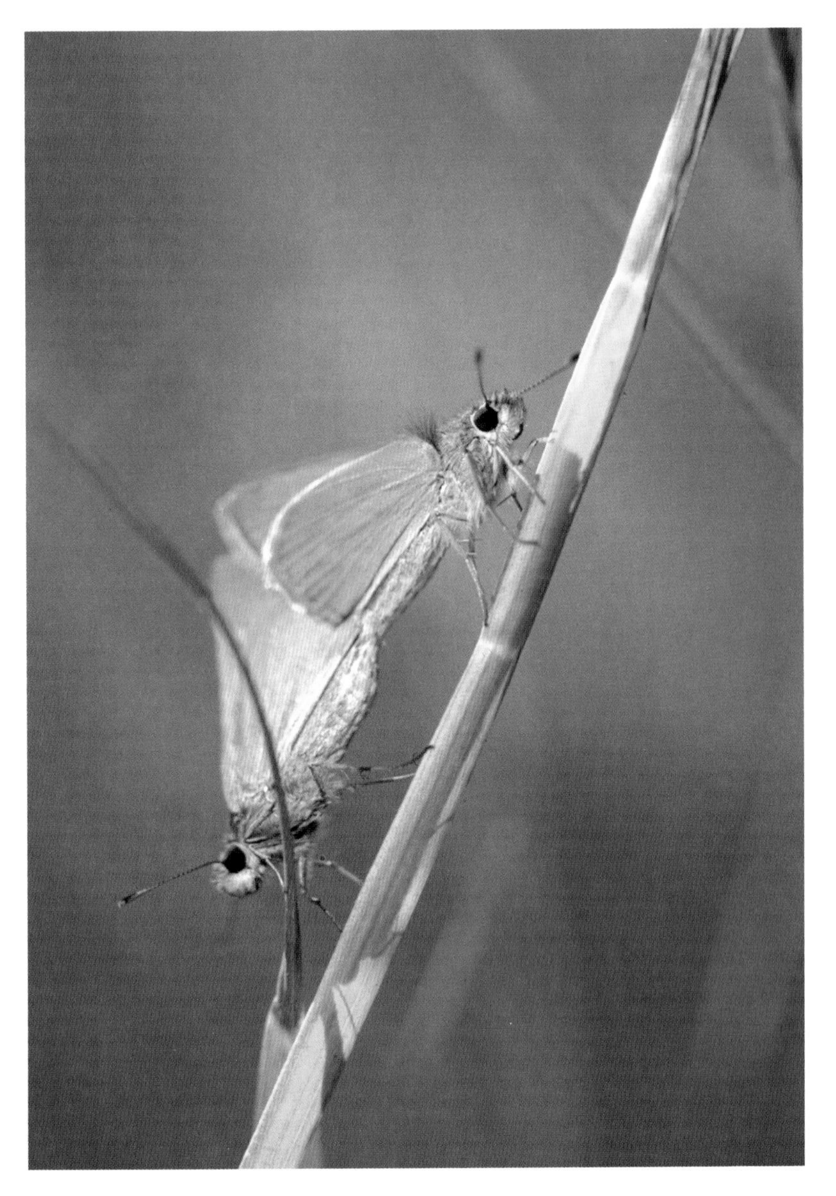

A male Julia's Skipper *(Nastra julia)* extends his scent scales, or "brushes," as part of the mating ritual.

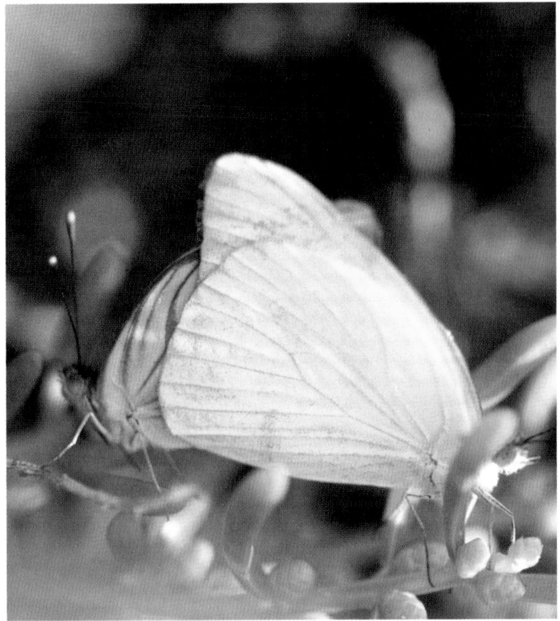

Mating of the Great Southern White *(Ascia monuste)*

flying if temperatures are below 40 degrees Fahrenheit and cannot fly well until they bring their body temperatures to between 60 and 108 degrees, with the optimum body temperature between 77 and 110 degrees. In order to raise the temperature of the muscles that control their wings and legs, they must absorb radiation from the sun or their surroundings or produce the heat themselves.

Butterflies exhibit six basic positions to obtain and regulate this heat: dorsal, lateral, body basking, dorsally closed, conduction, and "shivering." In early mornings and during cool periods, butterflies can often be seen basking in the sunlight with wings spread wide. This dorsal basking is the most common position used by most species. A butterfly basking with wings outspread in this manner snaps them together quickly if a cloud or other shadow passes over its body. This conserves heat while making the insect less conspicuous.

Lateral basking is used by some butterflies such as the Large Wood-Nymph (*Cercyonis pegala*), Hackberry Emperor (*Asterocampa celtis celtis*), Goatweed Leafwing (*Anaea andria*), and Sulphurs. They keep their wings closed and practically lie down on their sides, exposing the undersurface of the folded wings to the full rays of the sun. This tactic also helps eliminate a shadow, which might be a giveaway to a predator.

Satyrs (family Satyridae), the Falcate Orangetip (*Anthocharis midea*), and Blues are primarily body baskers, opening their wings only wide enough for their bodies and base of wings (usually black or dark-colored) to receive some of the sun's rays. A similar method is termed dorsally closed basking and is seen mostly in the Skippers (family Hesperiidae). Here the hindwings are open with the forewings only partially opened, funneling heat to the body.

In conduction basking, the butterfly chooses a rock, large leaf, the ground, or a dark-colored twig near the ground for perching, not only to absorb the heat from the sun but also to take advantage of the heat rising from the object on which it is resting. This method is used by almost all species, with wings held at their usual position.

Some species, especially those that overwinter or emerge early in the season, such as the Mourning Cloak (*Nymphalis antiopa*), Question Mark (*Polygonia interrogationis*), Eastern Comma (*P. comma*), and Red Admiral (*Vanessa atalanta rubria*), will "shiver" or rapidly vibrate the wings to warm the body before flying.

Butterflies are just as sensitive to excessive heat as they are to cold. On summer days when the temperature becomes very high, they attempt to prevent overheating by closing their wings dorsally over the back. They may also turn the head and abdomen directly toward the sun; this edge effect keeps the sun from striking the larger wing surface and helps keep the insect cooler. If the butterfly is still too hot, it will take cover in a shaded area.

Roosting

As late evening approaches and the air begins to turn cooler, butterflies seek a resting place for the night. Apparently, the major environmental factor that initiates roosting behavior is the decreasing radiation from the setting sun; when clouds pass over the sun before an approaching storm even on a hot summer day, butterflies will seek shelter and assume the roosting position.

For many species the selection of a roosting site is preceded by a short period of extremely active and erratic flight. During this period the butterfly spirals several feet above the vegetation, then dips down to a leaf or grass blade, testing it for suitability. It usually samples several sites before finally settling on one, most often choosing a fairly stable site with no other vegetation close by that would knock the insect from its perch if blown about by high winds. Often the perch is the dead and leafless stem of an herb, perhaps chosen because such a site is not ordinarily visited by ants, one of the butterfly's worst enemies during periods of immobility. Many butterflies will bask in the sun as long as possible and will sometimes use the basking perch as their roosting site for the night. This has been termed "vesper warming," and the Least Skipper (*Ancyloxypha numitor*), Eastern Tailed-Blue (*Cupido comyntas texana*), and Pearl Crescent (*Phyciodes tharos*) are noted for such behavior.

Other roosting positions are used, each species of butterfly habitually using a certain type. Some butterflies, such as the Black Swallowtail (*Papilio polyxenes*), will continue to feed late, then roost on westward-facing slopes or the outermost leaves of westward-facing plants in order to benefit from the last rays of the sun. Other butterflies, such as the Gemmed Satyr (*Cyllopsis gemma gemma*) and some of the Yellows, choose roosting sites that are shaded from the setting sun and will therefore be illuminated by the rising sun; such sites offer earlier warming the following morning and enable the butterfly to begin feeding sooner. However, where the accumulation of dew is great, even butterflies with east-facing roosting sites have a long wait. Until the sun reaches their wings, dispelling the moisture, the insects remain immobile, unable to fly.

Some species of butterflies randomly select sites each night or may use the site of one type for a few nights and then switch to a different type. During cool spring and autumn nights, undersides of large-leaved trees such as oak (*Quercus* spp.) or hickory (*Carya* spp.) are favorite sleeping spots for the Pipevine Swallowtail and Spicebush Swallowtail (*Pterourus troilus*), especially if the tree branches overhang a little-used road. Heat is trapped beneath the trees, with the temperature commonly several degrees higher beneath the branches. This extra heat provides a much warmer bed for the insects. On hot nights during the summer months, though, they seek the cooler outside tips of the upper branches.

Overwintering

Different species of butterflies have evolved their own special devices for surviving the cold of winter. In the Rio Grande Valley and along the coast where it remains relatively warm all winter and flowering plants are present, many resident butterflies do not have a diapause, or "resting period," and continue to fly and breed throughout the year. Other species that live and breed in a more northerly part of the state where freezing temperatures regularly occur must pass the winter in another life stage, as either egg, larva, pupa, or hibernating adult.

Several of the Hairstreaks overwinter in the egg stage, while most of the Skippers and some of the Blues overwinter in the larval stage. Swallowtails (family Papilionidae), Longwings (family Nymphalidae), Whites (family Pieridae), and Sulphurs spend the winter in the chrysalis or pupal stage, the exception being the species that wander into the state from Mexico or Florida and may breed but are not considered residents. Some butterflies, such as the Mourning Cloak, Hackberry Emperor, Red Admiral, Question Mark, and Painted Lady (*Vanessa cardui*), pass the winter in a semihibernating state, tucked away behind loose bark, in narrow cracks of buildings, or in hollows of posts or trees. They are able to survive the freezing cold of winter by thickening the blood (haemolymph) with certain natural substances (glycerol, sorbitol, or alcohol) that act in a manner similar to that of antifreeze in a car radiator. During this time of severe cold, a butterfly's metabolic rate becomes noticeably slowed; instead of continually eating, the insect is sustained mainly from stored body fats. Any period of warmth and sun, however, will see the butterfly out partaking of oozing sap, partially thawed fruit, or any substance containing amino acids.

The Monarch is an exception and cannot endure freezing in any stage of its life cycle. For its survival, it migrates to warmer climates in Mexico, where it lives out the winter resting on certain roosting trees in areas that have been used by migrating butterflies for hundreds of years.

Long-Distance Flights

Other species of butterflies travel long distances as well, both northward and southward and at different times of the year. The Painted Lady, Common Buckeye, Cloudless Sulphur (*Phoebis sennae*), Dainty Sulphur (*Nathalis iole*), Little Yellow (*Pyrisitia lisa lisa*), Gulf Fritillary (*Agraulis vanillae incarnata*), and Great Southern White (*Ascia monuste*) all move northward each spring, rearing many broods along the way. With the shortening day length of fall, some of the existing population heads back southward, often into areas of nonfreezing temperatures; the majority of the insects do not make a return flight and are killed by advancing freezes. The ones that do return as far as the semitropical climate continue to reproduce with no winter diapause. By spring the population will have again built to such numbers that it must disperse in search of available food plants, so a portion of the population once again heads northward.

A few species of butterflies, such as the Checkered White (*Pontia protodice*), are notorious for long-range dispersal. Dispersal movements, or emigration, differ from migration in being random and erratic in nature, whereas migration follows a habitual and predictable pattern. Checkered Whites exhibit a prime example of uncertain behavior, for they may be encountered almost anywhere in the United States during one breeding season and then not seen in the same area again for several years. In the southernmost portion of the state the Snout (*Libytheana carinenta*) may become extraordinarily abundant after drought-breaking rains in the fall and disperse hundreds of miles. Rio Grande Valley residents such as the Julia Longwing (*Dryas iulia*), Zebra Longwing (*Heliconius charithonius vazquezae*), Amymone (*Mestra amymone*), Yellow Angled-Sulphur (*Anteos maerula*), and Ruddy Daggerwing (*Marpesia petreus*) are sometimes sighted in the Panhandle during these late summer and early fall months.

Dangers to Caterpillars and Chrysalides

Butterflies, as do all other living creatures, have their enemies. In the larval, plant-eating stage, one of the worst and most common pests is the introduced fire ant (*Solenopsis invicta*). This was not always such a serious danger, but in recent years fire ants have spread to the extent that rare is the yard that does not have its bed or two of them. Often the ants form their mound in an out-of-the-way place, and a lot of damage can be done to eggs, caterpillars, and chrysalides before the ants are finally noticed. Once a caterpillar or chrysalis is found, a group of ants will congregate and entirely consume the "future" butterfly in only a few minutes. It is time well spent to walk about a butterfly garden at least once a week to check for these ferocious beasts. Especially look along branches and leaves of the food plants.

First-brood caterpillars are especially susceptible to various fungi during the early spring months, when there are many cloudy days, an abundance of rain, and sudden drops in temperature. Plentiful stands of the food plant in protected areas will best ensure survival of the larvae during this time.

Spring rains sometimes cause plants to die because of excessive moisture. This loss is a normal occurrence and something the larvae can cope with. Usually, they simply move to nearby areas of plants that survived in drier habitats.

Parasitism of the caterpillar, as well as the eggs and pupae, by various species of tiny flies and wasps is a common hazard. The gardener has little control in this situation. If a plentiful source of the food plant is available, there are usually enough healthy larvae left to continue the breeding cycle. The amount of parasitism varies from brood to brood and season to season, so the population as a whole usually levels out.

The worst predation of butterfly caterpillars is probably by birds and spiders. Caterpillars are the staple food of young birds in the nest and are eaten quite readily by the adult as well. Several species of spiders lurk among plant foliage, taking many caterpillars, especially the smaller, early-stage ones. Caterpillars have evolved various methods of protection from their enemies, such as camouflaged coloration, spines, or distastefulness, but still the loss is staggering.

Dangers to Adults

Even after butterflies have emerged and are on the wing, their lives are still fraught with many dangers. The first broods to emerge in the spring are susceptible to late freezes and sudden thunderstorms. The insects can sometimes survive by finding a crack beneath tree bark or by snuggling low to the ground among dried grasses to escape freezing temperatures.

They are not so lucky during really windy rainstorms. During turbulent weather, adult butterflies most commonly seek shelter beneath large leaves. If high winds accompany the rain, the leaves are turned over, exposing the butterflies to the driving rain. Also, the wind often loosens the butterfly from its hold on the leaf, forcing it to the ground, where it is either drowned or beaten to death.

Praying mantises (*Mantis* spp.), various spiders (order Araneae), red wasps (order Hymenoptera), carabid beetles (family Caraidae), ambush bugs (family Phymatidae), assassin bugs (family Reduciidae), robber flies (family Asilidae), dragonflies (order Odonata), anoles and green snakes (order Reptilia), and birds (order Aves) all readily feed on butterflies. Birds frequently catch butterflies on the wing; other predators lie in wait or stalk the nectaring insects. The large webs of the orb-weaving spiders are another great hazard to butterflies on the wing. Nighttime predation by such animals as raccoons (order Carnivora), opossums (order Marsupialia), mice and flying squirrels (order Rodentia) are a major cause of mortality in some species. During this roosting period, the butterflies are quiescent and totally defenseless against any kind of disturbance.

During the summer months the single most common disaster to butterflies is loss of nectar plants. Nectar sources frequently die as a result of drought, mowing, or herbicides. Also, a poorly planned garden that leaves a gap with no plants in flower usually results in either the butterflies dying or having to seek food elsewhere.

Early freezes in the fall take a toll on the adult butterfly population. Some manage to survive the cold by hiding in well-protected spots and will be seen on the wing for several weeks after the first frost or two. With the exception of the species that migrate or hibernate, extremely cold weather eventually kills both the nectar plants and the butterflies. Often larval food plants are killed before the feeding larvae reach a stage in which they can overwinter successfully.

Methods of Protection

Eggs, once they are laid, are at the mercy of the weather, parasitic wasps, and predators ranging from fire ants to beetles to birds. The best the mother butterfly can do to protect her future young is to conceal the eggs as well as possible and to choose a well-protected place on or near a healthy food source. Commonly, eggs are of a pale color, either whitish, greenish, or pale yellow, and are generally deposited on the foliage. Eggs deposited at the base of a leaf or along stems often become much darker in color after a day or so, blending more readily with their surroundings and becoming less conspicuous to predaceous eyes. Some butterflies use a strategy of wide dispersal of the eggs as a measure of safety, laying only one egg per plant on widely separated plants. Other species use "overabundance," laying a hundred or more eggs in one cluster.

The most vulnerable period of a butterfly's life is during the larval or caterpillar stage, and it is then that greatest mortality occurs. The extent of their built-in defenses against diseases is not known, but their obvious weapons against predator attack are numerous.

As with eggs, concealment is one of the major visual defenses for caterpillars. They are often pale green, as in the case of the Cabbage White and Lyside Sulphur (*Kricogonia lyside*), and remain immobile and hidden beneath a leaf except when feeding. They may be spotted, blotched, or striped to blend with the stems and branches of their food plants. The lengthwise blackish and reddish-brown striping of the Common Buckeye make it practically indistinguishable from the brownish striations of the branches of agalinis (*Agalinis* spp.), one of its food sources.

The art of appearing as something else is one of the most fascinating strategies evolved by larvae. First instar caterpillars of some of the Swallowtails along with the Viceroy (*Limenitis archippus*) and Red-spotted Purple (*L. arthemis*) appear as nothing more than fresh bird droppings and are overlooked by all but the most observing eye. Last instar larvae of some of the Swallowtails such as the Spicebush, Eastern Tiger Swallowtail (*Papilio glaucus*), Two-tailed Tiger Swallowtail (*P. multicaudata*), and Palamedes Swallowtail (*P. palamedes*), have large, conspicuous eyespots on the humped thorax that give the frightening impression of a snake's head.

The Red Admiral, Painted Lady, and Variegated Fritillary (*Euptoieta claudia*) use silk to form a leaf- or flower-nest in which to hide. They eat, rest, and occasionally even form the chrysalis within these shelters. Caterpillars of the Texas Emperor (*Asterocampa clyton texana*), Theona Checkerspot (*Thessalia theona*), Janais Patch (*Chlosyne janais*) and Bordered Patch (*C. lacinia adjutrix*) are gregarious in the first instars. By remaining clustered together, they are offered some protection by sheer numbers. If disturbed, they initiate a fright aspect by twitching, jerking, or rearing their heads in unison.

An assassin bug (family Reduviidae) threatens a Northern Oak Hairstreak *(Satyrium favonius ontario)*.

Osmeteria of the Eastern Black Swallowtail *(Papilio polyxenes asterius)*

Larvae of the Gulf Fritillary, Zebra Longwing, Mourning Cloak, and Common Buckeye, among others, are covered with branched spines that, although completely harmless, appear fearsome enough to cause would-be predators to have second thoughts. The egg-laying tubes of some parasitic wasps are not long enough to reach the body of the larva because of the spines, so parasitism is not possible. The larvae of the Monarch, Pipevine Swallowtail, and Queen (*Danaus gilippus thersippus*) have long, floppy tentacles that wave about menacingly when they are crawling or feeding. Swallowtail larvae possess a most formidable chemical defense—a scent gun. Molest one of these critters, and a two-pronged apparatus suddenly springs up out of the thorax, releasing a strong, almost overpoweringly obnoxious scent. These fleshy horns, called osmeteria, are usually bright orange or reddish. The predator attacking one of these larvae experiences a double whammy: the shock of a large, bright object suddenly appearing and a totally unexpected spraying with a horrible odor.

And of course, some caterpillars are just plain bad tasting. Having eaten plants that contain poisonous chemicals, the larvae retain the poisons, making themselves unpalatable to predators. Such terrible-tasting caterpillars do not bother to hide. They are usually brightly or strikingly colored and remain in the open, conspicuously and confidently munching away. Birds that feed on a bad-tasting caterpillar usually do not make a second mistake.

Once caterpillars have selected the site for pupation and the change from larvae to pupae has been completed, the pupae are unable to move from the site. Because they are so vulnerable, butterfly pupae have evolved extremely variable shapes and colors that help them hide from their many enemies. Pupae of the Longwings have odd-shaped extensions that break up the overall outline, making them look more like a dead, tattered leaf than a tasty morsel; the chrysalides of the Cloudless Sulphur are either green or brown, perfectly matching the stems and foliage of Partridge-pea (*Chamaecrista fasciculata*), its food plant. The pale green shell of the Monarch is decorated with a partial band of black and gold with occasional golden flecks, which blends perfectly with its leaf support. Pupae of the Gulf Fritillary and the Question Mark have light-reflecting silver or gold spots or splashes, which help them blend into the sunlight-and-shadow areas of the foliage. The black-speckled yellow chrysalis of the Janais Patch would be most conspicuous if placed against green foliage, but when attached among the pale stalks of its larval food plant, it is much less noticeable to predators. The intricate blackish and brownish mottling of Swallowtail pupae resemble jagged bits of wood or bark, appearing as an extension of a leaf rachis or a broken branch.

Because adult butterflies are also food for a number of predators, they, too, have evolved many different protective devices to survive. While we view the intricate colors, shapes, and patterns on a butterfly's wing with awe and wonder, they have been perfected by that particular insect through a long process of natural selection for a protective rather than aesthetic purpose. One of the most dramatic protection methods is cryptic or camouflage coloration and patterning, which allows the butterfly to blend into its surroundings. Frequently, the wings have the shape or pattern

Cloudless Sulphur (*Phoebis sennae*)

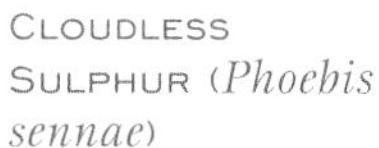

Cloudless Sulphur (*Phoebis sennae*)

Larvae of the Palamedes Swallowtail *(Papilio palamedes)* actually resemble a snake's head.

of objects such as broken pieces of bark, the mottling of pebbled ground, or the ribbing and coloration of dead leaves. One of the more common butterflies exhibiting this protective resemblance is the Goatweed Leafwing. When not feeding, the Goatweed Leafwing's usual resting place is on the ground or the trunk of a tree. The tannish or pale brown coloring and venation of the closed wings so resemble a dead leaf or a patch of bare ground that the butterfly seems to disappear the moment it comes to rest.

Wing edges of the group of butterflies known as Anglewings (*Polygonia* spp.) are conspicuously angled, cut, or scalloped. These irregular wing shapes, along with a dull brownish or grayish coloration and contrasting markings, obliterate their outline, helping them escape detection.

A disruptive pattern is commonly used by species such as the Zebra Swallowtail (*Eurytides marcellus*) and Zebra Longwing. When viewed alone, the sharply and distinctively striped wing pattern is readily visible, but when one of these butterflies comes to rest in dappled shade, it becomes lost among the sunshine and shadows. Large contrasting borders and bands on the Bordered Patch and Red Admiral and the broad yellow bands on the upper surface of the Giant Swallowtail (*Papilio cresphontes*) break up the overall color and shape. With this type of disruptive coloration, the insect may be seen but is not easily recognized for what it is. Eastern Tiger Swallowtails are dramatically visible when gathered around a mud puddle or nectaring at a clump of roadside flowers, but when they glide among tree branches or shrubbery, the yellow and black striping enables them to disappear in the filtered sunlight.

Butterflies that exhibit dull or cryptic coloration on the undersurface of their wings often have a contrastingly bright upper surface. Some have prominent, eyelike spots to frighten prospective predators. Such coloring is used by the Common Buckeye, whose prominent eyespots are concealed by the brownish, camouflaging pattern of the undersides. When disturbed, the Common Buckeye suddenly opens its wings and takes off in nervous, erratic flight, startling the would-be predator with brilliant colors and large "eyes" suddenly flashed in its face.

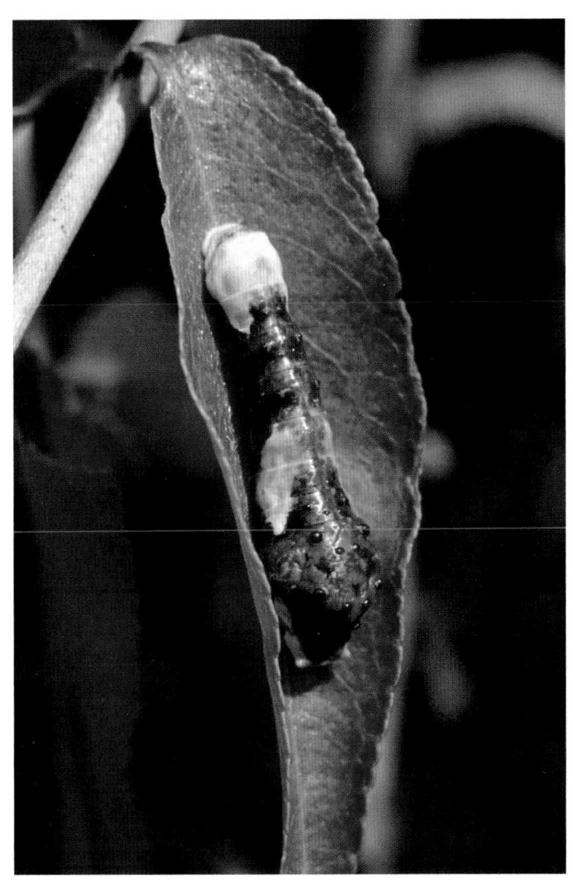

The larva of the Giant Swallowtail (*Papilio cresphontes*) appears to be bird droppings.

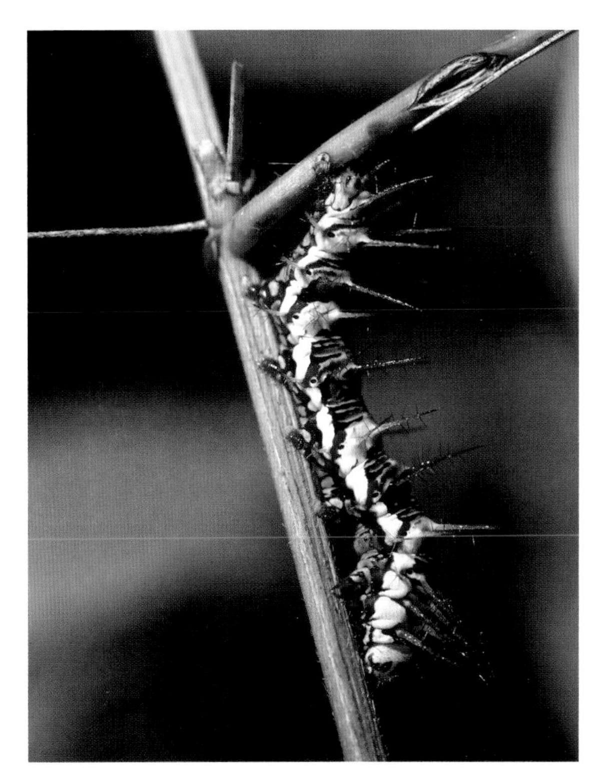

Fierce, branching spines of Julia Longwing (*Dryas iulia*) deter many predators.

(*Left*) Longitudinal striping helps this Common Buckeye (*Junonia coenia*) caterpillar blend in with branches.

The vivid green coloring of this Cloudless Sulphur (*Phoebis sennae*) larva enables it to blend into the surrounding foliage.

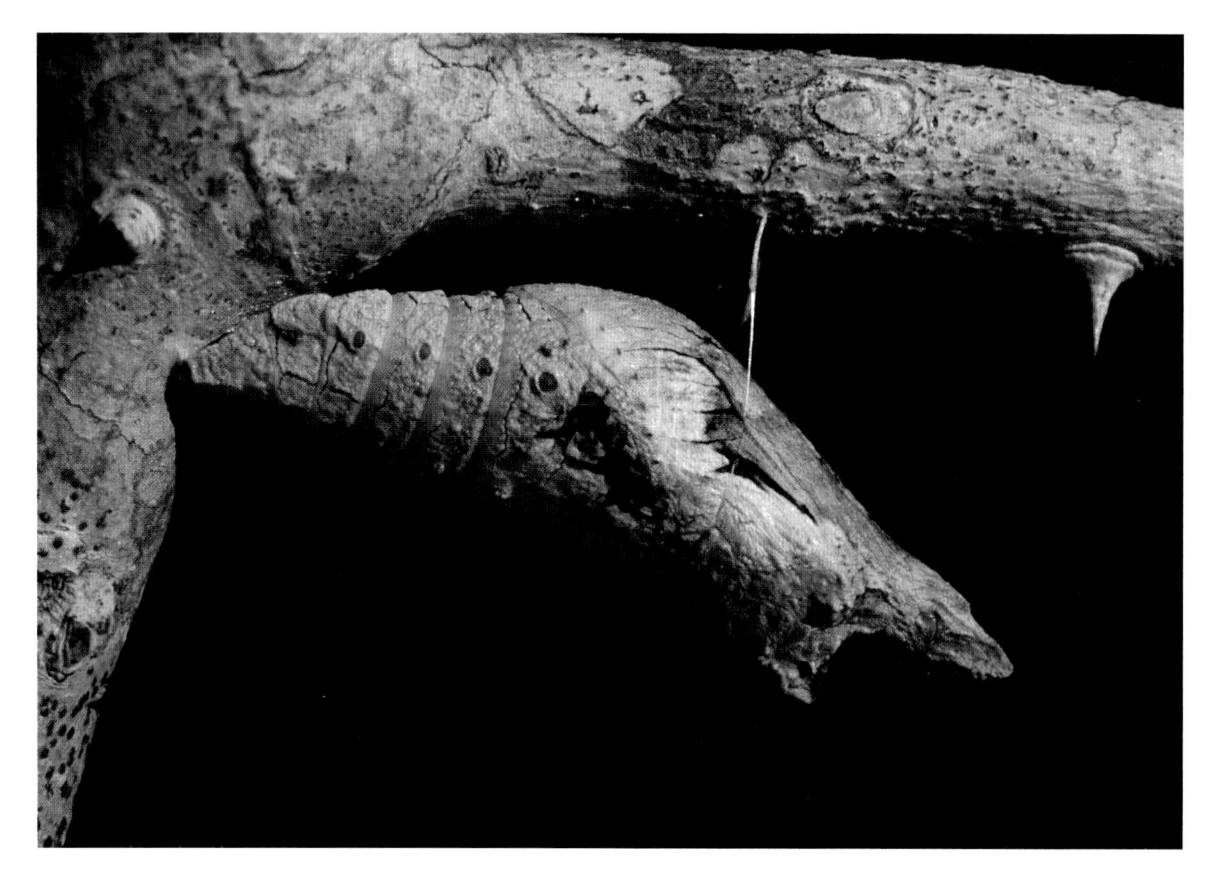

Branchlike pupa of the Giant Swallowtail *(Papilio cresphontes)*

Some butterflies have brightly colored spots or body parts that draw the attention of predators away from the vital sections of the body. The red spots on the lower wings of most of the Swallowtails and orange-tipped upper wings of the Arizona Sister (*Adelpha eulalia*) and Falcate Orangetip serve this purpose. The constantly moving "tails" and bright eyespots of the Hairstreaks make it most difficult for predators to distinguish head from tail. Close inspection often reveals portions of the wings missing near these false "parts," as if a bird or other predator had aimed for what it took to be a head and was left with a mouthful of wing instead.

Such butterflies as the Monarch, Queen, Gulf Fritillary, Pipevine Swallowtail, and Polydamas Swallowtail (*Battus polydamas*) rely on their showy appearances to attract the attention of predators by "thumbing their noses." This warning, or aposematic, coloration is found in butterflies genuinely dangerous or unpalatable to would-be predators. These insects retain poisonous alkaloids or toxins from the larval food plants on through the pupal stage and are quite indigestible. The toxins usually cause a bird or lizard to vomit, become disoriented, and otherwise feel quite ill. Usually one taste is enough. Monarchs and Queens, for instance, feed on members of the Milkweed Family (Asclepiadaceae), which contain cardiac glycosides, or heart poisons. Milkweed plants produce these chemicals as protection against herbivores, but larvae of the Monarch and Queen have developed the ability to tolerate and store these poisonous chemicals, making both the larvae and the adults poisonous to predators.

Some quite harmless and palatable butterfly species have evolved colors and patterns that mimic the unpalatable ones, thereby deriving protection because would-be predators mistake them for the bad guys and leave them alone. For instance, the "edible" Viceroy has developed coloration and wing venation much like that of the Monarch and is rarely molested. The Viceroy feeds primarily on willow (*Salix* spp.) and cottonwood (*Populus* spp.) and retains only small amounts of poisonous properties from its larval food plants. Mimics of the noxious Pipevine Swallowtail are many, including other Swallowtails such as the Black and Spicebush. In the eastern half of the state there are occasional black female forms of the usually yellow Eastern Tiger Swallowtail that look quite similar to the poisonous Pipevine. Even though the Red-spotted Purple has no tails and is not even a close relative, its general color is such that it can easily be mistaken for the noxious Pipevine, giving the insect an edge in its struggle for survival.

Home gardeners have many opportunities to provide plants for butterflies.

2 *Creating a Butterfly Garden*

As the natural habitat of butterflies is being drastically altered and in many instances destroyed entirely, there is much that the home gardener can do to take up the slack by providing these creatures with new areas where they can breed, find food, and lay eggs for future generations. In providing for the needs of butterflies, the gardener has not only the satisfaction of contributing to the continuation of the butterfly population but also the truly phenomenal pleasure of having these beautiful insects around. Watching the butterflies around us and trying to give them what they want sharpen our own senses of observation and awareness, so we cannot help learning something new. In such gardening, there is a constant seeking of new ideas, of new ways of doing things, and of offering new plants for the butterflies to try.

Butterfly gardening can be as simple or as complex as you want to make it. For the first time, you will be compelled to look at your garden through the eyes of others—the butterflies—and to consider their needs along with your own. In many instances, you will let their choices be first. To truly garden for butterflies, you must ask what they would prefer and, to the best of your ability, try to provide it. Usually the butterflies' preferences can be blended in quite nicely with your own garden plans so that the final effect is satisfactory to both.

As with any other type of gardening, planting to attract butterflies is simply understanding and working with the land. Plants selected for your garden should depend upon the interpretation of the land as well as other environmental factors of your locale. These plants should not only do well in your area but also be useful to the butterflies. The purpose of butterfly gardening is to attract the most species in the greatest numbers in a given space. This concept has been referred to as "butterfly production management" and is surely descriptive of what you will be trying to accomplish. Such a management program, if carried out with care and thoughtful planning, should be most gratifying.

A THOUGHTFUL PLAN CAN SATISFY BOTH GARDENERS AND BUTTERFLIES.

To accomplish such intensive site management, you must know which species of butterflies can be attracted to your garden, which larval food plants they prefer, and which nectar plants grow best in your area. From a good field guide, learn everything possible concerning the local species of butterflies, their habits, and their microhabitats. No matter how intensely you want to garden for butterflies, start with the most common species that use the most easily provided food plants; then gradually work toward attracting the rarer and more exotic insects. And no matter how hard you try, you cannot consistently attract breeding butterflies that do not already exist in your part of the state. It will do no good to introduce their food plant—they most likely will not come. It is far better to concentrate on attracting the species already existing in your region. All areas of Texas have several species of big, beautiful butterflies easily enticed to the garden with numerous smaller species just as interesting and intriguing in their own special ways.

Understanding Plant Terminology

As with almost all forms of wildlife, food is by far the most significant influence in a butterfly's life. Two stages in its life require two different types of food: the caterpillar stage when it eats only vegetative growth, and the adult stage when nectar is the primary food source. To provide these needs, consider three types of plants for the garden: "native," nonnative, and cultivated plants. There is

A BUTTERFLY GARDEN CAN BE PLANTED IN THE FRONT YARD.

surely nothing wrong in having plants from anywhere in the world, but it is important to understand a plant's natural range and the correct terminology to describe it.

In searching for the proper plants for a butterfly garden, you are frequently going to find plants and seeds advertised and sold as "wild, native plants and/or seeds." These terms are often misleading, for there is a definite difference between a plant being wild and being native, and even more so in being indigenous, which is rarely mentioned or stressed. In selecting plants that will thrive the very best in your garden, choose plants that are both native and indigenous—those that are growing in and well adapted to the exact area in which they will be planted.

A wild plant is any plant that has not been domesticated, meaning not having been developed by genetic manipulation. The problem is the term "native." A native plant is a wild plant growing within a prescribed area, be it natural or human defined. The borders of Texas are human defined, and if a plant grows within these boundaries, it is classified as a native plant. The term "indigenous" defines a native plant that has become adapted to a particular soil or habitat. This term is no longer accurate when a plant has been taken from the area where it has become environmentally adapted and is placed outside its natural range. For instance, a plant with a natural range west of the Pecos River that has been moved to East Texas is no longer an indigenous plant unless it naturally occurs there as well. When moved, it may still be called a native but, in actuality, becomes the same as a nonnative, no different from a plant from South America or Mexico.

The term "nonnative" is used here to define a native plant that does not naturally occur within the borders of Texas but actually grows quite well here. It is simply a plant that has been brought from its natural distribution and across boundary lines. Some plants from foreign countries, adjacent Mexico, or adjoining states of New Mexico or Louisiana are frequently used in garden plantings and grow beautifully in this state.

Cultivated plants are those that have been genetically manipulated to improve some physical feature, such as more blooms, longer blooming period, or lower growth habit. The relatively new term "cultivar" is often used in the name or description of these plants. Derived from the words *cultivated* and *variety*, in its truest sense it means a variation of an already-cultivated plant that has been manipulated and "improved" further, creating a new form or variety.

Larval Food Plants

Plants used by the female butterfly for egg deposition and later eaten by the larvae are often different species than the plants used as nectar sources by the adults. During the larval stage, the butterfly eats mostly vegetation. The flowers that will be so important to it for nectar later are totally useless to it now, unless of course, it eats the flowers, which some species do.

The female butterfly finds plants on which to lay her eggs by rather complicated and sophisticated chemical detection. She usually flies slowly from plant to plant, hovering around the leaves or stems, "smelling" and "tasting." Often she scratches the surface of a leaf to get a better smell or taste to determine the chemical content. Scents of some plants such as cabbage (*Brassica oleraceae* var. *capitata*), dill (*Anethum graveolens*), common fennel (*Foeniculum vulgare*), parsley (*Petroselinum crispum*), or cherry (*Prunus* spp.) are so strong the butterflies can detect them from a great distance and have no problem finding them. Other plants, such as Erect Pipevine (*Aristolochia erecta*) and the thin, grasslike flax (*Linum* spp.), must be diligently searched for among the tangled stems of grasses and other plants.

Some butterflies are very specialized in their larval food choices, using only plants within a few closely related families. In other instances, only one genus or species within a plant family is selected as a larval food source. Specialization goes even further in that in many situations the young larvae eat only buds or young fruit; if these are not available, the young hatchlings starve to death. Other species are not so particular: More than seventy-five species of plants have been reportedly used by the Spring/Summer Azure (*Celastrina ladon*), and more than ninety species are reportedly used by the Gray Hairstreak (*Strymon melinus franki*). The Painted Lady (*Vanessa cardui*) prefers thistle (*Cirsium* spp.) but will use more than one hundred other plants.

Butterflies such as the Blues and Hairstreaks (family Lycaenidae), even though they use a wide assortment of plants, are real stay-at-homes, living and breeding within a few hundred yards from where they were born.

A BUTTERFLY GARDEN ALONG A WALKWAY

They use only the plants available to them in that area. Not straying from their original breeding places, they form small local colonies sometimes with a distance of several miles to the next populated area of the same species.

Some species of butterflies, such as the Hackberry Emperor (*Asterocampa celtis celtis*), Question Mark (*Polygonia interrogationis*), American Lady (*Vanessa virginiensis*), and Red Admiral (*Vanessa atalanta rubria*), breed throughout the state because their food plants are available throughout. Others, such as the Zebra Swallowtail (*Eurytides marcellus*), will be found breeding only in the eastern portion of Texas because the larvae feed only on pawpaw (*Asimina* spp.) and that is the only place the plant occurs in the state. The Chisos Metalmark (*Apodemia chisosensis*) occurs in West Texas because its food plant, Havard's Plum (*Prunus havardii*), grows only on the caliche soils of that area. Butterfly species such as these have evolved a very intricate relationship with their food plants.

For egg laying, almost all species of butterflies more readily use native plants, but occasionally the cultivated sorts are chosen—especially if the wild species are not as abundant or in as good condition. Fortunately for the gardener, many of these wild plants are very appealing and add much to the overall scheme of the garden. Wild Black Cherry (*Prunus serotina*), the food of the Red-spotted Purple (*Limenitis arthemis*), and willow (*Salix* spp.) and elm (*Ulmus* spp.), used by the Viceroy (*L. archippus*), are all popular landscaping trees. Many trees, shrubs, and vines—such as

This sunny garden is protected from the wind by a charming picket fence.

Sassafras (*Sassafras albidum*) and Spicebush (*Lindera benzoin*) for the Spicebush Swallowtail (*Pterourus troilus*); Wafer-ash (*Ptelea trifoliata*) for the Eastern Tiger Swallowtail (*Papilio glaucus*); and the various passionflower vines (*Passiflora* spp.) used by the Gulf Fritillary (*Agraulis vanillae incarnata*), Julia Longwing (*Dryas iulia*), and Zebra Longwing (*Heliconius charithonius vazquezae*) are attractive in the garden.

Some larval food plants are seasonal, flourishing only briefly in the spring. Others may not be available until summer or fall. The most commonly used or most preferred plant may have been killed by an early frost or is not available due to other climatic conditions such as drought or flooding. Both the season and the availability of young, tender growth greatly influence the choice of the female toward egg-laying sites. As a result, eggs may be laid on one species of plant in early spring; then eggs of the second brood may be laid on an entirely different plant as it comes into season. The Common Buckeye (*Junonia coenia*), for instance, uses toadflax (*Nuttallanthus* spp.) and paintbrush (*Castilleja* spp.) for the first spring egg laying, plantain (*Plantago* spp.) for the summer broods, and agalinis (*Agalinis* spp.) and Yellow False Foxglove (*Aureolaria flava*) for the fall broods. Not only are such diverse choices of food plants to be expected within a particular area but some species of butterflies use different food plants in different regions. For example, in East Texas the Mourning Cloak (*Nymphalis antiopa*) commonly uses elm, but in West Texas the caterpillars are more often found on the leaves of willow, although both trees occur throughout the state.

Some species of butterflies are inconsistent in their breeding areas, and there are times when a species that has previously used a food plant in your area or garden will be conspicuously absent one year. This does not mean this particular species is gone from the area for good. Just give it time and another year—it may be back in even greater numbers. This unpredictable fluctuation is one of the things that makes butterfly gardening exciting. After a while you may begin to refer to time in your garden as "the year we had so many Common Buckeyes," "the year the Blue-eyed Sailors (*Dynamine dyonis*) were here," or "the year the Black Swallowtails ate all the parsley—and the fennel and the dill."

Keep Them Growing

In order for the entire life cycle of butterflies to progress normally in the garden, the caterpillars must have good-quality food and plenty of it at all times. Even though it is possible to have an abundance of larval food plants and still not have butterflies, there is surely a greater possibility of attracting them to the garden if the food plants are present. Besides the trees, shrubs, and perennials selected and planted in permanent locations, extra seeds or potted plants of some annual species left from spring planting can be kept on hand for continual planting throughout the season. Parsley and dill for Black Swallowtails (*Papilio polyxenes*) can be planted every two to six weeks during the breeding or flight season by scattering the seeds in open areas throughout the flower beds. Fill planters or pots with young plants, and place them around the porch or patio, rotating the containers as the foliage is eaten down. After the larvae have eaten a large portion of the potted plants, the hungry little creatures can be moved to larger plants, leaving the eaten plants to form new growth. During the summer months there are almost no native plants available that Black Swallowtails can eat, so their summer broods depend almost entirely on cultivated members of the Parsley Family (Apiaceae). For other species of butterflies that need young, tender foliage, Nasturtium (*Tropaeolum majus*) or hollyhocks (*Alcea* spp.), beans (*Phaseolus* spp.), and even cabbage can be planted at regular intervals.

Occasionally pinch or trim back portions of small shrubs and perennials to ensure continual new growth. Do not trim severely if a plant is less than three years old, for its root system may not be able to tolerate the loss of foliage. Vines such as passionflower and pipevine (*Aristolochia* spp.) and the Common Balloon-vine (*Cardiospermum halicacabum*) should be kept well watered and growing, especially during the summer months.

When fertilizing any plant, it is best to use a product applied to the soil. Any liquid fertilizer applied to the plant completely contaminates both the foliage and nectar, making them useless to the butterfly. If the product is purchased, make sure it contains no insecticide, as some fertilizers do.

Become a Caterpillar Lover

A "worm" to most folks is an ugly, repulsive creature and something to be squashed immediately upon sight. This attitude is most unfortunate, for none of the caterpillars of North American butterflies bite, sting, or spit in your eye. Not only are they perfectly harmless but when examined closely, they are among the most beautifully and intricately colored of all wild creatures. Many of them rival their final winged stage in markings and coloration.

To develop an appreciation of caterpillars, you need to become familiar with them and be able to recognize them. If plants for the larvae are being provided in your garden, it will certainly be important to distinguish the "good guys" from the "bad guys." Unfortunately, there is not space in this book to cover thoroughly the caterpillar stage of butterflies, but three new guides to caterpillars have recently been published (see the bibliography) that can be most helpful. These guides have excellent photos along with descriptions of the caterpillars, but there is a drawback. Butterfly larvae go through several molts, or skin changes, and often come out with a totally new look during that particular instar. Most field guides usually describe (or show) only the way the larvae appear in the last instar, which is often completely different from the earlier stages.

The lush, cottage-border approach in this garden allows the gardener to "hide" less attractive food plants at the back.

It would be good to keep a small notebook with descriptions of the caterpillars you find. Take pictures of them if possible. Identify the plant a caterpillar is feeding on. Then look in the index in this book or indexes in butterfly field guides; nearly always there will be an additional index of plants. See which butterfly larvae feed on the plant where you found the caterpillar. Check the description of the butterfly to see if its range includes your area. If so, then you most likely have something to get excited about.

A number of butterfly larvae are easily identified since they feed only on very specific plants. These plants, such as the milkweeds, passionflowers, pipevine, or members of the Mustard Family (Brassicaceae), contain volatile oils or poisons that are usually totally unacceptable to most other insects. If you find caterpillars eating any of these plants, they are almost certain to be butterfly larvae.

Some literature indicates that all larvae of butterflies are a serious problem of garden ornamentals or forage crops. As a general rule this is not true. Taken as a group, neither the adult butterflies nor their larval stages could be considered pests, and they have become innocent casualties of humankind's indiscriminate use of pesticides and herbicides in its war against a few noxious insects and unwanted plant growth. In North America only the imported Cabbage White (*Pieris rapae*) and the native Checkered White (*Pontia protodice*), whose caterpillars feed on cultivated members of the Mustard Family; Orange Sulphur (*Colias eurytheme*), which feeds on Alfalfa (*Medicago sativa*); and the Gray Hairstreak, which feeds on Cotton (*Gossypium hirsutum*), could be considered real offenders.

If nothing seems to match from all the comparisons and if the larva is not doing extensive damage, leave it alone. If not a butterfly larva, it could possibly be the caterpillar of some big, beautiful moth. Keep checking to see what is happening; then one day perhaps it will be in its last instar and recognizable as a really special butterfly you have been trying to attract.

Nectar Plants

Nectar plants, which attract adult butterflies, are found in both the native, nonnative, and cultivated categories. To attract the greatest number of butterflies to your yard, do not hesitate to mix the native with the nursery-obtained plants; in almost all instances they grow happily side by side. Just be sure the moisture, sun, and drainage needs are similar for each bed. Fortunately, cultivated plants have been genetically developed so they have a wide tolerance for differences in soils, amounts of moisture, heat, and the like. They have also built up resistance to pests and diseases that usually mean death to a wild plant out of its native range. You can usually plant this nursery-grown stock with confidence that it will generally perform well in your garden when given proper care.

There are some situations, however, where plants bred for largeness of flowers, hardiness, or other special feature have lost in the process their former ability to produce copious nectar. Petunias (*Petunia* spp.) are a good example. Most petunias available at nurseries today are stocky, variously colored, early flowering, and full of blossoms—but they are also scentless and have practically no nectar. The petunias of fifty years ago were rangy and mostly bluish, pinkish, or white in color, but they were intensely fragrant and bore an abundance of sweet nectar butterflies would fight for.

Flower Visitation

In working out his monumental system of classification, the eighteenth-century Swedish botanist Carl von Linné (Linnaeus) named every plant part in order to describe plants properly and consistently. He discovered that many flowers were endowed with structures not directly associated with the reproductive aspects of the plant, yet the structures were often situated very near the reproductive parts. These structures either produced or contained a wet, sweet fluid. As Linnaeus could determine no purpose for their existence, he gave the fluid the name "nectar"—the drink of the Greek gods—and the parts of the flowers producing or containing the fluid he named "nectaries."

In most species of plants, this grouping of special cells or nectaries is located within the flower and known as floral nectaries or nuptial nectaries since they aid in pollination. When these cells are situated on parts of the plant other than the flowers, such as stems, leaves, or bracts, they are called extrafloral nectaries or extranuptial nectaries. It is the floral nectaries that are most attractive to butterflies.

Floral nectaries vary widely in their appearance, sometimes very striking in shape and color and also very conspicuous in their placement on the flower. They can be found on all parts of the flower—the pistil or stamen or along any part of the petals or sepals. Some nectaries are very simple in structure; others, more elaborate and in the form of raised or enlarged rings or ridges of tissue, usually at or near the base of the petals.

The nectaries in short-tubed flowers, such as Butterfly Bush (*Buddleja davidii*) or the lantanas (*Lantana* spp.) and verbenas (*Verbena* spp.), are situated at the base of the tube with the nectar readily available. The shorter spurred violets (*Viola* spp.) also allow usage by a great number of butterflies. Here the nectar is produced by nectaries on stamens that extend into the spur and allow the sweet juice to flow into the spur, which acts as a storage tank or storage jug. Flowers such as columbine (*Aquilegia* spp.) and larkspur (*Delphinium* spp.) have nectaries formed in long, conical petals that sometimes taper to a hooked point. Only a butterfly with a very long proboscis can extract nectar from these longer nectaries.

Although there are some instances of openly exposed nectaries, many flowers have evolved protection for the nectar. Some of the more common methods include drooping flowers, narrow basal tubes and passages where water cannot enter, nectaries hidden deep within the flower, and the nectary being covered with thin flaps of tissue or a fringe of hairs. Such arrangements not only protect the nectar against evaporation but also ensure that it does not rapidly thicken or crystallize. They also prevent the nectar from being diluted by dew or washed away by rain. As an added measure of protection, some plants close the petals during cloudy or rainy weather and at night.

Nectar

The original purpose of nectar production by plants is still not clearly understood, but nectar did not arise in connection with pollination. There is evidence that plants produced a nectarlike substance previous to the evolution of pollination that usually occurred on parts of the plant totally independent of the floral region. The type of nectar we are most familiar with today—the sweet stuff produced by the flowers themselves—is, in evolutionary time, relatively new. Flowers seemingly refined its use and placement, along with the development of pollen, as an added attraction to insects to aid in the reproductive process.

Regardless of what nectar is to a plant, for butterflies it is, in most cases, their major means of nourishment. This celestial beverage is the one thing that will draw the most butterflies to your garden, and the amount and its quality determine whether they keep coming back day after day. And nectar means flowers. So, for the gardener wanting to attract butterflies, it can be put in quite simple

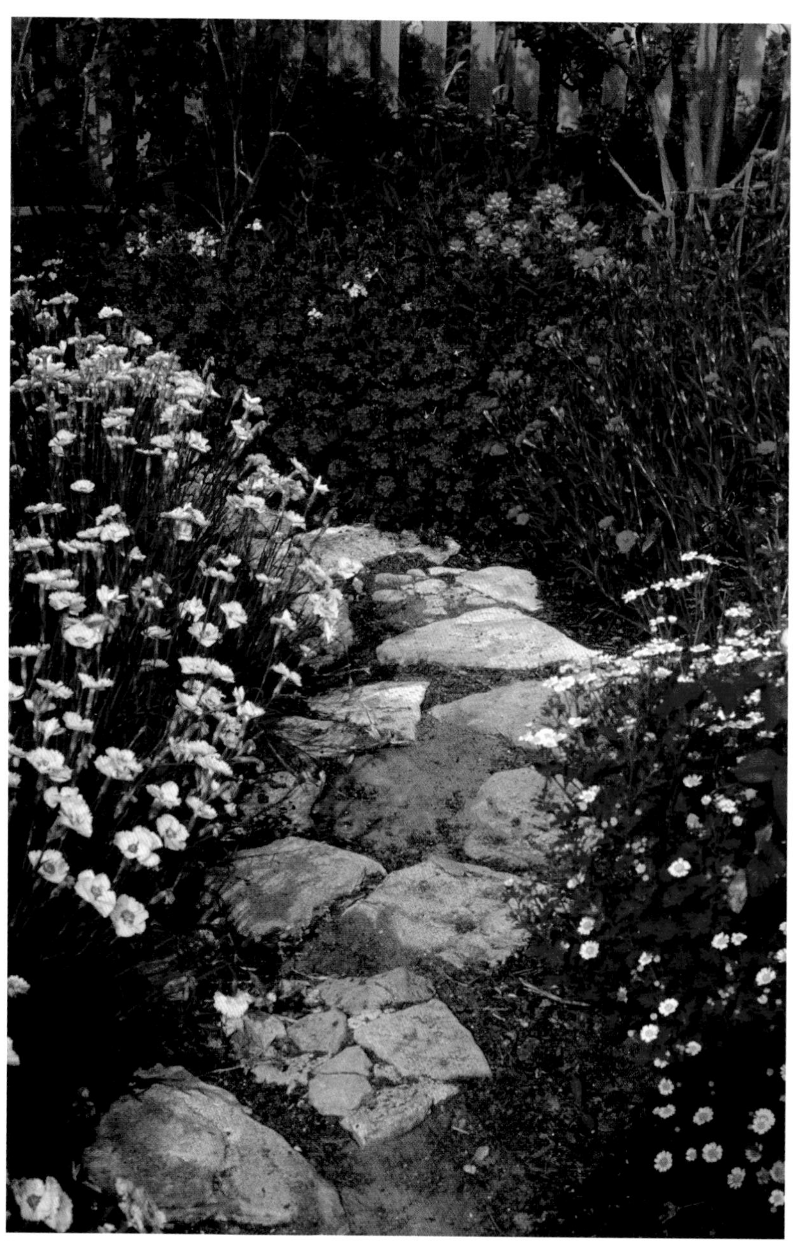

PINK AND PURPLE COLORS ANCHOR THESE BORDERS OF LOW-GROWING PERENNIALS.

terms: flowers equal nectar equals butterflies.

What is nectar? Basically, it is sugar water, generally containing from 25 to 40 percent sugar and 60 to 70 percent water. Some nectar, however, may contain as little as 8 percent sugar and some as much as 76 percent. Along with the sugars (the primary ones are glucose, fructose, and sucrose or saccharose) are trace amounts of amino acids, proteins, organic acids, phosphates, vitamins, enzymes, and flavonoids. The flavonoids produce the various scents of nectar.

Nectar is generally colorless but may occasionally have a slight yellowish, amber, or greenish coloring. Often it is noticeably scented or possesses a definite flavor. Anyone who has broken the end from a flower of Old-fashioned Petunia (*Petunia axillaris*), or honeysuckle (*Lonicera* spp.) or wild verbena, and licked the sweet juice from along the stamens and pistil and out of the "tube" can attest to the sweetness and distinct flavor of their nectar.

Nectar is such a powerful attractant that in some plants it has replaced visual attractants. Flowers yielding abundant nectar are frequently less showy or conspicuous than flowers with less nectar. For instance, the gaudy-colored, dinner-plate-sized flowers of some hibiscus (*Hibiscus* spp.) yield less nectar than a one-half-inch pale lavender lilac (*Syringa* spp.) blossom.

Different plants secrete optimal amounts of nectar at different times of the day, and sugar content can also fluctuate considerably. Concentration of nectar in the early morning is often very low in some plants. As the day advances, the intensity of sunshine, rising temperatures, or reduction of humidity may cause the sugar percentage to increase up to four times. In some cases, the nectar is the sweetest at the same time it is most abundant. In other plants, the sugar quality remains constant throughout the day, with variations in the amount of nectar produced. In still other situations, the amount of nectar produced is constant, while the amount of sugar varies. Usually the best nectar is produced at midday and is the sweetest when the sun is shining, the temperature increasing, and the humidity decreasing.

The variability of the nectar produced and its sugar content are primary reasons butterflies use certain plants in one part of the day, then switch to other plants at other periods. Studies have shown that in general, butterflies seem to prefer nectar with somewhat low sugar concentrations but high in nitrogen-rich amino acids. A butterfly may prefer the more diluted nectar to prevent water loss to the body; also, the thicker nectar may clog the proboscis and make gathering difficult.

Various environmental factors, such as temperature, humidity, wind, day length, sunlight, soil, and the health of the plant, greatly affect production or secretion of nectar. These factors vary in importance for different plants and at different times of the year, making for difficulty in determining optimum conditions for good nectar flow for a particular plant at any specific time.

Some plants produce nectar at quite low temperatures, while others do not begin secretion until the temperature is fairly high. Many plants flowering in very early spring produce nectar when the temperature is too cold for butterflies to fly. Other plants do not open blooms until the temperature is well into the eighties or nineties. If all other factors are favorable, a general rule is a cool night followed by a clear, hot day initiates the most abundant nectar flow.

Wind plays an important role in nectar production, depending upon the circumstances. Strong, cold winds are almost always detrimental, but mild, warm, drying winds after periods of rain are usually quite beneficial.

Length of daylight always affects the amount of nectar secretion, since this is an integral factor in the plant's reproduction cycle. The period of flowering and seed-set are intricately timed with day length, so the production of nectar as an extra pollination attraction during this time is very important.

In most instances, the amount of sunlight plants receive has a direct bearing on the amount of flower nectar produced. To attract the optimal number of butterflies to the garden, place plants in full sun. One of the major reasons butterflies prefer the open, sunny places is that the warmth of the sun promotes greater nectar production in the flowers. Therefore, the more sun and warmth received by the plants, the more nectar produced, providing more food for the butterflies. In almost all situations in which the same species of plants known to be readily used by butterflies are placed with some in sun and some in shade, the plants in the sun will be actively worked by the butterflies and the shaded plants virtually ignored. Butterflies may use the shaded plants to some extent when they are the number-one choice for nectar preference, but the amount of nectar obtained is not nearly as great as if the plants were in the sun. Often the butterflies do not find them worth the effort of exploration.

Type and condition of the soil have a direct bearing on the amount of nectar produced and its sugar content. Such soil conditions as fertility, moisture, and pH may affect not only the growth of the plant, especially if it is a transplanted wilding, but also the secretion of nectar and its quality. A well-grown plant with lush foliage or even adequate blooming does not necessarily mean that

A BUTTERFLY GARDEN SHELTERED BY BAMBOO FENCING

maximum nectar secretion is taking place or that the nectar is of the best quality. In many instances, when a wild, ordinarily heavily producing nectar plant is transplanted from one region to another, the nectar production is drastically lowered and the plant hardly used by butterflies. Very little research has been done on the relation of soils to maximum nectar production and quality, so the best thing the gardener can do is use native plants that are from as similar environments as possible, supplemented with cultivated stock such as Pentas (*Pentas lanceolata*) and Butterfly Bush or the abelias (*Abelia* spp.) and zinnias (*Zinnia* spp.), which are known to produce abundant and high-quality nectar under almost any conditions. But if native stock is transplanted, take a sample of soil and try to duplicate it as closely as possible in the garden.

The general health of a plant is important for nectar secretion. If a plant becomes stressed from lack of moisture to the point of even the slightest sign of wilting, nectar production stops immediately. The amount of moisture that perennials, shrubs, and trees have received months before has a direct bearing on nectar production. Often it is the late fall and winter rains that ensure abundant spring and early-summer nectar. Naturally, a good gardening policy is to keep the plants in the best growing condition by proper fertilizing and adequate moisture at all times, even during dry winter months.

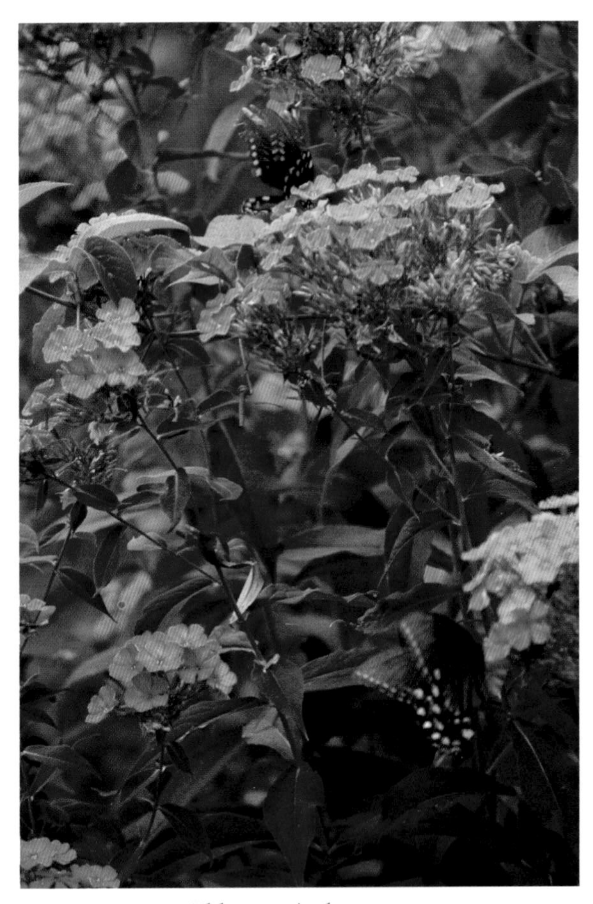

Summer Phlox (*Phlox paniculata*)

Color

Although butterflies can see in the full color spectrum, with observation it is quite obvious that some butterflies fly readily to flowers of one color while passing by an entire bed of flowers of another color. Some species of butterflies definitely have favorite colors. Butterflies in general find little attraction in flowers of greenish-blue to blue-green color. Many butterflies use orange flowers, and others fly most often to red flowers, but by far the most favored flower colors (when available) to most species of butterflies occur in the purple, pink, yellow, and white ranges, with true blues next best.

However, there are factors other than color that may influence a butterfly's choice of where it will dine. Many flowers possess visual guide marks or lines called nectar guides, which show insect visitors the direct route to the food. In looking over a patch of flowers, the butterfly naturally chooses the ones most clearly marked. It is much the same as if we had a choice of two places to go with the same objective at the end, but with a map drawn to one location and with just an address for the other. If our reward was to be exactly the same upon arrival at either location, we would most likely choose the one with the easy-to-follow map. Butterflies are no different, and many flowers have developed such maps. Sometimes these markings are very conspicuous to our own eyes, but in other cases they are marked in the ultraviolet range of colors and are visible only to the butterfly.

The guide marks used by some flowers are in the form of a ring of a lighter or darker color at the base of the petals, such as seen in morning glories (*Ipomoea* spp.), frogfruits (*Phyla* spp.), blue-eyed grasses (*Sisyrinchium* spp.), phlox (*Phlox* spp.), and verbenas. Other flowers, especially the tubular ones like Desert Willow (*Chilopsis linearis*), Lemon-mint (*Monarda citriodora*), Fragrant Devil's-claw (*Proboscidea louisianica fragrans*), and most salvias (*Salvia* spp.), penstemons (*Penstemon* spp.), and agalinis, have developed elaborate patterns of dots, splashes, or large solid patches of contrasting colors on the lower portion of the corollas that point to the nectar source. Still others are conspicuously marked with contrasting lines, as in the Spring Beauty (*Claytonia virginica*) and violets.

In many instances, flowers that appear solidly colored to us are distinctively patterned to a butterfly. Again, as in the patterns on a butterfly's wing, ultraviolet light plays a role. These ultraviolet markings are more common on the flowers than the patterns we can see. A number of flowers have a combination of nectar guides, both ultraviolet absorbing and visible to us.

The large blotches at the base of petals and on the lip of some tubular flowers also act as a visible landing platform for the butterfly. For the insect, this mark is an attraction, but it must be investigated a little further. With the scent-sensitive tips of the antennae the butterfly probes, and with the feet it taps the petals, until all senses tell it, "I am standing near sugar." Out rolls the proboscis, and the butterfly begins to imbibe the sweet juices.

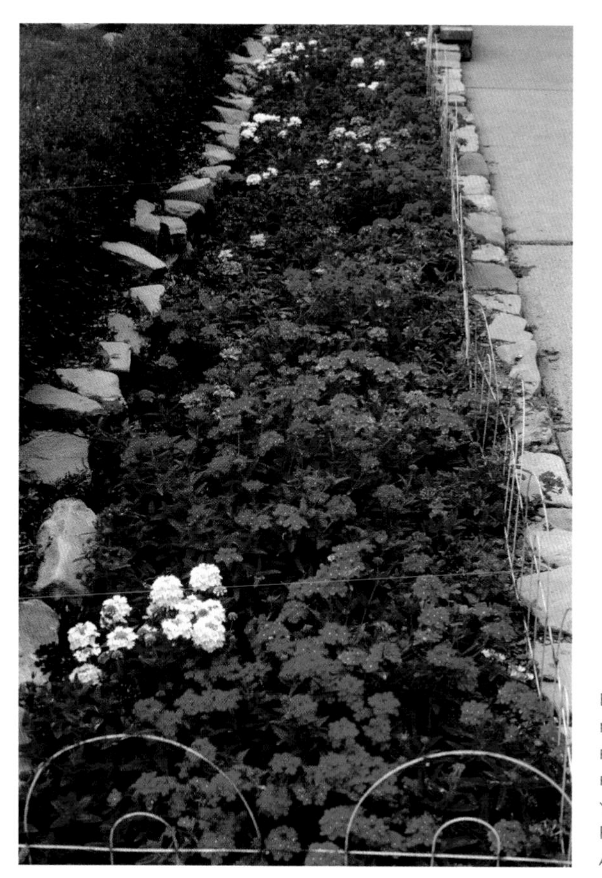

Butterflies fly most often to flowers in the purple, pink, and yellow ranges. Red flowers are also popular.

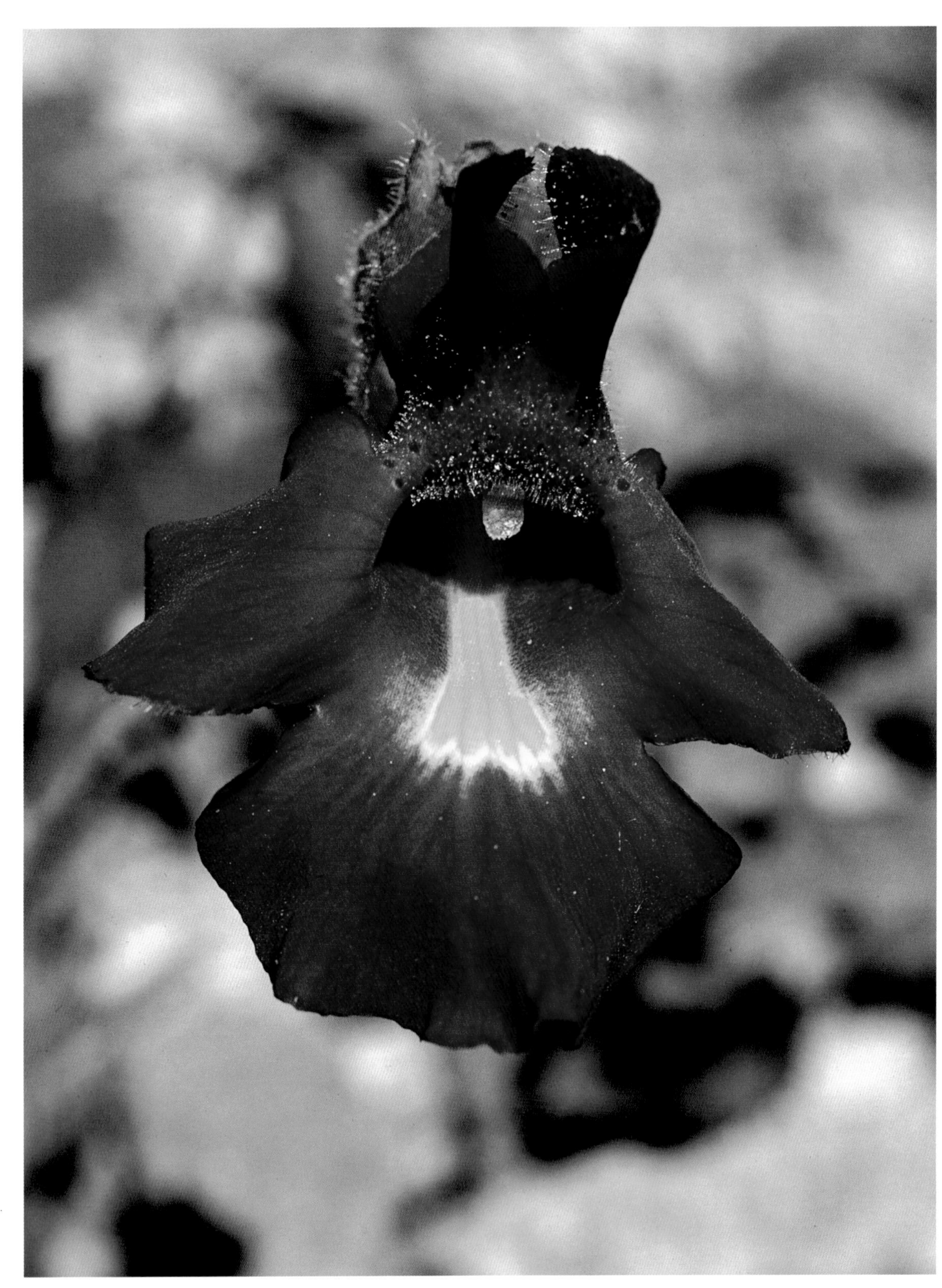

FRAGRANT DEVIL'S-CLAW (*Proboscidea louisianica fragrans*). A FLOWER'S GUIDE MARKS AFFECT NECTAR GATHERING.

Fragrance

In locating nectar, butterflies use the fragrance and scent markings of flowers as much as or more than the visual markings. The ultraviolet and other visual marks are specially scented, and even flowers that bear no visual nectar guides at all usually have the approach to the nectar marked by a fragrance guide. Often these scent marks smell quite different from the other parts of the flower. Even the food source itself, nectar, as well as the nectaries, is also usually distinguished by a stronger or entirely different scent.

After a certain length of time, both the fragrance and the color of the nectar guide change. This can easily be observed by comparing a freshly opened Red Buckeye (*Aesculus pavia*) flower or a bluebonnet (*Lupinus* spp.) flower with one that has been open long enough to be pollinated. The once strongly aromatic guide becomes almost scentless and changes from a striking yellow to a dullish red. These scent and color changes act almost as warnings to insects that either the flower has been pollinated or the pollen is aged and no longer viable and the nectar production diminished. It does not want to be bothered anymore. It is now withering or making fruit—not nectar.

Flower Shape

One important factor to a butterfly selecting flowers on which to feed is the flower shape. Some flowers may be abundant nectar producers, but because the nectar is situated at the bottom of long tubes or covered by stiff flaps or hairs, the nectar remains unavailable to all except butterflies possessing a very long proboscis. This inaccessibility definitely limits the number of species that can make use of such flowers.

The actual size of the flower, in most instances, is also important, as is its placement on the stem. The large trumpets of morning glories are usually worked by the larger, longer-tongued Swallowtails (family Papilionidae) and Fritillaries (family Nymphalidae). However, because the flower throat is widely spreading, some of the longer-tongued Skippers (family Hesperiidae) alight on the rim, then crawl deep within the blossom until the proboscis can reach the nectar. Long-tubed flowers of *Salvia* and *Penstemon*, which are arranged vertically along the plant stem, make nectar gathering difficult and are mostly used by larger butterflies that do not alight but feed while continually beating their wings.

The size of the flower clump is also significant in a butterfly's nectaring choice. A few solitary flowers, no matter how rich in nectar, are rarely as attractive as a cluster of numerous smaller flowers. Although butterflies are forced to take nectar where they can get it, their preference is a large supply easily obtained. Small, short-tubed flowers with wide, flat rims, with the flowers grouped in clusters, and with many clusters on the plant are ideal. Flowers in the Phlox Family (Polemoniaceae) and Verbena Family (Verbenaceae) are good examples, and almost any species from either group never fails to attract the butterflies. Many members of the Aster Family (Asteraceae) are excellent, as are most species of the Mint Family (Lamiaceae). Such a flower arrangement provides a good perch for the insects while taking nectar, and the

CHERRIES *(Prunus* spp.) ATTRACT THIS GOATWEED LEAFWING *(Anaea andria)*.

close clustering of the flowers saves much-needed energy that would otherwise be spent in flying from one solitary flower to another.

OTHER ATTRACTANTS

Not all butterflies are attracted solely to flower nectar and readily partake of liquid from such things as tree sap, honeydew, overripe or rotting fruit, dung, carrion, urine, and mud. Natural nectar, sap, and overripe fruit contribute much-needed protein to their diets, while dung, urine, and carrion provide amino acids. Salts, especially necessary to the males of some species, are usually obtained from mud puddles or the edges of streams, bogs, or seeps.

In order to get the species more attracted to these types of fluids into a chosen location, you must garden a little differently. Fortunately, it is not costly and can be quite easy. Since a tree that exudes sap is generally an injured or diseased tree, trying to deliberately furnish this particular substance as an attractant is not recommended. The butterflies that feed on tree sap readily take other liquids that the gardener can more easily provide.

In place of natural tree sap, try the process called sugaring. Make an elixir absolutely irresistible to butterflies by combining in a blender one pound of brown sugar, approximately one-half cup of dark, strong molasses, very old fruit of some sort, and a can of beer. Overripe bananas (*Musa* spp.)—the mushy kind with black skins—and very soft, squashy, even fermented peaches (*Prunus* spp.) are good choices. Instead of (or in addition to) the beer, some rum can be added; if neither of these is available, add a dash of artificial rum, banana, or peach flavoring. Leave this mixture thick enough to spread but with as much liquid as possible—remember that butterflies are after the liquid. Let the concoction ferment for half a day or overnight in an uncovered container; then brush it onto tree trunks or posts, placing it from ground level to shoulder height. Even better, pour it out into a large, shallow dish and place it on the ground. Goatweed Leafwings (*Anaea andria*), Hackberry Emperors, Common Buckeyes, Question Marks, Red-spotted Purples, and Mourning Cloaks generally go crazy over this mixture. If there is a wooded area near or adjacent to the flower garden, spread some of the bait on a log or stump in an opening with dappled shade. Then watch some of the less common species such as the Large Wood-Nymph and various Satyrs gather in mixed groups to sample the feast.

To attract the fruit-drink lovers who prefer the "real thing," place fruit on mulched ground or in a shallow dish in a spot in the sun or a semishaded area of the garden. Over-

PLACE FRUIT BOWLS AT THE OPEN EDGE OF A FLOWER BORDER.

ripe or damaged fruit can often be obtained quite cheaply or for free from grocery stores or fruit stands. Peaches, bananas, pears (*Pyrus* spp.), and both wild and cultivated plums (*Prunus americana*) and persimmons (*Diospyros* spp.), will be considered quite tasty. Hackberry Emperors, Question Marks, Goatweed Leafwings, and Mourning Cloaks especially love pears and persimmons and often congregate by the dozens on the ground to feed on the fallen fruit. Occasionally, Monarchs (*Danaus plexippus*) and Gulf Fritillaries can even be found enjoying the juice. Outer rinds of watermelon (*Citrullus vulgaris*) and cantaloupe (*Cucumis melo cantalupensis*) along with the juices are well liked by several species. Place thick sections or halves in dishes, or bore holes in the sides and hang the fruit by wires or strings from low limbs of trees.

Crushed black grapes (*Vitis* spp.) or various berries mixed with a small amount of honey or old molasses and allowed to begin fermentation are readily imbibed. A bit of added beer or rum draws an even larger crowd. The sippers apparently cannot get enough of this tasty brew and will crawl around the fruit, fluttering their wings erratically.

An excessively juicy mixture is best piled in a shallow, earth-colored pottery dish and placed at the edge of a flower border. Have several dishes scattered about the garden for best results. If some of the dishes are not being used, move them to other locations until the butterflies find them.

If fruit is to be placed in the flower border, be sure the area is sufficiently large for the butterflies to feed. Since they are eating on the ground, the feeding space needs to be large enough and clear enough to allow a quick takeoff in case of danger. Provide an area no smaller than three feet by five feet and completely free of plants. If a basking area has been provided, one end of this should do quite nicely as a feeding area.

Some butterflies do not regularly use the true nectar produced by flowers yet commonly feed on a substance that greatly resembles nectar in appearance. Known as "honeydew" or "leaf honey," this sweet, sticky substance is often found covering the surfaces of tree leaves, your car, and lawn furniture during summer months. It is a tree sap by-product excreted by aphids, scale insects, and leafhoppers. Butterflies such as the Snout (*Libytheana carinenta*), Hackberry Emperor, Mourning Cloak, and Painted Lady feed on this "false nectar."

If space permits, manure to be used in the flower beds can be piled in a sunny, out-of-the-way corner of the garden for the nourishment of certain species. Try to put the pile where it can be watered frequently, as it is only the liquid the insects seek. The amino acids produced by manure sometimes draw butterflies by the dozens. Raw meat that is too old for table use works very well.

Groups of butterflies are often seen feeding at the margins of small puddles or excessively moist areas. Sometimes they congregate in large numbers; sometimes there will be only two or three. The grouping may be of only one species or a mixture of several species. In almost all instances, the group contains only males. Occasionally, a female will join the group but usually neither stays as long nor appears very interested in the available liquid. The communal gathering of butterflies at such areas is aptly referred to as a "drinking club" or "puddling club." When a butterfly discovers an especially attractive site, others passing by notice the action and stop to join. Once the insects begin their drinking, they become so absorbed in the process they can be easily approached.

The habit of feeding at puddle margins is not necessarily related to water requirements; the insects are seeking salts or amino acids. Frequently, the puddles or moist areas are from water contaminated or polluted by dung, urine, or dead animals, which provide the nutrients needed by the insects. The importance of these salts to butterflies is evident in their continuing to probe certain puddles long after the soil has become almost dry, instead of visiting areas of fresh, uncontaminated water.

To provide these salts that the butterflies seem to love so much, a simple salt and sand area can be constructed. Begin, as with placing fruit, by choosing an open area of the flower border or an area close by. Be sure to position the area where it can be kept moist. If the puddling area is to be within a border, provide an area large enough and open enough for quick flight takeoff. The next step is to construct a form of short one- by two-inch boards in whatever shape desired to form a corral-type area approximately two by two feet square. The boards may be laid on top of the ground, or the soil may be scooped out and the top of the boards placed flush with the ground. The boards do not have to be nailed together if placed in the ground.

Line the form with heavy plastic, cutting to fit the top of the frame. White plastic is less unsightly in case rain should expose the edges. The plastic helps keep the sand moist as well as keeps the salt out of the soil around nearby plants. Fill the form with sand, the kind sold at lumberyards or businesses that mix concrete. After placing the sand in the form, water it down; add more sand if needed after settling. Next, sprinkle sea salt over the top of the sand, mixing it in well; then barely wet it down again. Rock salt may be sprinkled on the sand before the form is completely filled and then another inch or so of sand sprinkled over the salt. Do not make the area overly salty: a ratio of approximately one-half to three-fourths cup of salt to one gallon of sand should be about right. A bit of manure, urine, or beer sprinkled over the top will make it even more attractive. Top it off with a thin layer of pea-sized gravel if desired.

For the salts and minerals to be used by the butterflies, they must be available in the form of liquid. Therefore, the sand bed must be kept moist at all times. If you have a drip system in your garden, the ideal placement of the sand bed is with the bed near one of the drip emitters. Instead of the usual emitter, insert a length of small tubing that reaches the center of the sand bed. Attach the emitter to the free end of the tubing and place near the center of the sand. Cover the tubing and emitter with sand, if desired. Also, the sand bed can be built under or around a barely dripping faucet in a sunny area.

ZEBRA SWALLOWTAILS *(Eurytides marcellus)* PUDDLING

Basking

Butterflies spend a lot of time basking in the sun, and they bask more often in early spring and fall. Especially in the morning, butterflies can be seen resting with wings outspread, soaking in the warmth. The smaller butterflies frequently sit perfectly still with wings outspread on the flowers from which they are feeding. Larger butterflies usually choose an exposed area in the sun for basking, such as a leaf, stick, log, or open ground.

To provide basking areas in the garden, place a decorative log or rock in a more open area of the border among the plants. Pine needles (*Pinus* spp.) or other freshly fallen leaves spread thickly over a three- by five-foot area, with a rock or log added, also work quite well. The dark-colored leaves rapidly soak up the sun's rays, providing warmth to the butterfly from below as the sunlight does from above. In placing the basking material, whether leaves, rocks, or logs, be sure that it fully catches the early-morning sun. Another basking area facing west will be welcomed by butterflies flying late in the day, especially during the cooler fall months. A convenient place for warming up will allow them a few more minutes for feeding—much needed to see them through the lengthening, cooler nights.

Hibernation

Some species of butterflies, such as the Red Admiral, Mourning Cloak, and Goatweed

Cracks in fence posts provide hibernation areas for butterflies.

Leafwing, pass the winter in hibernation as adults. Cracks in logs, fence posts, loose boards, or tree bark are all potential hibernating sites. Remember the use of such places by insects when tidying up the garden in the fall; leave them in place if not too unsightly.

If your garden does not already have such sites, construct special places for overwintering butterflies. Simply tack rough cedar boards or large slabs of bark in protected places on fences or on the side of the house or garage. Place the boards or slabs vertically, leaving one side not completely nailed down, with a crack where the butterfly can wedge itself inside. Place the bark at various heights, beginning a few inches off the ground.

Brush piles or a shelter of stacked logs can be built. Lay the logs in layers, with each layer of logs pointing in a different direction, forming a crisscross design. Leave the bark on the logs—the loosened edges will make ideal hibernating sites.

A well-planned garden provides food for caterpillars and butterflies as well as room to fly.

3 A Planting Plan

One objective of a butterfly garden is that it be properly established and, therefore, easily maintained with little disturbance to the butterflies. A gardener should be able to enjoy the butterflies in the garden and derive pleasure from the efforts to attract them, not overwhelmed and discouraged by a lot of constant work. With this thought in mind, plan your garden correctly in the beginning, choosing plants specifically used by butterflies, and do the planting properly. The following pages are intended to help both in the planning and planting of such a garden and in maintaining it with a minimum of effort.

Know Your Area

There are a few things you should do before ever picking up a shovel or visiting a nursery. First, give some thought to the area and region in which you live (see the end sheet map of regions of Texas). Make notes in a sturdy notebook about the amount of rainfall, general soil type, and first and last freeze dates for your area (a local nursery or state extension office can help). Also study the hardiness zone map for Texas, and note in which of the Texas zones you live. Such ecological factors are going to determine to a great extent the general species of flowers that will thrive best in your garden.

After jotting down the general physiographic features of your locality, describe your immediate habitat—is it rural, suburban, or urban? Getting even more specific, define the minihabitat of your existing garden or property. Is it an open yard or lawn with practically no trees or shrubs, an area with straight borders along the property lines, a large and formally landscaped garden with shrubs, or

A BUTTERFLY GARDEN CAN BE LANDSCAPED IN ANY NUMBER OF ATTRACTIVE WAYS.

an area of well-landscaped but informal beds? Briefly describe your present garden in the notebook. This information will provide an understanding of where you already stand in regard to the possibilities of attracting butterflies, and it will be a great help in formulating future plans.

To begin the actual plans for planting, draw in your property boundaries on a fairly large sheet of graph paper. Make the outside boundaries as large on the paper as possible but in as nearly correct proportions as you can make them. The easiest and most accurate way of doing this is to refer to the original land plat and house plans. Make a note on the graph sheet of the scale being used. This drawing will be referred to frequently, and knowing the scale is important. Sketch in the house, including porches, patios, garage, or any other attached structures. All that is needed here is an outline of the outside dimensions. Make heavier lines in the house outline to indicate placement of all doors and windows.

The next step is to take the sheet of graph paper, a couple of pencils, a good eraser, a long tape measure, and a buddy to hold one end of the tape, and go out into the yard to continue this base model. An on-site drawing of these features is most important, for it is amazing how much can be forgotten when sitting inside the house. Walking about the yard, and using the tape measure at all times to keep things to scale, begin sketching in any existing structures, such as a tool shed, swimming pool, fountain, gazebo, benches, birdbath, play equipment, walkways, walls, fences, and hedges. Sketch in all outside water faucets or outlets, and include overhead (or underground) power lines, poles, or utility structures. Continue to draw in all flower beds or borders, vegetable garden, areas of ground cover, and the like. If there are objects on neighboring properties affecting your property, such as buildings, trees, or water runoff, locate these on the drawing.

Now make a second tour of the yard, and draw in all specimen trees or shrubs. On an extra sheet of paper or in the notebook, make notes about these plants. Such notes will be important when planning for the addition of butterfly-attracting plants. If some plants are not performing well where planted, perhaps they can be moved to a more desirable location or given to a neighbor, to be replaced with special butterfly plants; mark any such plants.

As you walk about making notes on the existing vegetation, also note which areas receive full sun all day long, sun only part of the day, or full shade during the growing season. The amount and areas of sun and shade are very important when planning new flower beds. Jot down notes about the soils, whether they are sand, clay, or loam. Pay special attention to the drainage of the garden area, and note any problems of erosion or water standing for long periods after rains. Make note of excessive wind tunnels and whether windbreaks would be helpful. Locate areas of the yard that are viewed from inside the house or from much-used areas, such as a porch or patio. Mark the sight lines of objects that would be better screened or completely hidden from view, whether on your property or the neighbor's. Using dash marks, sketch in areas where new beds can be put in or already existing beds can be enlarged or extended.

If you generally like your existing beds and think that only a few new plants can be added, then make as many notes as possible on problem areas in the yard. Small patches of lawn that are not doing well, small areas between the house and walkways that are hard to keep, a problem area between garage and street—all of these are potentially new areas for butterfly plants. Other such areas include those outside the garden fence (perhaps there is a vacant lot there), in rarely used alcoves, behind small buildings such as storage or tool sheds, or around woodpiles or compost bins. The area between the sidewalk in front of your house and the street is a possibility, but only if you live in a low-traffic area; you certainly do not want to entice butterflies to an area, only for them to end up in the grills of speeding vehicles. Possibly large boulders could be scattered in an open area formed by a circular driveway, with the entire area planted in a wild "meadow" or perhaps more contained with beds and meandering walkways.

Finally, on this map define the present lawn area. In the notebook, describe its status concerning health, looks, and workability.

Before planning the addition of any new beds or any new plants to already existing beds, you need to know exactly what plants you already have growing and where they are. Draw a diagram of each bed on a separate sheet of graph paper, placing all the perennials in the proper spots and drawing in areas where annuals are usually placed. Use the tape measure so that you can draw all of these to scale as nearly as possible, but use a large scale since you have only one bed per sheet and are going to need all the room you can get. This should complete the base map and give a good guide from which to do any future planning.

Now, spend some time studying both the overall yard sketch and each individual bed with its existing plants. It is vitally important that you know where all present plantings and possible new sites are located, as this information will form the basis for decisions about the type and quantity of any new plantings.

Regardless of how much or how little you eventually decide to do in the garden, well laid-out plans, of both present and future plantings, will save much time and expense later. Such plans, which may seem frustrating and time-consuming at first, will ensure much greater success in attracting butterflies into the area you plan to provide for them. With a thorough knowledge of your yard, you can place future plants in the locations best suited to their needs, guaranteeing thriving, healthy plants, which in turn means less work and more butterflies attracted.

Later, if more advice is needed in further rearranging the garden, or if help is needed with a serious landscaping or building problem, the scale drawing will save much time and cost in landscaping consulting fees. It might be wise to make several copies of the maps. Covered with a sheet of acetate or clear plastic to prevent dirty fingerprints or smudging of the drawings, they may be easier to work from if taped to a stiff cardboard backing. If you want to combine the various phases of the base plan with drawings of the beds, make each drawing directly onto a sheet of acetate. Use a different colored pen or pencil for each phase so each phase of the total garden can be clearly seen when the sheets are put together.

Decision Time

Now that a map has been made of what already exists in your garden, along with notes and sketches of all possible places for new plants or beds, it is time to take a long, hard look at the situation and make some decisions. It is important not to get carried away in your enthusiasm here, for gardening to attract butterflies is meant to be a leisurely and enjoyable experience, not an added chore. In these preliminary stages, consider who will be maintaining the garden and how much time and effort will be spent in the maintenance. Neglected plants are not healthy plants, and it is better by far to have one small, glorious bed of verbenas (*Verbena* spp.) or zinnias (*Zinnia* spp.) than to have two or three long borders of sickly or scantly flowering plants that are too much trouble or too expensive to take care of. Such a planting will be of little attraction or use to butterflies.

Give serious thought to the amount of money available to spend on the garden. Do not forget that plants must be purchased from time to time, as well as watering hoses, fertilizer, mulching materials, tools, and the like.

And this is the time to decide whether you want to plant only larval food plants, only nectar plants, or both. Studies have shown that female butterflies choose to lay their eggs in areas with abundant nectar sources, so a

Special butterfly-attracting plants can fill problem spaces in a garden.

combination of plants will attract the most species of butterflies.

After the decisions have been made about the time and money you are willing to invest and the type of attracting program you would like, it is time to begin formulating future plans. Make sure these decisions are written down on the plan sheet as a constant reminder of your intentions, and do not let enthusiasm carry you beyond reality.

New Plans

As you begin redesigning your garden, always try to see the finished product through the eyes of a butterfly. It is important to keep in mind that to attract and retain the interest of butterflies, you will need to keep the plantings uncomplicated, uncluttered, and in masses of individual species and color.

It will be worthwhile early on in the planning to visit local nurseries and garden centers to see the materials available for terracing, paths, edgings, fencing, patios, containers, and the like. Not only will this provide an idea of what is available to work with in the garden, but nurseries often will have displays with the products in use, suggesting ideas for your own yard. Driving through various neighborhoods and visiting arboretums and botanical gardens may offer new ideas and inspiration.

Now, referring back to the sketches of the yard and flower beds, begin drawing on new sheets, doing the same as before, except this time drawing in the actual work to be done. Again, draw to scale. First, make note of all structures and plants to be taken out. Then, locate on the map the places for additional trees or shrubs you plan to install. If a windscreen is needed, draw it in if it is to be provided by fencing or a living hedge. Draw in any future trellises, walkways, decorative edgings, or retaining walls. If driftwood or rocks for focal interest are to be added, these should be shown on the plans, to scale.

Go to the yard often while making new plans, and keep measuring. Try to visualize the final results of what is being planned. Use a water hose to outline new beds so actual measurements can be taken to get a more realistic idea of the size needed. Also, look at the proposed changes from inside the house to be sure irreparable mistakes are not being made. It would be most unfortunate to plant a tree or place a trellis in front of a window or glass door, blocking views of the future garden full of beautiful flowers and butterflies.

First-choice areas for any new plantings should, of course, be designed for maximum sunlight. The sun is the major factor around which the entire life cycles of most butterflies revolve. They choose the sunniest beds of flowers from which to gather nectar and the sunny sides of trees and shrubs on which to lay their eggs. They seek sunny areas in which to mate, and they bask in the sun to control the warmth of their bodies. They also use the sun for orientation during flight. With the major needs of the butterflies in mind, make full use of all areas of the garden where the most sun is available.

Do not expect to completely finish the plans for the garden in one try. You will want to start with the general drawings, then continue to refine them until you feel that you have the best choices and largest number of butterfly plants possible for your area and your garden.

No matter how many new beds and borders you would like to add to the garden, it is equally important to provide the butterflies with space. Butterflies like to sail and glide, to sample and soar, and must have enough room to escape their enemies. So do not fill the entire yard with tall shrubs or the beds with tall, herbaceous plants. Rather, through careful selection, create plenty of openings and an overall sense of spaciousness.

A small garden space can be challenging, but even the tiniest city lot has the potential for attracting and rearing butterflies. Even though the limited space means limited possibilities, it can still attract many species of butterflies if carefully planned. A look of largeness and naturalness can be obtained even in the small garden by curving the outlines of the beds and staggering heights of the plants in an undulating fashion rather than having perfectly straight and flat beds, which give a feeling of smallness, stiffness, and formality.

TRAILING LANTANA *(Lantana montevidensis)* WORKS WELL IN SMALL SPACES AS A SOLID PLANTING.

If the garden space is small and surrounded with a high fence, keep the beds along the fence line and leave the center of the garden open. Or perhaps add only one small, irregularly shaped bed containing low-growing flowers such as single zinnias or petunias (*Petunia* spp.) bordered with pansies (*Viola* spp.) or Sweet Alyssum (*Lobularia maritima*) toward the center of the open area. A solid bed of low, reclining lantanas (*Lantana* spp.) or verbenas would be a real butterfly feast table yet would not interfere with their flight from one border to another.

Should the sunniest areas for new beds be on sloping ground, you are a good step ahead, for staggering plants for height is much easier.

By using plants of approximately the same height, a solid, slanting sheet of color can be created. Otherwise, the plants can still be broken up in various heights for even more interest. Any kind of boulder or piece of driftwood usually appears its loveliest when viewed from a sloping angle and blends in with this type of planting more naturally.

Slopes so steep as to cause drainage or erosion problems can be terraced. This is more trouble in the beginning, but if done properly, it will be as self-sustaining after a year or so as any other type of planting. Terracing can be done with railroad ties, bricks, treated half-log edging, or stones. Use the material best suited to your house and the time and money you care to invest. Shapes of the beds should offer ease of maintenance and the best possible plant exposure to the butterflies.

In planning new beds or enlarging already existing ones, the question of width is important. In some instances, such as in a small yard, there may be no choice. But if plenty of space is available, then two factors should be considered. First, the beds should be wide enough to provide plenty of spreading space for both perennials and annuals. Second, the beds should be kept narrow enough for easy maintenance. Usually a bed six to eight feet wide is about right. Any narrower, and you will have to forget about having flowering shrubs fronted with lower-growing plants. Any wider, and it will be almost impossible to do the clipping, bug inspection, plant division, or fertilizing that will be needed from time to time.

An alternative is to have two or more smaller beds with a very narrow walkway between them. This is an excellent idea if you are planning an extensive attracting program using a lot of beds or intend to photograph the butterflies. Other possibilities include making a narrow walkway by placing a simple, narrow strip of flat stones on top of the mulching material or simply leaving a walkway by spacing the plants wider apart.

If there is a low spot or sink area in the yard or lawn that has been a real eyesore or problem area, consider it now a blessing. Instead of trying to get grass to grow there, plan to keep the grass pulled or cut back and keep the area deliberately wet. A flat stone or two can be placed around the edge of the area, with beer, fermented fruit juice, sugar water, or honey poured on the area from time to time. This may very well become the favorite gathering place for members of the local butterfly puddling club. It may never be as alluring as the ruts of a country road after a rain or ordinary barnyard muck, but if such places are not readily available to the butterflies of the neighborhood, then your little spot may prove a much-needed substitute.

Butterflies need protection from strong winds while feeding. If there is a constant strong breeze across your property channeled by open fields or heavy street traffic, consider the possibility of minimizing or breaking up this wind flow. Perhaps the house or another building could be used as partial screening, extending the protection with the use of lattice panels, wooden fencing, or plantings of trees and shrubs. Using larval food plants or good nectar producers as the extensions would serve a dual purpose in this situation.

Look around the garden, and note the route the butterflies will most likely take in leaving the garden. If it will be directly out into street traffic, a fence or hedge may be necessary. Such a structure will force the butterflies to fly up, putting them above the traffic. Other possibilities are inexpensive yet attractive: cane fencing, lattice panels, or tall-growing perennials or annuals. Another possibility is to attach chicken wire to lattice panels and then plant vines such as Queen's Wreath (*Antigon leptopus*), or passionflower (*Passiflora* spp.) or morning glory (*Ipomoea* spp.), to cover the wire. Ideally, a combination of fencing and plants would be better, for some plants might not reach a workable height until late in the season, whereas the fencing would give immediate benefit.

In redesigning the garden, give some thought to your own enjoyment of the plants as well as their use by the butterflies. Place some of the really fragrant plants such as Flowering Mimosa (*Albizia julibrissin*), Beebrush (*Aloysia gratissima*), Sweet Almond Verbena (*A. virgata*), Butterfly Bush (*Buddleja davidii*), Summer Phlox (*Phlox paniculata*), Old-fashioned Petunia (*Petunia axillaris*), or lilacs (*Syringa* spp.) close to a bedroom window, a porch or patio, or the driveway. If there are particular sections of the house where large portions of your at-home time is spent, try to make the scenes from the windows or glass doors of these rooms colorful and arresting. If there is a butterfly-attracting border beneath a kitchen window but not visible from inside, hang some baskets from the roof overhang to be viewed from the window. Butterflies will fly to these as well as visit the flowers in the bed below.

Constructing paths and walkways from one part of the garden to another, either for utility or aesthetics, may be a consideration. Perhaps it would be better to note that paths

Staggering plants for height is one option for gardens on sloping ground.

are wanted and draw in a sketch of possible sites when making drawings. Do not make any plans for actual construction of paths or walks until at least after the first season, and after the second season would probably be even better. This will give you the time and opportunity to see where the "lines of least resistance" are and which routes you naturally take when moving about in the garden. It will also give you time to see where underground water lines are needed and have them permanently installed. When constructing paths, keep them simple and with thought concerning durability and low maintenance.

If you have large acreage with a portion naturally wooded, perhaps on the edge of town or in the country, it may be possible to attract species of butterflies that rarely come to the more open, sunny areas. These butterflies, while seen less often, are very beautiful, and their shyness makes them of special interest. If the wooded area is close enough to be incorporated in the garden planting, enhance the area with choice larval and nectar trees and shrubs. Even if the wooded area extends somewhat away from the already existing beds, the area can be brought closer by planting the space between the woods and the beds with well-chosen trees, shrubs, and native flowering plants used by butterflies.

For instance, the woodlands could be extended with Flowering Dogwood (*Cornus florida*), Redbud (*Cercis canadensis*), Hercules'-club Prickly-ash (*Zanthoxylum clava-herculis*), Sassafras (*Sassafras albidum*), Tulip Tree (*Liriodendron tulipifera*), Spicebush (*Lindera*

This walled garden and the trees in it protect butterflies from strong winds.

benzoin), Red Bay (*Persea borbonia*), or pawpaw (*Asimina* spp.)—all important larval food plants. In front of the trees, rich, nectar-source shrubs such as Buttonbush (*Cephalanthus occidentalis*), Virginia Sweetspire (*Itea virginica*), New Jersey Tea (*Ceanothus americanus* var. *pitcherii*), or azaleas (*Rhododendron* spp.) and lantanas could be planted. Be sure to include such vines as Woolly Pipevine (*Aristolochia tomentosa*), Carolina Jasmine (*Gelsemium sempervirens*), and passionflower.

Even if the wooded area is too far from the garden to act as an extension of the borders and beds, there are many butterfly-attracting plants that can be used to enhance the edges of the woods. This is an excellent place for many larval food plants that may not fit into the garden scheme, yet the adult butterflies will readily nectar at your garden flowers after being reared in the wooded or "weedy" area.

Choosing Plants

In order to select the proper butterfly-attracting plants for the garden, you first have to know which butterflies can be expected there. You need to know the species found specifically in the region in which you live, especially in choosing larval food plants. Before deciding which plants to add, look at the butterflies listed in this book and their ranges. Also study the larval food plant list for the region of the state in which you live, for this gives a more complete listing of larval food plants as well as the butterflies that use them.

Referring to the plants profiled here along with the more general larval and nectar plant list provided, make notes of favorite nectar and larval food plants of the butterflies for your area. Study the descriptions of the plants given here as well as those in gardening catalogs; then decide which ones you would like to try. Choose more species than can be used in order to give yourself more versatility when planning the garden, for the exact plants wanted or needed may not be immediately available.

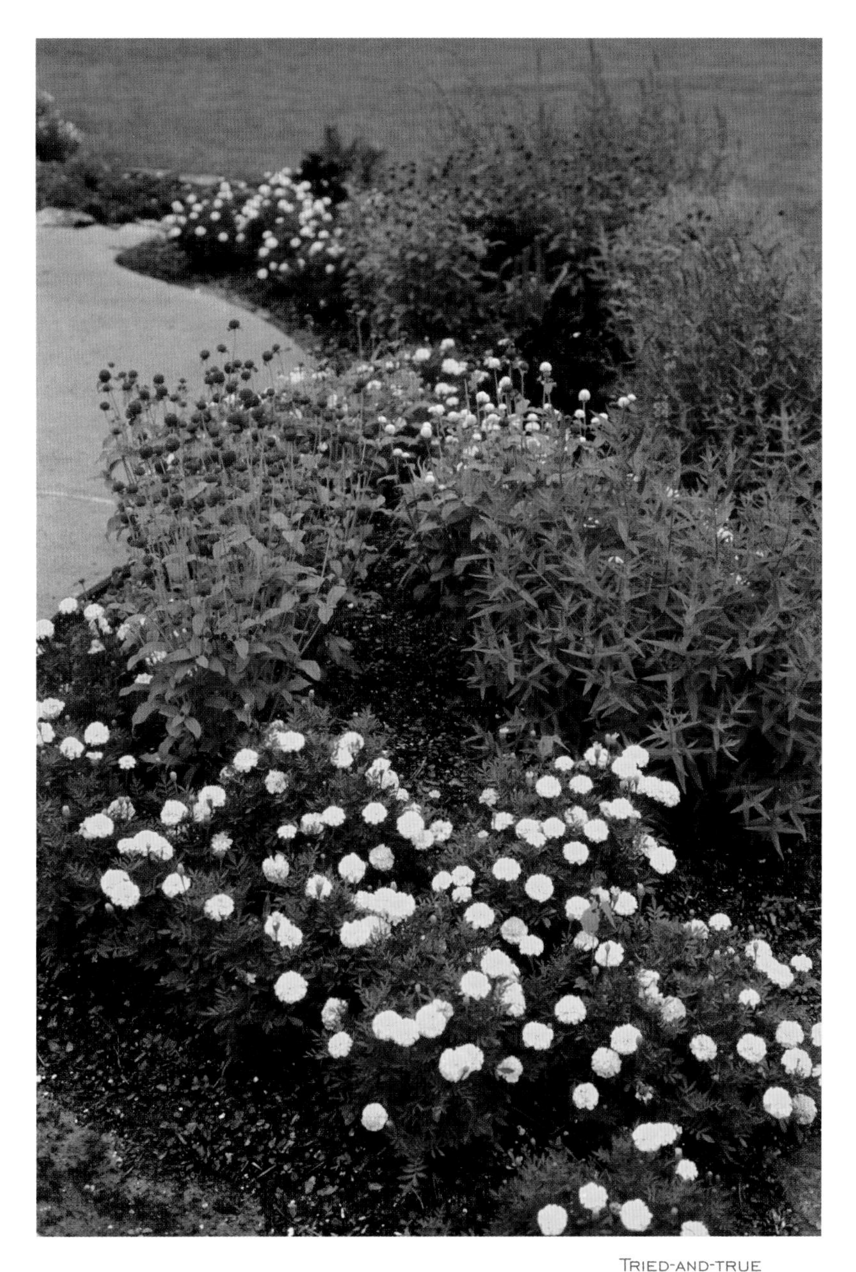

Tried-and-true annuals, such as the marigolds (*Tagetes* spp.) and Globe Amaranth (*Gomphrena globosa*) shown here, are good choices for the beginning gardener.

If a long-term attracting program is planned using both larval and nectar plants, but your time and energy for garden work are limited, it would be best to start with only two or three of the nectar- and larval-attracting trees and shrubs during the first year. Perhaps add one or two choice species of perennials to already existing beds or borders, and fill the additional open spaces with annuals. Even if there is no space in the borders for new plants, make no more new beds than you can easily mulch and water during the summer months. Instead, continue to improve all existing beds and, as time allows and following the original plan, properly prepare more new beds or enlarge already existing ones.

The season in which a garden is started will greatly influence the choice of plants. Ideally, a garden should be started in late summer or early fall, but a planting for butterflies can be made at any time. However, if your garden was started in either spring or summer, by the time fall arrives, perhaps you will have an even better understanding of what you want growing in the permanent garden.

Unless the existing flower beds already have really good, loamy soils, do not plan to set out perennials the first spring or summer. Perennials are generally long-lived, and much thought and care should go into their placement in properly prepared, permanent beds or borders. Instead, use annuals, which generally require minimal care, and spend time making plans for putting in new beds and improving the old ones. This will also allow an entire growing season to study plants thriving in your area and evaluate them for suitability to your personal garden site.

Also, if not already an accomplished gardener, you do not want to waste time and money on selecting hard-to-grow seeds or plants. For spring and summer the first year, stick to the common, tried-and true annuals, such as Mexican Sunflower (*Tithonia rotundifolia*), Shepherd's-needle (*Bidens alba*), or single-flowered marigolds (*Tagetes* spp.) or zinnias, which are easily grown from seed or can be readily obtained at a local nursery. Include some of the Old-fashioned Petunias if the seeds or plants can be found. Use these annuals for wonderful groupings and drifts to make great splashes of color. There is enough variety in height and growth in annuals to make exceptional garden displays, as well as enough color and flower shape to please butterflies. Annuals can certainly be the backbone of the planting for the first season, and some of them are such butterfly favorites that space should be allowed for them somewhere in the garden every year.

As long as the flowers are good nectar producers, the butterflies do not care whether the plants are annuals or perennials; just be sure there are lots of them. As previously stated, butterflies do not like a little bit of this and a little bit of that. They are attracted to large groupings of the same color and fragrance. So whatever you plant, do it en masse. When butterflies find large plantings of good nectar-producing flowers, they will continue to feed there until the flowering period has ended.

To cut down on yearly planting and maintenance, soils should be worked into topnotch condition and borders established as quickly as possible with good nectar-producing perennials. Well-established borders start flowering earlier in the spring, and this is important. Even in the central portion of the state, some butterflies are often out in February, so early nectar sources for them are vital.

If you have some established beds and are able to put out only one or two species of herbaceous perennials the first year, purchase a few good, fail-proof ones such as Summer Phlox, Showy Bergamot (*Monarda didyma*), or liatris (*Liatris* spp.) or verbenas. Save the big bucks for a couple of choice trees or shrubs, such as Butterfly Bush, Beebrush, Chaste Tree (*Vitex agnus-castus*), or lantanas, for planting in the fall. Add one or two choice but easy larval food plants such as Tropical Milkweed (*Asclepias curassavica*), dill, and passionflower vines.

In choosing plants for butterfly usage, pay special attention to length of the bloom period, especially for shrubs and perennials. Many times such plants flower only briefly; if the space in your garden is at a premium, plants having the longest flowering periods are certainly the most beneficial. On the other hand, many perennials may be trimmed back or perhaps naturally die back after flowering. In such a case, use them near plants that produce most of their vegetative growth and come into bloom either earlier or later.

Timing flowering periods is crucial for the butterfly gardener. A continuous production of flowers is absolutely mandatory, for a garden without nectar-producing flowers is going to mean a garden without food for the butterflies. Constant bloom is best guaranteed by always having at least one or two good old reliables such as Mexican Sunflower, Pentas (*Pentas lanceolata*), or zinnias and lantanas. Then, if some of the new plants being tried do not work well, there will still be something to draw in the butterflies.

PLACE RANGING PLANTS, SUCH AS COLUMBINE (*Agutlegta* spp.), IN FRONT OF TREES TO CREATE A WOODED, PROTECTED AREA FOR WOODLAND BUTTERFLIES.

Many of the plants purchased will probably be from local nurseries. Generally, the plants are sold in containers; if buying a native species, look the plants over carefully. Most stock sold as container plants is still in the pot or can in which it was raised. If lifted from the container, these plants should have a solid ball of earth filled with many fine rootlets showing on the outside of the soil ball. Occasionally, you may find "container" plants, especially native species, that have recently been dug from a field or roadside and indiscriminately crammed into cans or pots for quick sale. Whenever you find plants treated this way, refuse to buy them. Furthermore, let the nursery know that the reason you are not buying the plant is that it appears to have been dug from the wild, under unknown circumstances, instead of being properly propagated. Rarely do such plants survive, because of the stress of being improperly cared for. There are many nursery folks who are propagating their own native stock, and these places should be sought out and patronized. The importance of obtaining native plants from appropriate sources cannot be stressed enough.

Some firsthand information could possibly help in making final choices of plants for the garden. In-depth information is now available on the Web for almost every plant in the state. To learn even more about how a particular plant will grow in your area, visit and talk with knowledgeable and trustworthy nursery personnel about a plant's good and bad features. Local landscape architects and designers are gold mines of knowledge about certain plants. They are the ones who draw the plans for the home owner, purchase the plants, and guarantee the plants' survival after planting. Talk to them about the species that do best in your area or how they deal with the ones that are harder to grow. Have a list of chosen plants with you at all times so you can ask about specific ones. To avoid any confusion when discussing plants and to prevent mistaken purchases, use both the scientific and common names of the plant.

Visit arboretums, botanical gardens, city parks, university campuses, and trial gardens, and talk with the gardening personnel there. At the same time, compare butterfly usage in areas where a wide range of plants are growing and flowering together. Drive around town, and make notes of plants you see and like. Anytime you are in the country, observe the wild plants that butterflies are using. If the butterflies are avidly nectaring on a plant a few miles from your house, chances are they will readily use it if it is growing in your yard.

Study the seed and plant catalogs thoroughly. Instead of treating them simply as a list of available plants, use them as textbooks. The descriptions given for each plant should give a good idea of whether it will grow in your area and in your particular garden habitat. Pay special attention to the soil and moisture requirements for each plant, and choose the ones with requirements most closely matching what you can provide.

One important advantage to a diverse planting of both native and cultivated species is that it provides natural checks and balances that help keep unwanted insects and diseases under control. Generally, native species suffer from fewer diseases than cultivated stock, so when both native and cultivated plants are grown together, a single disease or species of insect is less likely to annihilate the garden.

No matter what time of year a permanent garden is started or the amount of work first put into it, a richly productive garden is not going to be accomplished in one year. It requires patience and a lot of trial and error and will need to be developed over two or three seasons or even longer. Each garden is unique. Therefore, you will need to keep working with various plants until you find those that grow best in the area you have provided as well as those for which butterflies show a particular fondness. Eventually, when your own favorites have been effectively combined with what the butterflies need and want, you will have a garden that is truly beautiful and functional—for both you and the butterflies.

Preparing Beds

Butterflies do not like disturbance or radical changes in their feeding area, so the sooner shrubs and perennials can be planted and arranged satisfactorily, the better. And since the plants should remain relatively undisturbed once established, proper bed preparation is very important. For new garden beds, first mark off the outlines of the desired shapes; then spade the soil to a depth of several inches within the entire outlined areas. Most gardeners are finding that in many situations, raised beds allow easier gardening and healthier plants. Raised beds greatly aid in drainage and can help alleviate problems with heavier soils.

If time and care are taken in preparing the beds correctly, they will be a joy to work with. If you skimp at all, however, trying to keep the beds going will be a continual frustration, with weak and diseased plants and an unwanted expense in replacing dead ones.

No matter whether old beds are being rejuvenated or totally new beds being created, some plants such as Purple Nutgrass (*Cyperus rotundus*), Yellow Nutgrass (*C. esculentus*), Common Bermuda Grass (*Cynodon dactylon*), St. Augustine Grass (*Stenotaphrum secundatum*), and Johnson Grass (*Sorghum halepense*)

Purple Coneflowers *(Echinacea purpurea)* are dependable nectar producers.

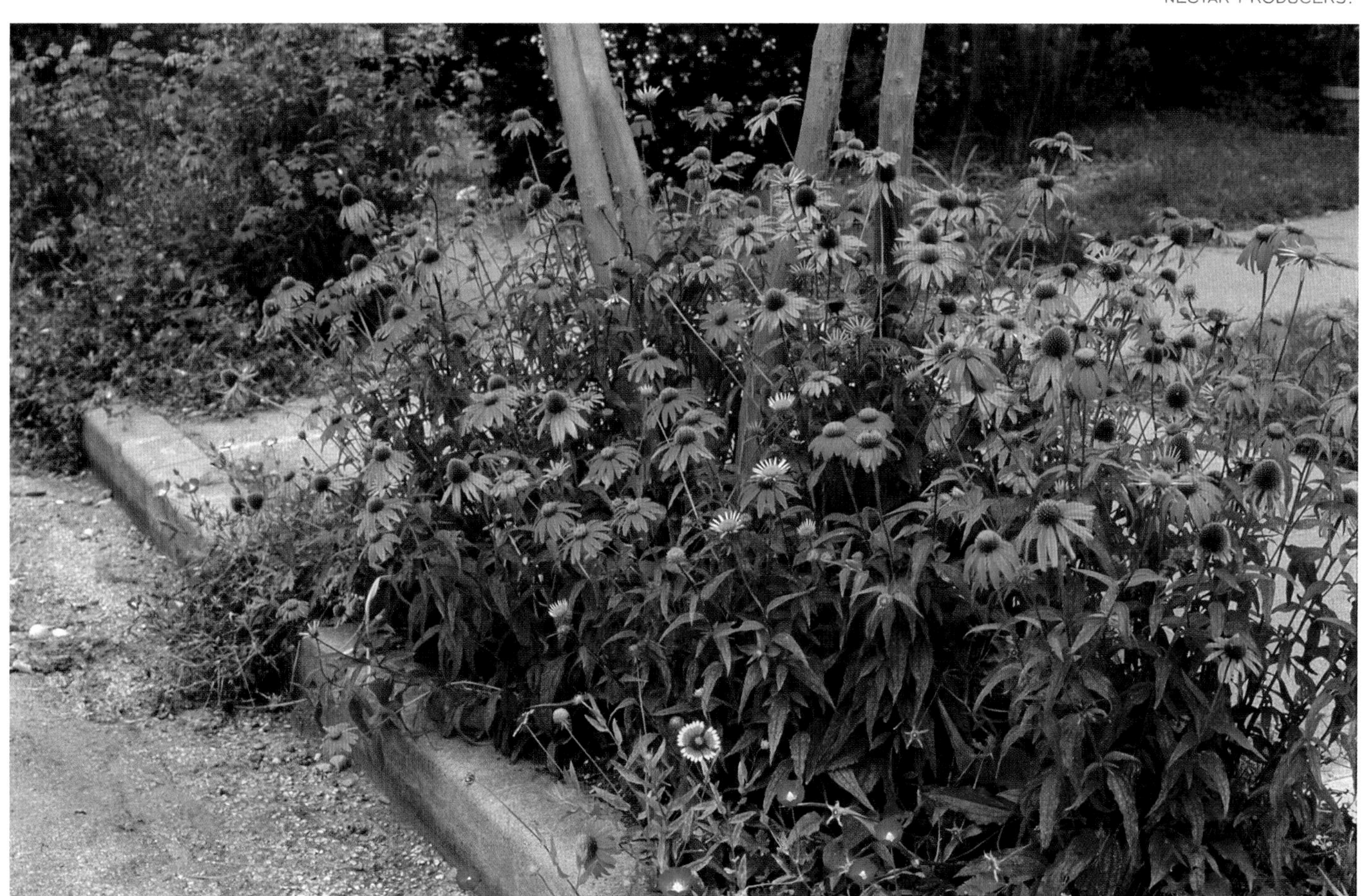

or herbaceous plants such as Common Dayflower (*Commelina communis*) and most garden mints (*Mentha* spp.) should absolutely never be left in the beds in any form. Even if the soil has to be screened, get every tiny piece of these plants, roots, or rhizomes out of the beds. This simply cannot be emphasized enough. If any portion of these plants is left in the beds and even if buried a foot or more deep, the plants will resprout and are almost impossible to get rid of. And these beasts cannot be eradicated by "pulling" or shallow digging—such attempts simply break them off at the roots. When this is done, the plant responds by producing a multitude of new shoots deep underground, resulting in an even greater number of plants. In the end, the garden may well be abandoned—to even try to eradicate these plants will be an overwhelming task for most gardeners.

There are three major considerations when planning and preparing the beds: soil, water, and mulch. These are all equally important for the health of the plants and for ease of future maintenance.

Soil

Soil is the growing medium and basic source for nutrients readily available to plants. From soil comes the makings of foliage, flowers, and nectar, and you cannot properly prepare the soil in a bed until you understand the requirements of the plants you plan to use there.

If a completely native garden is the objective, not too much will be needed for the soil. Using native plants in a soil richer than their normal growing soils will affect the sugar content and amount of nectar produced. This may be a good thing, or it may not. Watch the natives closely; if they are being heavily used by butterflies in the wild but are not being used in your garden setting, move the plants into soils as close to native soils as possible. However, if both native and cultivated plants are planned to be used in the same bed, an enrichment of the soil should be tried.

Whether native or cultivated, some plants demand deep, dry, well-drained sand, while others want seepy muck. For the plants that require a very particular type of growing situation, plan to prepare a special bed for them. The majority of plants, including many natives, grow well in moderately rich, well-drained soils of a loose or porous texture. Creating the perfect planting medium in which most plants will grow their absolute best takes time. Regardless of the type of soil you start with, whether deep sand, hard caliche, or the tightest of clays, organic matter is the best amendment available to enrich the soil. Organic matter not only helps hold moisture in the soil but slowly releases nutrients as the matter decomposes, thus providing the plants a steady source of the good things they need for healthy growth.

In the beginning stages of bed preparation, add plenty of organic matter to the beds. Sphagnum moss, barnyard or horse manure, old mushroom growing medium, and compost are all effective. Work in well-rotted sawdust, chopped corncobs, cotton burrs, peanut shells, pecan shells, or rice hulls. Half-deteriorated pine needles and bark chips are excellent. Visit the local fresh produce market where vendors shell peas in the summer and pecans in the fall, and get the discarded hulls. In some cities such as El Paso, Austin, and

Houston, organic composting programs are available that provide compost to the home gardener. If you have nothing but last year's leaves, run the lawn mower over them a couple of times to shred them finely, and then add them to the spaded soil. In South Texas, vegetable trimmings can possibly be gathered from the fields after harvest, composted, then eventually added to the soil. Add some sand and garden gypsum if the soil is heavy and full of clay particles.

If you suspect your soil is drastically lacking in something, do a soil test. Then add whatever is specifically needed, if anything. Finally, sprinkle in bonemeal, blood meal, and cottonseed meal. After you have added every good thing you can think of to make the soil richer, loamier, and better drained, turn the soil once more, mixing everything together thoroughly, and then level the bed. To settle the soil in freshly prepared beds, soak it down thoroughly and completely. Usually more soil will need to be added to bring the bed back to the desired height.

After plants are in and before adding the mulch, consider adding some earthworms. A most important thing to remember here is that plants do not eat dirt—they drink moisture. Earthworms do eat dirt, along with an enormous amount of ground litter. Their castings (excrement), when dissolved, enrich the soil and can then be taken up by the plants. During this ingesting and casting process, earthworms moderate the soils' pH to neutral; they also aerate or loosen the soil, making it more friable, increase moisture absorption and stability, and bring the deeper-buried nutrients closer to the soil surface, liberating them in a more soluble form that plants can use. When adding the earthworms to the beds, just sprinkle them on top of the ground—they will work their way into the soil. There is simply nothing better for continual aeration and enrichment of soils.

In creating the garden, if cost becomes a factor and a choice has to be made between some "extra" items, such as paving or edging materials and a load of manure for the beds, go for the manure. The health and survival of every plant depend to a large extent on the medium in which it is growing, so you simply cannot afford to skimp on good soil. It will be no problem adding the extras the following season, but once beds are planted, it would be a major undertaking to completely redo the soil.

Water

Supplemental water can come from several sources, the most common the tap (treated water) or rain barrels (collected water). If using all native plants is planned, then it will be best to install rain-collecting barrels. Many native plants will not respond well to tap water or may die altogether.

Whichever type of water is used, the most efficient and least troublesome way of providing necessary moisture to the plants is with a watering system of some sort. Watering systems range from simple to elaborate. One simple system, of course, is the familiar watering hose, which can be so frustrating, aggravating, and time-consuming that the watering may not be done adequately, if at all. On the other end of the spectrum is the elaborate (and expensive) network of underground lines complete with timers for automatically turning the water on and off.

One common method for watering gardens is placing one or more sprinklers about the yard, generally in spots where some of the water hits the flower beds. Sprinklers waste a tremendous amount of moisture because of evaporation, and they should never be used in the garden except in very early morning, allowing the plants to become completely dried off before starting nectar production. If sprinklers are used in late evening and foliage is not completely dry before nightfall, you are risking invasion of all kinds of fungus diseases to the plants.

No sprinkler of any kind should ever be used to water a butterfly garden. Most sprinklers have to run from four to six hours to saturate the soil well enough to benefit the plants, and butterflies cannot feed from flowers while a sprinkler is spraying them with water. The water also dilutes the nectar, or washes it away entirely, so the butterflies will have to wait until the nectar flow is back to normal, which may take several hours. Even if the sprinklers have been used in the early-morning hours, the nectar will have been affected. As often as gardens need to be watered during the hottest summer weather, the butterflies may be without food for a day or so at a time. This is too long. They will go somewhere else for their nectar.

One system that is not very expensive and works quite well is the drip system. There are several brands of drip systems to choose from, but basically they are all the same. A solid, flexible plastic pipe is run underground from the faucet to the beds; on top of the ground in the bed area are short lengths of much smaller flexible tubing running off the main pipe. At the end of each short length of tubing is an emitter or dripper, which allows the water to drip slowly to thoroughly saturate the soil. These systems do not work as well in the western half of the state due to lime buildup in the openings.

The best system of all is the round, seeping or soaker hoses from which the water gently and steadily oozes out into the soil. They work on the same principle as the drip system except minute holes are prepunched at regular intervals, so water comes out in a continuous stream instead of a drop at a time as in the drip system. The soaker hose is less expensive than a drip system, is less trouble to install, and does not have a problem with clogged holes (a frequent occurrence with emitters). These hoses come in different grades and quality; if you opt for this system, getting quality hoses with good fittings is highly recommended. They will last for many years, and a hose with sturdy, nonleaking connections is worth every penny of the few extra dollars of initial cash outlay.

To install, after plants are situated within the beds, attach the hose to a faucet, then curve it back and forth from front to back of the bed all the way from faucet to end of bed. After the soaker hoses are satisfactorily installed (making sure the entire bed is being reached with the water), cover them with a deep layer of loose mulch. Covering with mulch hides the hoses from view, conserves moisture, and protects the hoses from deer, squirrels, and mice—animals with a special liking for plastic. Flexibility and light weight make the hoses easy to install, move, or rearrange later if needed, without damaging the plants. When a soaker hose has been covered over with mulch, it is very easy to overwater the plants. Keep a hand digger or flat stake stuck in a bed and frequently check the soil moisture. To check, pull the digger or stake to one side. If the soil is satisfactorily moist to an approximate depth of six inches, then discontinue watering.

Once the plants in the beds have become deeply rooted and are growing well and the beds are properly mulched each year, there may no longer be a need for an intensive watering program. Most likely only an occasional watering during the hottest parts of the year will be necessary. But until that time, a good, reliable watering system is one of the best investments to be made to assure survival and good nectar production of the butterfly plants.

Mulch

For easiest upkeep once the shrubs and perennials have been planted, applying a good mulch on the beds is the nicest thing you can do for yourself—and the plants. Not only does the proper mulch eliminate practically all weeding, watering, and fertilizing but it also keeps the soil from cracking due to moisture evaporation and helps keep plant roots at a more even temperature, which encourages healthier growth. Also, a proper mulch prevents the erosion of good topsoil during heavy rains and stops water from splashing back onto the foliage, which could spread soilborne diseases. Soil-splattered plants are unsightly, and if the plants are low, the dirt may cover the flowers, contaminating the nectar to the extent it cannot be used by butterflies. Furthermore, an organic mulch is food for earthworms, which are constantly adding nutrients to the soil.

Do not add deep, permanent mulch to the beds until all plants are up and growing, especially if annuals have been used in combination with perennials. Until this time, some weeding may be necessary. If possible, bury the weeds pulled out. Often the extracted plants can simply be turned upside down and stuffed back into the holes from which they were dug. If this is not possible, at least turn them upside down (exposing the roots so there will be no danger of rerooting) and leave them on the flower bed to act as a thin mulch until a deeper one can be applied. As the plants deteriorate, they add more good nutrients to the soil. A little soil sprinkled on top of the uprooted plants and a good watering a few days after they have sufficiently dried out will speed the rotting process. If the pulled plants cannot be left on the bed, by all means add them to the compost pile. However, plants that have already gone to seed should never be left on the beds, for all you will be doing then is planting more.

Many materials can be used as a permanent mulch. Often, the same product used to enrich the soil can be used as mulching material when applied in depth and placed on top of the bed. Organic mulches are popular, and those from wood or vegetable by-products are especially desirable. These attractive mulches decompose slowly, continually enriching the soil. If the organic matter is relatively fresh or new when applied, it is best to add a good nitrogen fertilizer because the mulch has a tendency to deplete the soil of nitrogen while decomposing.

Sawdust is often free for the hauling from a local sawmill. Whether the sawdust is hauled yourself or bought, select from the oldest, most

Pine needles make an attractive and effective mulch.

rotted piles possible. The older the sawdust is when applied, the faster it deteriorates into the soil. To avoid any nitrogen loss from using sawdust, add a little blood meal or cottonseed meal to the soil before applying the mulch.

Pine needles make a beautiful mulch, as does shredded pine bark. Mushroom compost is excellent where available and is usually relatively inexpensive since you generally haul it yourself.

Do not let any of the leaves in the yard get away. Rake them into low piles, run the lawn mower over them a time or two, and then spread about the beds and around the plants. If all of the leaves cannot be used that winter, keep them in bags until spring. By then, the leaves that were placed on the beds in the fall will have partially decomposed and can be worked into the soil as humus. Then shred the saved leaves, applying fresh mulch for the summer.

Be constantly on the lookout for tree-trimming trucks. These folks usually shred their trimmings and are always looking for a close, convenient dumping spot; your yard is probably much nearer than the city garbage heap. They are usually happy to deposit their chipped trimmings in your yard or driveway. These make an attractive top dressing for the beds and are slower to deteriorate than leaves.

In the agricultural sections of the state, cottonseed hulls or burrs are available as are ground corncobs, peanut shells, rice hulls, and sunflower seed hulls. If possible to obtain, a mixture of materials is less likely to compact on the beds, and it provides better aeration and water penetration. All of these materials have a natural, attractive appearance and last well when applied to a three-inch depth. Ideally, a three-inch organic mulch should be spread over a one-inch layer of organic fertilizer, such as well-rotted horse or chicken manure.

Wood products or vegetable by-products for mulching are not as available in the western, northern, or central portion of the state and, where they are, are usually very expensive. There are ways around this, however, and by thoughtful preplanning, this should not deter anyone from having beds of good rich soil that will support a number of beautiful native and cultivated flowering shrubs and perennials. After the soil has been enriched as much as possible, and this may mean working with small areas at a time, lay the soaker hoses and put in the plants. Lay a thick covering of newspapers all over the bed. Ordinarily, newspapers do not work well, especially in the eastern part of the state, because they do not allow rain to soak through to the roots. But in the western half, the plants will benefit more from a few thorough soakings from the soaker hoses with the newspaper mulch conserving the moisture than from the few and infrequent rains. After the newspapers are put down, they can be covered with thin, flat stones, or a layer of shredded bark.

A mulch of decomposed granite or other crushed rock is often used. This helps conserve moisture, protect plant roots from extreme heat and cold, and prevent erosion. A native rock is often chosen, giving the garden a more natural appearance that blends with the surrounding landscape.

Using Native Plants

The use of native plants in the home landscape has been advocated for years, but only the true wildflower lover has taken the message to heart. In the more eastern states, gardeners have created beautiful woodland gardens using native plants such as Mayapple (*Podophyllum peltatum*), Bleeding Heart (*Dicentra eximia*), Jack-in-the-pulpit (*Arisaema triphyllum*), and trilliums (*Trillium* spp.). Only in the last few years has the term "wildflower gardening" come to be meaningful for the more southwestern states. We should take special interest in it, for nature has been more than generous in providing beautiful and varied species of trees, shrubs, and herbs. Texas has more than five thousand species of vascular plants (this includes ferns and grasses), many of them producing showy flowers or other features worthy of including in gardens. It is time to take a good look at these natives, learn their ways and growth needs, and claim them as our heritage.

Using these plants in our landscaping to attract butterflies has both practical and aesthetic advantages. Using native plants that have been observed to be useful to butterflies gives the gardener a better chance of bringing the insects into the garden and at the same time brings the natives closer so that their beauty can be more fully enjoyed and appreciated. To attract butterflies, you must plant the things they prefer, and certainly they are familiar with and have developed a liking for many of the natives. To combine the best-loved natives with the best-loved cultivated species is further guiding the garden toward becoming a butterfly paradise.

As an added bonus, most native herbaceous plants are perennials or self-sowing biennials or winter annuals. Once they are established, an abundance of plants is provided each year, both for your own space and usually with enough left over for sharing. The native species are already well adapted to the soil and climate in which they are growing. If planted in a similar situation in the garden, they are the least-demanding plants in cost and amount of care. Native plants also have the advantage of giving a more natural

PLANT A WILDFLOWER MEADOW IF SPACE PERMITS.

appearance to the garden. The flowering season often can be extended in the garden, both earlier and later, by using these hardier native species.

Planting Native Seeds

Study the wildflowers in your area, decide on a few that are wanted in the garden or in a naturalized area, and obtain seeds from a source as nearby as possible, even if it means collecting the seeds yourself. Even though a plant may have an extensive range, seeds collected from plants in the immediate area will have the highest germination rate and will grow and bloom better than plants from seeds collected elsewhere. Plants become adapted to soils, rainfall, and climatic conditions; when introduced into different habitats, they generally do not grow or flower as well.

Before collecting or buying seeds, very carefully study the area where you intend to plant. Make notes about type of soil, amount of moisture, and amount of sun or shade available. Most wildflowers, especially the ones planted for butterflies, need at least six to eight hours of sun each day, so select the site carefully.

Make note of the plants already growing there. If the area is covered in King Ranch Bluestem (*Bothriochloa ischaemum* var. *songarica*), Guinea Grass (*Panicum maximum*), Common Bermuda Grass, or other solid covering, this covering will need to be removed. Native plants generally cannot compete with such persistent nonnative species. Fall planting of seeds is generally best, as most species need to germinate in the fall in order to have developed strong root systems by spring. Some species need the winter's cold and moisture to break dormancy.

Following are some general hints on planting select wildflower seeds, whether in a scattered naturalized planting or as grouped individual species in the border. These suggestions provide the best way to ensure a successful showing of healthy plants with masses of nectar-producing flowers for the butterflies.

To prepare garden beds for planting, the first step is to move back the existing mulch and gently loosen the soil. For a large, naturalizing area, soil should be broken up according to its type: three to four inches if the soil is tight clay, one to two inches if sandy loam. If the ground is exceptionally dry, soak it thoroughly and wait two or three days before tilling.

Seeds can be planted in rows, in groups of separate species, or in your own special mix broadcast by hand. If the seeds are very small, it is helpful to mix them with dry builder's sand to prevent their clumping. After sowing, cover the seeds with a thin layer of soil. The general rule here is to cover no deeper than three or four times the diameter of the seed. One easy method for covering seeds is to flip over a rake and use the smooth edge to rake the soil over the seeds. This method covers a large percentage of the seeds.

Next comes a most important step. In order for the seeds to survive after germination, they must be in direct contact with the soil, so it is very important to firm the seedbed after covering. If a roller is not handy, the area can be firmed by gently patting with the back of a flat shovel. If nothing else is handy, walk on the area in flat-soled shoes, or lay a wide board down and walk on the board. Small areas can be pressed down by hand.

After the soil has been completely firmed, gently but thoroughly soak the entire area. All seeds need moisture to germinate, so keep the bed moist until the seedlings are up. Also, do not let the young seedlings dry out after sprouting. Do not overwater, but keep the area moist until all plants are well established and growing.

Do not overfertilize wildflowers. Heavy fertilizing usually results in plant death or an extraordinary amount of foliage but no flowers. If the wildflowers have been planted in a mass or in an uncultivated area, probably no fertilizing will ever be necessary. If the plants are in a border where more lush growth is desired, use a very weak fertilizer solution occasionally through the growing season. As a general rule, however, wild plants bloom much better if nothing richer than a light top dressing of organic compost and manure is used.

If the area is large or meadowlike and the vegetation becomes unruly or too unsightly late in the season, instead of mowing or shredding, do as much as possible with a scythe. This gets the vegetation down and the ripened seeds in contact with the soil but leaves the stems intact. Any chrysalides attracted to the stems will most likely remain attached and survive. Many larvae seek shelter at the base of plants, and stems left standing fairly high afford some winter protection.

Different species of wildflowers respond to different maintenance techniques. Some species respond better to mowing in the fall; others, to mowing in the spring. Others can tolerate year-round mowing or even burning off, while others cannot. Disking and the time of year it is done also have effect. Each of these may be favorable to one kind of plant but fatal to another. If an area is being naturalized, try dividing it into sections and experiment with one of the preceding techniques to see what happens. Most likely there will be a dominant species for each section, even if the entire area was originally planted with exactly the same mixture of species.

Seeds of wildflowers can be collected along the roadside. Always obtain permission if on park land or private land.

If the wild or untamed area is already begun but now you want to add some special butterfly plants, you certainly do not want to retill the area and lose the species already established. The simplest way to plant now is to choose spots with the least vegetation, scratch the earth as much as possible with a three-pronged hand digger, scatter seeds into the disturbed area, and barely cover with fine soil. Gently step on the area to firm the soil, and then water thoroughly.

One of the beauties of wildflowers is their ability to find their own niches—the places they like best. If something is found coming up in a different part of the garden from where it was originally planted, you might want to leave it there to see how it will do. If the conditions are conducive to the seed's sprouting, then the adult plant will most likely grow well also.

Get to know what the baby seedlings of the newly planted species look like. Study the seedbed constantly; if several plants look similar, chances are they are what you planted. Become familiar with these babies so if they are found in a different part of the garden, they will be recognized and not removed as weeds. They can then either be nurtured or moved to a more desirable area.

Include one or two species of the taller native grasses in your garden plantings. Many of the Skippers (family Hesperiidae) are grass feeders in the larval stage, so they need the native grasses to complete their life cycles in the garden. It is unusual to find any of the common native grasses in a nursery, but a couple of small clumps should be easily obtained from a landowner or from a construction site. One or two good-sized clumps can be divided before placing in the borders, providing plenty of starters. There are beautiful grasses all over the state, and the Skippers use many of them. Choose clump-forming ones (not those that spread by rhizomes or runners), such as Little Bluestem (*Schizachyrium scoparium*), Yellow Prairie Grass (*Sorghastrum nutans*), Broad-leaf Wood-oats (*Chasmanthium latifolium*), Hairy Grama (*Bouteloua hirsuta*), Sideoats Grama (*B. curtipendula*), Blue Grama (*B. gracilis*), and Southwestern Bristle Grass (*Setaria scheelei*). When clumps of grasses such as these are used in the flower border among flowering plants, the graceful foliage and seed heads add a most unusual interest to the plantings.

Native Grass Lawn

Unless you already have a well-established lawn of some special turf grass such as zoysia (*Zoysia* spp.), or St. Augustine or Common Bermuda, there is a wonderful alternative, at least for all except the extreme eastern portion of the state: Buffalo Grass (*Buchloe dactyloides*). This native, sun-loving grass requires less mowing and has fewer other maintenance problems than the more popular lawn grasses and also offers yet another opportunity for attracting the Green Skipper (*Hesperia viridis*).

Staying low to the ground and spreading by aboveground runners or stolons, Buffalo Grass rarely reaches more than eight inches in height. It is a tough, slow-growing, disease-resistant grass and, once established, needs practically nothing except admiration. Occasional mowing will help it become thicker and lusher, but toward the end of the season it should be allowed to become a little taller so it can blow and ripple in the wind. The small stems and soft green coloring of this grass give a natural, rustic look to the yard and never overwhelm the flower borders as the massive, dark green sheets of most lawn grasses do.

Male and female flowers of this grass are on separate plants. Female flowers are inconspicuous and remain low to the ground, later forming small burrlike seed pods. Male flowers are larger, comblike, and held on erect stems extending above the foliage. These later turn brown, remain for some time, and are very attractive.

A lawn of Buffalo Grass should be started in much the same way as preparing flower beds. Remove all existing vegetation by tilling no deeper than two inches; then rake or sift the soil. If the area was a weedy one, let the soil remain bare until any seeds remaining in the soil have sprouted. These sprouts should be eradicated either by hand or by using an herbicide with no soil-residual activity. Since it is much easier to eliminate weeds and other grasses before planting, start with as clean an area as possible.

Not only can a lawn of Buffalo Grass be used to complement your plantings but the lawn itself is a perfect place for some special plants that the butterflies will find enticing. After the lawn area has been properly prepared but before planting the Buffalo Grass seeds, randomly scatter bulbs, plants, or seeds of low-growing native wildflowers all over the area to be planted. With a bulb planter or trowel, dig a hole large enough for the bulb or roots to be covered completely; then plant the grass seeds as recommended.

Most of the smaller plants from bulbs have foliage that is somewhat grasslike and usually remains aboveground for only a few weeks. And even while it is green and growing, it blends in with the grass and is not distracting. By using only low-growing species and choosing the ones to plant according to their bloom period, you can have some plants in flower almost the entire season. Or make it an early-spring spectacular of wildflowers before the grass really gets going. Or plant only "rain lilies" (*Cooperia* spp., *Habranthus* spp., or *Zephyranthes* spp.), and have a lawn covered with white or gold after each rain.

There are three entirely different plants that have the general name "rain lily"—all of them beautiful and perfect for use in a Buffalo Grass lawn. In the genus *Cooperia* the flowers are mostly white or pinkish-white (yellow in some species), open in the afternoon, and remain open during the night. The flowers will begin closing the following day in the heat of the sun. In *Habranthus* and *Zephyranthes* the flowers are generally golden-yellow, often streaked with coppery-red. A thick colony of any of these makes an absolutely stunning sight a day or so after a rain, and while the flowers are open, the air is filled with an almost intoxicating fragrance.

For early-spring color, try using blue-eyed grasses (*Sisyrinchium* spp.), Yellow Star-grass (*Hypoxis hirsuta*), Spring Beauty (*Claytonia virginica*), Blue Funnel-lily (*Androstephium coeruleum*), and False Garlic (*Nothoscordum bivalve*)—the latter, one of the few flowers the Falcate Orangetip (*Anthocharis midea*) will take the time to visit for nectar. Following these will be Celestials (*Nemastylis geminiflora*), Golden-eye Phlox (*Phlox roemeriana*), and the wild onions (*Allium* spp.). There are many species of wild onions in the state, ranging from white to dark rose to yellow in color and with bloom periods from early spring to frost. Many of them produce wonderfully

A LAWN OF BUFFALO GRASS *(Buchloe dactyloides)* IS A PERFECT BACKDROP FOR ATTRACTING BUTTERFLIES TO YOUR PLANTINGS.

fragrant flowers that bear an abundance of nectar and are much used by many species of butterflies. Choose the ones already growing in your area or as near to the same conditions as possible.

The native, ground-hugging Erect Pipevine (*Aristolochia erecta*) is a must for any native grassy area. Its grasslike leaves are one of the favored larval food plants for the Pipevine Swallowtail (*Battus philenor*) and possibly the Polydamas Swallowtail (*B. polydamas*).

For the best effect in the lawn-garden, choose only really low-growing plants so that the flowers either mingle with the grass or barely peek over the top. Leave these little jewels to reseed themselves, and if the lawn is not mowed too frequently, they will quickly naturalize and form showy colonies.

Adopt a Weed

Many food plants used by larvae are plants that can only be referred to as downright "weeds": those plants having totally undesirable characteristics or with very few desirable ones. And it is true that some of these plants might stretch the patience of a gardener having only a little space, but there are numerous others that perhaps could be given an area in the garden "just for the butterflies." Just cultivate the attitude of mind that these plants are not really weeds but future butterflies.

A patch of Heart-leaf Stinging-nettle (*Urtica chamaedryoides*) may not be what you want growing up front among the zinnias, but a healthy stand of it at the back of the border and out of harm's way will ensure many generations of Red Admirals (*Vanessa atalanta rubria*). There will never be a bright show of flower color from evax (*Evax* spp.) or cudweed (*Gnaphalium* spp.), but their nondescript leaves and flower heads provide larval food and a fluffy pupating nest for the American Lady (*Vanessa virginiensis*).

Conversely, some weeds such as Woolly Croton (*Croton capitatus*), food plant of the Goatweed Leafwing (*Anaea andria*), are very attractive. Woolly Croton can be used in a mixed border, where its silvery foliage blends beautifully with the brighter green of other plants. Ferny-leaved Partridge-pea (*Chamaecrista fasciculata*), eaten by the Cloudless Sulphur (*Phoebis sennae*), makes an interesting contrast to coarser-leaved cultivated plants. Many of the clump-forming native grasses become lovely accents in the flower border and are the main food source of many species of Skippers.

Check the list of larval food plants in chapter 8, and choose three or four of the weedy sort that grow in your area. Then, give some thought to adopting one or more of these waifs into your garden. If you do not want them in any of the flower beds or borders, consider areas along and on the outside of a fence or building. Perhaps use them to hide the compost heap or winter's brush pile. If they already exist in the garden area, simply allow them their space by not mowing them down. No special care is needed for these wildings, for they are generally extra tough, having developed under such adversities as always being unloved and unwanted.

Seed and Plant Sources

To obtain some of the special plants recommended in this book (especially those needed for larvae), check with local nurseries first. If they do not carry the plants you need, ask if they can obtain the material for you. Make sure the material they offer or can obtain has been propagated from your local region. Do not accept "native" material actually grown in Utah or California; this was tried, first-hand, and it does not work. Excellent native, nectar-producing plants from one region of the state can (and usually do) change in the quality and quantity of nectar when transplanted into another region. If a choice has to be made, you will be much better off using a locally grown second-choice plant rather than a first-choice plant that has been propagated and grown in other climates.

If plants native to your region are wanted or needed that are not available at your local nurseries, here are some suggestions. Whenever you travel and see massive construction going on, stop and ask permission to dig the plants needed. Contact the local offices of the Highway Department and the county commissioners in your area to learn of sites where road work is in progress or planned for the future. The county agent may know of land being cleared. If you travel county, ranch, or farm roads, you will frequently find fence rows in the process of being cleaned or cleared of the existing vegetation. Permission to remove plants from such sites is usually readily granted.

Local contractors or builders are another source for obtaining plants. Usually they know months in advance where large construction is to be done, and the digging of a few plants beforehand is of little concern to them.

Some cities have "plant rescue teams" that stay in contact with local contractors and dig choice plants before they are lost to construction. Usually these plants are placed in arboretums, city gardens, and zoos. Check with local garden clubs to find out about such rescue groups in your area. Garden clubs are often good sources for native plants and seeds. Many of them have sales or exchanges throughout the year and would welcome your participation. Various other organizations have sales and exchanges, usually during early spring or late fall. A good source of information for such events is the newsletter of the Native Plant Society. At least one of the gardening magazines that specializes in covering the state has a letter section that can be of great benefit in trying to locate such unpopular and little-known plants as nettle and cudweed.

Do not forget friends, relatives, and even total strangers. Always be on the lookout for places to visit in rural areas as possible sources for needed species. Rare is the landowner who will deny the privilege of taking a few plants generally considered weeds. If an extra-good spot of wildflowers is found, the landowner is almost always delighted to share some seeds. Do not be shy about knocking on doors. When driving down a road and a much-needed plant is sighted in someone's yard, stop and visit. Rarely will you walk away empty-handed.

For just such times, you should always be prepared. Make up a collecting basket or box, and always have it handy. In the basket include scissors, clippers, a pocketknife, cheesecloth or nylon netting, various sizes of plastic and paper bags, twist ties, gloves, flagging, metal or plastic tabs or ties (for labeling), a

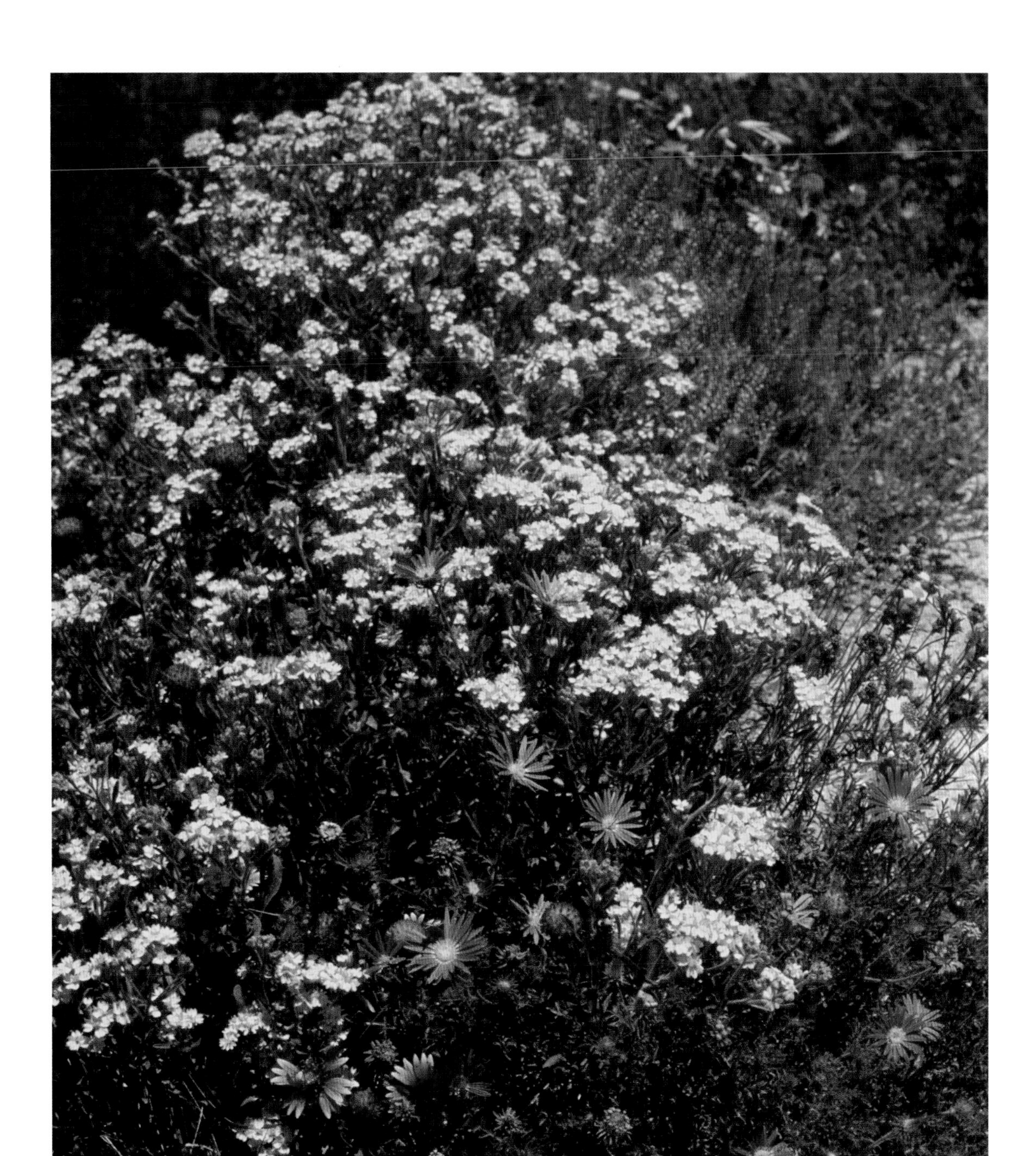

A NATURALIZED BUTTERFLY GARDEN.

waterproof pen, a small notebook, and transparent tape. If space in the car trunk permits, always carry a sharpshooter shovel, several thin cardboard boxes, a bundle of newspapers, and a couple of jugs of water (tightly sealed). Now you are ready for gathering seeds and plants when the opportunity arises. **Do not forget or neglect to ask permission before collecting, and do not collect on public land.**

It is very easy, especially when the opportunity to dig plants from a large area arises, to take things simply because they are pretty or might attract butterflies or because they are being destroyed. In such a situation, be firm and be selective. Before digging one plant, consider such things as the plant's growth habit, length of bloom period, how it will fit into your landscaping scheme, and, most important, how often and by how many species of butterflies it is used. This is not to say that new species should not be continually tried, but serious thought should be given to the planting space available in your garden so you do not fill it up with less-than-choice species.

One of the easiest methods of transplanting the smaller, more shallow-rooted plants from the wild is to spread a layer of soil in the bottom of a thin cardboard box and, as the plants are dug, compactly place the plants in the box until full. Set the box in a rolled-down plastic bag, and water the plants well. Then, partially close the plastic bag until ready to plant. Extra-heavy or double paper sacks can be used in the same way, either cutting or rolling the tops down before placing the plants inside.

Some native plants are more easily started from seeds than by transplants. If an especially attractive plant is seen along the roadside, perhaps one of exceptional color or robustness, tie flagging tape around its base and also tie some of the tape near the base of some nearby object, such as a fence post or small bush. Give the plant time to set seeds; then go back and gather them. Store the seeds in small paper bags (never plastic) in a cool, dark place or in the refrigerator until proper planting time.

If purchasing wildflower seeds, buy only those packaged as individual species. Be very specific in your selections, choosing only the species that will do well in your garden situation and have the flowers with characteristics that butterflies like, namely, nectar, fragrance, a good landing platform, and favored colors. Packaged mixes contain few species that are butterfly favorites; although they provide an array of colorful flowers, butterflies will ignore them.

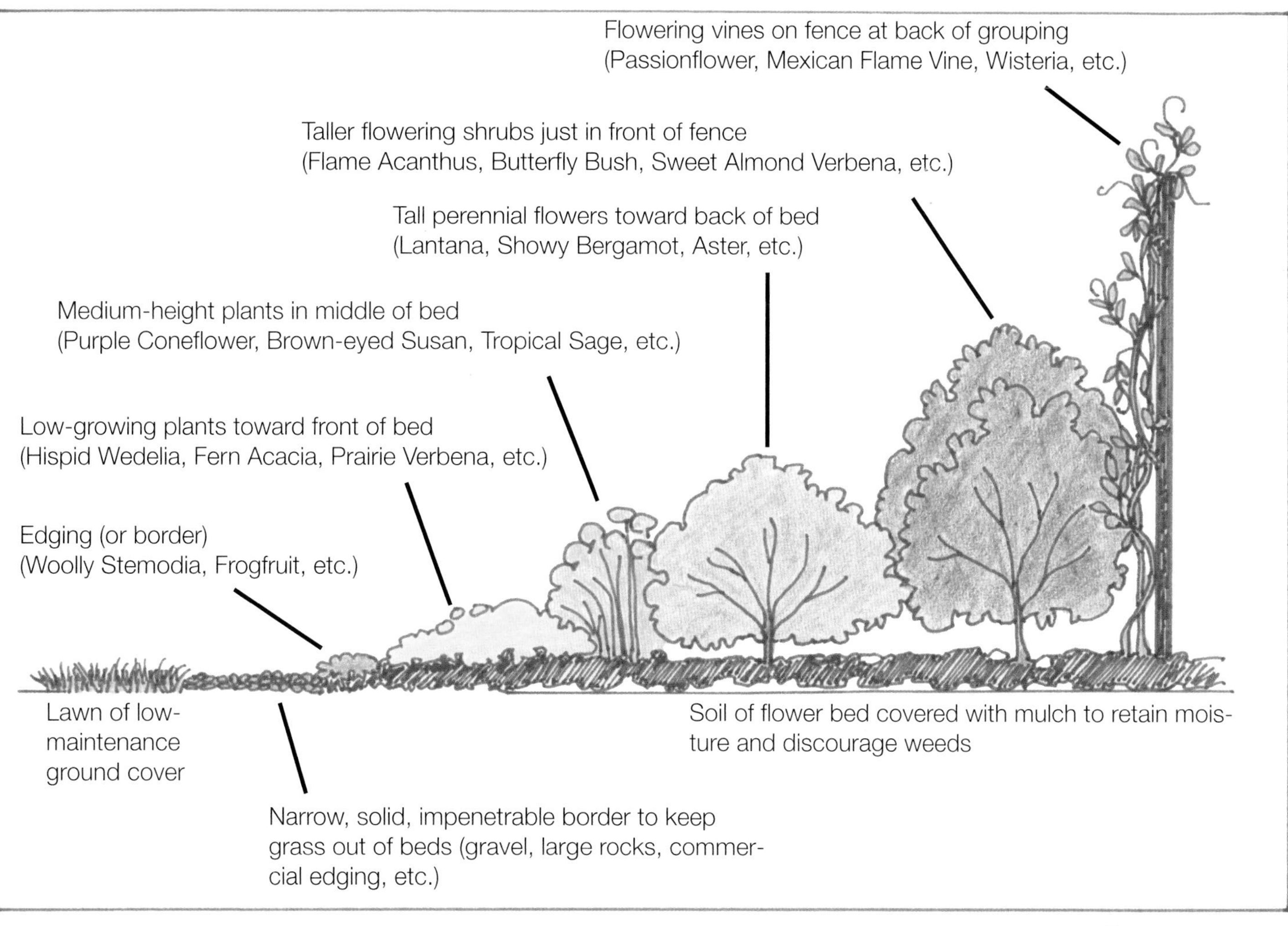

Typical border arrangement

Landscape Plans

Here are sample garden designs for inspiration. Use your imagination and personal preferences in substituting plants and materials, adapting these ideas to fit your yard and area. When choosing plants, refer to the complete larval food plant list as well as the general nectar plant list—the plants listed are all known to be used by the butterflies in the area where they are given.

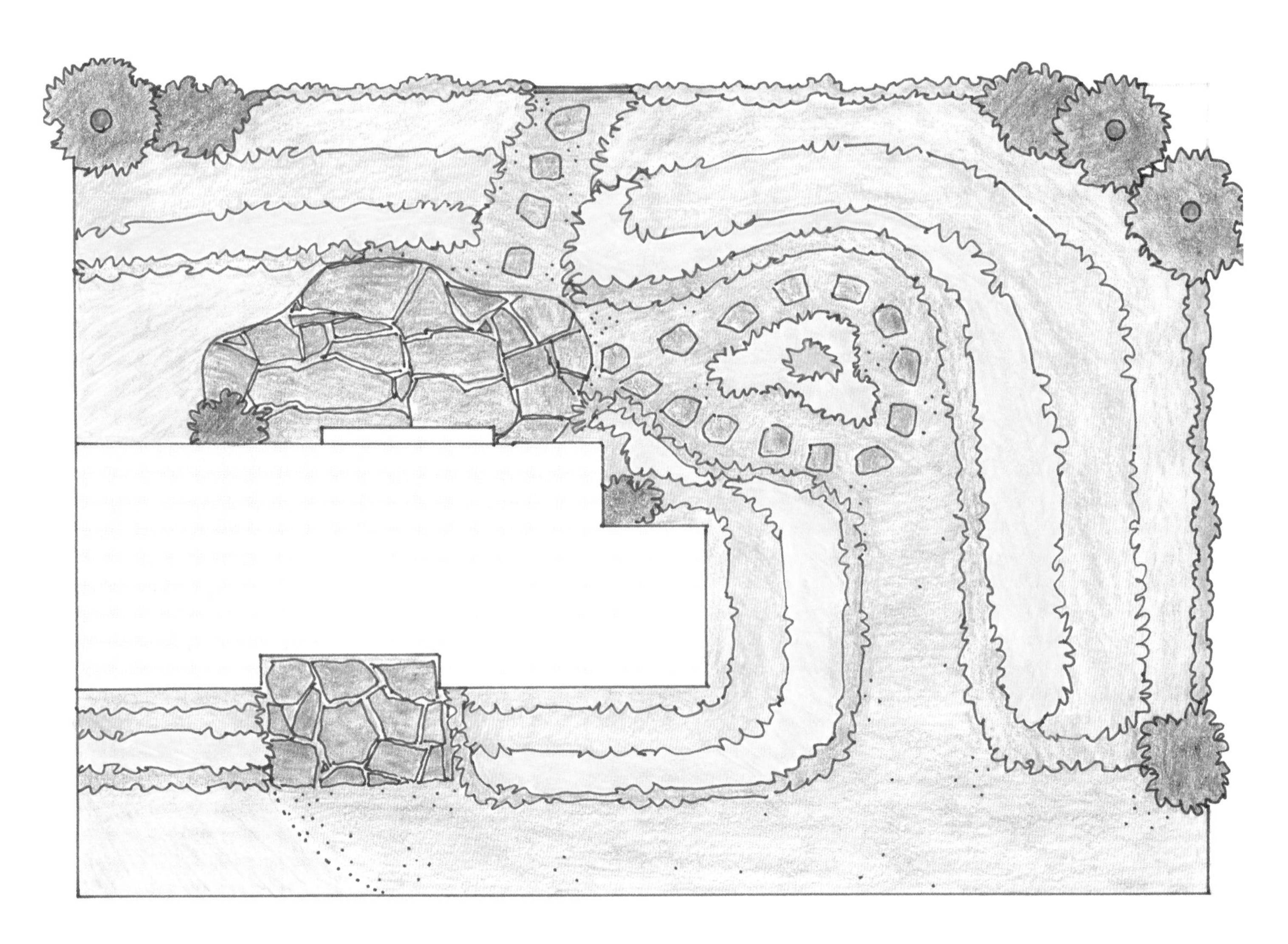

HIGH PLAINS/ ROLLING PLAINS (Region 1)

Abilene, Amarillo, Big Spring, Lubbock, Midland, Wichita Falls

(L) indicates larval food plant
(N) indicates nectar plant

Trees

American Elm (L)
(*Ulmus americana*)
Black Locust (L)
(*Robinia pseudoacacia*)
Chaste Tree (N)
(*Vitex agnus-castus*)
Chinquapin Oak (L)
(*Quercus muhlenbergii*)
Cottonwood (L)
(*Populus deltoides*)
Desert Willow (N)
(*Chilopsis linearis*)
Flowering Mimosa (N)
(*Albizia julibrissin*)
Hackberry (L)
(*Celtis laevigata*)
Honey Mesquite (L) (N)
(*Prosopis glandulosa*)
Little-leaf Mulberry (L)
(*Morus microphylla*)
Redbud (L) (N)
(*Cercis canadensis*)
Red Mulberry (L)
(*Morus rubra*)
Western Soapberry (L)
(*Sapindus saponaria* var. *drummondii*)

Shrubs

Agarita (N)
(*Mahonia trifoliolata*)
Barbados-Bird-of-Paradise (N)
(*Caesalpinia pulcherrima*)
Black Dalea (L)
(*Dalea frutescens*)
Butterfly Bush (N)
(*Buddleja davidii*)
Cat-claw Acacia (L)
(*Acacia greggii*)
Chickasaw Plum (N)
(*Prunus angustifolia*)

Eastern Red Cedar (L)
(*Juniperus virginiana*)
False Indigo (L) (N)
(*Amorpha fruticosa*)
Fragrant Lilac (N)
(*Syringa vulgaris*)
Fragrant Sumac (L)
(*Rhus aromatica*)
Roemer's Acacia (L)
(*Acacia roemeriana*)
Smooth Yucca (L)
(*Yucca glauca*)
Trailing Lantana (N)
(*Lantana montevidensis*)
Trifoliate Orange (L) (N)
(*Poncirus trifoliata*)
Wafer-ash (L)
(*Ptelea trifoliata*)
Wright's Acacia (L)
(*Acacia greggii* var. *wrightii*)
Yellow Bells (L) (N)
(*Tecoma stans*)

Vines (on trellis/fence)

Climbing Milkweed Vine (L) (N)
(*Funastrum cynanchoides*)
Drummond's Virgin's Bower
(*Clematis drummondii*)
Mexican Flame Vine
(*Pseudogynoxys chenopodioides*)

Tall Herbs (3–5 feet)

Argentina Verbena (N)
(*Verbena bonariensis*)
Cosmos (N)
(*Cosmos bipinnatus*)
Golden Crownbeard (L) (N)
(*Verbesina encelioides*)
Hollyhock (L) (N)
(*Alcea rosea*)
Maximilian Sunflower (N)
(*Helianthus maximiliani*)
Mexican Sunflower (L) (N)
(*Tithonia rotundifolia*)
Plains Sunflower (L) (N)
(*Helianthus petiolaris*)
Purple Marsh-fleabane (N)
(*Pluchea odorata*)
Showy Bergamot (N)
(*Monarda didyma*)
Showy Milkweed (L) (N)
(*Asclepias speciosa*)
Summer Phlox (N)
(*Phlox paniculata*)

Medium Herbs (2–3 feet)

Brown-eyed Susan (N)
(*Rudbeckia hirta*)
Cardinal Flower (N)
(*Lobelia cardinalis*)
Cherry Sage (N)
(*Salvia greggii*)
Copper Globe-mallow (L)
(*Sphaeralcea angustifolia*)
Fern Acacia (L)
(*Acacia angustissima* var. *hirta*)
Globe Amaranth (N)
(*Gomphrena globosa*)
Goldenrod (N)
(*Solidago* spp.)
Heath Aster (L)
(*Symphyotrichum ericoides*)
Ironweed (N)
(*Vernonia* spp.)
Liatris (N)
(*Liatris* spp.)
Old-fashioned Petunias (N)
(*Petunia axillaris*)
Purple Coneflower (N)
(*Echinacea purpurea*)
Scarlet Globe-mallow (L)
(*Sphaeralcea coccinea*)
Tropical Milkweed (L) (N)
(*Asclepias curassavica*)
Tropical Sage (L)
(*Salvia coccinea*)
Thread-leaf Groundsel (N)
(*Senecio flaccidus*)
Woolly Paper-flower (N)
(*Psilostrophe tagetina*)

Low Herbs (1–2 feet)

Butterfly Weed (N)
(*Asclepias tuberosa*)
Evax (L)
(*Evax* spp.)
Flax (L)
(*Linum* spp.)
Gregg's Mistflower (N)
(*Conoclinium dissectum*)
Huisache Daisy (N)
(*Amblyolepis setigera*)
Paintbrush (L)
(*Castilleja* spp.)
Peppergrass (L)
(*Lepidium* spp.)
Prairie Verbena (N)
(*Glandularia bipinnatifida*)
Society Garlic (N)
(*Tulbaghia violacea*)
Sweet Sand-verbena (N)
(*Abronia fragrans*)
Tahoka Daisy (N)
(*Machaeranthera tanacetifolia*)
Trailing Lantana (N)
(*Lantana montevidensis*)
Two-leaved Senna (L) (N)
(*Senna roemeriana*)

Edging (to 1 foot)

Bayou Violet (L)
(*Viola sororia*)
Bristle-leaf Dyssodia
(*Dyssodia tenuiloba*)
Dogweed (L) (N)
(*Thymophylla* spp.)
Fringed Pipevine (L)
(*Aristolochia fimbriata*)
Frogfruit (L) (N)
(*Phyla* spp.)
Garden Pansy (L)
(*Viola x wittrockiana*)
Nasturtium (L)
(*Tropaeolum majus*)
Plains Zinnia (N)
(*Zinnia grandiflora*)
Scarlet Pea (L)
(*Indigofera miniata*)
Sweet Alyssum (L) (N)
(*Lobularia maritima*)

Plants beneath Trees

Drummond's Wax-mallow (N)
(*Malvaviscus drummondii*)
Fern Acacia (L)
(*Acacia angustissima* var. *hirta*)
Spring Elbow-bush (N)
(*Forestiera pubescens*)

Flagstone

Decomposed Granite

Turf

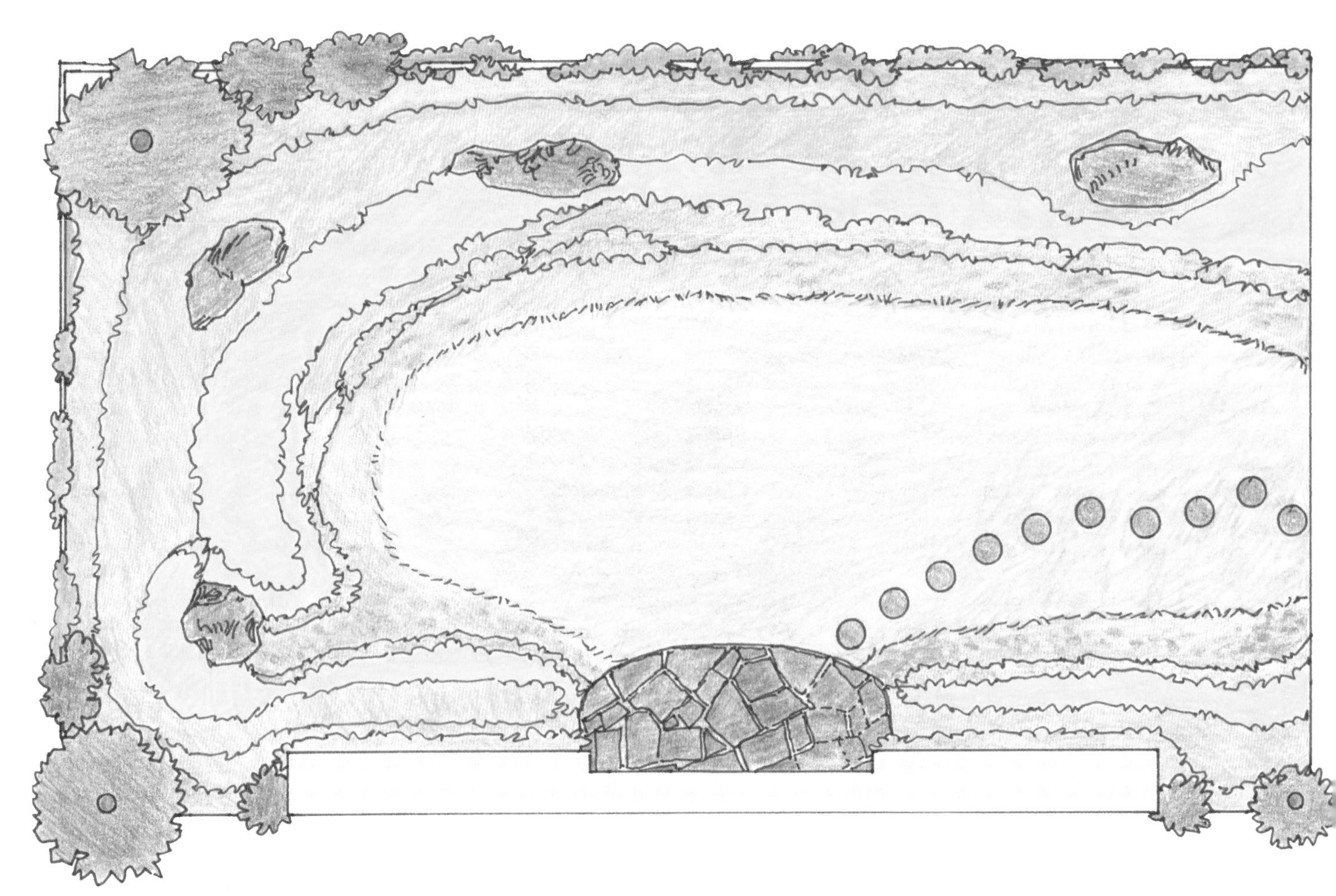

GRASSLANDS/
PRAIRIES/
SAVANNAHS
(REGION 2)

Dallas, Bryan/College Station, Fort Worth, Goliad, Gonzales, Paris, Sherman, Waco

(L) indicates larval food plant
(N) indicates nectar plant

Trees

American Elm (L)
(*Ulmus americana*)
Ash (L)
(*Fraxinus* spp.)
Black Locust (L)
(*Robinia pseudoacacia*)
Bur Oak (L)
(*Quercus macrocarpa*)
Carolina Basswood (L) (N)
(*Tilia americana* var. *caroliniana*)
Chaste Tree (N)
(*Vitex agnus-castus*)
Chinquapin Oak (L)
(*Quercus muhlenbergii*)
Eastern Red Cedar (L)
(*Juniperus virginiana*)
Flowering Mimosa (L) (N)
(*Albizia julibrissin*)
Hackberry (L)
(*Celtis laevigata*)
Hercules'-club Prickly-ash (L)
(*Zanthoxylum clava-hercules*)
Honey Mesquite (L) (N)
(*Prosopis glandulosa*)
Little-leaf Mulberry (L)
(*Morus microphylla*)
Mexican Plum (L) (N)
(*Prunus mexicana*)
Northern Catalpa (L)
(*Catalpa speciosa*)
Redbud (L) (N)
(*Cercis canadensis*)
Red Mulberry (L)
(*Morus rubra*)
Rusty Blackhaw (N)
(*Viburnum rufidulum*)
Sassafras (L)
(*Sassafras albidum*)
Southern Red Oak (L)
(*Quercus falcata*)
Western Soapberry (L)
(*Sapindus saponaria* var. *drummondii*)
Wild Black Cherry (L) (N)
(*Prunus serotina*)
Willow (L)
(*Salix* spp.)

Shrubs

Argentina Senna (L)
(*Senna corymbosa*)
Arkansas Yucca (L)
(*Yucca arkansana*)
Barbados Bird-of-Paradise (N)
(*Caesalpinia pulcherrima*)
Black Dalea (L)
(*Dalea frutescens*)

Butterfly Bush (N)
(*Buddleja davidii*)
Cherry Sage (N)
(*Salvia greggii*)
Common Beebrush (N)
(*Aloysia gratissima*)
Common Buttonbush (N)
(*Cephalanthus occidentalis*)
Downy Prickly-ash (L)
(*Zanthoxylum hirsutum*)
False Indigo (L) (N)
(*Amorpha fruticosa*)
Flame-leaf Sumac (L)
(*Rhus copallina*)
Fragrant Lilac (N)
(*Syringa vulgaris*)
Illinois Bundleflower (L)
(*Desmanthus illinoensis*)
Mexican Buckeye (L) (N)
(*Ungnadia speciosa*)
Redroot (L)
(*Ceanothus herbaceus*)
Spring Elbow-bush (N)
(*Forestiera pubescens*)
Sweet Almond Verbena (N)
(*Aloysia virgata*)
Texas Kidneywood (L)
(*Eysenhardtia texana*)
Trailing Lantana (N)
(*Lantana montevidensis*)
Wafer-ash (L)
(*Ptelea trifoliata*)
West Indian Lantana (L) (N)
(*Lantana camara*)
Wright's Snakeroot (N)
(*Ageratina wrightii*)

Vines (on trellis/fence)

Climbing Milkweed Vine (L) (N)
(*Funastrum cynanchoides*)
Common Balloon-vine (L)
(*Cardiospermum halicacabum*)
Groundnut (L)
(*Apios americana*)
Least Snoutbean (L)
(*Rhynchosia minima*)
Mexican Flame Vine (N)
(*Pseudogynoxys chenopodioides*)
Purple Passionflower (L)
(*Passiflora incarnata*)
Red-fruited Passionflower (L)
(*Passiflora foetida*)
Virginia Butterfly-pea (L)
(*Centrosema virginianum*)
Wisteria (L)
(*Wisteria* spp.)
Woolly Pipevine (L)
(*Aristolochia tomentosa*)
Yellow Passionflower (L)
(*Passiflora lutea*)

Tall Herbs (3–5 feet)

Golden Crownbeard (L) (N)
(*Verbesina encelioides*)
Hollyhock (L)
(*Alcea rosea*)
Ironweed (N)
(*Vernonia* spp.)
Liatris (N)
(*Liatris* spp.)
Maximilian Sunflower (N)
(*Helianthus maximiliani*)
Mexican Sunflower (L) (N)
(*Tithonia rotundifolia*)
Plains Sunflower (L) (N)
(*Helianthus petiolaris*)
Purple Marsh-fleabane (N)
(*Pluchea odorata*)
Summer Phlox (N)
(*Phlox paniculata*)
Tooth-leaved Goldeneye (N)
(*Viguiera dentata*)
Virginia Frostweed (L) (N)
(*Verbesina virginica*)
Yellow Prairie Grass (L)
(*Sorghastrum nutans*)

Medium Herbs (2–3 feet)

Agalinis (L)
(*Agalinis* spp.)
Broad-leaf Wood-oats (L)
(*Chasmanthium latifolium*)
Brown-eyed Susan (N)
(*Rudbeckia hirta*)
Butterfly Weed (N)
(*Asclepias tuberosa*)
Cardinal Flower (N)
(*Lobelia cardinalis*)
Cherry Sage (N)
(*Salvia greggii*)
Copper Globe-mallow (L)
(*Sphaeralcea angustifolia*)
Canna (L)
(*Canna x generalis*)
Heath Aster (L) (N)
(*Symphyotrichum ericoides*)
Late-flowering Eupatorium (L) (N)
(*Eupatorium serotinum*)
Mealy Sage (N)
(*Salvia farinacea*)
Purple Coneflower (N)
(*Echinacea purpurea*)
Rootstock Iresine (L)
(*Iresine rhizomatosa*)
Scarlet Globe-mallow (L)
(*Sphaeralcea coccinea*)
Texas Abutilon (L)
(*Abutilon fruticosum*)
Tropical Milkweed (L) (N)
(*Asclepias curassavica*)
West Indian Lantana (N)
(*Lantana camara*)

Low Herbs (1–2 feet)

Cudweed (L)
(*Gamochaeta* spp.)
Fern Acacia (L)
(*Acacia angustissima* var. *hirta*)
Flax (L)
(*Linum* spp.)
Hispid Wedelia (L) (N)
(*Wedelia acapulcensis* var. *hispida*)
Paintbrush (L)
(*Castilleja* spp.)
Pigeon-wings (L)
(*Clitoria mariana*)
Two-leaved Senna (L)
(*Senna roemeriana*)
Violet Ruellia (L)
(*Ruellia nudiflora*)

Edging (to 1 foot)

Fringed Pipevine (L)
(*Aristolochia fimbriata*)
Frogfruit (L) (N)
(*Phyla* spp.)
Garden Pansy (L)
(*Viola x wittrockiana*)
Nasturtium (L)
(*Tropaeolum majus*)
Powderpuff (L)
(*Mimosa strigillosa*)
Prairie Verbena (N)
(*Glandularia bipinnatifida*)
Sweet Alyssum (L) (N)
(*Lobularia maritima*)

Plants in Lawn

Blue-eyed Grass (N)
(*Sisyrinchium* spp.)
Bluets (N)
(*Houstonia* spp.)
Copper Lily
(*Habranthus tubispathus*)
False Garlic (N)
(*Nothoscordum bivalve*)
Prairie Rain Lily
(*Cooperia* spp.)
Spring Beauty (N)
(*Claytonia virginica*)
Texas Centaury
(*Centaurium texense*)
Texas Dutchman's-breeches (L)
(*Thamnosma texana*)
Wild Onion (N)
(*Allium* spp.)
Yellow Star-grass (N)
(*Hypoxis hirsuta*)

Flagstone

Decomposed Granite

Turf

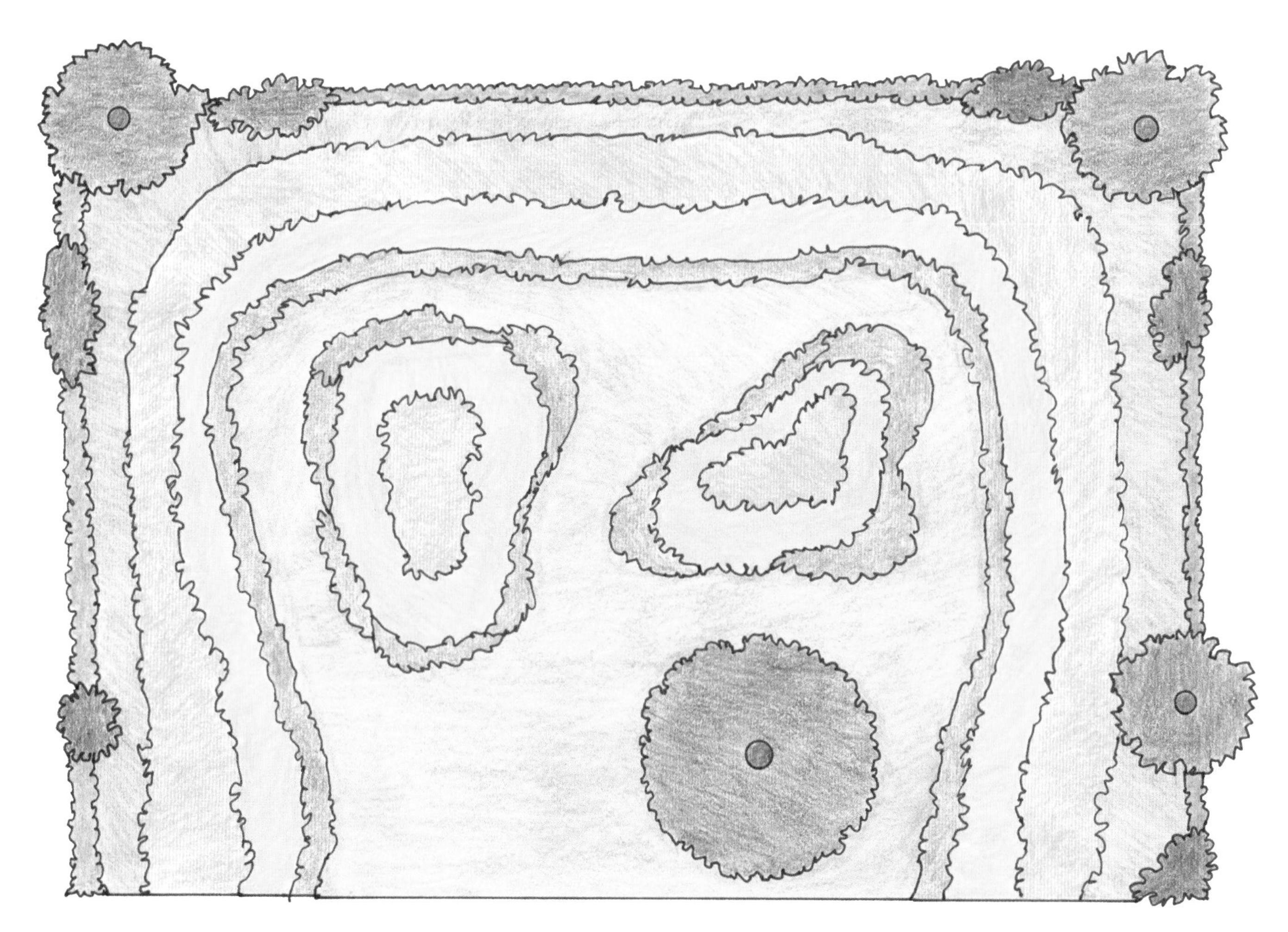

PINE/HARDWOODS
(REGION 3)
Mount Pleasant, Longview, Tyler, Nacogdoches, Lufkin, Huntsville, Beaumont

(L) indicates larval food plant
(N) indicates nectar plant

Trees

Ash (L)
(*Fraxinus* spp.)
Black Locust (L)
(*Robinia pseudoacacia*)
Carolina Basswood (L) (N)
(*Tilia americana* var. *caroliniana*)
Camphor Tree (L)
(*Cinnamomum camphora*)
Carolina Silverbells (N)
(*Halesia carolina*)
Chaste Tree (N)
(*Vitex agnus-castus*)
Eastern Red Cedar (L)
(*Juniperus virginiana*)
Elm (L)
(*Ulmus* spp.)
Flowering Dogwood (L)
(*Cornus florida*)
Flowering Mimosa (L) (N)
(*Albizia julibrissin*)
Hackberry (L)
(*Celtis laevigata*)
Hercules'-club Prickly-ash (L)
(*Zanthoxylum clava-herculis*)
Mexican Plum (L) (N)
(*Prunus mexicana*)
Oak (L)
(*Quercus* spp.)
Pine (L)
(*Pinus* spp.)
Red Bay (L)
(*Persea borbonia*)
Red Mulberry (L)
(*Morus rubra*)
Redbud (L) (N)
(*Cercis canadensis*)
Sassafras (L)
(*Sassafras albidum*)
Sweet-bay (L)
(*Magnolia virginiana*)
Tulip Tree (L)
(*Liriodendron tulipifera*)
Western Soapberry (L)
(*Sapindus saponaria* var. *drummondii*)
Wild Black Cherry (L) (N)
(*Prunus serotina*)

Shrubs

Azalea (N)
(*Rhododendron* spp.)
Butterfly Bush (N)
(*Buddleja davidii*)

Common Buttonbush (N)
(*Cephalanthus occidentalis*)
Coral Bean (L)
(*Erythrina herbacea*)
Deerberry (L)
(*Vaccinium stamineum*)
False Indigo (L) (N)
(*Amorpha fruticosa*)
Fragrant Lilac (N)
(*Syringa vulgaris*)
Glossy Abelia (N)
(*Abelia x grandiflora*)
Highbush Blueberry (L)
(*Vaccinium corymbosum*)
Illinois Bundleflower (L)
(*Desmanthus illinoensis*)
New Jersey Tea (L) (N)
(*Ceanothus americanus* var. *pitcherii*)
Parsley-leaved Hawthorn (L) (N)
(*Crataegus marshallii*)
Spicebush (L) (N)
(*Lindera benzoin*)
Sumac (L)
(*Rhus* spp.)
Tall Pawpaw (L)
(*Asimina triloba*)
Trailing Lantana (L) (N)
(*Lantana montevidensis*)
Trifoliate Orange (L)
(*Poncirus trifoliata*)
Virginia Sweetspire (N)
(*Itea virginica*)
Wafer-ash (L)
(*Ptelea trifoliata*)
Wax-myrtle (L)
(*Morella* spp.)

Vines (on trellis/fence)

Climbing Hempweed (N)
(*Mikania scandens*)
Climbing Milkweed Vine (L) (N)
(*Funastrum cynanchoides*)
Common Balloon-vine (L)
(*Cardiospermum halicacabum*)
Groundnut (L)
(*Apios americana*)
Least Snoutbean (L)
(*Rhynchosia minima*)
Purple Passionflower (L)
(*Passiflora incarnata*)
Southern Hog-peanut (L)
(*Amphicarpaea bracteata*)
Virginia Butterfly-pea (L)
(*Centrosema virginianum*)
Wisteria (L)
(*Wisteria* spp.)
Woolly Pipevine (L)
(*Aristolochia tomentosa*)
Yellow Passionflower (L)
(*Passiflora lutea*)

Tall Herbs (3–5 feet)

Argentina Verbena (N)
(*Verbena bonariensis*)
Bergamot
(*Monarda* spp.)
Blue Mistflower (N)
(*Conoclinium coelestinum*)
Golden Crownbeard (L) (N)
(*Verbesina enceiloides*)
Goldenrod (N)
(*Solidago* spp.)
Hollyhock (L)
(*Alcea rosea*)
Ironweed (N)
(*Vernonia* spp.)
Joe-Pye Weed (N)
(*Eutrochium fistulosum*)
Liatris (N)
(*Liatris* spp.)
Mexican Sunflower (L) (N)
(*Tithonia rotundifolia*)
Purple Coneflower (N)
(*Echinacea purpurea*)
Purple Marsh-fleabane (N)
(*Pluchea odorata*)
Slender-leaf Mountain-mint (N)
(*Pycnanthemum tenuifolium*)
Swamp Milkweed (N)
(*Asclepias incarnata*)
Virginia Frostweed (L) (N)
(*Verbesina virginica*)
Yellow Prairie Grass (L)
(*Sorghastrum nutans*)

Medium Herbs (2–3 feet)

Agalinis (L)
(*Agalinis* spp.)
Broad-leaf Wood-oats (L)
(*Chasmanthus latifolium*)
Brown-eyed Susan (N)
(*Rudbeckia hirta*)
Butterfly Weed (N)
(*Asclepias tuberosa*)
Cardinal Flower (N)
(*Lobelia cardinalis*)
Cherry Sage (N)
(*Salvia greggii*)
Common Canna (L)
(*Canna indica*)
Flamingo Plant (L)
(*Jacobinia carnea*)
Globe Amaranth (N)
(*Gomphrena globosa*)
Liatris (N)
(*Liatris* spp.)
Partridge Pea (L) (N)
(*Chamaecrista fasciculata*)
Pentas (N)
(*Pentas lanceolata*)
Old-Fashioned Petunias (N)
(*Petunia axillaris*)
Tropical Milkweed (L) (N)
(*Ascepias curassavica*)
Woolly Croton (L)
(*Croton capitatus*)

Low Herbs (1–2 feet)

Cudweed (L)
(*Gamochaeta* spp.)
Evax (L)
(*Evax* spp.)
Fern Acacia (L)
(*Acacia angustissima* var. *hirta*)
Flax (L)
(*Linum* spp.)
Green Milkweed (L) (N)
(*Asclepias viridis*)
Pigeon-wings (L)
(*Clitoria mariana*)
Spring Bittercress (L)
(*Cardamine rhomboidea*)
Violet Ruellia (L)
(*Ruellia nudiflora*)

Edging (to 1 foot)

Blue-eyed Grass (N)
(*Sisyrinchium* spp.)
Fringed Pipevine (L)
(*Aristolochia fimbriata*)
Frogfruit (L) (N)
(*Phyla* spp.)
Garden Pansy (L)
(*Viola x wittrockiana*)
Nasturtium (L)
(*Tropaeolum majus*)
Peppergrass (L)
(*Lepidium* spp.)
Powderpuff (L)
(*Mimosa strigillosa*)
Scarlet Pea (L)
(*Indigofera minuata*)
Sweet Alyssum (L) (N)
(*Lobularia maritima*)
Violets (L)
(*Viola* spp.)

Flagstone

Decomposed Granite

Turf

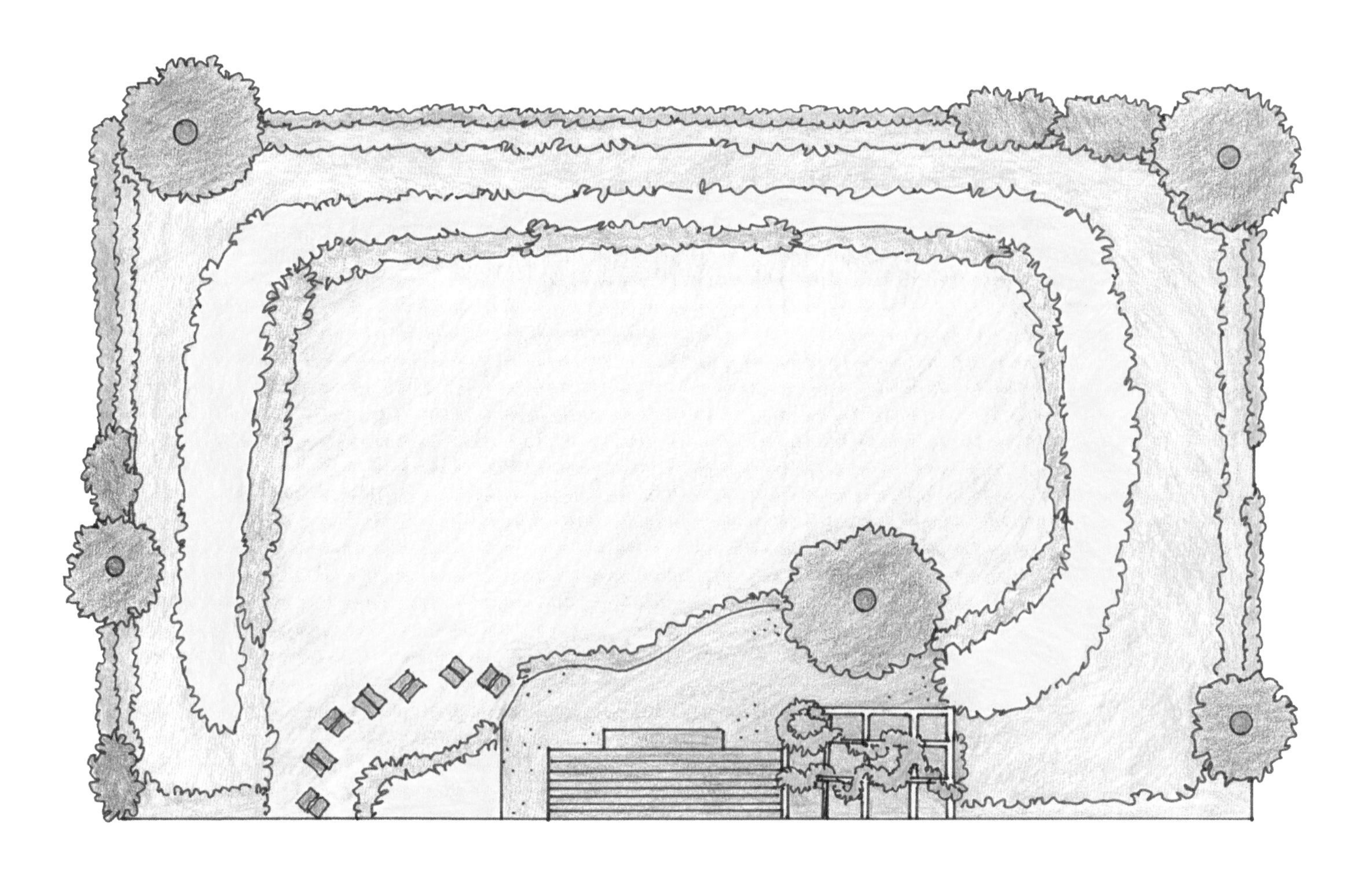

COASTAL PRAIRIES/ MARSHES/ BEACHES (REGION 4)

Corpus Christi, Galveston, Houston

(L) indicates larval food plant
(N) indicates nectar plant

Trees

Anacahuita (N)
(*Cordia boissieri*)
Ash (L)
(*Fraxinus* spp.)
Carolina Basswood (L) (N)
(*Tilia americana* var. *caroliniana*)
Chaste Tree (N)
(*Vitex agnus-castus*)
Elm (L)
(*Ulmus* spp.)
Flowering Dogwood (L)
(*Cornus florida*)
Flowering Mimosa (N)
(*Albizia julibrissin*)
Hackberry (L)
(*Celtis laevigata*)
Mexican Plum (L) (N)
(*Prunus mexicana*)
Oak (L)
(*Quercus* spp.)
Orange (L) (N)
(*Citrus* spp.)
Pine (L)
(*Pinus* spp.)
Red Bay (L)
(*Persea borbonia*)
Red Mulberry (L)
(*Morus rubra*)
Redbud (L) (N)
(*Cercis canadensis*)
Sassafras (L)
(*Sassafras albidum*)
Sweet-bay (L)
(*Magnolia virginiana*)
Wild Black Cherry (L) (N)
(*Prunus serotina*)

Shrubs

Argentina Senna (L)
(*Senna corymbosa*)
Azalea (N)
(*Rhododendron* spp.)
Barbados Bird-of-Paradise (N)
(*Caesalpinia pulcherrima*)
Barbados Cherry (L)
(*Malpighia glabra*)
Bougainvillea (N)
(*Bougainvillea glabra*)
Butterfly Bush (N)
(*Buddleja davidii*)
Cenizo (L)
(*Leucophyllum frutescens*)
Common Buttonbush (N)
(*Cephalanthus occidentalis*)
Coral Bean (L)
(*Erythrina herbacea*)

Downy Prickly-ash (L)
(*Zanthloxylum hirsutum*)
Drummond's Wax-mallow (L)
(*Malvaviscus drummondii*)
Dwarf Screwbean (L)
(*Prosopis reptans*)
False Indigo (L) (N)
(*Amorpha fruticosa*)
Flame-leaf Sumac (L)
(*Rhus copallina*)
Lime Prickly-ash (L)
(*Zanthloxylum fagara*)
New Jersey Tea (L) (N)
(*Ceanothus americanus* var. *pitcherii*)
Parsley-leaved Hawthorn (L) (N)
(*Crataegus marshallii*)
Sweet Almond Verbena (N)
(*Aloysia virgata*)
Trifoliate Orange (L)
(*Poncirus trifoliata*)
Virginia Sweetspire (N)
(*Itea virginica*)
Wafer-ash (L)
(*Ptelea trifoliata*)
Wax-myrtle (L)
(*Morella* spp.)

Vines (on trellis /fence)

Alamo Vine (N)
(*Merremia dissecta*)
Asian Pigeon-wings (L)
(*Clitoria ternate*)
Climbing Hempweed (N)
(*Mikania scandens*)
Climbing Milkweed Vine (L) (N)
(*Funastrum cynanchoides*)
Common Balloon-vine (L)
(*Cardiosperma halicacabum*)
Drummond's Virgin's Bower (L)
(*Clematis drummondii*)
Groundnut (L)
(*Apios americana*)
Least Snoutbean (L)
(*Rhynchosia minima*)
Purple Passionflower (L)
(*Passiflora incarnata*)
Queen's Wreath (N)
(*Antigonon leptopus*)
Red-fruited Passionflower (L)
(*Passiflora foetida*)
Salt-marsh Morning Glory (N)
(*Ipomoea sagittata*)
Snapdragon Vine (L)
(*Maurandya antirrhiniflora*)
Virginia Butterfly-pea (L)
(*Centrosema virginianum*)
Wild Cow-pea (L)
(*Vigna luteola*)
Wisteria (L)
(*Wisteria* spp.)
Woolly Pipevine (L)
(*Aristolochia tomentosa*)
Yellow Passionflower (L)
(*Passiflora lutea*)

Tall Herbs (3–5 feet)

Argentina Verbena (N)
(*Verbena bonariensis*)
Cardinal Flower (N)
(*Lobelia cardinalis*)
Golden Crownbeard (L) (N)
(*Verbesina encelioides*)
Goldenrod (N)
(*Solidago* spp.)
Gulf Vervain (N)
(*Verbena xutha*)
Joe-Pye Weed (N)
(*Eutrochium fistulosum*)
Powdery Thalia (L)
(*Thalia dealbata*)
Purple Marsh-fleabane (N)
(*Pluchea odorata*)
Swamp Milkweed (N)
(*Asclepias incarnata*)
Texas Abutilon (L)
(*Abutilon fruticosum*)
Virginia Frostweed (L) (N)
(*Verbesina virginiana*)
Wild Bergamot (N)
(*Monarda fistulosa*)

Medium Herbs (2–3 feet)

Agalinis (L)
(*Agalinis* spp.)
Betony-leaf Mistflower (N)
(*Conoclinium betonicifolium*)
Blue Mistflower (N)
(*Conoclinium coelestinum*)
Broad-leaf Wood-oats (L)
(*Chasmanthium latifolium*)
Brown-eyed Susan (N)
(*Rudbeckia hirta*)
Cherry Sage (N)
(*Salvia greggii*)
Lemon Beebalm (N)
(*Monarda citriodora*)
Liatris (N)
(*Liatris* spp.)
Pentas (N)
(*Pentas lanceolata*)
Pickerel Weed (N)
(*Pontederia cordata*)
Pigeonberry (L) (N)
(*Rivina humilis*)
Purple Coneflower (N)
(*Echinacea purpurea*)
Shrimp Plant (L)
(*Justicia brandegeana*)
Tropical Milkweed (L) (N)
(*Asclepias curassavica*)
Violet Ruellia (L)
(*Ruellia nudiflora*)

Low Herbs (1–2 feet)

Hispid Wedelia (L) (N)
(*Wedelia acapulcensis* var. *hispida*)
Lindheimer's Tephrosia (L)
(*Tephrosia lindheimeri*)
Wright's False Mallow (L)
(*Malvastrum aurantiacum*)

Edging (to 1 foot)

Frogfruit (L) (N)
(*Phyla* spp.)
Garden Pansy (L)
(*Viola x wittrockiana*)
Powderpuff (L)
(*Mimosa strigillosa*)
Prairie Verbena (N)
(*Verbena bipinnatifida*)
Scarlet Pea (L)
(*Indigofera miniata*)
Spreading Sida (L)
(*Sida abutifolia*)
Sweet Alyssum (L) (N)
(*Lobularia maritima*)
Violets (L)
(*Viola* spp.)
Woolly Stemodia (L)
(*Stemodia lanata*)

Flagstone

Decomposed Granite

Turf

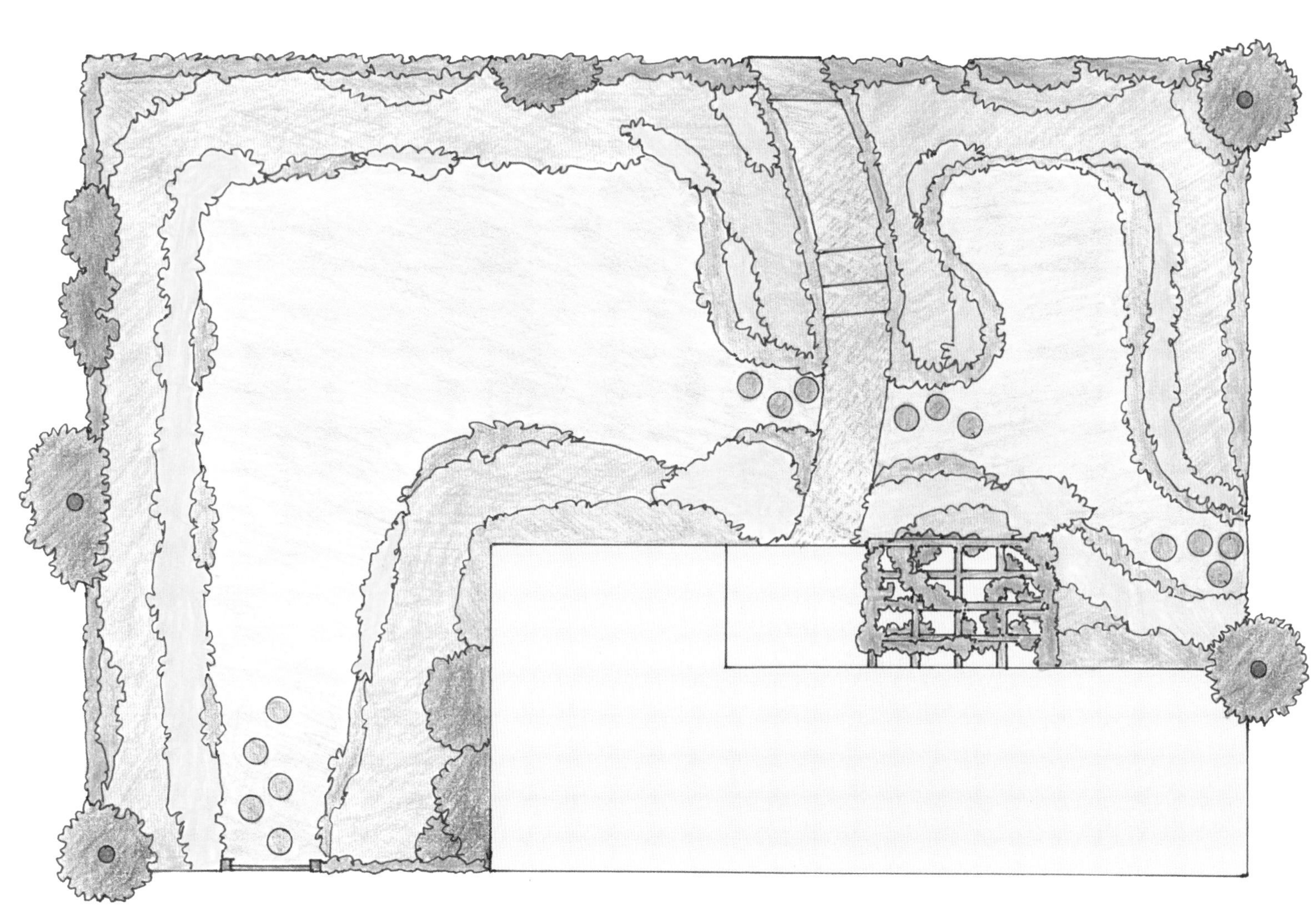

CHAPARRAL PLAINS/RIO GRANDE VALLEY (REGION 5)

Alice, Crystal City, Eagle Pass, Edinburg, Harlingen, Kingsville, Laredo, McAllen, Mission, Pharr

(L) indicates larval food plant
(N) indicates nectar plant

Trees

Anacahuita (N)
(*Cordia boissieri*)
Black Locust (L)
(*Robinia pseudoacacia*)
Cedar Elm (L)
(*Ulmus crassifola*)
Chaste Tree (N)
(*Vitex agnus-castus*)
Desert Willow (N)
(*Chilopsis linearis*)
Hackberry (L)
(*Celtis* spp.)
Honey Mesquite (L) (N)
(*Prosopis glandulosa*)
Orange (L) (N)
(*Citrus* spp.)
Retama (N)
(*Parkinsonia aculeata*)
Western Soapberry (L)
(*Sapindus saponaria* var. *drummondii*)

Shrubs

Acacia (L)
(*Acacia* spp.)
Argentina Senna (L)
(*Senna cormybosum*)
Barbados Cherry (L)
(*Malpighia glabra*)
Bougainvillea (N)
(*Bougainvillea glabra*)
Butterfly Bush (N)
(*Buddleja davidii*)
Carlowrightia (L)
(*Carlowrightia* spp.)
Common Beebrush (N)
(*Aloysia gratissima*)
Coral Bean (L)
(*Erythrina herbacea* var. *arborea*)
Cenizo (L)
(*Leucophyllum frutescens*)
Crucita (N)
(*Chromolaena odorata*)
Desert Lantana (N)
(*Lantana achyranthifolia*)
Drummond's Wax-mallow (L) (N)
(*Malvaviscus drummondii*)
Flame Acanthus (L) (N)
(*Anisacanthus quadrifidus* var. *wrightii*)

Mexican Bird-of-Paradise (N)
(*Caesalpinia mexicana*)
Southwest Bernardia (L)
(*Bernardia myricifolia*)
Spiny Hackberry (L)
(*Celtis pallida*)
Sweet Almond Verbena (N)
(*Aloysia virgata*)
Texas Kidneywood (L)
(*Eysenhardtia texana*)
West Indian Lantana (L) (N)
(*Lantana camara*)
Yellow Bells (L) (N)
(*Tecoma stans*)

Vines (on trellis/fence)

Chinese Wisteria (L)
(*Wisteria sinensis*)
Climbing Hempweed (N)
(*Mikania scandens*)
Climbing Milkweed Vine (L) (N)
(*Funastrum cynanchoides*)
Common Balloon-vine (L)
(*Cardiospemum halicacabum*)
Drummond's Virgin's Bower (L)
(*Clematis drummondii*)
Least Snoutbean (L)
(*Rhynchosia minima*)
Mexican Flame Vine (N)
(*Pseudogynoxys chenopodioides*)
Queen's Wreath (N)
(*Antigonon leptopus*)
Red-fruited Passionflower (L)
(*Passiflora foetida*)
Slender-lobed Passionflower (L)
(*Passiflora tenuiloba*)
Snapdragon Vine (L)
(*Maurandya antirrhiniflora*)
Tropical Balloon-vine (L)
(*Cardiospermum corundum*)
Virginia Butterfly-pea (L)
(*Centrosema virginianum*)
Yellow Passionflower (L)
(*Passiflora lutea*)

Tall Herbs (3–5 feet)

Argentina Verbena (N)
(*Verbena bonariensis*)
Croton (L)
(*Croton* spp.)
Golden Crownbeard (L) (N)
(*Verbesina encelioides*)
Hollyhock (L)
(*Alcea rosea*)
Mexican Sunflower (L) (N)
(*Tithonia rotundifolia*)
Virginia Frostweed (L)
(*Verbesina virginica*)
White-flowered Plumbago (L)
(*Plumbago scandens*)

Medium Herbs (2–3 feet)

Awnless Bush-sunflower (N)
(*Simsia calva*)
Betony-leaf Mistflower (N)
(*Conoclinium betonicifolium*)
Broad-leaf Wood-oats (L)
(*Chasmanthium latifolium*)
Brown-eyed Susan (N)
(*Rudbeckia hirta*)
Cherry Sage (N)
(*Salvia greggii*)
Clammyweed (L)
(*Polanisia dodecandra*)
Croton (L)
(*Croton* spp.)
Flamingo Plant (L)
(*Jacobinia carnea*)
Canna (L)
(*Canna x generalis*)
Goldenrod (N)
(*Solidago* spp.)
Gregg's Mistflower N)
(*Conoclinium greggii*)
Liatris (N)
(*Liatris* spp.)
Mexican Oregano (N)
(*Poliomintha longiflora*)
Partridge-pea (L)
(*Chamaecrista fasciculata*)
Pentas (N)
(*Pentas lanceolata*)
Pigeonberry (L)
(*Rivina humilis*)
Rio Grande Dicliptera (N)
(*Dicliptera sexangularis*)
Shrimp Plant (L)
(*Justicia brandegeana*)
Skeleton-leaf Goldeneye) (N)
(*Viguiera stenoloba*)
Trailing Boneset (N)
(*Fleischmannia incarnata*)
Tropical Milkweed (L) (N)
(*Asclepias curassavitica*)
Tropical Sage (N)
(*Salvia coccinea*)
Willow-leaf Aster (L) (N)
(*Symphyotrichum praealtum*)
Wright's Abutilon (L)
(*Abutilon wrightii*)

Low Herbs (1–2 feet)

Fern Acacia ((L) (N)
(*Acacia angustissima* var. *hirta*)
Hispid Wedelia (L) (N)
(*Wedelia acapulcensis* var. *hispida*)
Ruellia (L)
(*Ruellia* spp.)
Upright Pipevine (L)
(*Aristolochia erecta*)

Edging (to 1 foot)

Brittle-leaf Dyssodia (L) (N)
(*Dyssodia tenuiloba*)
Common Dogweed (L) (N)
(*Thymophylla pentachaeta*)
Prairie Verbena (L) (N)
(*Verbena bipinnatifida*)
Dwarf Crownbeard (N)
(*Verbesina nana*)
Fringed Pipevine (L)
(*Aristolochia fimbriata*)
Frogfruit (L) (N)
(*Phyla* spp.)
Nasturtium (L)
(*Tropaeolum majus*)
Powderpuff (L)
(*Mimosa strigillosa*)
Purple Bush-bean (L)
(*Macroptilium atropurpureum*)
Scarlet Pea (L)
(*Indigofera miniata*)
Sweet Alyssum (L) (N)
(*Lobularia maritima*)

Flagstone

Decomposed Granite

Turf

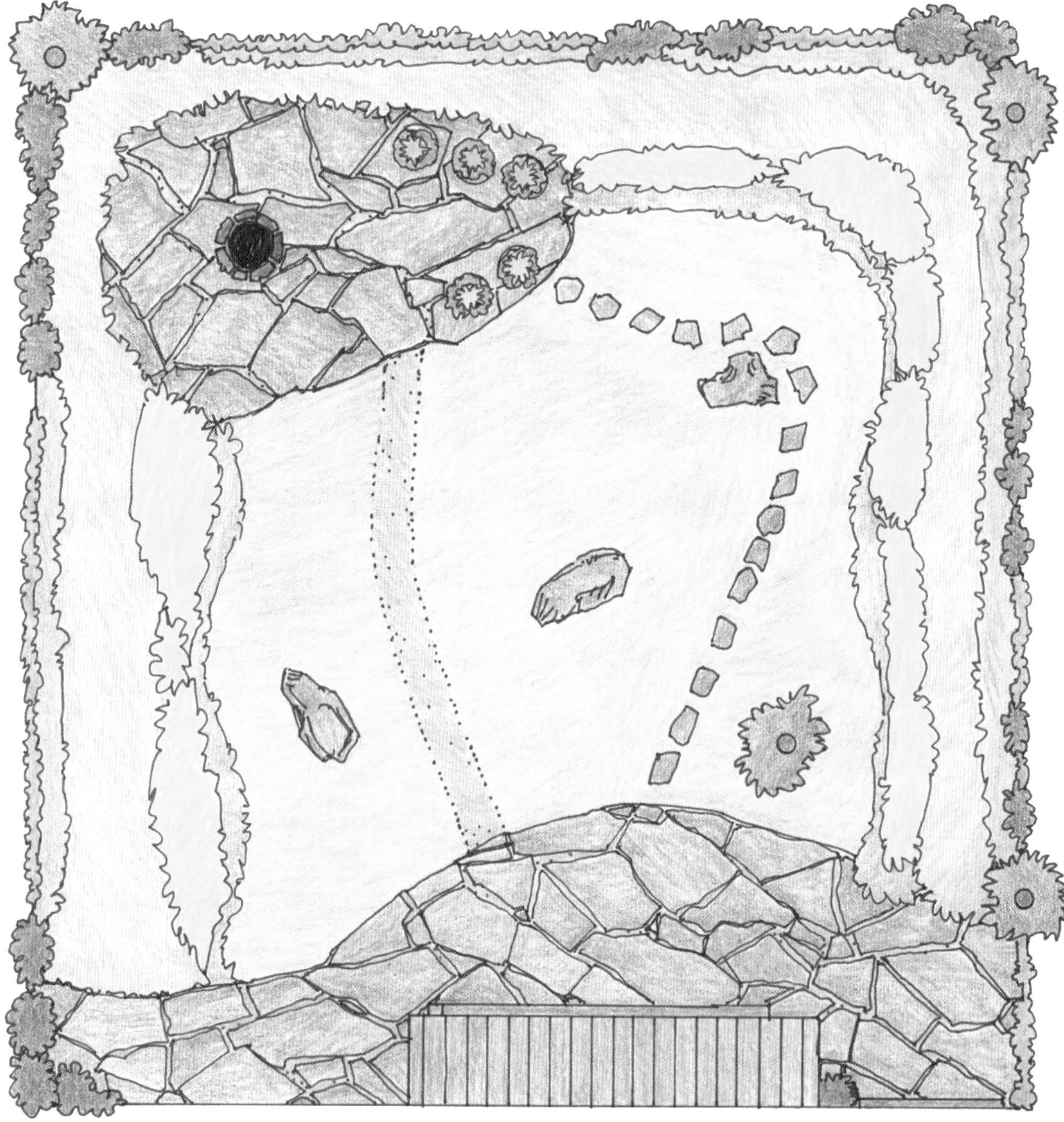

EDWARDS PLATEAU (REGION 6)

Killeen, Austin, San Antonio, Del Rio, Odessa, Midland, San Angelo

(L) indicates larval food plant
(N) indicates nectar plant

POTS

Use plants from the Medium Herbs or Low Herbs lists bordered by the lower-growing plants from the Edging list.

 *Trees*

Anacahuita (N)
(*Cordia boissieri*)
Anacacho Orchid Tree (N)
(*Bauhinia lunarioides*)
Ash (L)
(*Fraxinus* spp.)
Ashe Juniper (L)
(*Juniperus ashei*)
Carolina Basswood (L) (N)
(*Tilia americana* var. *caroliniana*)
Cedar Elm (L)
(*Ulmus crassifolia*)
Chaste Tree (N)
(*Vitex agnus-castus*)
Desert Willow (N)
(*Chilopsis linearis*)
Flowering Mimosa (N)
(*Albizia julibrissin*)
Golden-ball Lead Tree (L) (N)
(*Leucaena retusa*)
Hackberry (L)
(*Celtis laevigata*)
Honey Mesquite (L) (N)
(*Prosopis glandulosa*)
Live Oak (L)
(*Quercus virginiana*)
Mexican Orchid Tree (N)
(*Bauhinia mexicana*)
Mexican Plum (L) (N)
(*Prunus mexicana*)
Red Bay (L)
(*Persea borbonia*)
Redbud (L) (N)
(*Cercis canadensis*)
Retama (N)
(*Parkinsonia aculeata*)
Spanish Oak (L)
(*Quercus buckleyi*)
Western Soapberry (L)
(*Sapindus saponaria* var. *drummondii*)
Wild Black Cherry (L) (N)
(*Prunus serotina*)

 *Shrubs*

Agarita (N)
(*Mahonia trifoliolata*)
Barbados Bird-of-Paradise (N)
(*Caesalpinia pulcherrima*)
Barbados Cherry (L)
(*Malpighia glabra*)
Butterfly Bush (N)
(*Buddleja davidii*)
Cenizo (L) (N)
(*Leucophyllum frutescens*)
Common Beebrush (N)
(*Aloysia gratissima*)
Common Buttonbush (N)
(*Cephalanthus occidentalis*)
Desert Lantana (N)
(*Lantana camara*)
Downy Prickly-ash (L)
(*Zanthoxylum hirsutum*)
Drummond's Wax-mallow (L)
(*Malvaviscus drummondii*)
False Indigo (L) (N)
(*Amorpha fruticosa*)
Flame Acanthus (L) (N)
(*Anisacanthus quadrifidus* var. *wrightii*)
Havana Snakeroot (L) (N)
(*Ageratina havanensis*)
Lime Prickly-ash (L)
(*Zanthloxylum fagara*)
Mexican Buckeye (L) (N)
(*Ungnadia speciosa*)
Redroot (L)
(*Ceanothus herbaceus*)
Red Yucca (N)
(*Hesperaloe parviflora*)
Southwest Bernardia (L)
(*Bernardia myricifolia*)
Spiny Hackberry (L)
(*Celtis pallida*)
Spicebush (L)
(*Lindera benzoin*)
Spring Elbow-bush (N)
(*Forestiera pubescens*)
Sweet Almond Verbena (N)
(*Aloysia virgata*)
Texas Kidneywood (L) (N)
(*Eysenhardtia texana*)

Trailing Lantana (N)
(*Lantana montevidensis*)
Wafer-ash (L)
(*Ptelea trifoliata*)
Wright's Snakeroot (N)
(*Ageratina wrightii*)
Yellow Bells (L) (N)
(*Tecoma stans*)

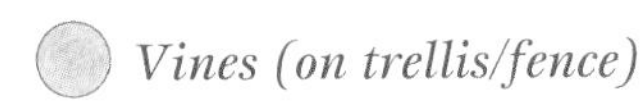

Vines (on trellis/fence)

Climbing Milkweed Vine (L) (N)
(*Funasrum cynanchoides*)
Common Balloon Vine (L)
(*Cardiospermum halicacabum*)
Drummond's Virgin Bower (L)
(*Clematis drummondii*)
Groundnut (L)
(*Apios americana*)
Least Snoutbean (L)
(*Rhynchosia minima*)
Mexican Flame Vine (N)
(*Pseudogynoxys chenopodioides*)
Purple Passionflower (L)
(*Passiflora incarnata*)
Queen's Wreath (N)
(*Antigonon leptopus*)
Red-fruited Passionflower (L)
(*Passiflora foetida*)
Slender-lobed Passionflower (L)
(*Passiflora tenuiloba*)
Snapdragon Vine (L)
(*Maurandya antirrhiniflora*)
Virginia Butterfly-pea (L)
(*Centrosema virginianum*)
Wisteria (L)
(*Wisteria* spp.)
Yellow Passionflower (L)
(*Passiflora lutea*)

Tall Herbs (3–5 feet)

Argentina Verbena (N)
(*Verbena bonariensis*)
False Wissadula (L)
(*Allowissadula holosericea*)
Golden Crownbeard (L) (N)
(*Verbesina encelioides*)
Hollyhock (L)
(*Alcea rosea*)
Joe-Pye Weed (N)
(*Eutrochium fistulosum*)
Mexican Sunflower (L) (N)
(*Tithonia rotundifolia*)
Purple Marsh-fleabane (N)
(*Pluchea odorata*)
Russian Sage (N)
(*Perovskia atriplicifolia*)
Scarlet Bouvardia (N)
(*Bouvardia ternifolia*)
Summer Phlox (N)
(*Phlox paniculata*)
Swamp Milkweed (L) (N)
(*Asclepias incarnata*)
Tooth-leaved Goldeneye (N)
(*Viguiera dentata*)
White-flowered Plumbago (L)
(*Plumbago scandens*)
Wild Bergamot (N)
(*Monarda fistulosa*)
Yellow Prairie Grass (L)
(*Sorghasturm nutans*)

Medium Herbs (2–3 feet)

Broad-leaf Wood-oats (L)
(*Chasmanthium latifolium*)
Brown-eyed Susan (N)
(*Rudbeckia hirta*)
Cardinal Flower (N)
(*Lobelia cardinalis*)
Cherry Sage (N)
(*Salvia greggii*)
Canna (L)
(*Canna x generalis*)
Globe Amaranth (N)
(*Gomphrena globosa*)
Gregg's Mistflower (N)
(*Conoclinium dissectum*)
Ironweed (N)
(*Vernonia* spp.)
Liatris (N)
(*Liatris* spp.)
Mealy Sage (N)
(*Salvia farinacea*)
Mexican Oregano (N)
(*Poliomintha longiflora*)
Partridge-pea (L)
(*Chamaecrista fasciculate*)
Pentas (N)
(*Pentas lanceolata*)
Pigeonberry (L) (N)
(*Rivina humilis*)
Porterweed (N)
(*Stachytarpheta* spp.)
Prairie Agalinis (L)
(*Agalinis heterophylla*)
Purple Coneflower (N)
(*Echinacea purpurea*)
Scarlet Globe-mallow (L)
(*Sphaeralcea coccinea*)
Shrimp Plant (L) (N)
(*Justicia brandegeana*)
Skeleton-leaf Goldeneye (N)
(*Viguiera stenoloba*)
Tropical Milkweed (L) (N)
(*Asclepias curassavica*)
Violet Ruellia (L)
(*Ruellia nudiflora*)

Low Herbs (1–2 feet)

Cedar Sage (N)
(*Salvia roemeriana*)
Cudweed (L)
(*Gamochaeta* spp)
Damianita (N)
(*Chrysactinia mexicana*)
Fern Acacia (L)
(*Acacia angustissima* var. *hirta*)
Flax (L)
(*Linum* spp.)
Hispid Wedelia (L) (N)
(*Wedelia acapulcensis* var. *hispida*)
Paintbrush (L)
(*Castilleja* spp.)
Prairie Verbena (L) (N)
(*Verbena bipinnatifida*)
Spectacle Pod (N)
(*Dithyrea wislizenii*)
Tahoka Daisy (N)
(*Machaeranthera tanacetifolia*)
Two-leaved Senna (L)
(*Senna roemeriana*)
Woolly Paperflower (N)
(*Psilostrophe tagetina*)

Edging (to 1 foot)

Bayou Violet (L)
(*Viola sororia*)
Frogfruit (L) (N)
(*Phyla* spp.)
Garden Pansy (L)
(*Viola x wittrockiana*)
Gregg's Dalea (L) (N)
(*Dalea greggii*)
Lindheimer's Tephrosia (L)
(*Tephrosia lindheimeri*)
Plains Zinnia (N)
(*Zinnia grandiflora*)
Prickle-leaf Dogweed (N)
(*Thymophylla acerosa*)
Sweet Alyssum (L) (N)
(*Lobularia maritima*)
Western Peppergrass (L) (N)
(*Lepidium alyssoides*)
Woolly Stemodia (L)
(*Stemodia lanata*)

Plants in Lawn

Annual Pennyroyal
(*Hedeoma acinoides*)
Blue-eyed Grass (N)
(*Sisyrinchium* spp.)
Bluets (N)
(*Houstonia* spp.)
Buffalo Grass (L)
(*Buchloe dactyloides*)
Cut-leaf Germander (N)
(*Teucrium cubense* var. *laevigatum*)
Cut-leaf Gilia
(*Giliastrum incisum*)
Drummond's Wild Onion (N)
(*Allium canadense*)
False Garlic (N)
(*Nothoscordium bivalve*)
Flax (L)
(*Linum* spp.)
Funnel Lily
(*Androstephium coeruleum*)
Lady Bird's Centaury
(*Centaurium texense*)
Prairie Celestials
(*Nemastylis geminiflora*)
Prairie Rain Lily
(*Cooperia* spp.)
Spring Beauty (N)
(*Claytonia virginica*)
Spring Evax (L)
(*Evax verna*)
Texas Dutchman's-breeches (L)
(*Thamnosma texana*)
Texas Toadflax (L)
(*Nuttallanthus texanus*)
Upright Pipevine (L)
(*Aristolochia erecta*)
White Milkwort
(*Polygala alba*)

Flagstone

Decomposed Granite

Turf

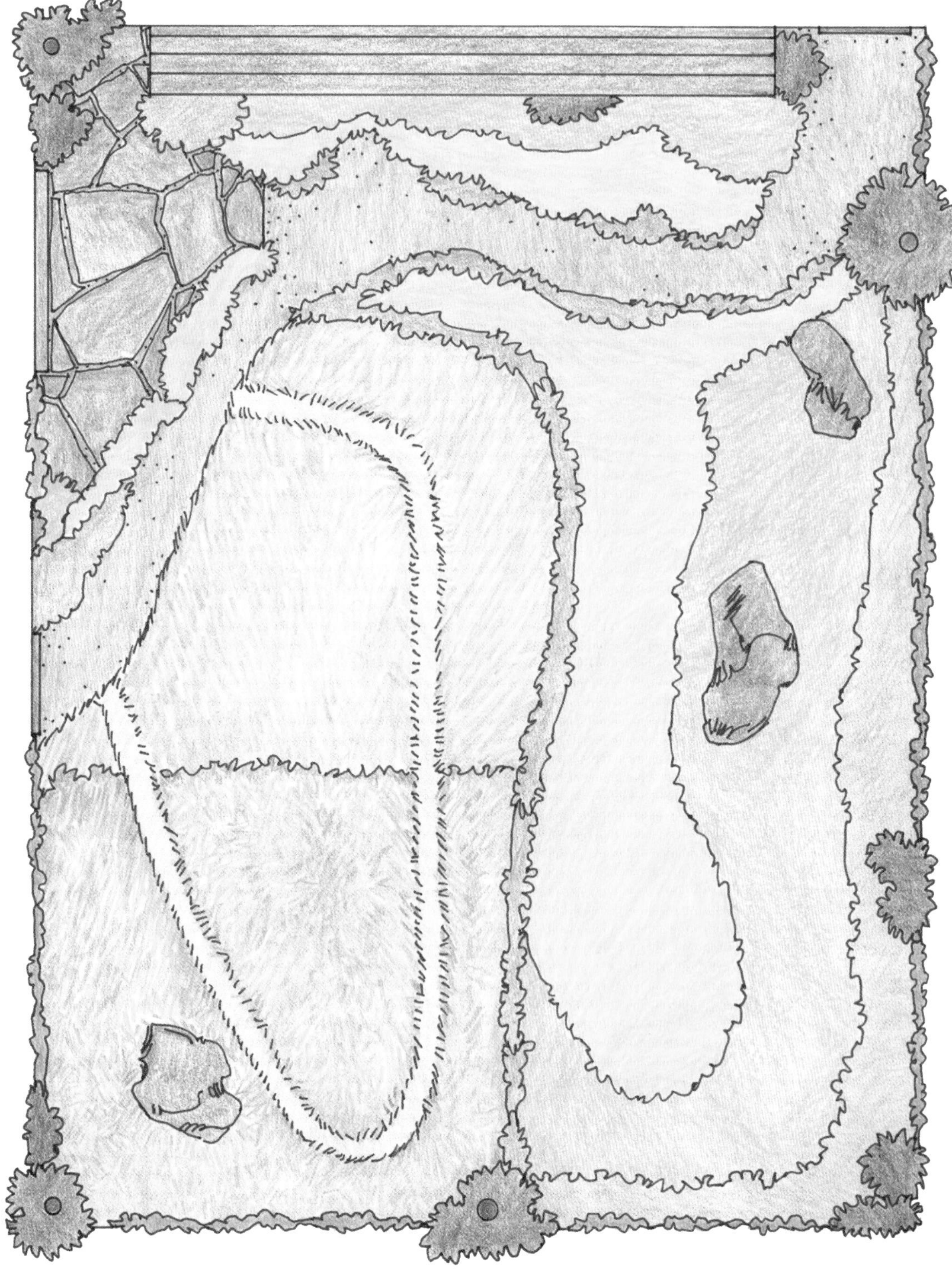

TRANS-PECOS
(REGION 7)

Alpine, Balmorhea, El Paso, Fort Davis, Fort Stockton, Marathon, Presidio, Terlingua, Van Horn

(L) indicates larval food plant
(N) indicates nectar plant

Trees

Alligator Juniper (L)
(*Juniperus deppeana* var. *deppeana*)
Chaste Tree (N)
(*Vitex agnus-castus*)
Choke cherry (L) (N)
(*Prunus virginiana*)
Cottonwood (L)
(*Populus deltoides*)
Desert Willow (N)
(*Chilopsis linearis*)
Golden-ball Lead-tree (L) (N)
(*Leucaena retusa*)
Hackberry (L)
(*Celtis laevigata*)
Honey Mesquite (L) (N)
(*Prosopis glandulosa*)
Little-leaf Mulberry (L)
(*Morus microphylla*)
Oak (L)
(*Quercus* spp.)
Redbud (L) (N)
(*Cercis canadensis*)
Retama (N)
(*Parkinsonia aculeate*)
Western Soapberry (L)
(*Sapindus saponaria* var. *drumondii*)
Wild Black Cherry (L) (N)
(*Prunus serotina*)

Shrubs

Big Bend Silverleaf (L)
(*Leucophyllum minus*)
Black Dalea (L)
(*Dalea frutescens*)
Cat's-claw Mimosa (N)
(*Mimosa aculeaticarpa* var. *biuncifera*)
Cenizo (L)
(*Leucophyllum frutescens*)
Common Beebrush (N)
(*Aloysia gratissima*)
Common Buttonbush (N)
(*Cephalanthus occidentalis*)
Desert Ceanothus (N)
(*Ceanothus greggii*)
False Indigo (L) (N)
(*Amorpha fruticosa*)

Four-wing Saltbush (L)
(*Atriplex canescens*)
Mexican Buckeye (L)
(*Ungnadia speciosa*)
Mexican Orange (N)
(*Choisya dumosa*)
New Mexico Locust (L)
(*Robinia neomexicana*)
Red Yucca (N)
(*Hesperaloe parviflora*)
Smooth Yucca (L)
(*Yucca glauca*)
Texas Carlowrightia (L)
(*Carlowrightia texana*)
Texas Kidneywood (L) (N)
(*Eysenhardtia texana*)
Wafer-ash (L)
(*Ptelea trifoliata*)
Willow-leaf Baccharis (L) (N)
(*Baccharis salicifolia*)
Wright's Snakeroot (N)
(*Ageratina wrightii*)
Yellow Bells (L) (N)
(*Tecoma stans*)

Vines (on trellis/fence)

Climbing Milkweed Vine (L) (N)
(*Funastrum cynanchloides*)
Drummond's Virgin's Bower (L)
(*Clematis drummondii*)
Slender-lobe Passionflower (L)
(*Passiflora tenuiloba*)
Snapdragon Vine (L)
(*Maurandya antirrhiniflora*)
Texas Snout-bean (L)
(*Rhynchosia senna* var. *texana*)

Tall Herbs (3–5 feet)

Argentina Verbena (N)
(*Verbena bonariensis*)
False Wissadula (L)
(*Allowissadula holosericea*)
Golden Crownbeard (L) (N)
(*Verbesina encelioides*)
Goldenrod (N)
(*Solidago* spp.)
Hollyhock (L)
(*Alcea rosea*)
Mexican Sunflower (L) (N)
(*Tithonia rotundifolia*)
Purple Marsh-fleabane (N)
(*Pluchea odorata*)
Scarlet Bouvardia (N)
(*Bouvardia ternifolia*)
Summer Phlox (N)
(*Phlox paniculata*)
Tooth-leaved Goldeneye (N)
(*Viguiera dentata*)

Medium Herbs (2–3 feet)

Aster (L) (N)
(*Symphyotrichum* spp.)
Butterfly Weed (N)
(*Asclepias tuberosa*)
Cardinal Flower (N)
(*Lobelia cardinalis*)
Cherry Sage (N)
(*Salvia greggii*)
Chocolate Daisy (N)
(*Berlandiera lyrata*)
Globe Amaranth (N)
(*Gomphrena globosa*)
Lemon Beebalm (N)
(*Monarda citriodora*)
Liatris (N)
(*Liatris* spp.)
Mealy Sage) (N)
(*Salvia farinacea*)
Pigeonberry (L) (N)
(*Rivina humilis*)
Prairie Flax (L)
(*Linum lewisii*)
Purple Coneflower (N)
(*Echinacea purpurea*)
Skeleton-leaf Goldeneye (N)
(*Viguiera stenoloba*)
Texas Milkweed (L)
(*Asclepias texana*)
Thread-leaf Groundsel)N)
(*Senecio flaccidus*)
Tropical Sage (N)
(*Salvia coccinea*)

Low Herbs (to 1–2 feet)

Cedar Sage (N)
(*Salvia roemeriana*)
Croton (L)
(*Croton* spp.)
Damianita (N)
(*Chrysactinia mexicana*)
Fern Acacia (L)
(*Acacia angustissima* var. *hirta*)
Gregg's Coldenia
(*Coldenia greggii*)
Gregg's Dalea (L) (N)
(*Dalea greggii*)
Gregg's Mistflower (N)
(*Conoclinium dissectum*)
Hispid Wedelia (L) (N)
(*Wedelia acapulcensis* var. *hispida*)
Huisache Daisy (N)
(*Amblyolepis setigera*)
Paintbrush (L)
(*Castilleja* spp.)
Prairie Verbena (N)
(*Verbena bipinnatifida*)
Sand-verbena (N)
(*Abronia* spp.)
Tahoka Daisy (N)
(*Machaeranthera tanacetifolia*)
Two-leaved Senna (L)
(*Senna roemeriana*)
Western Peppergrass (L) (N)
(*Lepidium alysssoides*)
Woolly Paper-flower (N)
(*Psilostrophe tagetina*)

Edging (to 1 foot)

Common Dogweed (L) (N)
(*Thymophylla pentachaeta*)
Desert Marigold (N)
(*Baileya multiradiata*)
Desert Zinnia (N)
(*Zinnia acerosa*)
Dwarf Germander (N)
(*Teucrium lacinatum*)
Frogfruit (L) (N)
(*Phyla* spp.)
Limoncillo (N)
(*Pectis angustifolia*)
Plains Zinnia (N)
(*Zinnia grandiflora*)

Plants for Buffalo Grass Lawn

Bluebells
(*Campanula rotundifolia*)
Blue-eyed Grass (N)
(*Sisyrinchium* spp.)
Bluets (N)
(*Hedyotis* spp.)
Copper Lily (N)
(*Zephyranthes longifolia*)
False Garlic (N)
(*Nothoscordum bivalve*)
Flax (L)
(*Linum* spp.)
Needle-leaf Gilia
(*Gilia rigidula* var. *acerosa*)
Prairie Rain Lily (N)
(*Cooperia* spp.)
Slender Shellflower
(Southwestern Pleatleaf)
(*Nemastylis tenuis*)
White Milkwort
(*Polygala alba*)
Wild Onion (N)
(*Allium* spp.)
Yellow-flowered Onion (N)
(*Allium coryi*)

Plants for Natural Area

Blue-flowered Flax (L)
(*Linum lewisii*)
Mealy Sage (N)
(*Salvia farinacea*)
Paintbrush (L)
(*Castilleja* spp.)
Prickle-leaf Dogweed (N)
(*Thymophylla acerosa*)
Spotted Beebalm (N)
(*Monarda punctate*)
Wright's Verbena (N)
(*Verbena wrightii*)

Flagstone

Decomposed Granite

Turf

Eastern Black Swallowtail *(Papilio polyxenes esterius)*

4 An Instant Butterfly Garden

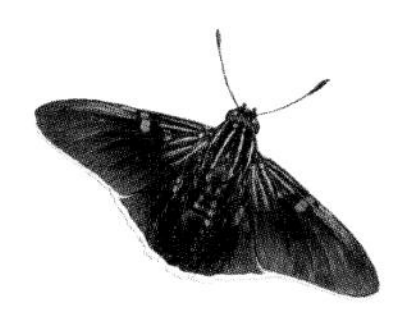

If you are really excited about the prospect of attracting butterflies, but it is too late in the season to dig beds or set out trees and shrubbery, there is a way you can still make a butterfly garden almost instantly—use plants in containers.

Site Possibilities

Before heading for the nearest plant dealer, take a few minutes to give some thought to the possibilities that your yard may offer. Next, following some of the same steps used in creating a new garden, make a plan of how much space you have and the amount of time and money you want to spend.

Start by walking around your house and yard, noting areas where pots, containers, or hanging baskets can be used. There are many ways to make a small space become a haven for butterflies, so give every nook and cranny a good, long look. Because the choice of location for a container grouping is as important as for planting beds or borders, consider only open, protected locations in full sun. An area close to a water faucet or the end of a watering hose would be really handy, as potted plants need watering more often than bedded plants.

Following are a few site possibilities to keep in mind as you walk about, but do not be limited by these. No yard is exactly like another or used in the same way, so let your imagination soar. Choose what will work for your yard while providing the butterflies what they need.

1. A sunny wall of the house, garage, or tool shed where half baskets could be attached or full baskets hung from the roof edge

2. A strip of trellis fastened to a sunny wall for climbing larval food plants and a grouping of pots or other containers on the ground beneath
3. One corner of the yard, or an area where a corner can be "made" with lengths of bamboo fencing and a grouping of containers arranged in staggered heights
4. Outermost edges of uncovered porches, decks, or patios
5. Wooden or metal railings of stairways, balconies, or porches where pots or baskets can be attached
6. The immediate area around a mailbox post
7. Wooden fences, pillars, or posts where hanging baskets can be fastened in a staggered arrangement
8. Old wrought-iron or metal tables, benches, or chairs or wooden picnic tables that are no longer used and where pots can be arranged
9. The edge of a driveway or walk or on steps or stairways where containers can be placed in a pleasing arrangement yet still allow passage
10. Window boxes on the sunny sides of the house with baskets hanging above

Containers

Now that some garden sites have been selected where containers would work well, next consider the containers themselves. Gather all the hanging baskets, pots, or planting containers you possess, and set them out where they are easy to see and work with. Old wash pots, battered buckets, and bushel baskets add character to the planting. Sturdy baskets from the house can be used. They will last only one season but are lovely when a potted plant is placed inside. Look around for such objects as concrete blocks or short square or round tiles, such as those used for drainage or chimney flues, that can be stood on end and filled with soil.

Picturesque old stumps or pieces of driftwood can be incorporated into the grouping, not only as effective focal points but as convenient basking areas for the butterflies. If the arrangement of pots is to be on or near the ground, a salt or puddling area can be added. Make plans for this in one of the groupings.

In making container selections, keep in mind that metal or plastic containers have a tendency to become exceptionally hot when placed in the sun; plant roots may be burned from the overheated soil. Wooden or clay pots keep the roots much cooler, although clay is very porous, with rapid loss of soil moisture. Plants in clay pots will most likely need more frequent watering. One way around this problem is to set one clay pot inside a larger clay pot, filling the area between the two pots with sphagnum moss or sand. If this filler is kept moist, it not only reduces evaporation but helps keep the plant roots several degrees cooler. Plants in plastic pots also benefit from the coolness of being placed in larger clay pots lined with moist sphagnum.

Any type of wooden container is probably best for these instant plantings, but they have the drawback of being heavier than either plastic or clay. If the containers are very large, swivel rollers can be screwed to the bottom. This not only makes them easier to move around but raises them off the ground and prevents rotting of the wood.

A CHARMING CONTAINER OF OLD-FASHIONED PERTUNIAS *(Petunia axillaris)*

Wonderful configurations can be made by stacking concrete blocks. If the gray color of the blocks is intrusive, they can be painted with a flat paint of soft, subtle earth colors or concealed with rustic split logs, cork sheeting, rough cedar boards, or old bricks. The blocks should be stacked so that there are holes for filling with soil.

Measure the width and depth of the containers selected for the plantings and the overall area where they are to be placed. Put as much of the future arrangement together as possible; then make a list of things that you need to obtain. Finally, make a list of the more common, tougher plants known to be readily used by butterflies. Include height and bloom period of each plant. Once at the nursery, ask for advice about the number of plants needed for each container you have.

Selecting and Arranging Plants

The choice of plants for this instant garden will be limited to what the nurseries have in stock at the time, but there should be enough available to attract butterflies within hours after the plants are brought home and arranged.

In making the selection for these containers and baskets, it is important here as in a regular garden planting to choose a few species and have many containers of the same plant rather than a half-dozen different pots of a dozen different species. Let the size, shape, and placement of the containers break up the monotony of the planting and be of interest to people. At the same time, the mass of a single nectar source will be of most interest to the butterflies.

If hanging baskets are being planned in the new arrangement, occasionally nurseries will offer baskets already planted with Pentas (*Pentas lanceolata*), Sweet Alyssum (*Lobularia maritima*), or verbenas (*Verbena* spp.) and lantanas (*Lantana* spp.). If planting them yourself, wire baskets already lined with sphagnum moss are best. After purchasing needed plants, baskets, extra sphagnum moss, and soil, you are now ready to begin planting and making final arrangement of containers in the space chosen.

For container planting, the very best potting soils should be used, and this can become expensive when filling several large containers. To cut down on the amount of soil needed, place in the bottom of the container a large plastic pot upside down or sealed, empty plastic jugs. This takes up some of the space yet leaves plenty of soil for the plants. It also lightens the container considerably, making it easier to move if necessary.

In planting the baskets, before adding any soil, place a plastic drip dish (the kind ordinarily used beneath pots) in the bottom of the basket. This acts as a reservoir and cuts down on the amount of watering required. Make sure the moss around the sides is extra thick; if it is not, the soil will be washed away during continual summer watering. Extra sphagnum may be needed to secure the plants that are planted on the sides of the basket and to close any gaps after the plants are in place.

Food plants for some larvae can easily be grown in some of the containers. Place four to six plants of parsley (*Petroselinum crispum*) for the Black Swallowtail (*Papilio polyxenes*) in a low container to be placed in a prominent spot. Or plant a few of the taller-growing common fennel (*Foeniculum vulgare*) or dill (*Anethum graveolens*) in the middle of a pot (also for the Black Swallowtail); then plant parsley as an edging or use a flowering plant such as Sweet Alyssum or Nasturtium (*Tropaeolum majus*). A large pot or two of such a planting will add a ferny touch to an otherwise all-floral arrangement and make the overall area more attractive. Place a small trellis or a strong, decorative, many-twigged shrub branch in some of the pots, and plant

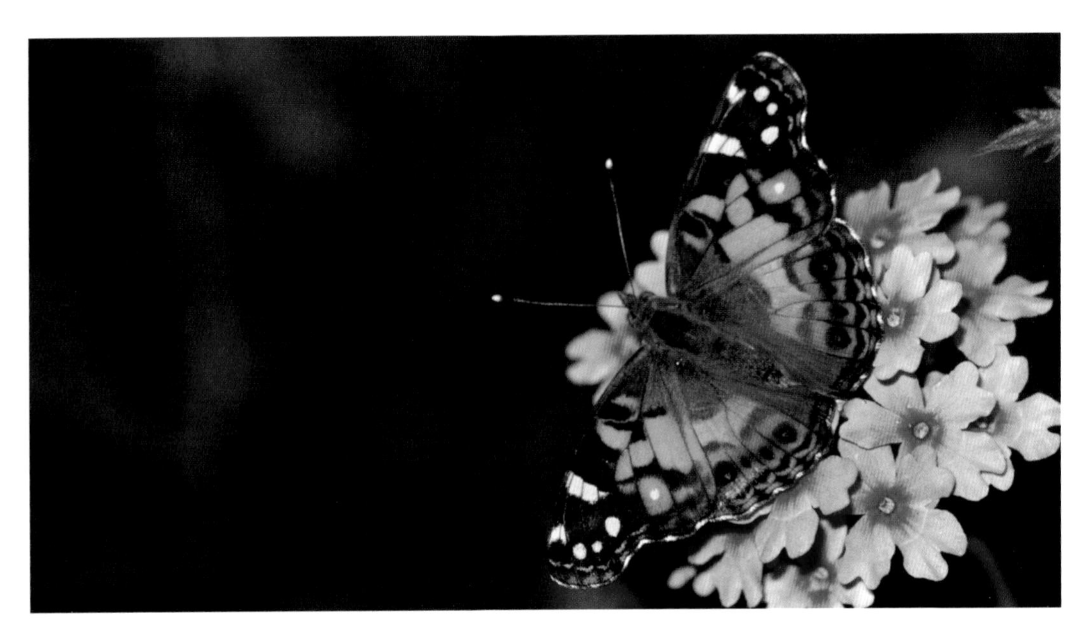

VERBENAS (*Verbena* spp.) WORK WELL IN CONTAINER GARDENS, HERE WITH AMERICAN LADY (*Vanessa virginiensis*).

climbing vines of pipevine (*Aristolochia* spp.) for the Pipevine Swallowtail (*Battus philenor*) or passionflowers (*Passiflora* spp.) for the Gulf Fritillary (*Agraulis vanillae incarnata*) and Variegated Fritillary (*Euptoieta claudia*).

Every container of plants, whether purchased already potted or planted yourself, should be mulched with small- to medium-sized bark chips. Bark mulch prevents rain from washing soil from around the roots and, more important, helps conserve moisture. Never use sphagnum for mulching, since it dries out very quickly when exposed to air. When dry, the moss acts as a sponge, drawing water away from the roots of the plant.

If gardening in one of the more arid regions of the state, place shallow terra-cotta or plastic pans or drip trays beneath the pots to conserve liquid that may run out of the pots when water or fertilizer is being applied. Attached to the bottom of hanging baskets, pans or trays prevent dripping on the plants underneath and provide a longer period of available moisture. In rain-prone East Texas, excellent and immediate drainage may be required. Making large drainage holes and raising containers off the ground by placing them on thin stones may be necessary.

In the final arranging, for both the health of the plants and for attracting the attention of butterflies, plants should be placed at varying heights. Place taller-growing plants to the back or middle of the group. Some plants, such as lantana and verbena, spread outward and should be placed low and to the front or side of the grouping. Different heights can easily be obtained by placing the containers on old bricks, flat stones, tiles, overturned chipped or cracked pots, or old stumps.

RED-FRUITED PASSIONFLOWERS (*Passiflora foetida*) ARE STUNNING CLIMBING VINES.

A word of caution here: Once they are arranged, do not move the plants about simply on a whim. Butterflies are very habitual; once a butterfly has found a food source, it comes back to it regularly until the flower is no longer in bloom. If the plant is moved very far away, the butterfly tends to become a little put out and may discontinue feeding. This is another reason to keep plants in a grouping composed of several plants of the same species, and the plants of each type all of one color. This way, if a container of a certain plant in the most conspicuous spot stops flowering well, it can be exchanged for a better-flowering one from the back of the grouping. Only the pots, not a species, will be changed, and the butterflies will not be as leery of the difference in appearance.

One excellent plant that should be included in this type of planting is the nonnative Tropical Milkweed (*Asclepias curassavica*). Not only will this plant bloom throughout the summer and into fall but if brought indoors and placed in a sunny position, it will con-

MALACHITE *(Siproeta stelenes biplagiata)* AND LANTANA *(Lantana* spp.)

tinue blooming through the winter. On warm, sunny days (when the hibernators might be out) it can be carried (or rolled) outside. For this continuation of bloom, it is best if the plant is not allowed to form seedpods. If possible, have several pots of this plant and keep some of them trimmed back for new, lower growth and more bloom.

At the end of the flowering season, if there were any perennials in your container planting, do not let them go to waste. Empty the containers, separate the plants into small divisions, and plant these in the permanent border. There will most likely be enough plants to make a nice showing the next season and provide an ample nectar source for the butterflies.

MAINTENANCE

Plants in containers need more attention than the same species planted in the garden. This does not mean a lot of time-consuming work must be done, but it does mean the plants must be checked at least once a day. Constant monitoring of the moisture, feeding, and rearranging will be necessary to ensure the plants remain healthy and full of flowers.

Since the objective is a mass of plants in a very small area, the plants will be overcrowded aboveground, and the roots underground will have little room in which to expand. There is going to be much competition between the plants within each pot, and the soil will quickly be depleted of both nutrients and moisture. And again, the caution to water the plants at soil level and not in an overall "sprinkling" fashion. Here, as in the garden, nectar is the attractant and must not be diluted or washed away.

For newly planted containers, a root stimulator should be added at the time of planting. After two weeks a regular feeding regime should be started using a good, all-around fertilizer. A good combination is to apply a slow-release fertilizer at the time of planting, and then once weekly apply a weak solution of liquid fertilizer. If a dry fertilizer is used, water before and after adding the granules. The absolute best is a weekly addition of earthworm pellets. It is imperative that container plants receive adequate feeding, for without added food they quickly sicken and die.

Watering is the biggest chore for container gardening, but an absolutely necessary one. It usually has to be done daily, and during July and August twice a day may not be too often. It is not the easiest thing to do, either, especially for hanging baskets. The simplest arrangement is to have the containers as close to a faucet as possible and to install a hose-reel on the house or a nearby fixture; then the hose can be conveniently unrolled for watering and rolled back up when finished. A stout bamboo pole firmly wired to the last two or three feet of the hose allows the top of the hanging baskets to be reached without a stepladder. If there are only a few containers, the watering problem can be solved by simply dropping a handful of ice cubes into the containers each day.

Keep all spent flowering heads clipped to ensure continual flowering. Especially watch native species, for if these are left to form ripe seeds, most species will stop flowering. Some plants, such as lantana and verbena, benefit from clipping portions of the plants back to near the base, thus encouraging new growth and another round of flowering.

Constructing a Dry Bed

The term "dry bed" is used here to describe a temporary flower bed where the soil in the bed does not come in contact with the ground. It is essentially a bottomless box lined with plastic sheeting and placed on top of concrete or some other impermeable surface. This alternative might be right for you if you have neither the time nor the inclination to care for numerous flower beds or hanging baskets. Or it may be that your yard is too shaded for butterflies and the only really open, sunny spot is a driveway or a patio.

Constructing a dry bed is relatively simple. Select an area on a hard surface such as a concrete driveway, a wide walkway, or a brick or stone patio or porch. Begin to lay out the building materials, which can be railroad ties, rough cedar logs, sections of short half logs, concrete blocks, old bricks, or stones. Place them in the bed configuration best suiting your needs and space. Designing the bed is the same as if you were digging one in the garden, except in this case you are placing it on top of a hard surface.

After the bed is outlined to your satisfaction, be sure it is built at least eight to ten inches high; twelve to fourteen inches would be even better. Keep in mind that the bed will act exactly as a pot or container and must be filled with extra-rich potting soil; ordinary

Zebra Swallowtail *(Eurytides marcellus)* and Garden Zinnia *(Zinnia elegans)*

garden soil will not do. If the building materials make it necessary for the bed to be deep, the bottom portion can be partially filled with small plastic pots turned upside down to conserve soil. If the bed has been placed where drainage direction is important, heap up gravel or small round stones in the bottom of the bed site on the opposite side of the needed direction of drainage flow, slightly raising and slanting the bed on one side.

Once the bed is at the desired height and its drainage material is arranged, completely line the interior of the bed with an extra-heavy sheet of plastic, preferably clear, placing it on top of the drainage material. Smooth the sheet into all the corners to conform to the shape of the bed as much as possible. Do not split or cut the plastic to make it conform in shape, but fold it tightly and tape if necessary. After you have made this liner fit as well as possible, punch a few drainage holes in the bottom with an ice pick or a small nail.

Place a one- or two-inch layer of coarse gravel, pine bark, or wood chips in the bottom of the bed. Add a two-inch layer of sphagnum moss for moisture retention. Next, begin to fill the bed with potting mix. The soil will be somewhat fluffy, so, to prevent settling, pour the soil out in layers, ever so gently patting down between layers. After the bed is filled with the soil, roll back the excess plastic around the edges and tuck it in neatly, leaving it about an inch below the top of the walls. Trimming excess plastic can be done at this time.

Before the actual planting, place the potted plants in a pleasing position; then, stand back and visualize what they will look like in a month or so. If some of the plants are the trailing type, place these toward the front or the outer edges; arrange taller plants either in the center or at the back. As the plants are being placed, water them in with the addition of a root stimulator. After all planting is done, spray the entire area gently and briefly just to settle the soil and clean off the foliage.

If there are gaps in the building material or around the edges that allow the plastic to be seen, stuff moist sphagnum moss gently into the cracks. Cover the entire top of the bed with pine bark or dark-colored wood chips. This greatly benefits in conserving moisture, which is quickly lost in this type of planting.

As this bed is truly an artificial one, having no bottom contact with garden soils, the combination of growing plants and evaporation depletes the soil moisture very quickly. Treat the bed as a large potted or container plant. Do not neglect its daily watering and weekly fertilizing.

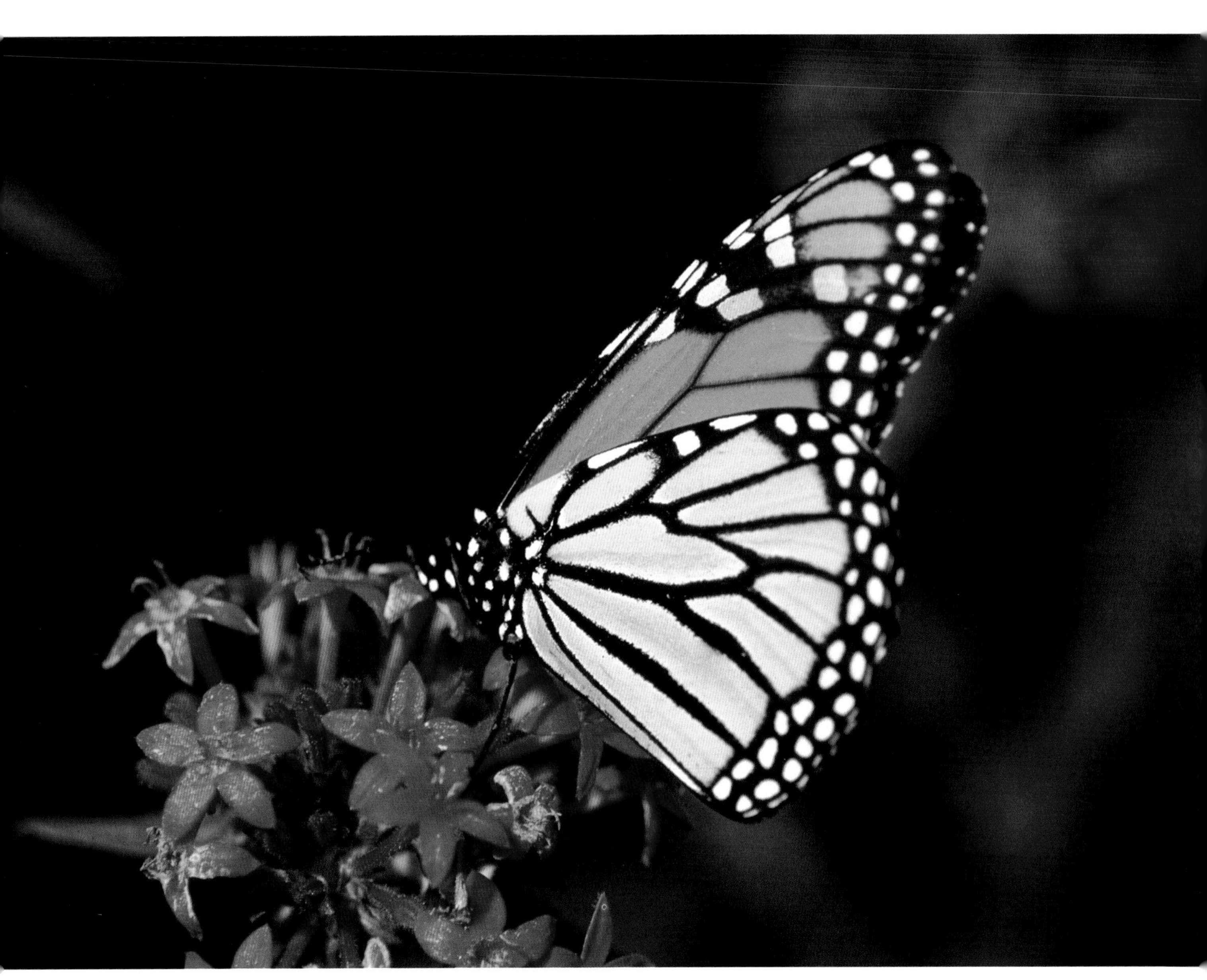

Female Monarch *(Danaus plexippus)* on Pentas *(Pentas lanceolata)*

5 Let Nature Do It: Butterfly-Friendly Pest Controls

Establishing a garden to attract certain caterpillars and adult butterflies is also going to invite other insects that may not be so welcome. In the natural world there are no "good" bugs or "bad" bugs. What we may occasionally consider obnoxious insects have their place and function in nature the same as the insects we admire or think lovely—it is simply a matter of the way we are "seeing" things at the moment. In the following discussion, while I speak of the bad guys and show methods of removing them in order to better establish a garden for a particular insect, in this case the butterfly, in no way am I passing judgment on any living thing. Personally I can, and do, appreciate the fortitude, mystery, and beauty of an aphid (family Aphididae) or Japanese beetle (*Popillia japonica*) as much as the most gorgeous of butterflies.

Clearly, however, there are times in butterfly gardening when some species of insect becomes so numerous or damaging that it is detrimental to your purpose, and some sort of action needs to be taken. If the usual methods of wide-scale eradication were applied, it is quite likely the butterfly larvae would also be eliminated in the process. To keep the garden balanced in favor of the desired creatures, certain garden practices and a system of organic pest control may be needed.

The best defense against any kind of pest insect or disease is to have the healthiest plants possible. Bugs first go for weak, malnourished plants. If the plants in your garden are provided with a well-balanced diet through compost-rich soils, adequate moisture, and the sunshine needed for strong growth, they will be less stressed and therefore less susceptible to harmful attack.

A pest outbreak is often triggered by the

HEALTHY PLANTS ARE THE BEST DEFENSE AGAINST PESTS. THESE LANTANA *(Lantana* spp.*)* BENEFIT FROM GOOD AIR CIRCULATION AND AN EFFECTIVE MULCH.

use of a potent, quick-release fertilizer or one of the miracle growth sprays that initiates tender, succulent plant growth. Plants are far less likely to be bothered if they are allowed to maintain a natural, steady growth pace by the use of manure or compost in the soil and slowly deteriorating mulches.

Plants being grown out of their natural ranges or habitats may have a hard time adjusting and may be wide open to all kinds of insect attack. Plants that continually attract insect invasion should be promptly removed and replaced with tougher, more hardy species.

Give plants in the flower beds plenty of room to breathe. Good air circulation helps prevent many fungus diseases, such as mildew, crown rot, and black spot. Keep the garden clean by immediately removing severely infected plants, either by burning them or by burying them deeply in a new flower bed in the making.

Many and various bacteria can be found in good, rich compost, along with the molds and fungi that live in the soil. Both are providing various antibiotics to plants for fighting diseases, so use lots of compost in the beds. If some action must be taken, treat only the plants showing actual damage, and apply treatment only to the areas affected.

There are many alternatives to indiscriminate use of chemicals. For the health and safety of the butterflies and their larvae, apply a system using only natural controls, integrating physical controls, biological controls, and controls through the use of plants themselves.

Companion Plants

Some plants apparently thrive and are at their healthiest when placed in close proximity to certain other plants. The various benefits of such companion planting are not easy to explain. Perhaps it is due to certain nutrients being brought from a lower depth by one plant and thus made available to a more shallow-rooted plant. Perhaps it is shade provided to the roots of one plant by another during a critical growth period, or maybe one plant benefits from a certain combination of room and light allowed by certain other plants. Or perhaps it is simply that plants like to be near their friends.

No exact, scientific explanation has been offered for the generally healthier effects obtained from such planting, but it does seem to work. Not only do plants appear to be larger and produce more flowers and fruit when placed in combination with certain other species but they are noticeably less stressed when placed in small groups of their own kind, instead of being separated with one here and one there. It is not difficult to see the difference between the health of a solitary plant lost in the maze of a garden border and that of several plants of the same species placed in groups.

An example of the interactions of such plantings is within members of the Fabaceae or Bean Family, such as Groundnut (*Apios americana*), White Clover (*Trifolium repens*), or bluebonnets (*Lupinus* spp.) and any of the peas (*Pisum* spp.) and beans (*Phaseolus* spp.). Leguminous plants form a symbiotic relationship with certain bacteria in the soil that live in small nodules formed along the roots. These nodules produce nitrogen and other beneficial substances that are released into the soil and used by nearby plants, creating healthier growing conditions.

The benefit of companion planting will be most noticeable when using predominantly native species in the landscape. Most cultivated species have been manipulated to adapt to various soils and environments, but the natives do best in as close to original growing conditions as possible, that is, with others of like kind or certain associated species.

Along with planting natives in groups of the same species, it is equally important to plant them near their natural neighbors. To learn this association, pay special attention when driving the countryside or hiking. Notice which plants are more commonly found in close association to the plants you have or are planning for your garden. These associate plants may not be known for producing abundant nectar, but if there is space in your garden, and if the plant is not too offensive, plant a couple close to the prized nectar plant.

Probably as important as providing the associate plants is making sure the soil in the garden beds is the same as or as close to those of the native habitat as possible. Since plants obtain their liquid food from the nutrients released from the soil, the correct "diet" of dirt is extremely important.

Repellent Plants

Although the reason behind the obvious benefits of placing certain companion plants in close proximity to others may remain a mystery for the time being, there is another planting combination that is more easily understood. Repellent planting uses certain plants that, either by odor or the release of chemicals through their roots, ward off attack by certain insects. Scattering these plants among the butterfly-attracting plants helps keep undesired insects away.

Some members of the Lily Family (Liliaceae) are among the most useful repellent plants, with Society Garlic (*Tulbaghia violacea*) and garlic (*Allium sativum*) being especially potent deterrents. Many members of the genus *Allium*, or onions, both wild and cultivated, keep nearby plants free of insect pests, and the very attractive large, showy clusters of flowers are a great nectar source, drawing in numerous butterflies. As an added bonus for the garden, all of the alliums can be eaten, except perhaps some of the ornamental ones.

The foliage of Four-o'clock (*Mirabilis jalapa*), as well as milkweed (*Asclepias* spp.), rue (*Ruta* spp.), parsley (*Petroselinum crispum*), dill (*Anethum graveolens*), common fennel (*Foeniculum vulgare*), and anise (*Pimpinella anisum*), is poisonous to many insects. When these plants are used among other plants, most insects tend to leave the nonpoisonous plants alone as well. Larvae of the Monarch (*Danaus plexippus*), Queen (*D. gilippus thersippus*), and Black Swallowtail (*Papilio polyxenes*) have evolved methods for ingesting, storing, and using certain plant poisons, but fortunately most of the harmful insects have not. Other plants with foliage toxic to certain insects include Common Flax (*Linum usitatissimum*), Wormwood (either *Artemisia absinthium* or A. *stelleriana*), and Borage (*Borage officinalis*), petunias (*Petunia* spp.), larkspurs (*Delphinium* spp.), and geraniums (*Pelargonium* spp.), especially white-flowered ones.

Many insects are repelled by the pungent odor of Nasturtium (*Tropaeolum majus*). On the other hand, Nasturtium attracts almost all species of aphids. This can be good if the plants are placed where they can act as a trap crop for the aphids, keeping the insects away from other plants as well as making them easier to destroy.

Feverfew (*Chrysanthemum parthenium*) is a plant known and used since early times for insect control. Insects in general do not like its pungent foliage and avoid it. Nettles (*Urtica* spp. and *Tragia* spp.), important larval food plants for the Red Admiral (*Vanessa atalanta rubria*), are also valuable both as companion plants and as repellent plants. Nettles greatly increase the potency of herbs while repelling several plant invaders. Tansy (*Tanacetum vulgare*) controls ants, which move aphids from one plant to another, and marigolds (*Tagetes* spp.) help control nematodes (phylum Nematoda) in the soil. The tall, small-flowered Mexican Mint-marigold (*T. lucida*) evidently exudes a repellent from the root system in greater potency than other marigold species.

As far as I know, no in-depth research has been done on mixing repellent plants into a concentrated planting of butterfly plants. Most of the repelling qualities of these plants come from the odors emitted through their foliage, especially when the foliage is crushed, and these odors may either make it hard or impossible for butterflies to find the nectar

The foliage of petunias *(Petunia* spp.) is toxic to some insects.

plants or make the general area so disagreeable the butterflies will not stay around. When using repellent plants, constantly observe the actions of the butterflies in the garden. If you suspect the repellent plants are doing more harm than good, remove them or move them to an area where they are not often disturbed.

Physical Controls

For some of the larger insects, hand removal may be the simplest and safest method. Fill a can or jar with a mixture of water topped with kerosene, and drop the insects into the liquid as you remove them from the plants. Grasshoppers (order Orthoptera) are easily picked from plants at night. To capture sow bugs and pill bugs (family Asellidae), moisten the ground and lay a board on the moist area. In early mornings and late evenings, lift the board and dispose of the clusters of bugs that have gathered.

Slugs and snails (class Mollusca) are reportedly attracted to saucers of stale beer. If this does not work, use a straw or pine needle mulch around the plants. These soft-bodied mollusks do not care much for the rough texture of the mulch, finding it difficult to move across. Blood meal and bonemeal sprinkled on top of the soil (or mulch) is also a deterrent to slugs and snails as well as pill bugs, sow bugs, ants (family Formicidae), aphids, deer (*Odocoileus* spp.), rabbits (order Lagomorpha), and household cats (*Felis catus*). Wood ashes from the fireplace sprinkled thickly around plants act as a great deterrent to many pests. Other irritants to be strewn around plants include crushed dried hot peppers, finely crushed eggshells, camphor, powdered charcoal, builder's sand, and cedar shavings.

An Irish potato (*Solanum tuberosum*) sliced in half and buried about two inches beneath the soil's surface is a simple but effective trap for wireworms (larvae of click beetles [family Elateridae]). After a couple of days, lift the potatoes and destroy the worms.

Traps made from plastic jugs containing sugar water flavored with vanilla (*Vanilla planifolia*), sassafras (*Sassafras albidum*), lemon (*Citrus limon*), anise, or common fennel will attract many flying and jumping insects. Place these jugs near the infested plants. To trap grasshoppers, half fill a wide-mouthed jar with this mixture and bury the jar to the rim in the ground. Other trap baits to try are a crushed banana peel, a cup each of sugar and vinegar, and enough water to almost fill a gallon jug. A brew of mashed fruit, one or two cups of sugar, and some yeast for fermentation is especially attractive to Japanese beetles. Fill narrow-mouthed jugs no more than half full with these mixtures. Smaller insects crawl down into the jugs to get at the bait and drown, but butterflies usually do not enter.

There are several creatures generally considered helpful in most gardens that are definitely not desired in a butterfly garden. Most of these are avid predators on eggs or larvae, including those of moths and butterflies. Some predators that should not be encouraged to remain in a butterfly garden include most of the parasitic wasps (family Braconidae), ichneumons (family Ichneumonidae), trichogrammatids (family Trichogrammatidae), along with assassin bugs (family Reduviidae), stink bugs (family Pentatomidae), robber flies (family Asilidae), praying mantises (family Mantidae), and spiders (order Araneae).

Praying mantises are one of the most familiar insects sold as a biological pest control and for a general garden are highly recommended instead of pesticides. But they have no place in a butterfly garden. Praying mantises are one of the worst enemies of butterflies, hiding among the foliage and seizing the insect as it approaches the flower to nectar.

The little anoles (*Anolis carolinensis*), sometimes referred to as chameleons since they can quickly change their coloring to match the surrounding environment, readily prey on both butterfly larvae and adults, consuming all they find, except perhaps the Monarch and Queen.

Biological Controls

Along with some of the "bad" insects, there are others that are especially beneficial to a butterfly garden, and their presence should be both protected and encouraged. Just as some viruses, bacteria, and fungi help keep plants healthy, many insects, during one or more stages of their life, attack and devour other insects that are damaging the plants. One of the most beneficial is a small beetle variously

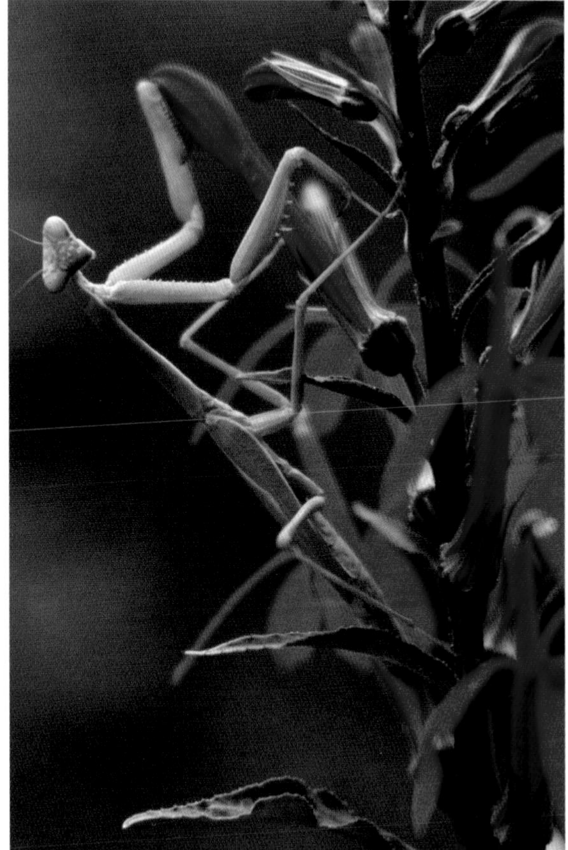

(left) A CRAB SPIDER (FAMILY THOMISIDAE) ATTACKS A CERAUNUS BLUE *(Hemiargus ceraunus astenidas)*.

(right) THE PRAYING MANTIS (FAMILY MANTIDAE) IS ONE OF THE BUTTERFLY'S WORST ENEMIES.

known as ladybug, lady beetle, ladybird, or aphid wolf (family Coccinellidae). Ladybug eggs, a bright yellow or orange, are attached in small clusters to plant leaves or stems, behind tree bark, or among debris scattered on the ground. The larvae are flat, warty, and somewhat carrot-shaped, their grayish-black coloring spotted with blue and orange. Both the larvae and the adults eat aphids, leafhoppers (family Cicadellidae), mealybugs (families Pseudococcidae and Eriococcidae), and scale insects (superfamily Coccidae). Each larva can devour up to four hundred aphids, and one adult ladybug may consume five thousand or more. Unfortunately, they just as readily consume the eggs of butterflies.

Green lacewings (family Chrysopidae) are beautiful, delicate, fairylike insects with clear, gauzy green wings and red eyes. Both larvae and adults suck the body fluids from aphids, mealybugs, common thrips (family Thripidae), spider mites (*Tetranychus* spp.), and cottony-cushion scales. They actually do a better job than ladybugs of keeping aphids under control. Ladybugs and lacewings can be purchased from some nurseries (see appendix). When ordering, be sure to purchase the ladybugs that are already conditioned to eat aphids.

Larvae of the firefly or lightning bug (family Lampyridae) are long, flat, and wormlike. They live on the ground under bark and in

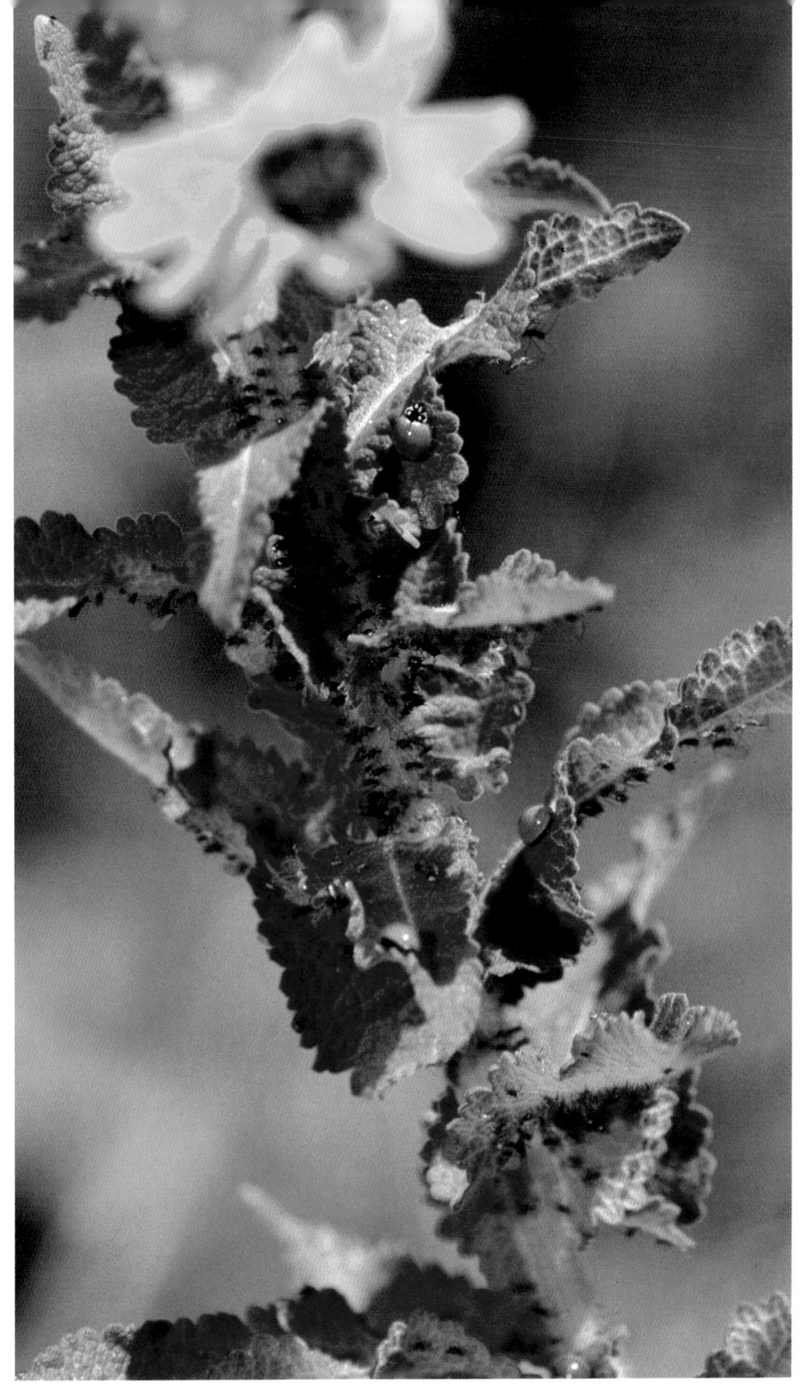

Ladybugs (family Coccinellidae) are beneficial insects that devour many common garden pests.

moist places and are voracious nighttime predators, feeding especially on slugs, snails, and cutworms (larvae of the noctuid moths—family Noctuidae).

And keep a toad (family Bufonidae) handy. Or two or three. A single toad can consume up to three thousand insects in a month, especially ants, sow bugs, and pill bugs. Having rather broad tastes, they also eat grasshoppers, cutworms, beetles, snails, and slugs.

Natural Insecticidal Controls

Because some plants have a natural repellent quality while growing, others (or sometimes the same ones) can be very effective when used as a spray, dust, or mulch. Some of these plants are not especially attractive, so they can be grown in out-of-the-way spots and gathered when needed.

The best preventive for garden problems is to keep a close watch on the plants and catch the insects or diseases before they really get established. If this is done, often only a light spraying of a small portion of the plant will be required. If the infected plant is a nectar plant, hand-spray with a small bottle sprayer in such a manner as to keep the spray away from the flowers, thereby not disturbing the nectar source of the butterflies. If the foliage of a larval food plant is being damaged, use the spray sparingly and only where no caterpillars are feeding.

For many of the following concoctions, using a blender is suggested, but do not use the household blender. If an old, no-longer-used machine is not available, simply mash the material until it is as mushy as possible. Chopped plant material can also be left to steep in warm water for several hours or placed in a jug and left in the sun for a couple of days. The liquid strained off is usually potent enough for use. Wear rubber gloves when chopping or blending the plants and also when spraying. Use a small plastic sprayer bottle or a small commercial hand sprayer in order to confine the spray to exactly where it is needed. Two or three such sprayings with a weak solution are much more desirable in butterfly gardening than one extra-strong dose, which may be lethal to everything that touches it for weeks.

Keep in mind that even though these insecticides are concocted from plants, they are very potent and should never be used indiscriminately. Use much caution, and apply them only to the pest-infected portions of the plants. Leave insecticides on plants only long enough to kill the pests; then thoroughly wash the plants with a soft dousing from the water hose.

- An excellent spray for spider mites can be prepared by chopping the leaves, stems, and spent flowers of the Flowering Tobaccos (*Nicotiana* spp.) in a blender with enough water to make a liquid. Strain and add enough water for a spray.
- An even better spray can be made by obtaining the strongest chewing tobacco on the market, combining one-half cup of the tobacco with three cups of water, and simmering (not boiling) on very low heat for twenty or thirty minutes. Remove from heat, and leave the tobacco soaking in the same water for a couple of days. When ready to apply, dilute with a little more water and then strain through a nylon stocking before using in a sprayer. Try burying unused chewing tobacco around plants to help control underground pests such as June bug (family Scarabaeidae) larvae, wireworms, nematodes, or cutworms.
- A tea of chopped Heart-leaf Stinging-nettle (*Urtica chamaedryoides*) makes a good spray for aphids. Use gloves and wear a long-sleeved shirt when gathering the nettle. Some people are highly allergic to the sting of this plant, which can cause large blisters on the skin. After the nettle has been gathered, chop in a blender. Remove to a glass jar or jug, completely

The poisonous compounds contained within the sap of the genus *Asclepias* naturally repel most insects.

cover the chopped nettle with water, cover the container tightly, and leave standing for several days or until the plant parts are well softened. When ready to use, strain and mix one part of the tea to seven parts water.

- The leaves of Common Elderberry (*Sambucus nigra* var. *canadensis*) contain oxalic acid from which a tea can be made to deter aphids.
- A mixture of one-half cup of soft soap such as Ivory liquid (read the label carefully—nonsoap detergents will not work) thoroughly mixed with two quarts of water is an excellent control for aphids. Soap is very strong; use a couple of mild applications instead of one strong one to lessen the risk of burning the plants. A potassium-based insecticidal soap can be obtained at most plant nurseries. This special mixture smothers its victims and is much less damaging to plants than household soaps. Remove any soap mixture from the plants with a gentle washing from the hose after the infestation is over, especially if used on *Asclepias* (notorious for attracting aphids). The Monarch and the Queen are constantly going to be needing the plants for egg laying. If there are eggs or larvae already on the plants, aphids can be kept under partial control by running your fingers along the stems and on the undersides of the leaves, crushing the soft-bodied insects. This must be done almost daily and will not completely eliminate the aphid colony but does keep them controlled until the butterfly larvae have all pupated. Then the plants should be thoroughly doused with the soap-and-water mixture to completely eliminate the entire aphid population.
- Nasturtiums in the border are the larval food for the Cabbage White (*Pieris rapae*) and Great Southern White (*Ascia monuste*) and a trap for aphids. When Nasturtium is ground in a blender in combination with Wormwood, Garden Sage (*Salvia officinalis*), and Chamomile (*Anthemis nobilis*), along with enough water to make a spray, it is very effective in the control of white flies (family Aleyrodidae) and aphids.
- Brew the chopped stems and leaves of Wormwood into a tea, and sprinkle the mixture on the ground and on young plants. The bitter taste repels slugs and snails.
- Also, use Wormwood tea in combination with the water from soaked Quassia (*Picrasma excelsa*) chips for an especially potent control. Quassia chips may be purchased at the drugstore or nurseries that carry natural control products. To make the spray, soak four ounces of chips in two gallons of water for several days. Then simmer slowly over very low heat for three or four hours; cool and strain before using. When combining this with the Wormwood spray, add one-half teaspoon of soft soap to a bottle full of the spray to make it stick to the plants better. Insects find the bitter taste of these brews completely unpalatable. Again, if used on any of the butterfly larval food plants, the spray should be thoroughly washed off as soon as the pest infestation is over.
- Chop citrus (*Citrus* spp.) peels in a blender with water, let stand for several hours, strain, and use on various insect pests. This is a good, all-around spray and may be tried on almost all chewing or juice-sucking

insects. It is very strong and, if not diluted sufficiently, can severely burn foliage.

- A mixture of three cloves of garlic, one medium onion, and a teaspoon of hot pepper combined in a blender with a quart of water is an excellent repellent for many insects, including aphids, thrips, and grasshoppers. Let the mixture stand for half an hour or so before straining; then use one part mixture to three or four parts water. These ingredients can also be mixed in a large jar of water and left to steep in the sun two or three days before using.
- Many gardeners have reported phenomenal success with an elixir called bug juice. To prepare this mixture, collect about a cup of the bugs causing the problem (or as many as possible in the case of small insects, such as aphids or mealybugs). Be sure to collect any that look weakened or sick. Place the insects in a blender, using the proportion of one-half cup of bugs to two cups of water, and blend until liquefied. If a blender is not available, mash the insects thoroughly, then add the water, and let set for several hours. Strain through a sieve or cheesecloth, retaining all the liquid possible. Dilute the juice with four to eight parts water, and then use as a spray.

  Any juice left over can be frozen for a year or more. In order to get enough insects for this juice, you will have to have a pretty bad infestation. By having some of the proper juice on hand from the previous summer, you can use at the first sign of the insects the following spring, before they become a real problem. This juice can also be prepared using either slugs, snails, or pill bugs; when the liquid is sprayed, poured, or sprinkled around the base of plants, it proves to be a powerful deterrent to these pests.
- For an infestation of mealybugs, dip a cotton swab in alcohol and apply to each insect seen as well as to the axil of each leaf, where the eggs and very young are hidden.
- If nematodes are a serious problem, consistent use of compost and natural fertilizer in the soils should eliminate them. Lime and fish fertilizer make a useful repellent. A teaspoon of sugar sprinkled into the planting hole before setting out plants such as annuals is also helpful.
- Stems and foliage of Common Fennel can be cut into two- or three-inch pieces and scattered in circles around plants as a thin mulch to keep snails away. Pill bugs and sow bugs may be controlled by sprinkling a weak lime solution (two pounds of lime dissolved in five gallons of water) around plants. Sprinkle cornmeal around plants where cutworm damage is apparent; cutworms love the meal, but they cannot digest it and will die from overeating.

The Heavies

If a really terrible insect infestation becomes established in the garden and various natural methods have been tried with no success, then perhaps a consultation with local nursery personnel might be needed concerning the best chemical pesticides to use. Whatever is chosen, try applying a slightly weaker solution than recommended the first time. Keep a close watch on the plants, and as soon as the pests appear to be gone, immediately and thoroughly wash the plant, reducing the chance of any larvae or adult butterflies being harmed.

Perhaps one of the most effective yet least toxic of the really serious insecticides is pyrethrin. This is an extract of the toxins from the dried flowers of a daisylike perennial formerly in the genus *Pyrethrum* but now known as *Chrysanthemum cineraiifolium*. Some seed and plant sources now offer seeds of this chrysanthemum, so gardeners can grow their own insecticidal plants, sprinkling the dried, crushed flowers around infested plants. Pyrethrin is especially effective against aphids, thrips, leafhoppers, and many beetles.

Sometimes pyrethrin is used mixed with such chemicals as ryonia and rotenone. Although they are not persistent in the environment, all of these other extracts are extremely toxic and nonselective, killing most things they come in contact with. Any such poison should be used with the greatest caution.

Systemics are the new rage in pesticides today. Systemic poisons are usually placed in the ground around a plant either as a liquid, a powder, or tablets. As the substance dissolves, it is taken up by the roots of the plant and carried into all plant parts. The poisons infiltrate the leaves and stems of the plants as well as the pollen and nectar of the flowers, meaning sure death not only to larvae but adult butterflies as well. Furthermore, if these chemicals are placed in the soil, they will, by the water movement and soil disturbance, be carried to other parts of the garden, eventually contaminating many more plants than the ones they were originally used on. It is my suggestion to leave the systemics at the store.

Another product that should never be used in a butterfly garden is *Bacillus thuringiensis*, commonly referred to as Bt. This bacterial pathogen works internally, causing paralysis and death within twenty-four hours to all butterfly larvae (and other critters) that eat even a small amount of any part of the plant on which the bacterium has been used. This product is sold in liquid, dust, and granular forms and is marketed under the brand names Dipel, Thuricide, Biotrol, Attact, Bactisphere, and Soilserv Bacillus Bait. Do not use any of these products in a butterfly garden.

Fire Ant Control

One of the worst enemies to butterfly larvae and chrysalides is the imported fire ant (*Solenopsis invicta*). Fire ants become established after flying into an area during mating flights, which occur anytime from spring through fall. They can also be brought to the garden in containers of nursery stock or on shared plants from neighbors. If you want to raise butterflies, you simply have to get rid of any and all fire ants anywhere near your plants.

To eradicate fire ants, there are safe, nonchemical ways as well as chemicals specific to fire ants. Fire ants continually move their larvae and the queen about within the mound and in tunnels to regulate their temperatures. Midmorning to early evening is the best time to treat the mound in the spring, for the ants will have the larvae near the top of the mound to receive the warmth of the sun. During the summer months they are near the surface only during the cooler mornings and late evenings. At these times the colony can be destroyed by pouring boiling water on the mound. Use about three gallons per mound, and be sure it is as hot as you can

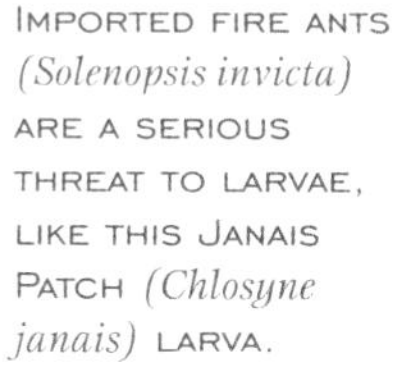
Imported fire ants *(Solenopsis invicta)* are a serious threat to larvae, like this Janais Patch *(Chlosyne janais)* larva.

possibly get it. Do not disturb the mound in any way before pouring the water on; in fact, walk up to the mound slowly and quietly. Any earth tremors can send the ants deep within their tunnel, carrying both the queen and the larvae. The queen has to be destroyed to completely eradicate a colony. Oils from citrus peels are very toxic to fire ants. Remove as much of the white inner lining of orange (*Citrus sinensis*), grapefruit (*C. maxima* × *paradisi*), or lemon as possible. Run the outer rind through a blender with enough water to blend nicely. Add more water, and pour this on the fire ant mound. A commercial pesticide containing d-limonene kills ants on contact. An insecticidal soap can also be blended with water and poured on the mound.

The toxins Pro-Drone and Logic Fire Ant Bait affect only fire ants. Scattered about the mound, the bait is picked up by the ants, carried underground, and fed to the young. The bait is a growth inhibitor; because the young do not develop properly, the colony is eventually eliminated. Although expensive, Amdro Ant Bait is very effective when scattered about the mound.

The latest and most promising weapon may be the imported phorid flies (family Phoridae) from South America. These tiny flies lay their eggs within the ants, and the young fly caterpillars slowly consume the ants' bodies. Eventually the ant colony is totally eradicated or significantly reduced.

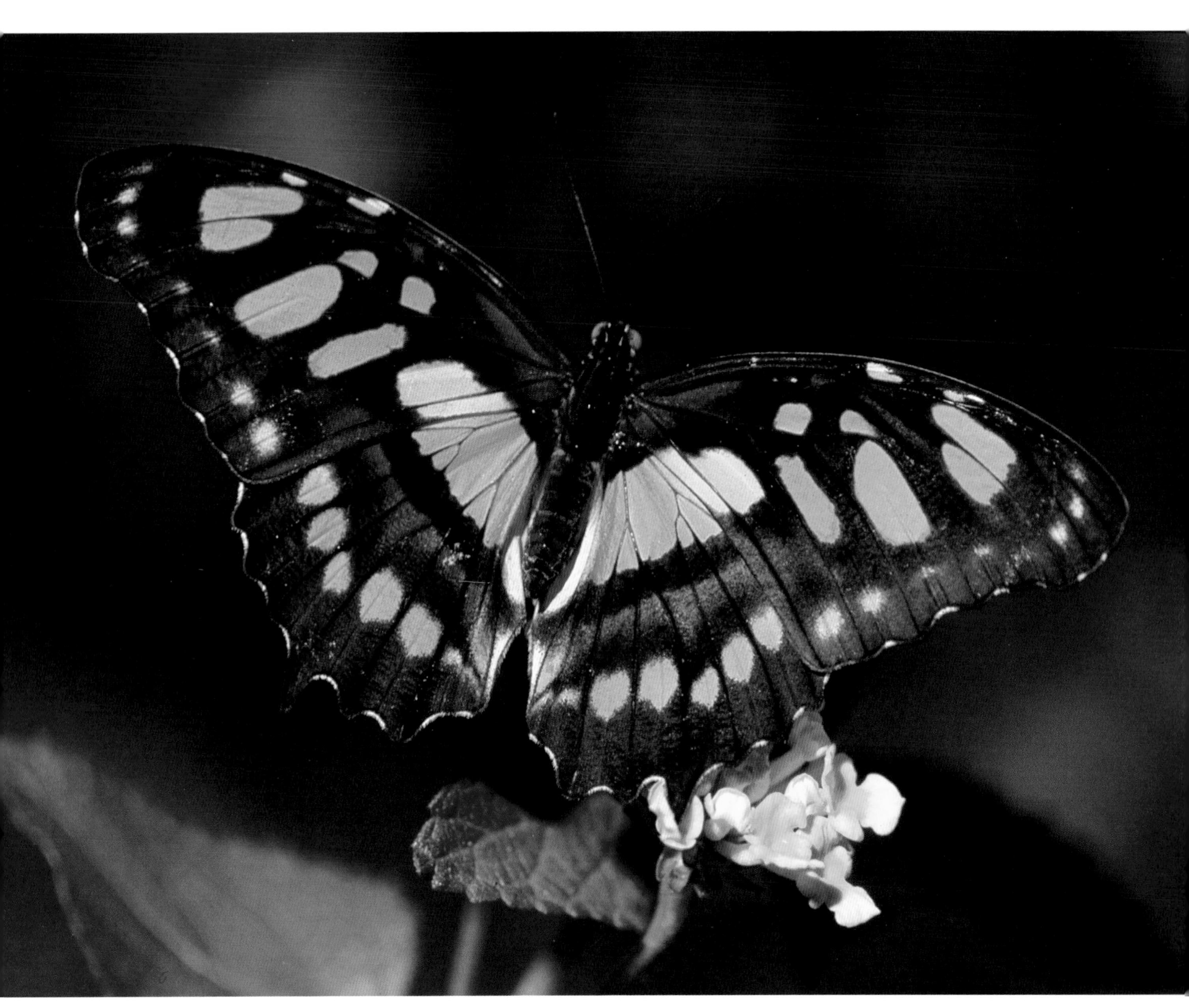

Malachite *(Siproeta stelenes biplagiata)*

6 A Special South Texas Garden

Gardeners in the semitropical area of the Rio Grande Valley have an excellent opportunity for attracting some of the rarest and most beautiful species of butterflies to be found in the state. Not only are there species that are permanent Texas residents only in the Valley area but there are some even less common species that come into South Texas from Mexico and Central and South America and will, under favorable conditions, remain and breed, often for several years in a row. This northward movement, added to the already abundant species consistently breeding here, gives this region the highest number of total species of any area in the state—almost three hundred.

There are probably several environmental factors that determine the residence and breeding status of these uncommon butterflies, but two of the most important are the occasional periods of freezing temperatures and the scarcity of larval food plants. A severe freeze may completely eliminate a population of these exotic species of butterflies, and other than providing protection through the planting of windbreaks, providing plenty of "hiding" places, and protecting both the larval and flowering plants, there is little that can be done to prevent the extermination of the butterflies during this time.

Many of the larval food plants for these more uncommon species of butterflies are native and have been almost entirely eliminated from the Valley area through clearing practices for the raising of food crops. In a few instances, they may never have been abundant members of the Valley flora. This shortage of food plants limits the range, distribution patterns, and number of individuals coming into the area, and females may not be able to locate a plentiful supply of the appropriate food plants on which to lay their eggs. Only in

THE NECTAR OF THE BRIGHT RED FLOWER CLUSTERS OF SHRIMP PLANT *(Justicia brandegeana)* WILL ATTRACT BUTTERFLIES TO THE GARDEN.

areas where an adequate source of the larval food plants are available will there ever be a chance of a colony becoming well established. By incorporating the necessary food plants in home gardens, the chances of getting colonies of these "special" butterflies to remain and breed are greatly increased.

Very few of the needed plants are going to be readily available at nurseries, so the best method of obtaining many of the plants listed here will be through gathering seeds, taking cuttings from native plants, or haunting native plant sales. Special orders for some of the non-natives can be placed through local nurseries. Visit the local chambers of commerce, and get names and phone numbers of local nature, birding, and garden clubs to find sources and methods of obtaining needed plants.

The information given here is in no way the "final word" on the butterflies to be found in the Rio Grande Valley. Each year as more naturalists and researchers become interested and involved in the butterflies in this region, new species are continually being added. And, as with all such work, the more that is learned about the distribution, habitats, and life cycles of a wildlife species, the status of their abundance is continually changed and updated. Take this then as a guideline—to be subtracted from and added to while more research is completed on these uncommon species.

The following list has been compiled from current literature, communications, and personal observations in the field. The butterflies listed are those that, at the present time, are known to breed only in the southernmost portion of the state, and only if their food plant is known. Many other species of butterflies stray or wander into the Valley area and perhaps breed there, but at present, their food plants are not known, so these species are not listed.

The abbreviation "spp." following a plant genus in this list indicates that there are more than one species of that genus occurring in the Valley, and the butterflies are known to or will probably use all or several of them.

Butterfly	***Food plant***
Aguna, Emerald (*Aguna claxon*)	Mexican Orchid Tree (*Bauhinia mexicana*)
Aguna, Gold-spotted (*Aguna asunder*)	Mexican Orchid Tree (*Bauhinia mexicana*)
Aguna, Tailed (*Aguna metophis*)	Mexican Orchid Tree (*Bauhinia mexicana*)
Angled-Sulphur, Yellow (*Anteos maerula*)	Senna (*Senna* spp.)
Astraptes, Flashing (*Astraptes fulgerator*)	Chaste Tree (*Vitex agnus-castus*) Coyotillo (*Karwinskia humboldtiana*)
Banner, Common (*Epiphile adrasta*)	Common Balloon-vine (*Cardiospermum halicacabum*) Little-fruit Supplejack (*Serjania brachycarpa*) Urvillea (*Urvillea ulmacea*)
Bluewing, Mexican (*Myscelia ethusa*)	Vasey Adelia (*Adelia vaseyi*)
Cracker, Guatemalan (*Hamadryas guatemalena*)	Purple Wings (*Dalechampia* spp.)
Crescent, Pale-banded (*Anthanassa tulcis*)	Rio Grande Dicliptera (*Dicliptera sexangularis*) Runyon's Ruellia (*Ruellia nudiflora* var. *runyonii*) Water-willow (*Justicia* spp.)
Daggerwing, Many-banded (*Marpensia chiron*)	Common Fig (*Ficus carica*)
Daggerwing, Ruddy (*Marpesia petreus*)	Common Fig (*Ficus carica*)
Emperor, Pavon (*Doxocopa pavon*)	Spiny Hackberry (*Celtis pallida*)
Emperor, Silver (*Doxocopa laure*)	Spiny Hackberry (*Celtis pallida*)
Flasher, Frosted (*Astraptes alardus*)	Coral Bean (*Erythrina herbacea*)
Flasher, Gilbert's (*Astraptes alector*)	Mexican Orchid Tree (*Bauhinia mexicana*)
Flasher, Two-barred (*Astraptes fulgerator azul*)	Chaste Tree (*Vitex agnus-castus*) Coyotillo (*Karwinskia humboldtiana*)

Butterfly	*Food plant*
Fritillary, Mexican (*Euptoieta hegesia merediana*)	Hierba del Venado (*Turnera diffusa*) Red-fruited Passionflower (*Passiflora foetida*)
Giant-Skipper, Manfreda (*Stallingsia maculosus*)	Manfreda (*Manfreda* spp.)
Greenstreak, Clench's (*Cyanophrys miserabilis*)	Retama (*Parkinsonia aculeata*)
Greenstreak, Goodson's (*Cyanophrys goodsoni*)	Pigeonberry (*Rivina humilis*)
Greenstreak, Tropical (*Cyanophrys herodotus*)	Climbing Hempweed (*Mikania scandens*) West Indian Lantana (*Lantana camara*)
Hairstreak, Aquamarine (*Oenomaus ortygnus*)	Soursop (*Annona globiflora*)
Hairstreak, Red-spotted (*Tmolus echion*)	West Indian Lantana (*Lantana camara*)
Hairstreak, Strophius (*Allosmaitia strophius*)	Barbados Cherry (*Malpighia glabra*)
Leafwing, Angled (*Memphis glycerium*)	Croton (*Croton* spp.)
Leafwing, Pale-spotted (*Memphis pithyusa*)	Croton (*Croton* spp.)
Longtail, Brown (*Urbanus procne*)	Common Bermuda Grass (*Cynodon dactylon*) Guinea Grass (*Panicum maximum*) Johnson Grass (*Sorghum halepense*) Rye Grass (*Lolium perenne*) St. Augustine Grass (*Stenotaphrum secundatum*)
Longtail, Mottled (*Typhedanus undulatus*)	Argentina Senna (*Senna corymbosa*) Christmas Senna (*S. pendula*) Coffee Senna (*S. occidentalis*)
Longtail, Teleus (*Urbanus teleus*)	Fringe-leaf Paspalum (*Paspalum setaceum* var. *ciliatifolia*) Guinea Grass (*Panicum maximum*)
Longwing, Banded Orange (*Dryadula phaetusa*)	Passionflower (*Passiflora* spp.)
Longwing, Crimson-patched (*Heliconius erato petiverana*)	Passionflower (*Passiflora* spp.)
Longwing, Isabella's (*Eueides isabella*)	Passionflower (*Passiflora* spp.)
Malachite (*Siproeta stelenes biplagiata*)	Lance-leaved Water-willow (*Justicia ovata* var. *lanceolata*) Runyon's Ruellia (*Ruellia nudiflora* var. *runyonii*) Runyon's Water-willow (*J. runyonii*) Shrimp Plant, Green (*Blechum pyramidatum*)
Mellana, Common (*Quasimellana eulogius*)	Guinea Grass (*Panicum maximum*)
Metalmark, Blue (*Lasaia sula*)	Screwbean Mesquite (*Prosopis reptans*)

Butterfly	***Food plant***
Metalmark, Curve-winged (*Emesis emesia*)	Mexican Bird-of-Paradise (*Caesalpinia mexicana*)
Metalmark, Falcate (*Emesis tenedia*)	Drummond's Virgin's Bower (*Clematis drummondii*)
Mimic-queen, Tiger (*Lycorea cleobaea*)	Common Fig (*Ficus carica*) Papaya (*Carica papaya*)
Patch, Banded (*Chlosyne endeis pardelina*)	Small-flowered Carlowrightia (*Carlowrightia parviflora*)
Patch, Rosita (*Chlosyne rosita*)	Dicliptera (*Dicliptera sexangularis*) Downy Water-willow (*Justicia pilosella*)
Peacock, Banded (*Anartia fatima*)	Lance-leaved Water-willow (*Justicia ovata* var. *lanceolata*) Rio Grande Dicliptera (*Dicliptera sexangularis*) Runyon's Water-willow (*J. runyonii*) Ruellia (*Ruellia* spp.)
Pellicia, Glazed (*Pellicia arina*)	Little-fruit Supplejack (*Serjania brachycarpa*)
Pixie, Red-bordered (*Melanis pixe*)	Guamúchil, Monkeypod (*Pithecellobium dulce*)
Purplewing, Dingy (*Eunica monima*)	Lime Prickly-ash (*Zanthoxylum fagara*)
Sailor, Blue-eyed (*Dynamine dyonis*)	Catnip Noseburn (*Tragia ramosa*)
Scrub-Hairstreak, Lantana (*Strymon bazochii*)	West Indian Lantana (*Lantana camara*) Scented Lippia (*Lippia graveolens*) White-flowered Lippia (*L. alba*)
Scrub-Hairstreak, Red-crescent (*Strymon rufofusca*)	Bladder-mallow (*Herissantia crispa*) Three-lobe False-mallow (*Malvastrum coromandelianum*)
Scrub-Hairstreak, Red-lined (*Strymon bebrychia*)	Common Balloon-vine (*Cardiospermum halicacabum*)
Scrub-Hairstreak, White (*Strymon albata*)	Texas Abutilon (*Abutilon fruticosum*) Wright's Abutilon (*A. wrightii*)
Scrub-Hairstreak, Yojoa (*Strymon yojoa*)	Heart-leaf Hibiscus (*Hibiscus martianus*)
Silverdrop, Broken (*Epargyreus exadeus*)	Senna (*Senna* spp.)
Silverspot, Mexican (*Dione moneta poeyi*)	Corky-stemmed Passionflower (*Passiflora suberosa*)
Sister, Band-celled (*Adelpha fessonia*)	Crucillo (*Randia rhagocarpa*)
Skipper, Brown-banded (*Timochares ruptifasciatus*)	Barbados Cherry (*Malpighia glabra*)

Butterfly	***Food plant***
Skipper, Double-dotted (*Decinea percosius*)	Common Bermuda Grass (*Cynodon dactylon*) Cut Grass (*Leersia* spp.) Rye Grass (*Lolium perenne*) St. Augustine Grass (*Stenotaphrum secundatum*)
Skipper, Falcate (*Spathilepia clonius*)	Purple Bush-bean (*Macroptilium atropurpureum*)
Skipper, Fawn-spotted (*Cymaenes odilia*)	Guinea Grass (*Panicum maximum*) Paspalum (*Paspalum* spp.)
Skipper, Glassy-winged (*Xenophanes tryxus*)	Drummond's Wax-mallow (*Malvaviscus drummondii*) Malva de Caballo (*Malachra capitata*)
Skipper, Guava (*Phocides polybius lilea*)	Guava (*Psidium guajava*) Strawberry Guava (*P. cattlelanum*)
Skipper, Hermit (*Grais stigmatica*)	Jopoy (*Esenbeckia berlandieri*)
Skipper, Malicious (*Synapte malitiosa*)	Guinea Grass (*Panicum maximum*)
Skipper, Mercurial (*Proteides mercurius*)	Least Snoutbean (*Rhynchosia minima*) Wild Cow-pea (*Vigna luteola*)
Skipper, Mimosa (*Cogia calchas*)	Black Mimosa (*Mimosa pigra*)
Skipper, Olive-clouded (*Lerodea arabus*)	Common Bermuda Grass (*Cynodon dactylon*)
Skipper, Pale-rayed (*Vidius perigenes*)	Bristle Grass (*Setaria* spp.) St. Augustine Grass (*Stenotaphrum secundatum*)
Skipper, Potrillo (*Cabares potrillo*)	Common Velvet-bur (*Priva lappulacea*)
Skipper, Purple-washed (*Panoquina lucas*)	Guinea Grass (*Panicum maximum*) Sugar Cane (*Saccharum officinarum*)
Skipper, Purplish-black (*Nisoniades rubescens*)	Sharp-pod Morning Glory (*Ipomoea cordatotriloba*)
Skipper, Starred (*Arteurotia tractipennis tractipennis*)	Croton (*Croton* spp.)
Skipper, Violet-banded (*Nyctelius nyctelius*)	Guinea Grass (*Panicum maximum*) Sugar Cane (*Saccharum officinarum*)
Skipper, Violet-patched (*Monca crispinus*)	Paspalum (*Paspalum* spp.)
Soldier (*Danaus eresimus*)	Beaked Cynanchum (*Cynanchum barbigerum*) Climbing Milkweed Vine (*Funastrum cynanchoides*)
Stripe-streak, Creamy (*Arawacus jada*)	Blue Potato Bush (*Solanum rantonnatii*)
Sulphur, Statira (*Aphrissa statira*)	Coin-vine (*Dalbergia ecastophyllum*) Senna (*Senna* spp.)

Butterfly	***Food plant***
Sulphur, Yellow-angled (*Anteos maerula*)	Senna (*Senna* spp.)
Swallowtail, Broad-banded (*Papilio astyalus*)	Lemon (*Citrus limon*) Lime (*C. aurantifolia*)
Swallowtail, Ornythion (*Papilio ornythion*)	Lime Prickly-ash (*Zanthoxylum fagara*)
Swallowtail, Ruby-spotted (*Papilio anchisiades*)	Lemon (*Citrus limon*)
Swallowtail, Thoas (*Papilio thoas*)	Lime Prickly-ash (*Zanthoxylum fagara*) Lemon (*Citrus limon*)
White, Giant (*Ganyra josephina*)	Downy Caper (*Capparis incana*) Jamaican Caper (*C. cynophallophora*)
White-Skipper, Turk's-cap (*Heliopetes macaira*)	Amantillo (*Abutilon trisulcatum*) Bladder-mallow (*Herissantia crispa*) Drummond's Wax-mallow (*Malvaviscus drummondii*)
Yellow, Barred (*Eurema daira*)	Sticky Joint-vetch (*Aeschynomene viscidula*)
Yellow, Boisduval's (*Eurema boisduvaliana*)	Christmas Senna (*Senna pendula*) Lindheimer's Senna (*S. lindheimeriana*)
Yellow, Dina (*Pyrisitia dina*)	Bitterbush (*Picramnia pentandra*) Goat Bush (*Castela erecta*)

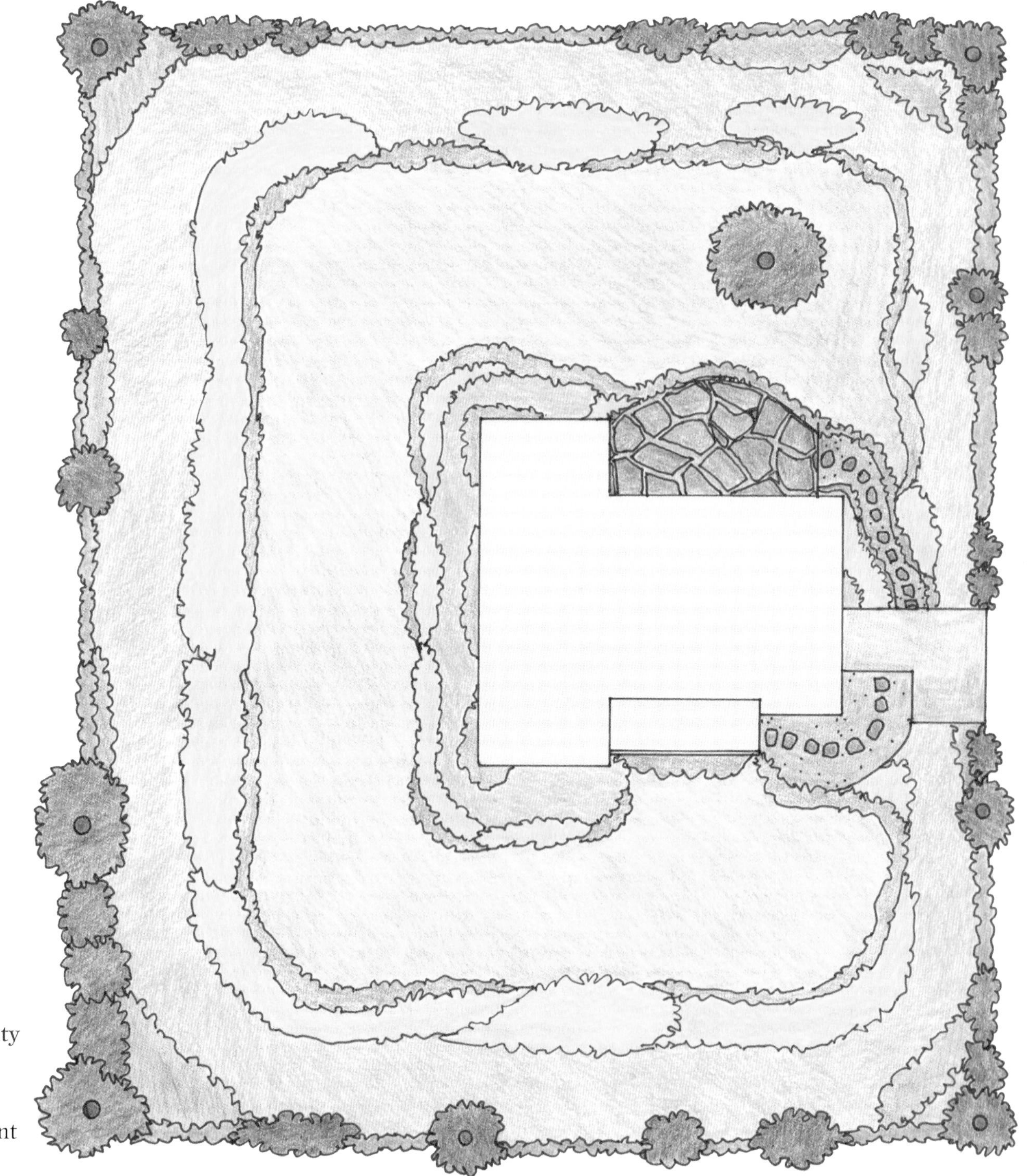

SPECIAL SOUTH TEXAS GARDEN

Brownsville, Harlingen, McAllen, Rio Grande City

(L) denotes larval food plant
(N) denotes nectar plant

Trees

Anacahuita (N)
(*Cordia boissieri*)
Chaste tree (N)
(Vitex agnus-castus)
Desert Willow (N)
(*Chilopsis linearis*)
Guamúchil, Monkeypod (L)
(*Pithecellobium dulce*)
Mexican Orchid Tree (L) (N)
(*Bauhinia mexicana*)
Orange/Lime (L) (N)
(*Citrus* spp.)

Shrubs

Argentina Senna (L)
(*Senna corymbosa*)
Barbados Cherry (L)
(*Malpighia glabra*)
Bougainvillea (N)
(*Bougainvillea glabra*)
Buttonbush (N)
(*Cephalanthus* spp.)
Candle-stick Senna (L)
(*Senna alata*)
Carlowrightia (L)
(*Carlowrightia* spp.)
Cenizo (L)
(*Leucophyllum frutescens*)
Christmas Senna (L)
(*Senna pendula*)
Coral Bean (L)
(*Erythrina herbacea*)
Coyotillo (L)
(*Karwinskia humboldtiana*)
Crucita (N)
(*Chromolaena odorata*)
Drummond's Wax-mallow (L) (N)
Malvaviscus drummondii)
Flame Acanthus (L) (N)
(*Anisacanthus quadrifidus* var. *wrightii*)
Guayacan (L)
(*Guajacum angustifolium*)
Jopoy (L)
(*Esenbeckia berlandieri*)
Lime Prickly-ash (L)
(*Zanthoxxylum fagara*)
Lippia (L) (N)
(*Lippia* spp.)
Vasey Adelia (L)
(*Adelia vaseyi*)
Vine Mimosa, Softleaf Mimosa (L) (N)
(*Mimosa malacophylla*)

Vines (on trellis/fence)

Climbing Hempweed (L) (N)
(*Mikania scandens*)
Climbing Milkweed Vine (L) (N)
(*Funastrum cynanchoides*)
Drummond's Virgin's Bower (L)
(*Clematis drummondii*)
Least Snout-bean (L)
(*Rhynchosia minima*)
Little-fruit Supplejack (L)
(*Serjania brachycarpa*)
Mexican Flame Vine (N)
(*Pseudogynoxys chenopodioides*)
Morning Glory (L) (N)
(*Ipomoea* spp.)
Queen's Wreath (N)
(*Antigonon leptopus*)Red-fruited Passionflower (L)
(*Passiflora foetida*)
Shrubby Blue Sage (N)
(*Salvia ballotiflora*)
Snapdragon Vine (L)
(*Maurandya antirrhiniflora*)
Tropical Balloon-vine (L)
(*Cardiospermum corundum*)

Tall Herbs (3–5 feet)

Argentina Verbena (N)
(*Verbena bonariensis*)
Croton (L)
(*Croton* spp.)
Golden Crownbeard (L) (N)
(*Verbesina enceloides*)
Frostweed (L) (N)
(*Verbesina* spp.)
Goldenrod (N)
(*Solidago* spp.)
Lindheimer's Senna (L)
(*Senna lindheimeriana*)
Mexican Sunflower (L) (N)
(*Tithonia rotundifolia*)
Purple Marsh-fleabane (N)
(*Pluchea odorata*)
White-flowered Plumbago (L) (N)
(*Plumbago scandens*)

Medium Herbs (2–3 feet)

Awnless Bush-sunflower (N)
(*Simsia calva*)
Betony-leaf Mistflower (N)
(*Conoclinium betonicifolium*)
Croton (L)
(*Croton* spp.)
Heath Aster (L) (N)
(*Symphyotrichum ericoides*)
Liatris (N)
(*Liatris* spp)
Mexican Oregano (N)
(*Poliomintha longiflora*)
Pentas (N)
(*Pentas lanceolata*)
Pigeonberry (L) (N)
(*Rivina humilis*)
Rio Grande Dicliptera (L)
(*Dicliptera sexangularis*)
Shrimp Plant (L)
(*Justicia brandegeana*)
Skeleton-leaf Goldeneye (N)
(*Viguiera stenoloba*)
Three-lobe False-mallow (L)
(*Malvastrum coromandelianum*)
Tropical Milkweed (L) (N)
(*Asclepias curassavica*)
Tropical Sage (N)
(*Salvia coccinea*)

Low Herbs (1–2 feet)

Hispid Wedelia (L) (N)
(*Wedelia acapulcensis var. hispida*)
Gregg's Keelpod (N)
(*Synthlipsis greggii*)
Huisache Daisy (N)
(*Amblyolepis setigera*)
Gregg's Mistflower (N)
(*Conoclinium dissectum*)
Runyon's Water-willow (L)
(*Justicia runyonii*)
Texas Abutilon (L)
(*Abutilon fruticosa*)
Trailing Lantana (L) (N)
(*Lantan montevidensis*)

Edging (to 1 foot)

Bristle-leaf Dyssodia (L) (N)
(*Dyssodia tenuiloba*)
Common Dogweed (L) (N)
(*Thymophylla pentachaeta*)
Desert Zinnia (N)
(*Zinnia acerosa*)
Frogfruit (L) (N)
(*Phyla* spp.)
Lindheimer's Tephrosia (L)
(*Tephrosia lindheimeri*)
Powderpuff (L)
(*Mimosa strigillosa*)
Prairie Verbena (L) (N)
(*Glandularia bipinnatifida*)
Scarlet Pea (L)
(*Indigofera miniata*)
Texas Stonecrop (L)
(*Lenophyllum texanum*)
Upright Pipevine (L)
(*Aristolochia erecta*)

Flagstone

Decomposed Granite

Turf

Theona
Checkerspot
(Chlosyne theona)

7 Butterfly Profiles

Following is a small sampling of the many species of butterflies to be found throughout the state. Some, such as the Gray Hairstreak (*Strymon melinus franki*) and the Eastern Black Swallowtail (*Papilio polyxenes asterius*), can be found in gardens during the entire growing season. Others, such as the Falcate Orangetip (*Anthocharis midea*) and Henry's Elfin (*Callophrys henrici*), can be found in only small specialized areas and for only short periods of time each year.

The butterflies shown and described here were chosen because they exhibit some unusual interest, because they show the variability of the state's insects, or simply because they are beautiful. There are many butterflies more common than some described, but space does not allow their inclusion.

No scientific order was followed in the arrangement of species within this section. Instead, they are placed generally by size—beginning with one of the largest butterflies, the Giant Swallowtail (*Papilio cresphontes*), and ending with one of the smallest, the Western Pygmy-Blue (*Brephidium exilis exilis*). In a few instances, similar-appearing species have been placed near one another for easy comparison.

The sets of information at the beginning and end of each description will help explain each butterfly's life cycle.

COMMON AND SCIENTIFIC NAMES: In most instances, the common names are the ones used in *Kaufman Field Guide to Butterflies of North America* by Jim P. Brock and Kenn Kaufman. Scientific names follow *A Catalogue of the Butterflies of the United States and Canada* by Jonathan P. Pelham. In some instances, secondary common names are given (in parentheses) for clarity.

FAMILY: The family shows the butterfly's particular place within the order Lepidoptera.

SIZE: Size of each species of butterfly varies from brood to brood, season to season, and region to region—sizes given here are from actual specimens or general literature.

BROODS: The exact number of broods remains unknown for many species. Even with some of the better-understood species, climatic factors and geography play a major role in the number of broods produced each year. The numbers given here are according to the best information available.

FLIGHT TIME: First months given are when that particular species flies, breeds, and can be expected to be seen on the wing after spending the winter in some other form. Dates given in parentheses ( ) refer to butterflies that spend part of their adult life cycle in hibernation during the coldest days of winter but can be seen flying about on warmer, nonfreezing days. They also indicate the period that particular butterflies spend the winter months in the Rio Grande Valley area.

OVERWINTERS: Resting or overwintering stages are spent in various forms specific to each species of butterfly. These are listed as egg, caterpillar (larva), chrysalis, or adult. The stage given in parentheses ( ) indicates the stage in which the butterfly spends the winter months in the warmer climate of the Rio Grande Valley area.

RANGE: For this book, the state has been divided into seven general regions (see map): High Plains/Rolling Plains (Region 1), Grasslands/Prairies/Savannahs (Region 2), Pine/Hardwoods (Region 3), Coastal Prairies/Marshes/Beaches (Region 4), Chaparral Plains and Rio Grande Valley (Region 5), Edwards Plateau (Region 6), and Trans-Pecos (Region 7). The region in which both the butterfly and its larval food plant(s) can be found is given by number. The first number(s) given is where the butterfly is known to consistently breed; numbers in parentheses ( ) are where the butterflies wander or emigrate and may occasionally breed but are not known to spend the winter in any form. "Throughout" indicates the butterfly breeds throughout the state.

Within each description there is a general, overall view of the butterfly and its characteristics and then the following information:

EGG: A general rather than in-depth description is given here because most eggs are so small that a microscope or strong hand lens is required to examine the features in detail.

CATERPILLAR: A caterpillar goes through several molts (discarding of outer covering) during this cycle, with some of the instars (stage between molts) often appearing quite unlike the previous one. The last instar is described here unless otherwise noted.

CHRYSALIS: The last, practically immobile, stage of the four stages of metamorphosis before the insect emerges as a fully fledged butterfly, and often very beautiful in its own right.

FOOD PLANTS: Some of the plants known to be eaten by the caterpillar (larva).

PARTS EATEN: Portions of plant eaten by the caterpillar, such as flower buds, flowers, young fruits, or foliage.

NOTE: An additional bit of information.

RELATED SPECIES: Butterflies or plants that are similar and may be of interest to the gardener.

Giant Swallowtail

(Papilio cresphontes)

Family: Swallowtail (Papilionidae)
Size: 4–6 inches
Broods: Two or more
Flight time: February–November (all year)
Overwinters: Chrysalis (adult)
Range: Throughout

Of the three largest butterflies found in North America, the Giant Swallowtail is the most common. The brownish-black upper surface is marked by two broad yellow bands of yellow spots meeting near the tip of the forewing; the lower surface is yellow splashed with black bands, veins, and borders. A row of blue iridescent crescents decorates the black band on the hindwing. Tails on the hindwings are long, spoon-shaped, and centered with elongated yellow spots.

Usually a strong, high flier, the Giant Swallowtail is easily enticed down to flowers, drinking from them long and thirstily. It is especially fond of Butterfly Bush (*Buddleja davidii*), Glossy Abelia (*Abelia* × *grandiflora*), and the lantanas (*Lantana* spp.), flying about leisurely from one bush to another. It also partakes of the fluids of mud, fruit juices, and manure.

Caterpillars commonly feed on leaves of trees in the Citrus Family (Rutaceae) and are referred to as an "orange puppy" or "orange dog." It is not hard to see the resemblance to a puppy as the caterpillar lies stretched out on a leaf in the sun. The caterpillar's osmeteria, or scent organs, are red and emit a most unpleasant odor when the caterpillar is handled.

Females are prolific egg producers and will usually lay four or five hundred eggs during their lifetime.

EGG: Usually pale green, but sometimes yellowish to orange; laid singly on new growth, generally near leaf tip; sometimes deposited on branch near young foliage.

CATERPILLAR: Brownish- or greenish-maroon with white or cream markings, a wide, cream-colored band across middle of body, and large white patch on rear; shiny, resembling fresh bird droppings; larva rests exposed on upper or lower sides of leaves or along young branches; makes no "nest" or shelter.

CHRYSALIS: Mottled in white, grays to black, tans to dark brown, usually camouflaged to match strata where attached, usually a piece of wood or bark; head attached upright or horizontally by silken mat and silken thread around middle; may not emerge for one to two years.

FOOD PLANTS: Hercules'-club Prickly-ash (*Zanthoxylum clava-herculis*), Lime Prickly-ash (*Z. fagara*), Downy Prickly-ash (*Z. hirsutum*), Wafer-ash (*Ptelea trifoliata*), and the cultivated grapefruit (*Citrus maxima* × *paradisi*), lemon (*C. limon*), lime (*C. aurantifolia*), sour orange (*C. aurantium*), sweet orange

GIANT SWALLOWTAIL

(*C. sinensis*), garden rue (*Ruta graveolens*), and fringed rue (*R. chalapensis*).
PARTS EATEN: Young to midmature foliage.
RELATED SPECIES: The Thoas Swallowtail (*Papilio thoas*) is difficult to distinguish from the Giant Swallowtail, especially on the wing. It is a rare resident only in the Rio Grande Valley but may occasionally wander throughout the state and as far north as Kansas. Two other similar swallowtails, the Ornythion (*P. ornythion*) and the Broad-banded (*P. astyalus*), are uncommon residents in the Valley area. The Thoas and Giant can best be identified by the yellow spot on the tails of the hindwings; the yellow spots are absent on the Ornythion and Broad-banded.

Eastern Tiger Swallowtail

(Papilio glaucus)

Family: Swallowtail (Papilionidae)
Size: 3½–5⅝ inches
Broods: Two or more
Flight time: February–November
Overwinters: Chrysalis
Range: 2, 3, 4, 5, 6

The Eastern Tiger Swallowtail has a most distinctive and descriptive coloring, primarily a soft, velvety yellow with the forewings displaying four large black stripes trailing downward from the upper or leading margin. The lower wing surface bears a row of several red dots set within a smudgy blue and black marginal band of the hindwing. An interesting dimorphism occurs in Eastern Tiger Swallowtails. Some of the females are black, mimicking the unpalatable Pipevine Swallowtail (*Battus philenor*). The black coloration usually occurs only from female eggs laid by a black mother. Male eggs from the same black mother produce yellow males.

The Eastern Tiger is a strong flier, at times flying very rapidly, at other times slowly soaring with wings at an angle. It often flies high among the trees, frequently along watercourses or forest borders. It visits flowers readily, sometimes barely clutching the flower with the feet and with wings fluttering rapidly, similar to other Swallowtails. At other times it may cling below the nectar source with wings spread flat, thoroughly working the flowers. When nectaring at especially rich sources such as Butterfly Bush (*Buddleja davidii*), Butterfly Weed (*Asclepias tuberosa*), azaleas (*Rhododendron* spp.), or lilacs (*Syringa* spp.), it closes its wings and crawls among the flowers, slowly extracting every minute drop of nectar. It may also occasionally be found

EASTERN TIGER SWALLOWTAIL. MIMIC BLACK COLORATION SHOWN AT RIGHT.

on carrion. Male Eastern Tigers are fond of wet areas and often gather in groups to obtain salts from the moisture. This puddling behavior is practiced by many species of butterflies.

Both the male and female participate in the courtship ritual, with much fluttering and flying about before landing and actually mating. During the courtship flight, the male releases a perfumelike pheromone that acts as an aphrodisiac to calm the female for mating. If the pair is disturbed during copulation, the female flies carrying the male.

EGG: Round, smooth, large for butterfly egg, yellow-green; laid singly on leaf of food plant.

CATERPILLAR: Young larva shiny, mottled in browns, with white saddle across back, much resembling a fresh bird dropping; mature larva green, the head portion enlarged, banded crosswise with a solitary narrow yellow band bordered by a black stripe, and with two yellow and black eyespots; eats at night; during the day rests on silken mat on top of leaf, with leaf edges pulled together to form a tentlike shelter.

CHRYSALIS: Yellowish to greenish or brownish; held upright or horizontally by silken mat and strand of silk, often formed on ground in leaf litter or low on trunk of tree.

FOOD PLANTS: Velvet Ash (*Fraxinus velutina*), Carolina Ash (*F. caroliniana*), Green Ash (*F. pennsylvanica*), White Ash (*F. americana*), Wild Black Cherry (*Prunus serotina*), Choke Cherry (*P. virginiana*), Mexican Plum (*P. mexicana*), Northern Catalpa (*Catalpa speciosa*), Carolina Hornbeam (*Carpinus caroliniana*), Cottonwood (*Populus deltoides*), American Snowbells (*Styrax americanus*), Sassafras (*Sassafras albidum*), Sweet-bay (*Magnolia virginiana*), Tulip Tree (*Liriodendron tulipifera*), Spicebush (*Lindera benzoin*), Wafer-ash (*Ptelea trifoliata*), and the cultivated Camphor Tree (*Cinnamomum camphora*), Fragrant Lilac (*Syringa vulgaris*), and apple (*Malus pumila*) and peach (*Prunus persica*).

PARTS EATEN: Foliage.

NOTE: The earliest known painting of an American butterfly is of the Eastern Tiger Swallowtail, supposedly painted by John White, a member of Sir Walter Raleigh's 1587 expedition to this country. The drawing was published in 1634 as a black-and-white woodcut. The Eastern Tiger Swallowtail is the state butterfly of Alabama.

RELATED SPECIES: The Western Tiger Swallowtail (*Papilio rutulus*) is very similar to the Eastern Tiger, bearing only one tail, but its range barely enters the state around El Paso. The Two-tailed Tiger (*P. multicaudata*) and Three-tailed Tiger (*P. pilumnus*) are similar to the Eastern Tiger in coloring but have more tails. Range of the Two-tailed is mostly in Regions 1, 6, and 7, while the Three-tailed is only in the lowermost portion of Region 5 with occasional sightings in the southwestern portion of Region 7.

Palamedes Swallowtail

(Papilio palamedes)

Family: Swallowtail (Papilionidae)
Size: 4 ½–5 ⅛ inches
Broods: Two or more
Flight time: February–November
Overwinters: Chrysalis
Range: 3, 4, 6

A large, slow-flying denizen of more moist wooded or semishaded gardens, this splendid Swallowtail sails among the shrubbery, casu-

Palamedes Swallowtail

ally taking nectar from many species of flowers. These Swallowtails often congregate in showy numbers where there are good stands of flowers such as Pickerel Weed (*Pontederia cordata*), Yellow Iris (*Iris pseudacorus*), Summer Phlox (*Phlox paniculata*), Argentina Verbena (*Verbena bonariensis*), or azaleas (*Rhododendron* spp.).

There are several Swallowtails with similar coloring and markings, but the Palamedes is distinguished by its large size, the upper surface of the hindwings bearing an unbroken yellow band, and a long, creamy-yellow to orange-red stripe on the lower surface of the wings along the base and parallel to the body. This stripe is large and easily seen as the insect is feeding, for usually in taking nectar, it does not rest on the flower but hovers above the blossom with fluttering wings, lightly grasping the petals with its feet. Another identifying marking is the large abdomen conspicuously striped in black and yellow.

EGG: Smooth, pale yellowish or greenish; laid singly on the undersurface of young leaf.

CATERPILLAR: Young larvae blotched brown and white, shiny; mature larvae green on upper portion of body, velvety brown or buff on lower side; portion of body behind head enlarged and with two black eyespots circled in orange, two smaller yellow spots behind these "eyes," rest of body covered with small blue dots circled in black; rests in a shelter constructed by making a bed of silk along the midrib on upper surface of a leaf and drawing the sides together.

CHRYSALIS: Mottled greenish; held upright by silken pad and silk thread around body.

FOOD PLANTS: Red Bay (*Persea borbonia*),

Sassafras (*Sassafras albidum*), Sweet-bay (*Magnolia virginiana*), and the cultivated avocado (*Persea americana*).
PARTS EATEN: Foliage.
NOTE: In some areas Palamedes Swallowtails roost communally at night with several taking shelter together high in the branches of tall Red Bay trees. More commonly associated with the swampy, steamy Big Thicket area, this butterfly can also be found in rare, local sites of the dry, rocky Hill Country and along the salt-laden Gulf Coast.

Pipevine Swallowtail

(Battus philenor)

Family: Swallowtail (Papilionidae)
Size: 2 ¾–5 inches
Broods: Five to six
Flight time: March–November (all year)
Overwinters: Chrysalis (adult)
Range: Throughout

The Pipevine is one of the most common, plainer, or less marked of the Swallowtails. The upper surface of the forewings is a soft velvety black, and the hindwings are overlaid with metallic blue or turquoise scales. The metallic sheen on the wings of the male is usually brighter and covers a larger area than on the female. Two rows of very narrow, somewhat inconspicuous, creamy to yellow dashes rim the margins of both wings but are usually more prominent on the hindwings. The lower surface of the hindwing has a wide row of large orangish-red dots bordered with bands of metallic blue scaling.

Retaining some of the poisonous properties of its larval food plant, this butterfly is most unpalatable to its bird and lizard predators. Three other Swallowtails—the Spicebush (*Pterourus troilus*), females of the Eastern Black (*Papilio polyxenes asterius*), and black females of the Eastern Tiger (*P. glaucus*)—along with the Red-spotted Purple (*Limenitis arthemis*), have evolved their coloring to closely resemble the Pipevine, obtaining protection through this mimicry. The dark-phased female of the Eastern Tiger Swallowtail is the closest in coloring and markings, but the Eastern Tiger can instantly be recognized by the bright orange spots on the upper surface of the hindwings near the tip of the abdomen, which are lacking in the Pipevine. Scent scales of the male Pipevine Swallowtail are in a slender, pocketlike depression along the inner edge of the hindwings and are quite noticeable on close inspection. Like the Monarch's (*Danaus plexippus*), the body of the Pipevine Swallowtail is very tough, enabling it to survive after being bitten or "tasted." As an added protection, glands within the orangish-red dots along the abdomen emit an acrid, unpleasant odor when the insect is molested.

PIPEVINE SWALLOWTAIL

The Pipevine Swallowtail has a swift flight but stops often to visit flowers. This butterfly seems to prefer pink to purple hues, although yellows and oranges are almost as readily used. Some of its favorite early-spring nectar sources are fruit tree blossoms, Butterfly Bush (*Buddleja davidii*), and azaleas (*Rhododendron* spp.). Summer will find it nectaring on Globe Amaranth (*Gomphrena globosa*), Summer Phlox (*Phlox paniculata*), Wild Bergamot (*Monarda fistulosa*), Butterfly Weed (*Asclepias tuberosa*), Mexican Sunflower (*Tithonia rotundifolia*), and various verbenas (*Verbena* spp.), lantanas (*Lantana* spp.), and liatris (*Liatris* spp.). Favorite fall sources include Joe-Pye Weed (*Eutrochium fistulosum*), Golden Crownbeard (*Verbesina encelioides*), and ironweed (*Vernonia* spp.).

During the day male Pipevine Swallowtails readily seek mud puddles and spend much time there.

EGG: Spherical, large, reddish-brown; laid singly or in small clusters of up to twenty on underside of host leaf, along leafstalk, or occasionally along stem of plant. Although each grouping usually consists of only a few eggs, almost every leaf and stem of the plant may be utilized.

CATERPILLAR: Young, brownish-colored larvae remain in small groups during the first two or three instars, dispersing as they become older; mature larva soft-bodied, ranging in color from almost orange to dark maroon; long, floppy tubercles extend along body, with two much longer ones on segment of body just behind head; feeds at night and during the day; makes no shelter.

CHRYSALIS: Greenish beneath with mottled yellowish patches on top portion; supported upright or horizontally from silken mat and silk strand around body.

FOOD PLANTS: All native species of *Aristolochia* are used as well as most cultivated spec0ies, especially Fringed Pipevine (*A. fimbriata*). The female occasionally lays eggs on the cultivar Elegant Pipevine (*A. elegans*), but young larvae will die on it. In the last instars (or almost-mature stages), the larvae can survive on this plant, although they do not like it much.

PARTS EATEN: Foliage or, in some instances, the entire aboveground portions of food plant.

NOTE: Large, loose groups of Pipevine Swallowtails often roost at night in a tree or large-leaved shrub, hanging from branches and undersides of leaves.

Eastern Black Swallowtail

(Papilio polyxenes asterius)

Family: Swallowtail (Papilionidae)
Size: 3 ¼–4 ¼ inches
Broods: Three or more
Flight time: February–November
Overwinters: Chrysalis
Range: Throughout

The Eastern Black Swallowtail is adapted to many situations within open country. This is a common butterfly of fields and meadows, cultivated farmland, parks, golf courses, and flower gardens. It seems to show no preference between dry uplands and moist marshes, as long as the area is open and not wooded.

A great lover of flowers, it likes to lazily drift among the plants, taking nectar and pausing frequently to bask with wings fully outspread. It is especially attracted to gardens that have both plentiful flowers and

Eastern Black Swallowtail

good stands of its favored cultivated larval food plants. Favored nectar plants include blossoms of fruit trees such as apple (*Malus pumila*) and peach (Prunus persica), along with various thistles (*Cirsium* spp.), zinnias (*Zinnia* spp.), and lantanas (*Lantana* spp.), Joe-Pye Weed (*Eutrochium fistulosum*), Butterfly Weed (*Asclepias tuberosa*), Summer Phlox (*Phlox paniculata*), and Blue Mistflower (*Conoclinium coelestinum*).

Female Eastern Black Swallowtails closely mimic the Pipevine Swallowtail (*Battus philenor*), gaining an advantage from the Pipevine's toxicity. Upper wing surfaces are mostly black with one or two rows of creamy to bright yellow spots bordering both wings, especially those of males. These bands of yellow spots are sometimes absent or very faint in the female. The rows of spots are separated on the hindwings by a narrow wash of metallic blue. Lower wing surfaces of both sexes are similar to those of the Spicebush Swallowtail (*Pterourus troilus*), being black with two rows of orange-red spots separated by a band of widely spaced smudged blue spots. A solitary large, black-centered red or orangish spot near the outer angle of each hindwing occurs on both upper and lower surfaces. Rows of small yellow dots line the black abdomen.

Eastern Black Swallowtails are often seen in moist areas or around mud puddles, methodically sucking up the moisture with its accumulated salts. To seek mates, a male patrols a chosen area or occasionally claims a perching place from which to fly out to inspect whatever passes by. He changes his perching site frequently, usually not using the same space more than two or three days. The female flies to a hilltop (or to the highest terrain around) to mate, with a male usually in pursuit. Once at the mating ground, the male and female flutter near one another briefly, then land, where they copulate. If the female lives longer than a week, she often mates a second time. The number of broods produced often depends on the availability of larval food plants.

EGG: Round to somewhat oval, smooth, cream to yellowish; laid singly on a flower bud or leaf of food plant.

CATERPILLAR: Young larva may be various shades of brown, or perhaps black and white with a wide white saddle across back; mature larva primarily pale green to whitish with wide, crosswise bands of black, the bands interspersed with yellow dots or slashes; larvae do not eat cast skin after molting, as is common with many species.

CHRYSALIS: In summer can be of camouflaging shades of green or tannish mottled with darker browns and blacks; overwintering fall form usually of camouflaged mottled dark brown or blackish; two short projections on head; supported upright from silk mat and silk strand around body.

FOOD PLANTS: Nuttall's Mock Bishop's-weed (*Ptilimnium nuttallii*), Ribbed Mock Bishop's-weed (*P. costatum*), Thread-leaf Mock Bishop's-weed (*P. capillaceum*), Spreading Scale-seed (*Spermolepis inermis*), Texas Dutchman's-breeches (*Thamnosma texana*), and the cultivated common fennel (*Foeniculum vulgare*), dill (*Anethum graveolens*), and parsley (*Petroselinum crispum*) seem to be favored food plants. Other natives less commonly used include Nuttall's Prairie Parsley (*Polytaenia nuttallii*), Anise-root (*Osmorhiza longistylis*), Forked Scale-seed (*Spermolepis divaricata*), Queen Anne's Lace (*Daucus carota*), Rattlesnake Weed (*D. pusillus*), Stalked Berula (*Berula erecta*), Spotted Water-hemlock (*Cicuta maculata*), Water Parsnip (*Sium suave*), Wild Celery (*Apium graveolens*), Wild Chervil (*Cryptotaenia canadensis*), Yellow Pimpernel (*Taenidia integerrima*), and the cultivated garden rue (*Ruta graveolens*), fringed rue (*R. chalapensis*), and parsnip (*Pastinaca sativa*).
PARTS EATEN: Flowers and immature seedpods preferred but will consume almost all aboveground parts of plant except tough stems.
NOTE: Queen Anne's Lace is often listed as a food source (and is included here), but I have yet to find a caterpillar on this rough plant.

Spicebush Swallowtail

(Pterourus troilus)

Family: Swallowtail (Papilionidae)
Size: 3–4 inches
Broods: Two or more
Flight time: March–November
Overwinters: Chrysalis
Range: 2, 3, 4, 6

The Spicebush is a beautiful butterfly with a majestic flight, covering ground quite rapidly when the need arises. Ordinarily, though, it flies about with an up-and-down motion, frequently sailing with motionless, outspread wings and veering around as if guided by the tails. It is usually quite wary, whether in flight or feeding, and not easily approached.

Often seen about edges of woodlands, along banks of streams, and in open fields and flower gardens, the Spicebush Swallowtail seeks out the lower-growing flowers. All the Swallowtails seem to prefer flowers that form heads or clusters, such as Butterfly Bush (*Buddleja davidii*), Joe-Pye Weed (*Eutrochium fistulosum*), Summer Phlox (*Phlox paniculata*), various lantanas (*Lantana* spp.), and liatris (*Liatris* spp.). When a Swallowtail finds choice plants, it remains for long periods, going from one flower cluster to another. While feeding, it continues to flutter its wings, lightly grasping the flowers with the feet. It has been suggested that the Swallowtails do this in order not to tip the flower with their heavier bodies and perhaps spill the nectar.

Spicebush Swallowtails are also great moisture lovers and frequently congregate

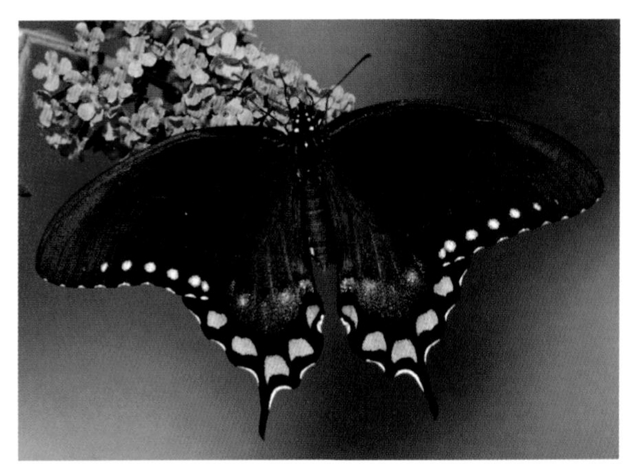

SPICEBUSH SWALLOWTAIL

around mud puddles and patches of moist dirt or along the edge of a stream.

In coloration this butterfly mimics the unpalatable Pipevine Swallowtail (*Battus philenor*), but the Spicebush Swallowtail differs in having a more complete row of large white or pale bluish spots along the edges of both fore- and hindwings on the upper surface (dots on the Pipevine Swallowtail are very small or nonexistent). Also, the iridescent sheen of the hindwings is reduced to a wide band on the Spicebush. In the female this sheen is a lovely blue; in the male it has a distinct greenish cast. A prominent round spot of orange-red occurs on the trailing margin of each hindwing near the outer angle, but these can rarely be seen in flight. There are two rows of orange or red dots on the lower surface of the hindwing, instead of one row as on the Pipevine Swallowtail.

EGG: Round, smooth, greenish-white; laid singly on underside of food plant leaf.

CATERPILLAR: Green above, with pale tan or grayish-beige beneath; rows of small blue dots cover body; two large yellow or orange eyespots on enlarged, "humped" portion of body behind the head, one pair the largest and with large black dots forming pupil of the "eye"; mature caterpillars large; larvae feed at night and rest during the day on a silken mat on upper surface of a leaf with sides of leaf pulled together.

CHRYSALIS: Mottled greenish to tannish or various browns; two short projections on head; attached upright, horizontally, or occasionally head downward, from silken mat and silken thread around body.

FOOD PLANTS: Red Bay (*Persea borbonia*), Sassafras (*Sassafras albidum*), Sweet-bay (*Magnolia virginiana*), Tulip Tree (*Liriodendron tulipifera*), Spicebush (*Lindera benzoin*), and the introduced Camphor Tree (*Cinnamomum camphora*).

PARTS EATEN: Foliage.

Zebra Swallowtail

(Eurytides marcellus)

Family: Swallowtail (Papilionidae)
Size: 2 ½–4 inches
Broods: Three or more
Flight time: February–November
Overwinters: Chrysalis
Range: 3

Descriptively named the "Zebra" Swallowtail from its distinctive coloration, this butterfly is the only native representative in North America of a group known as the "Kite" butterflies because of their long, slender tails. The triangular-shaped wings are sharply striped in black and greenish-white, beautifully accented with red and blue dots on both surfaces of the wings near the very long tails.

Zebra Swallowtail

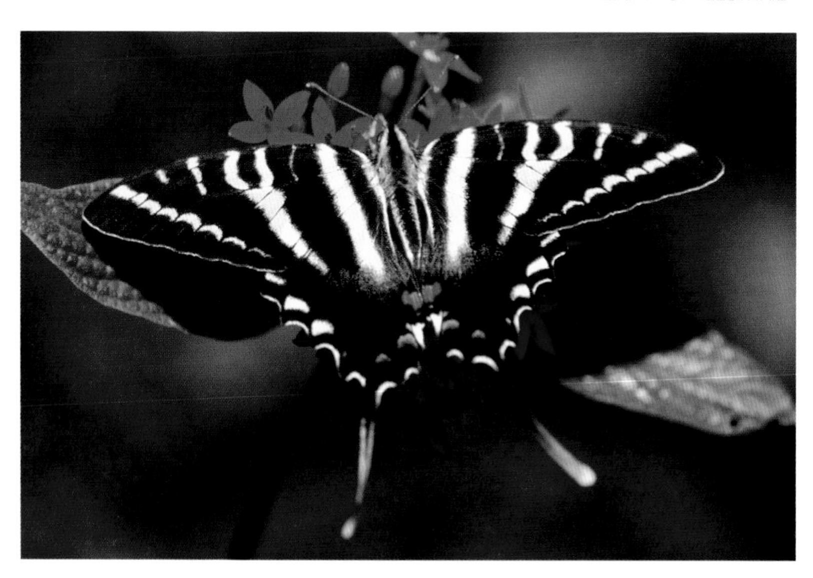

A brilliant red stripe occurs about midway on the lower hindwing.

This beauty remains low to the ground in flight. Although it flaps its wings slowly, because of its size it moves very quickly and with an erratic, bobbing motion. It does not remain long in one place but flits from flower to flower, sipping briefly. Males are especially fond of congregating in groups at mud puddles and will remain at some length, sipping the enriched moisture.

Zebras show a variation of color during the season. Brood relationships are extremely complex and in the past have been given different names by various authors. Early-spring forms have short wings and short tails and are the palest in coloration; this brood is the most numerous. Summer forms are larger, with darker coloring and longer wings and tails. Butterflies that emerge latest in the season are the largest and darkest of all. Some larvae from each brood hibernate until the next year, causing still more confusion in the coloring of those seen on the wing.

There is no mimicry involved with this species, and although they are palatable to all forms of predators, their worst enemy seems to be spiders, both the ones that lie in wait on flowers and the web builders.

This butterfly does not seem to adapt very well to urban development, needing wilderness conditions for breeding. However, if some larval food plants are provided in gardens near natural stands of pawpaw (*Asimina* spp.), the female readily uses them along with the wild plants for depositing eggs.

Zebra Swallowtails rarely emigrate, usually remaining in the area of the earlier stages of their life cycle.

EGG: Pale green; deposited singly on underside of food plant leaf.
CATERPILLAR: Pale green with crosswise rows of tiny black dots and narrow bands of pale yellow and blue; wider and darker bands of yellow, black, and blue occur on larger, "humped" portion of abdomen directly behind the head; some larvae may be mostly black, with narrow orange and white lines across the body and with the head solid black; rests on underside of a leaf or at base of plant; makes no shelter.
CHRYSALIS: Brownish to blackish, mottled in various tans, browns, and blacks, resembles piece of wood; rests upright from silken mat and silken strand around middle.
FOOD PLANTS: Tall Pawpaw (*A. triloba*) and Dwarf Pawpaw (*A. parviflora*).
PARTS EATEN: Foliage.
NOTE: Both the male and female often use the underside of pawpaw leaves for roosting at night and resting during inclement weather.

Arizona Sister (Eulalia Sister)

(Adelpha eulalia)

Family: Brush-footed (Nymphalidae)
Size: 3–5 inches
Broods: Two or more
Flight time: April–October
Overwinters: Caterpillar
Range: 1, 2, 6, 7 (5)

Usually seen only when gliding high among oak trees (*Quercus* spp.), the Arizona Sister spends much time patrolling moist canyons, gullies, or oak-covered hillsides and woodlands while seeking mates. They are fond of basking and frequently alight on sun-drenched rocks to soak up the much-needed warmth. While not an avid nectar feeder, if some rich source such as Buttonbush (*Cephalanthus occidentalis*), Texas Colubrina (*Colubrina texensis*), Butterfly Bush (*Buddleja davidii*), or Sweet Almond Verbena (*Aloysia virgata*) is sighted in the garden, they visit often and partake deeply of the sweet liquid. They also readily take nutrients from moist soils, fruit, and aphid honeydew.

The Arizona Sister is one of the largest and most gorgeous of our western species, with upper wings velvety dark brownish-black, the forewing with three large white spots and a large, startling orange-red patch near the tip. Hindwings are marked with a wide, slanting white band, tapering to a point near the lower edge of the wing. Below, the wings are of soft, muted colors, appearing "washed" or somewhat indistinct. A group of two short blue bands and one orange band occurs on upper wings, while the lower bears a distinct row of blue crescents along the margin and an interrupted blue line bordering the wide white band.

EGG: Surface of small, six-sided cells and covered in minute hairs, pale green; laid singly on margin of leaf near base of spine.

CATERPILLAR: Well-camouflaged, dark green on upperside and along sides, olive-brown beneath, or sometimes yellow-orange; both forms with four to six pairs of green, spiny, elongated knobs, the pair near head the longest; forms resting perch by extending frass pellets with silk from tip of leaf.

CHRYSALIS: Head with two short horns extending forward, pale brown or tannish, with fine, brown and metallic gold streaks and markings; hangs downward from silken mat.

FOOD PLANTS: The oaks, Bur (*Q. macrocarpa*), Emory (*Q. emoryi*), Gambel (*Q. gambelii*), and Plateau Live Oak (*Q. virginiana* var. *fusiformis*).

ARIZONA SISTER
(EULALIA SISTER)

PARTS EATEN: Young foliage on young trees or "sprouts."

NOTE: The Arizona Sister has a curious habit of finding formerly moist areas or places where some chemically rich substance once existed, then spitting a fluid on the area. By doing so, micro particles of nutrients are dissolved. The butterfly then "drinks" or "sucks up" the enriched moisture.

Monarch

(Danaus plexippus)

Family: Brush-footed (Nymphalidae)
Size: 3 3/8–4 5/8 inches
Broods: One to three
Flight time: March–May and September–December
Overwinters: Adult in Mexico
Range: Throughout

Monarch *(female)*

The Monarch is probably the best-known and most easily recognized butterfly in North America. The upper surface of the wings is a rich burnt-orange with black veins and borders, the borders liberally sprinkled with a double row of small white dots. The male also has a conspicuous black hindwing dot, which is a cluster of special scent scales. From these scales the male can emit a strong fragrance that attracts nearby females for mating. Females lack these black scent scales and are generally a little darker in color. The lower wing surface of both sexes is paler, duskier orange, with a black, white-dotted marginal border.

Moving in a slow, rather deliberate, soaring flight, Monarchs begin to return from their Mexico wintering grounds in early March, just as milkweeds (*Asclepias* spp.), the larval food plants, are showing fresh young growth. Females have mated before leaving their wintering grounds, and by the time they reach Texas, they are ready to begin depositing their eggs. A few males may be intermingled with the females, but they will soon die with few if any making a very long journey northward.

Monarch larvae feed on many species of *Asclepias* as well as some other genera of the Milkweed Family. Most members of this family are poisonous; the plants contain cardiac glycosides, or heart poisons. These chemicals are carried over from the larval stage to the adult butterfly, especially in the female, making the insect unpalatable to predators. However, some species of *Asclepias* and other related genera have very low toxin concentrations, leaving the adult butterfly relatively unprotected. Given a choice, a female Monarch finds the most poisonous plants on which to lay her eggs.

During the breeding season, the Monarch's favorite habitats are open fields, meadows, and flower gardens. It is a great lover of flower nectar and freely visits many different species. In the fall it seems especially fond of Blue Mistflower (*Conoclinium coelestinum*), Crucita (*Chromolaena odorata*), Virginia Frostweed (*Verbesina virginica*), Giant Ironweed (*Vernonia gigantea*), Mexican Sunflower (*Tithonia rotundifolia*), and various lantanas (*Lantana* spp.), goldenrods (*Solidago* spp.), and verbenas (*Verbena* spp.). In many instances, milkweed serves as both the larval food source and favored nectar source. Butterflies are one of the pollinators of milkweed, as can be observed from the not infrequent presence of pollen bundles dangling from their legs.

The Monarch has a strong, powerful flight and moves among flowering plants with much deliberation. Its sight is exceptional, and it is not easily approached, especially in the spring. On the southward migration during autumn, Monarchs are often tired and anxious to feed, so they can be observed more closely at this time.

EGG: Short, cone-shaped, with many lengthwise ridges and crosslines, creamy to pale green; usually laid singly on undersurface of leaf; if plant is large, more than one leaf often used.

CATERPILLAR: Conspicuously striped crosswise with narrow black, yellow, and white bands, and with two long, black, threadlike, fleshy filaments near the head and two shorter filaments near the rear; makes no shelter.

CHRYSALIS: Rounded, bluish-green with row of small gold dots and narrow bands of black and gold near top; hangs downward from silken pad.

FOOD PLANTS: Various members of the Milkweed Family, especially Antelope-horns Milkweed (*A. asperula*), Green Milkweed (*A. viridis*), Prairie Milkweed (*A. oenotheroides*), and the cultivated Tropical Milkweed (*A. curassavica*). Probably many other species are used, as well as Climbing Milkweed Vine (*Funastrum cynanchoides*) and Net-leaf Milkvine (*Matelea reticulata*). Butterfly Weed (*A. tuberosa*) is often listed as a food plant, but because of its roughness and a very low concentration of toxins, it is less commonly used.

NOTE: The fall Monarch migration is a familiar sight to most people, especially to those traveling during this period. The insects commonly move southward with cold fronts, similar to the movements of geese. In some areas at certain times, there are continuous, seemingly unending lines often a mile or more wide.

During these autumn movements, Monarchs spend the nights roosting in trees in large groups. Each evening near sunset they begin dropping from the sky to settle on lower tree leaves and branches. Incredibly, the same trees are used year after year. This phenomenon is as unexplainable as the Monarchs' eventual return to exactly the same spot and Oyamel Fir (*Abies religiosa*) trees where their ancestors had spent the previous winter in Mexico.

The Monarch is the official insect of Texas, listed in 1995.

RELATED SPECIES: The more coastal-occurring Queen (*Danaus gilippus thersippus*) and Soldier (*D. eresimus*) also feed on milkweed and related plants in the larval stage and are toxic to most predators. They do not migrate but spend the winter in the pupal or adult stage. The nonpoisonous Viceroy (*Limenitis archippus*), which belongs to another family and uses entirely different food plants in the larval stage, has evolved coloration similar to the Monarch's and thereby gains even more protection.

Viceroy

(Limenitis archippus)

Family: Brush-footed (Nymphalidae)
Size: 2 ½–3 ⅛ inches
Broods: Two or more
Flight time: April–November (all year)
Overwinters: Third instar caterpillar (adult)
Range: Throughout

Of the same genus as the Red-spotted Purple (*Limenitis arthemis*), which mimics the Pipevine Swallowtail (*Battus philenor*) for protection, the Viceroy is totally different in coloration and here in Texas, mimics the poisonous Monarch (*Danaus plexippus*). In the southeastern states it often mimics the various color phases of the Queen (*D. gilippus thersippus*), which is also toxic and distasteful and often may be more common than the Monarch. General coloration of the Viceroy, above and below, is a rich, russet-orange with conspicuously wide, black venation. A distinctive black line curves across the lower wings above the black-bordered margins. Both wings are bordered in wide, white-dotted black bands with a group of white dots near the tips of the forewings.

The Viceroy is very fond of a wide variety of flowers but is especially attracted to white-flowered ones such as Buttonbush (*Cephalanthus occidentalis*), Mountain-mint (*Pycnanthemum incanum*), and the fall-flowering vine Climbing Hempweed (*Mikania scandens*). While nectaring, the Viceroy usually keeps its wings partially expanded, differing from the Monarch, which usually feeds with its wings closed. Taking advantage of the sun's warmth, Viceroys can frequently be seen basking with half- or fully opened wings.

Flight consists of a series of rather rapid wing beats alternating with a period of gliding, enabling the insect to cover ground in a slow, erratic pattern. This butterfly prefers open fields, meadows, and sunny gardens but likes a bit of moisture. It is one of the most commonly seen species along open, sunny stream banks or along edges of marshes if there are flowering plants around. Not only does the Viceroy take nectar readily but it also sips moisture from sap, mud, rotting wood, fungi, dung, and insect honeydew.

EGG: Dome-shaped, flattened on bottom, covered in minute protrusions, pale green or

VICEROY

yellow; laid singly, usually on upper side of tips of young leaves of food plant; deposition in midafternoon.

CATERPILLAR: Mature larva mottled brown and olive, with creamy white saddle patch on the back, shiny, resembling fresh bird dropping; region behind head enlarged or "humped" with two short, feathery black horns.

CHRYSALIS: Mottled brownish and whitish, conspicuous rounded disk projecting from abdomen; first brood(s) hangs downward from silken mat; last (fall) brood spends winter in rolled-up leaf attached to food plant tree.

FOOD PLANTS: Black Willow (*Salix nigra*), Carolina Willow (*S. caroliniana*), Sandbar Willow (*S. exigua*), Cottonwood (*Populus deltoides*), Wild Black Cherry (*Prunus serotina*), the cultivated Weeping Willow (*S. babylonica*), Silver-leaf Poplar (*Populus alba*), apple (*Malus pumila*), and pear (*Pyrus communis*). Probably all species of willow are used as well as many other plants not listed here.

PARTS EATEN: First spring larvae often feed at night on catkins (inflorescences) of some tree species; later in the season larvae eat tips of leaves, preferably young ones.

NOTE: The Viceroy was previously thought to have derived all its protection from imitating the coloration of other poisonous species, but it is now known that willow, the preferred food plant of the Viceroy, also contains poisonous compounds, thus making the Viceroy somewhat unpalatable in its own right.

There are two subspecies in Texas—*L. a. watsoni* from the east to Central Texas and *L. a. obsoleta* in far West Texas (Big Bend area).

Red-spotted Purple

(Limenitis arthemis)

Family: Brush-footed (Nymphalidae)
Size: 2 ¼–4 inches
Broods: One to three
Flight time: March–November
Overwinters: Third instar caterpillar
Range: 2, 3, 4, 6, 7

Commonly seen flying along forest edges, woodland paths, and water courses, the Red-spotted Purple is a great lover of flowers and visits parks and gardens where good nectar sources are available. Not being particular about food choices, it can just as readily be seen feeding on sap, fruit, decaying wood or fungi, insect honeydew, dung, or dead animals.

When mating, the male does not patrol but chooses a perch in the open on trees or tall bushes and waits for the female to fly by.

Similar to the Pipevine Swallowtail (*Battus philenor*) in its plainness and lack of conspicuous markings, the Red-spotted Purple

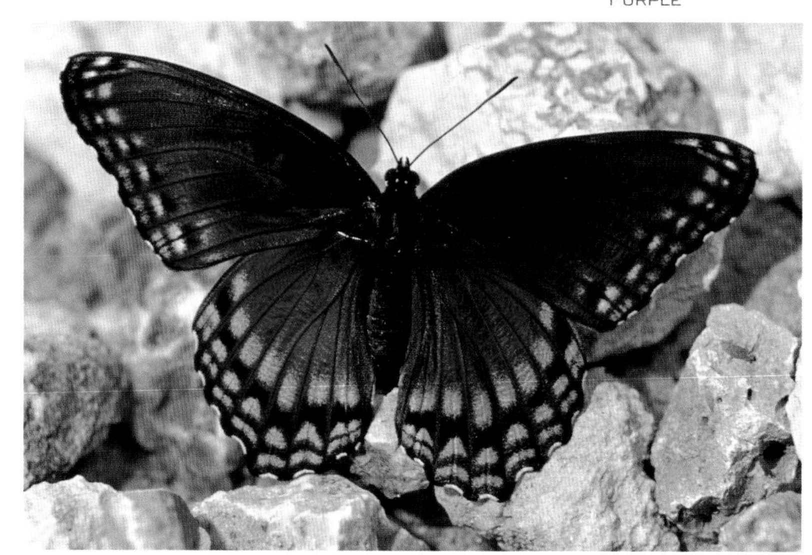

RED-SPOTTED PURPLE

is nevertheless very striking with a beautiful iridescent sheen of blue across the outer portion of the black wings on the upper surface. The blue is a bit darker and most noticeable on the basal portion of the hindwings. Faint, iridescent reddish patches near the tips of the forewings are visible under certain lighting. The undersurface is distinguished by a cluster of red markings near the body in addition to a row along the outer white- and black-banded margins. When trying to separate the two in the field, note the absence of tails on the Red-spotted Purple.

Even though it is in an entirely different family, the Red-spotted Purple has evolved to mimic the poisonous Pipevine Swallowtail and thus obtains protection from birds and other predators.

EGG: Dome-shaped, flattened on bottom, pitted, covered in minute protrusions, grayish- to pale green; laid singly on upperside tip of young leaves of food plant.

CATERPILLAR: Brownish, reddish, or greenish with white or cream saddle patch on the back, warty; area behind the head white, enlarged or "humped," and bearing two small, brushlike horns or bristles; eats both night and day; rests on top of leaf or along stem; makes no shelter.

CHRYSALIS: Yellowish-brown, mottled in dark greens and grays; hangs downward from silken mat.

FOOD PLANTS: Cottonwood (*Populus deltoides*), Carolina Hornbeam (*Carpinus caroliniana*), Eastern Hop-hornbeam (*Ostrya virginiana*), Wild Black Cherry (*Prunus serotina*), Choke Cherry (*P. virginiana*), Deerberry (*Vaccinium stamineum*), and the cultivated Silver-leaf Poplar (*Populus alba*), apple (*Malus pumila*), and pear (*Pyrus communis*).

PARTS EATEN: Foliage, preferably young or immature.

RELATED SPECIES: There are two subspecies in the state, with the eastern subspecies shown here (*L. a. astyanax*) ranging west as far as Real County and the westernmost subspecies (*L. a. arizonensis*) occurring west of the Pecos River.

Malachite

(Siproeta stelenes biplagiata)

Family: Brush-footed (Nymphalidae)
Size: 3 ¼–4 inches
Broods: Two or more
Flight time: October–December (all year)
Overwinters: Adult
Range: 5 (4, 6, 7)

A freshly emerged Malachite is one of the most beautiful of butterflies. Unfortunately, the green coloring of the wing scales fades in the sunlight and is not nearly so brilliant within a day or so after emergence. There

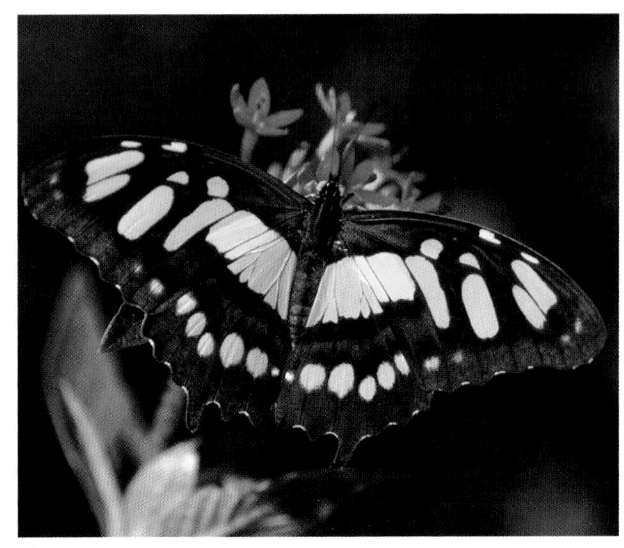

MALACHITE

is no mistaking the Malachite for any other, though, for the green and brownish markings are very distinctive. The dark chocolate or brownish-black on the upper wing surface is richly spotted and banded with dark jade or emerald-green. The lower wing surface is beautifully patterned in tawny-orange or rusty-brown, with large spots and wide bands of a lighter, pearly-green. A short but prominent tail is present on the hindwings. Females are usually paler in coloring than the males.

Malachites frequent flower gardens and can be seen even in cool weather visiting favorite nectar sources. For this beauty, have an abundance of Barbados Cherry (*Malpighia glabra*), Crucita (*Chromolaena odorata*), Queen's Wreath (*Antigon leptopus*), White-flowered Plumbago (*Plumbago scandens*), Tropical Sage (*Salvia coccinea*), lantana (*Lantana* spp.), and verbena (*Verbena* spp.) in the garden. In Mexico one of the best flowers for nectaring is Male Mujer (*Cnidoscolus palmeri*), a member of the Spurge Family (*Euphorbiaceae*). It is closely related to the more familiar Bull Nettle (*C. texanus*) and, like the Bull Nettle, is clothed with stinging hairs. This butterfly also makes use of various other substances as food sources, such as mud, dung, rotting fruit, and rotting leaf litter. At night, several adults roost communally on the undersides of leaves and branches on low shrubbery.

In seeking females with which to mate, the male perches on vegetation and waits for passing females or will occasionally patrol an area, slowly flying back and forth.

EGG: Dark green, laid singly or in small groups of two or three on lower surface of very young food plant leaves or plant seedlings.

CATERPILLAR: Head with two long, red or black horns curving backward; body velvety black with many branching spines.

CHRYSALIS: Two small horns on head, two rows of short, gold, spikey tubercules on body, pale yellowish-green, covered with powdery "bloom"; hangs downward from silken mat.

FOOD PLANTS: Violet Ruellia (*Ruellia nudiflora*), American Water-willow (*Justicia americana*), Runyon's Water-willow (*J. runyonii*), Hooker's Plantain (*Plantago hookeriana*), Red-seed Plantain (*P. rhodosperma*), Virginia Plantain (*P. virginica*), and Green Shrimp Plant (*Blechum pyramidatum*). Possibly Downy Water-willow (*J. pilosella*) and probably many other species of Ruellia are used.

PARTS EATEN: Foliage.

Mourning Cloak

(Nymphalis antiopa)

Family: Brush-footed (Nymphalidae)
Size: 2 ¼–3 ¾ inches
Broods: One
Flight time: January–May and September–December
Overwinters: Adult
Range: Throughout

Largest and most striking of the Anglewings, the Mourning Cloak has an extraordinary range, covering most of the Northern Hemisphere from Alaska to South America and straying to England, where it is known as the Camberwell Beauty.

The coloring and pattering of this butterfly are distinctive, the wings primarily a rich brownish-maroon overlaid with a shimmering purplish sheen. Wing margins are conspicuously angled and rimmed in a pale, velvety

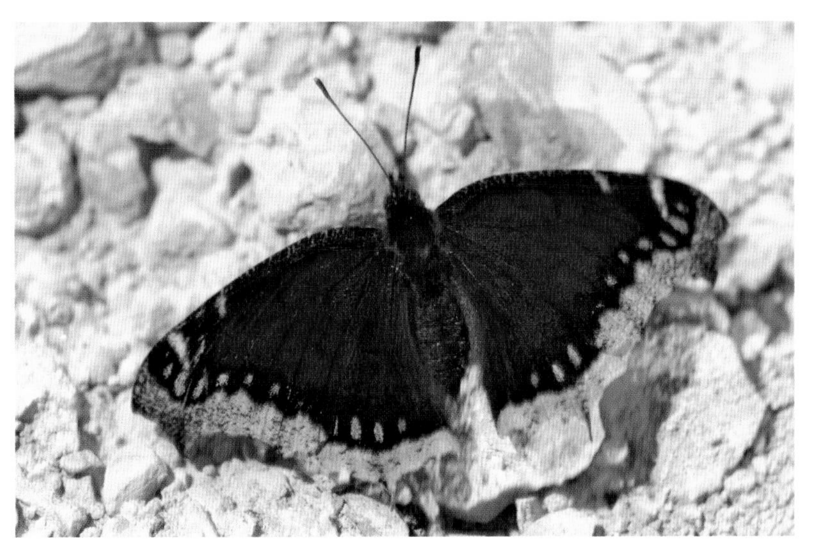

Mourning Cloak

yellow border with an inner row of large, brilliant blue spots. The undersurface of the wings is dull, grayish-black with an irregular, dirty cream or pale yellowish border.

Its flight is strong and erratic, but the insect will often remain on the ground until almost stepped on, then leap into the air in a circling flight, only to settle down again close to its original position. This flight is often accompanied by a rattling noise and then a loud conspicuous click as it closes its wings on alighting. The Mourning Cloak has a habit of resting on tree trunks or posts, head downward with wings closed over the back, but often basks on the ground or on low shrubbery in the sun with wings expanded. When resting on the ground with wings closed, it is almost impossible to see.

Mourning Cloaks are one of the hibernators but can often be seen flying around on sunny days during the winter even though there may be snow on the ground. They often come forth for a bit of tree sap and then retreat to a crack or crevice until the next "warm" day. They are one of the first butterflies to visit sap flows in the early spring, and trees that have been "tapped" by the Yellow-bellied Sapsucker (*Sphyrapicus varius*) are especially favored feeding sites. Mourning Cloaks generally are not frequent visitors to flowers, preferring open woodlands, but they will occasionally be seen around Frostweed (Verbesina microptera), Tooth-leaved Goldeneye (*Viguiera dentata*), or asters (*Symphotrichum* spp.), usually later in the year. They are especially fond of nectar from the flowers of various milkweeds (*Asclepias* spp.) and goldenrods (*Solidago* spp.). During the hot days of summer, Mourning Cloaks estivate, remaining dormant until fall. With cooler days they emerge to feed and store fats for winter hibernation.

EGG: Whitish, becoming darker before hatching; laid in one layer of two hundred or more, forming a wide band around twig of food plant tree.

CATERPILLAR: Velvety black with raised white dots and several rows of branching black spines or bristles on the body and with row of red spots along the back; caterpillars remain in groups until pupation, raising their heads and thrashing about in unison if disturbed; often leave the feeding area en masse to pupate.

CHRYSALIS: Mottled pale brown to grayish-black, covered with bluish-gray, powdery "bloom"; short "horns" on head, many tubercles on body; hangs downward from silken mat.

FOOD PLANTS: Black Willow (*Salix nigra*), Carolina Willow (*S. caroliniana*), Sandbar Willow (*S. exigua*), Arizona Cottonwood (*Populus fremontii*), Cottonwood (*P. deltoides*), Carolina Basswood (*Tilia americana*), Ameri-

can Elm (*Ulmus americana*), Slippery Elm (*U. rubra*), Western Hackberry (*Celtis occidentalis*), Net-leaf Hackberry (*C. laevigata* var. *reticulata*), Red Mulberry (*Morus rubra*), White Ash (*Fraxinus americana*), Berlandier's Ash (*F. berlandieriana*), Eastern Hop-hornbeam (*Ostrya virginiana*), and the cultivated Weeping Willow (*S. babylonica*), Siberian Elm (*U. pumila*), and Silver-leaf Popular (*Populus alba*), as well as the cultivated pear (*Pyrus communis*).
PARTS EATEN: Young foliage.
NOTE: The Mourning Cloak is our longest-lived butterfly, with some adults surviving up to almost a year.

Zebra Longwing

(Heliconius charithonius vazquezae)

Family: Brush-footed (Nymphalidae)
Size: 2 ¾–3 ¾ inches
Broods: Two or more (several)
Flight time: April–December (all year)
Overwinters: Chrysalis (adult)
Range: 2, 3, 4, 5, 6 (1, 7)

This butterfly belongs to a very small subfamily with most species in the Tropics.
The Zebra Longwing is our most distinctively marked and colored member, with bold black and yellow zebra stripes across the upper surface of the long, narrow wings. Two rows of yellow dots border the lower margins of the hindwings. The lower surface of both wings is much paler, with a cluster of bright crimson spots near bases of both wings. Banding is similar on the lower surface, but the yellow becomes more creamy-colored and the red is replaced with a beautiful rosy-pink patch decorating the tip of the hindwing.

ZEBRA LONGWING

The Zebra Longwing does not usually stray far from its place of emergence but may occasionally wander widely from spring to fall. Its flight is rather slow, with more sailing and drifting than flapping of the wings, but when disturbed or alarmed, it can move quickly and usually darts into low shrubbery. It prefers white, bluish, or purplish flowers such as Climbing Plumbago (*Plumbago scandens*), Crucita (*Chromolaena odorata*), Gregg's Mistflower (*Conoclinium dissectum*), Trailing Lantana (*Lantana montevidensis*), and Prairie Verbena (*Glandularia bipinnatifida*), as well as sand-verbena (*Abronia* spp.).

The Zebra Longwing and the Crimson-patched Longwing (*H. erato petiverana*) are the only two known breeding butterflies in the state with the ability to use pollen as a food source. Gathering minute amounts of pollen on the knobby tip of the proboscis, the butterfly releases a drop of fluid to dissolve the pollen; the insect is then able to drink the liquid in the usual manner. Pollen is extremely rich in protein, and this special food enables the female to lay an unusually large number of eggs—up to one thousand eggs over a long lifetime of three months or more.

Adults choose low shrubbery for roosting, with both males and females gathering in small groups and returning to the same site night after night. Longwings are thought to be the most intelligent among the butterflies, as shown by freshly emerged Longwings readily learning locations of good flower sources and communal roosting sites by association with the older insects.

The male chooses only a small territory for patrolling for females. He is also attracted to female pupae by scent; just before the female is ready to emerge, the male opens her shell with his abdomen and mates with the still unreleased female. He then deposits a repellent chemical or pheromone on the tip of the female's abdomen, which repels other males and thereby prevents her from mating again.

EGG: Ribbed, pale yellow, becoming darker; laid singly or in occasional clusters on very young terminal leaves.

CATERPILLAR: Pure white dotted with brownish-black, with six rows of branching, shiny black spines; larvae feed at night; make no shelter.

CHRYSALIS: Mottled tans and browns, dotted with gold and silver, several "winged" areas and covered with numerous short spines; hangs downward from silken mat.

FOOD PLANTS: The passionflowers, Purple (*Passiflora incarnata*), Bracted (*P. affinis*), Corky-stemmed (*P. suberosa*), Slender-lobe (*P. tenuiloba*), Red-fruited (*P. foetida*), and Yellow (*P. lutea*).

PARTS EATEN: Especially young leaves but occasionally buds, flowers, tendrils, and young fruits.

NOTE: Even though the Zebra Longwing may breed in certain areas of the state and raise one or more broods there, it neither emigrates southward nor overwinters in another form, so does not survive the first severe freezes.

Gulf Fritillary

(Agraulis vanillae incarnata)

Family: Brush-footed (Nymphalidae)
Size: 2 ½–3 ¾ inches
Broods: Two to several
Flight time: April–December (all year)
Overwinters: Chrysalis (adult)
Range: Throughout

The tropical or semitropical Gulf Fritillary is one of the most common yet spectacular butterflies found in the state. The upper surface of both wings is a bright tawny-orange with prominent black veins and markings, with a group of three silver dots near the body on the forewing. The lower surface of both wings is a soft, rich brown heavily splashed with silver spots and bars. A large patch of coral-pink in the basal portion of the forewing is usually visible, or at least partially so, at all times. To watch a group of these insects as they fly about, flashing their silver in the sun, is a breathtaking sight long remembered.

Genetically placed with the Longwings, which include the Zebra (*H. charithonius vazquezae*) and Julia (*Dryas iulia*), the Gulf Fritillary does not have the exceptionally long, narrow wings so characteristic of these other species.

The Gulf Fritillary is a fast flier but usually stays within a few feet of the ground while searching for nectar plants. It is quite addicted to flower visiting and works good

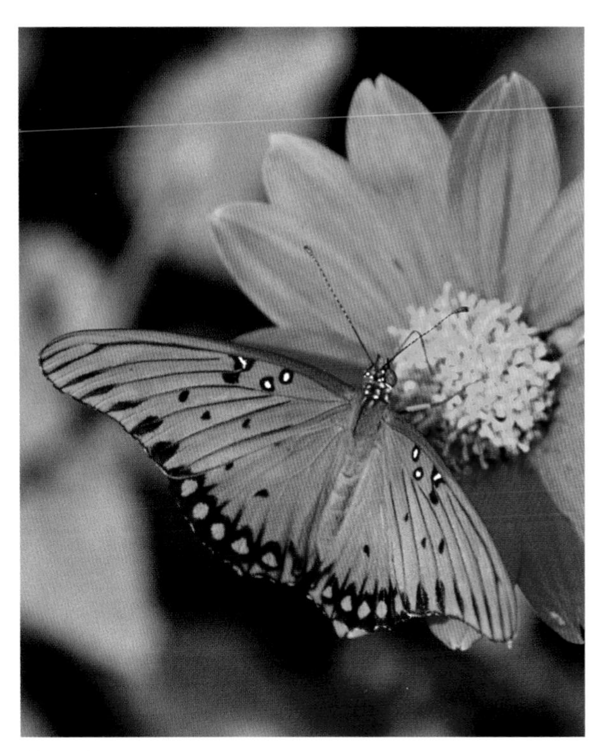

GULF FRITILLARY

nectar sources continually, hardly leaving the plants during the day. It is extremely fond of Butterfly Bush (*Buddleja davidii*), Mexican Sunflower (*Tithonia rotundifolia*), and various zinnias (*Zinnia* spp.), lantanas (*Lantana* spp.), thistles (*Cirsium* spp.), and verbenas (*Verbena* spp.). This butterfly is one of the few readily attracted to white flowers, and it also visits red-flowered ones. Both of these colors are the least used by most butterflies.

If there are several vines of passionflower (*Passiflora* spp.) in the garden or nearby available as a larval food source, there will be a continuous parade of Gulf Fritillaries during the entire season. Where both larval food and nectar sources are available, there is continual egg laying by the females, with great overlaps of emerging adults.

These insects contain toxic body juices from the poisonous larval food and generally go unmolested by predators. Spiders, however, are voracious predators on both the adults and larvae. For protection during the night, adults often roost in small groups near the ground on blades of grass or on the leaves and lower stems of herbaceous plants.

Adults do not stray far from the larval food plants for most of the year, although they occasionally migrate long distances northward. Breeding is not possible in these regions since *Passiflora* does not grow in such cold climates.

EGG: Oblong, ribbed, pale yellow, becoming golden-brown; laid singly on practically any part of the food plant but especially on the underside of a leaf.

CATERPILLAR: Head with two much longer backward-curving spines; body striped lengthwise in muddy maroon and bluish-black and covered with branching black spines; makes no shelter.

CHRYSALIS: Head with two short points; mottled in various browns and blacks; hangs downward from silken mat.

FOOD PLANTS: Almost all *Passiflora*, but the native species preferred. The evergreen cultivated Blue Passionflower (*P. caerulea*) is much utilized where available.

PARTS EATEN: Mostly leaves but sometimes buds, flowers, and young fruit.

NOTE: Where it overwinters as an adult, the Gulf Fritillary becomes greatly reduced in numbers during periods of severe cold. Due to the abundance of native food plants and favorable climatic factors most of the time, though, they are able to reestablish rather quickly.

The Gulf Fritillary is one of the fifteen butterflies now found in Hawaii; they arrived there in 1977 and probably made the journey from California. It is quite possible they were introduced in an attempt to control an invasion of *Passiflora* that was rampant at the time.

Goatweed Leafwing

(Anaea andria)

Family: Brush-footed (Nymphalidae)
Size: 2 3/8–3 1/4 inches
Broods: Two to several
Flight time: March–November (all year)
Overwinters: Adult
Range: Throughout

Belonging to a group commonly referred to as Anglewings or Leafwings, this large, robust butterfly, with an underwing coloring of softly mottled grays or purplish-browns, does much resemble a dried and withered leaf. Tips of the forewings are pointed; the hindwings are somewhat scalloped and with short tails. This irregularity, along with the Goatweed Leafwing's habit of resting on the ground with wings folded and at a decided slant, adds even more to its illusion of being a fallen leaf. In such a position it is very difficult to see, and it often flies from directly beneath your feet. If captured and handled, it usually "plays possum" by falling over on its side and pretending to be dead. Because it habitually rests on the ground or on tree trunks, when it takes flight, the bright red-orange color of its upper wing surface is quite unexpected, giving it an edge in escaping predators.

GOATWEED LEAFWING

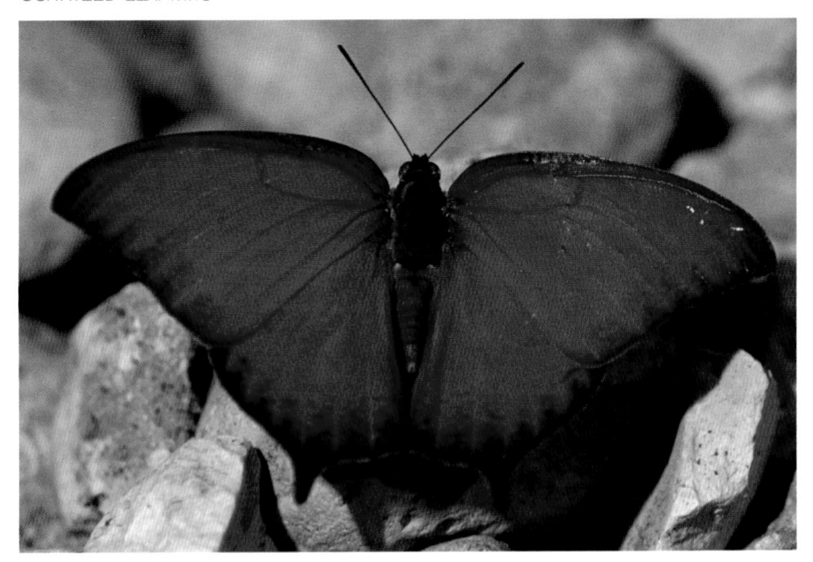

Food fare of the Goatweed Leafwing is usually made up of tree sap, juices of fruits, or moisture from dung, decaying wood, or rotting fungi. Having a rather short proboscis, the butterfly can gain nectar from only a few flower species. This butterfly is an occasional inhabitant of the garden, however, feeding from the more daisylike flowers. When not feeding, it frequently chooses a warm spot on a tree trunk, where it rests head downward.

Last-brood adult Goatweed Leafwings do not die during the cold winter months but become partially inactive, taking refuge inside unheated buildings, behind loose boards or tree bark, or in protected crevices of trees or posts.

EGG: Greenish-cream becoming reddish on top; laid singly on the underside of food plant leaf.

CATERPILLAR: Head with orange horns; body grayish-green, covered in minute points or bumps, tapering toward rear; in later instars, the larva pulls the lengthwise edges of a leaf together, making a loose tent open at both ends; when caterpillar uses smaller-leaved species of *Croton*, several leaves will be silked together, making a tentlike shelter but with top and bottom portions open.

CHRYSALIS: Blunt, thick, pale green mottled in white and various browns; hangs downward from silken mat.
FOOD PLANTS: Tropic Croton (*Croton glandulosus*), Texas Croton (*C. texensis*), Leatherweed Croton (*C. pottsii*), One-seeded Croton (*C. monanthogynus*), Silver-leaf Croton (*C. argyranthemus*), and Woolly Croton (*C. capitatus*), with the latter seemingly the most preferred.
PARTS EATEN: Foliage.
RELATED SPECIES: The similar but darker Tropical Leafwing (*A. aidea*) is an uncommon resident in the Rio Grande Valley and lower coastal area. A rare stray into the Valley is the Crinkled or Angled Leafwing (*Memphis glycerium*). Quite unlike any of these in coloration is the Pale-spotted Leafwing (*M. pithyusa*), which also finds its way occasionally into South Texas. Instead of the orange upper surface of other Texas Leafwings, the Pale-spotted is black above with blue-green iridescence and with a marginal line of white or blue spots on forewings.

Great Southern White

(Ascia monuste)

Family: White/Sulphur (Pieridae)
Size: 2 ¼–3 ⅜ inches
Broods: Many
Flight time: May–November (all year)
Overwinters: Adult
Range: 4, 5 (1, 2, 3, 6, 7)

This large, white butterfly lacks any prominent markings on the lower surface of the wings except the veins and some scaling, which are dusted in soft charcoal. The upper surface of wings is strikingly edged with dark gray or black half diamonds. Tips of the antennae of the Great Southern White are a lovely pale blue. Females are of two color forms, depending upon day length; the spring and summer broods, living in longer days, are dark gray, while the late summer and fall broods, living in shorter days, are white like the males.

GREAT SOUTHERN WHITE

Often when areas become overpopulated or food becomes scarce for other reasons, Great Southern Whites migrate northward. During migration flights the insects apparently continue on their course in a steady flight and in an unchanging direction, rarely stopping to nectar at flowers. Otherwise, both males and females sip flower nectar, with the females beginning to feed earlier in the day than the males.

Great Southern Whites are common at all times in the Rio Grande Valley area, where they breed almost throughout the year. They can readily be seen along edges of coastal marshes, on beaches, and in tidewater areas searching for nectar plants or Saltwort (*Batis maritima*), their larval food plant. Farther inland, they are readily observed in gardens or along roadsides during the warmer months, where they breed but do not winter in any form.

EGG: Elongated but wider in middle, ribbed, pale yellow; laid singly or in clusters of up to fifty on certain food plants; eggs can withstand inundation by salt water for short periods of time.

CATERPILLAR: Gray to brownish-green or yellow with several stripes of maroon, purplish-green, or dark gray, covered with black dots of various sizes and short hairs; makes no shelter.

CHRYSALIS: White or creamy with black dots and markings; rests upright from silken mat and silk strand around body.

FOOD PLANTS: Saltwort along coastline; inland a wide range of both native and cultivated members of the Mustard Family (*Cruciferae*), such as native Shepherd's-purse (*Capsella bursa-pastoris*), Virginia Peppergrass (*Lepidium virginicum*), Southern Marsh Yellowcress (*Rorippa teres*), and Clammyweed (*Polanisia dodecandra*). The cultivated Nasturtium (*Tropaeolum majus*) and Bird's Rape (*Brassica rapa*) are readily used, as well as cultivated broccoli (*B. oleraceae* var. *italica*), brussels sprouts (*B. o.* var. *gemmifera*), and radish (*Raphanus sativus*).

PARTS EATEN: Foliage.

NOTE: The Giant White (*Ganyra josephina*), also a resident of the Rio Grande Valley area, is similar but is larger (largest of North American Whites) and has a distinctive black spot on both surfaces of upper wings. The Florida White (*Glutophrissa drusilla*) is distinguished by having long upper wings conspicuously pointed at the tips. Also a resident of the Rio Grande Valley, it prefers shaded hardwoods to the open coastal areas.

Variegated Fritillary

(Euptoieta claudia)

Family: Brush-footed (Nymphalidae)
Size: 1 ¾–3 ⅛ inches
Broods: Three or more
Flight time: March–October (all year)
Overwinters: Chrysalis (adult)
Range: Throughout

The Variegated Fritillary is a tannish to tawny-orange and black butterfly, with black markings covering the upper surface of both wings in bars, dots, lines, and dashes to form a rather complex pattern. A somewhat nar-

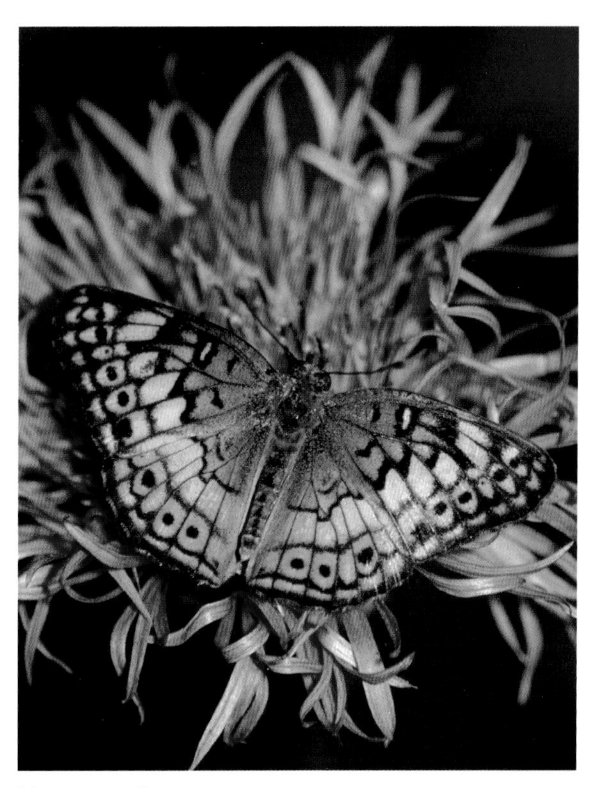

VARIEGATED FRITILLARY

row, zigzag black band marks the middle of both wings. Lower surfaces of both wings are marked with a mottling of white and brown, with a large orange area on the forewing near the body and a smaller patch on the hindwing. This insect does not have the silver markings on the lower wing surfaces as do the true Fritillaries.

Relationship to the true Fritillaries is shown through its use of violets and pansies (*Viola* spp.) as larval food plants. It also shows a close relationship to the Longwings with the larvae using passionflowers (*Passiflora* spp.) as a food plant. The Variegated Fritillary is known to use a wide assortment of plants other than these, and in some areas it is commonly found on flax (*Linum* spp.). In order to find the correct food plant, females rapidly move the forelegs about or scratch the plant's surface, testing and tasting for suitability.

These butterflies are widespread and usually not present in great numbers at any one time in a garden. They are fast fliers and rather far-ranging but readily come to gardens where a good supply of nectar plants is available, remaining longer if proper plants are available for egg deposition. Whether feeding, mating, or searching for plants on which to lay eggs, they usually fly close to the ground in a hovering or darting manner.

EGG: Ribbed, cream to pale green; laid singly on a leaf or stem of food plant.

CATERPILLAR: Striped in orange-red and white, the stripes with white and black dashes; six rows of black branching spines lengthwise of body, the front pair larger and pointing forward over the head; makes no shelter.

CHRYSALIS: Pearly- to silvery-white, sprinkled with black dots and gold-colored bumps; hangs downward from silken pad.

FOOD PLANTS: In Texas the genus *Linum* is apparently the most preferred, with almost all species being used. Other food plants include Purple Passionflower (*Passiflora incarnata*), Red-fruited Passionflower (*P. foetida*), Erect Spiderling (*Boerhaavia erecta*), Spreading Spiderling (*B. intermedia*), Scarlet Spiderling (*B. coccinea*), Purslane (*Portulaca umbraticola*), Whorled Nod-violet (*Hybanthus verticillatus*), the cultivated Garden Pansy (*Viola × wittrockiana*) and Blue Passionflower (*P. caerulea*), and probably any or all violets.

PARTS EATEN: Leaves, buds, petals, and young fruit.

NOTE: These butterflies occasionally stray as far north as Ontario and Quebec but do not breed there.

Cloudless Sulphur

(Phoebis sennae)

Family: White/Sulphur (Pieridae)
Size: 2 ¼–3 ⅜ inches
Broods: One to several
Flight time: February–December (all year)
Overwinters: Chrysalis (adult)
Range: Throughout

One of the largest and most common of the Sulphurs, this denizen of open, sunny places is distinctive and impressive when seen. The upper surface of the wings is clear, bright yellow, unmarked in males; the forewings have a solitary square or diamond-shaped spot near the middle of the wings in females. Some summer broods of Cloudless Sulphur females may be almost white, while fall broods have varying amounts of black markings along outer margins. The lower surface is yellow with small brownish markings; the female is darker and more heavily marked. Each wing bears one or two small whitish or translucent spots near the middle.

A very strong flier, the Cloudless Sulphur moves about the garden very rapidly. If approached too closely, it will fly for quite a distance before coming to rest again. These butterflies are frequent visitors to gardens and can be seen working flowers throughout the day. They are much more common, and come to the garden in greater numbers, during late summer and fall months.

Although not able to tolerate cold northern winters, Cloudless Sulphurs migrate northward each summer, rearing two or more broods where breeding conditions are favorable. By the end of summer, local southern populations may also have built to intolerable numbers, at which times they migrate northward in large groups. They may travel as far north as Canada. With the arrival of cooler temperatures many perish; however, others begin a return flight southward, where at least some of them overwinter. Their impressive flights can be seen in many areas each fall but are never as spectacular in numbers or frequency as the migratory flights of the Monarchs (*Danaus plexippus*).

CLOUDLESS SULPHUR

Since the Cloudless Sulphur is found in almost every open habitat, plants used for nectaring are very diverse. Some of the most common ones include Butterfly Bush (*Buddleja davidii*), Buttonbush (*Cephalanthus occidentalis*), Butterfly Weed (*Asclepias tuberosa*), Pentas (*Pentas lanceolata*), lantanas (*Lantana* spp.), and zinnias (*Zinnia* spp.). This is another of the butterflies that uses red-colored flowers extensively, often being seen around Showy Bergamot (*Monarda didyma*), Drummond's Wax-mallow (*Malvaviscus drummondii*), Tropical Sage (*Salvia coccinea*), and Cardinal Flower (*Lobelia cardinalis*).

As with other Sulphurs, the Cloudless likes to congregate around mud puddles and spend a lot of time taking moisture from such areas.

Males patrol their territory all during the day seeking females. Once a receptive female is found, the male approaches and touches her with his wings, using his scent brushes to release pheromones. Once mated, the joined pair usually flies off, the male flying and the female passive beneath.

EGG: Elongated, ribbed, white or cream, later turning orange or red; laid singly on flower bud or young leaf of food plant.

CATERPILLAR: Yellowish to greenish, striped along sides, and with rows of small blue-black dots across the back; sometimes may be orange-yellow, with narrow crosswise bands of greenish-gray to black, interspersed with rows of tiny black dots; makes shelter of folded and silk-tied leaves of food plant.

CHRYSALIS: Green, pink to brownish, usually camouflaged to match pupation site; rests upward from silken mat at rear and with silken strand around body.

FOOD PLANTS: Christmas Senna (*Senna pendula*), Maryland Senna (*S. marilandica*), Two-leaved Senna (*S. roemeriana*), Partridge-pea (*Chamaecrista fasciculata*), Delicate Sensitive-pea (*C. nictitans*), and the cultivated Argentina Senna (*S. corymbosa*), Coffee Senna (*S. occidentalis*), Sickle-pod Senna (*S. obtusifolia*), and Candle-stick Senna (*S. alata*).

PARTS EATEN: Buds, flowers, and young leaves.

NOTE: There are several intermediate, varying colorations and markings of the caterpillar. The greener form (which eats mostly foliage) seems to be the most common from Central Texas eastward. The yellow form (which eats mostly flowers) is more frequent westward. Any caterpillar found on a species of *Senna* is most likely one of the Sulphurs, Yellows, or Whites.

Question Mark

(Polygonia interrogationis)

Family: Brush-footed (Nymphalidae)
Size: 2 ¼–3 inches
Broods: Two or three
Flight time: February–November (all year)
Overwinters: Adult
Range: Throughout

Another of the Anglewing or "dead-leaf" butterflies, the Question Mark is not a frequent visitor to flowers but is easily attracted to mud, tree sap, insect honeydew, carrion, and rotting fruit. It is especially fond of spoiled fruit and actually becomes intoxicated if the fruit has fermented. The Question Mark spends most of its time along the edges of semishaded trails, in woodland openings, or along shrubby borders, preferring the shaded coolness to sunlit areas. During winter, adults take shelter in cracks or crevices or behind loose boards and tree bark, coming out to fly around during the warmer days.

The Question Mark usually rests with its wings folded above the back. If basking, it

QUESTION MARK

tips over until almost lying flat on the ground. When doing so, the Question Mark is camouflaged so well it is hard to distinguish this butterfly from rocky ground or fallen leaves, the type of area it prefers for resting.

The upper surface of the jagged-edged wings is brightly colored, being primarily orange with black dots, spots, mottling in the outer portion, and violet extending around the wings to form a narrow border. The lower surface is brownish with a violet-gray sheen; the hindwings are tailed and bear a small silver streak somewhat in the form of a printed question mark.

EGG: Slightly longer than wide, widely ribbed, pale green; laid singly or in groups of up to eight, attached either in columns to the upper surface or in chains hanging beneath lower surface or in rows along the margins of young leaves of the food plant; often deposited on plants near the food plant instead of on it.

CATERPILLAR: Mature larva mottled reddish-brown to black, with numerous orange-brown, branched spines covering the head and body; young caterpillars somewhat gregarious; makes no shelter.

CHRYSALIS: Yellowish to mottled grayish or brownish, spotted with gold or silver; several projections; hangs downward from silken mat.

FOOD PLANTS: American Elm (*Ulmus americana*), Cedar Elm (*U. crassifolia*), Winged Elm (*U. alata*), Slippery Elm (*U. rubra*), Sugar Hackberry (*Celtis laevigata*), Net-leaf Hackberry (*C. l.* var. *reticulata*), and the herbs False Nettle (*Boehmeria cylindrical*) and Heart-leaf Stinging-nettle (*Urtica chamaedryoides*). The cultivated Siberian Elm (*Ulmus pumila*) is sometimes used.

PARTS EATEN: Young foliage.

NOTE: There are two forms of the Question Mark, with the winter form described and shown here. In the summer form, the upper surface of the hindwings is almost black, and the undersurface is dark grayish-black with mottling of maroon and blue.

Large Wood-Nymph (Common Wood-Nymph)

(Cercyonis pegala)

Family: Brush-footed (Nymphalidae)
Size: 1 ¾–3 inches
Broods: One
Flight time: May–November
Overwinters: Newly hatched larva
Range: 1, 2, 3, 6

Although called a Wood-Nymph, this butterfly is not much of a true forest dweller. It much prefers brushy roadsides, woodland edges, trails, or even grassy meadows. It is a great lover of flowers and visits them often, seeming especially fond of Buttonbush (*Cephalanthus occidentalis*), Joe-Pye Weed (*Eutrochium fistulosum*), Virginia Frostweed (*Verbesina virginica*), and various milkweeds (*Asclepias* spp.), mistflowers (*Conoclinium* spp.), and thistles (*Cirsium* spp.). It will readily be found on fallen or fermenting fruit such as peaches (*Prunus* spp.), pears (*Pyrus* spp.), or persimmons (*Diospyrus* spp.).

Flight of the Large Wood-Nymph is usually short and with an appearance of being slow and weak, yet it is very erratic and extremely difficult to follow. The butterfly commonly sits fully exposed with folded wings on a leaf or branch. If disturbed, it will take off, often dropping into the grass or flying into thick shrubbery, where it alights on the underside

LARGE WOOD-NYMPH (COMMON WOOD-NYMPH)

of a leaf or a twig and is almost impossible to find again.

To see this butterfly with wings spread is an uncommon sight, for it seems a bit reluctant to show the rich, dark chocolate-brown coloring and bright yellow banding of the upper surface. Two large, blue-centered black dots decorate this wide yellow band. The lower surface is just as striking, with the wings mottled with fine, barklike striations. Forewings bear a wide pale yellow band decorated with two large eyespots of white or blue circled with black. Females are usually larger than the males, with a softer, lighter brown coloring and paler but larger eyespots. Both sexes have a series of smaller eyespots edging the hindwings.

To bask, the Large Wood-Nymph tilts the folded wings to one side, almost laying them flat. When warm enough on one side, it flips the wings over to warm the other side.

EGG: Elongated and larger in middle, deeply ribbed, white, cream, or yellow, becoming pale brownish, with brown or pink mottling; females deposit between two hundred and three hundred eggs singly on or near grasses in late summer.

CATERPILLAR: Pale yellowish or greenish, striped lengthwise with green and yellow lines and covered with short, fuzzy hairs; rear of caterpillar with two reddish tails; makes no shelter.

CHRYSALIS: Yellowish-green with various white or yellow bands, stripes, or lines; hangs downward from silken mat.

FOOD PLANTS: Various grasses, including Purpletop Tridens (*Tridens flavus*) and several of the bluestems in the genus *Andropogon*.

PARTS EATEN: Leaves.

NOTE: Males live only two to three weeks after emerging; females, which emerge up to a week after the males, may live several months, delaying egg laying as late in summer as possible. After emerging in late fall, the tiny larvae do not eat but immediately hibernate for the winter at the base of the food plant.

Southern Dogface

(Zerene cesonia)

Family: White/Sulphur (Pieridae)
Size: 2 1/8–2 3/4 inches
Broods: Two to several
Flight time: April–December (all year)
Overwinters: Chrysalis (adult)
Range: Throughout

In the more northern portion of its range and from a distance, the Southern Dogface can be mistaken for the Clouded Sulphur (*Colias philodice*), but a close inspection of the upper surface of the wings quickly distinguishes one from the other. The Southern Dogface is aptly named. Other Sulphurs have an almost even black border on the upper wing surface. The black in the border of the Southern Dogface is such that the remaining yellow of the wings forms a "dog face." A black dot within each yellow forewing patch forms the "eye" of the dog. The lower wing surface of the Southern Dogface is generally greenish-yellow, with a black-rimmed white dot on the forewings. Rosy or magenta scaling along the veins and along the wing margins of a fresh specimen is usually conspicuous. This rose coloring becomes even more noticeable in the winter form *rosa*. The forewing tips of the Southern Dogface are prominently pointed, in contrast to the rounded tips of the Clouded Sulphur.

SOUTHERN DOGFACE

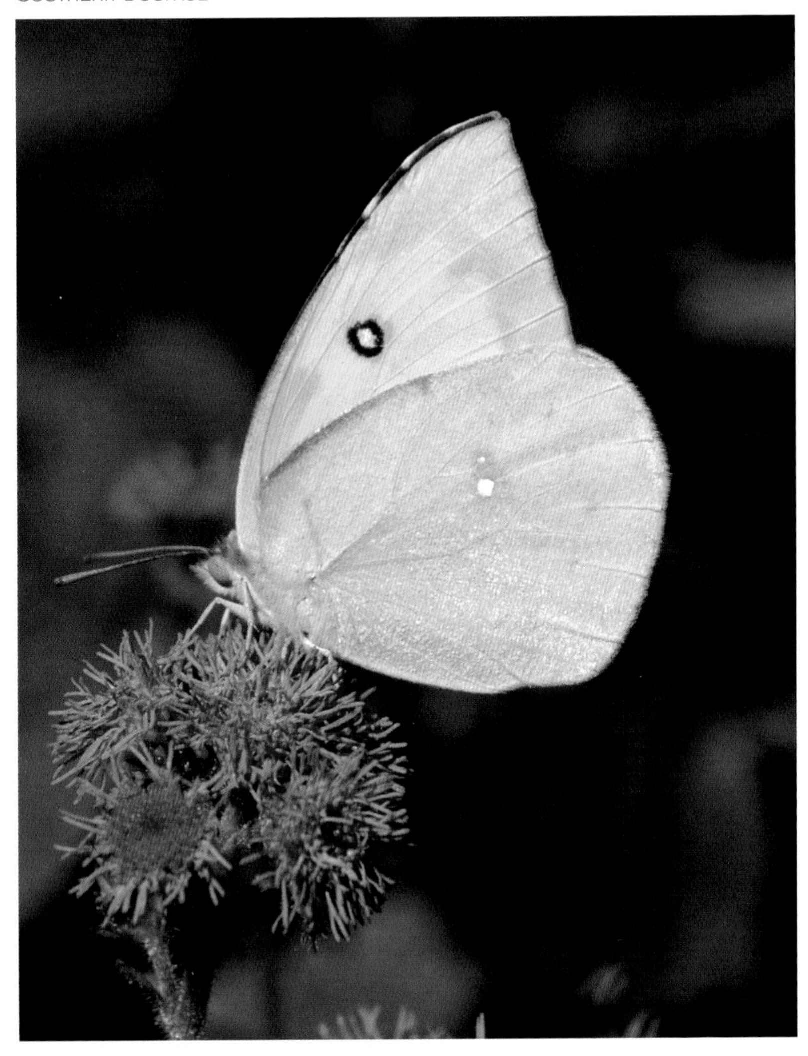

Flying swiftly, the Southern Dogface stops only briefly around each flower, although it spends much time in the garden sipping nectar. It is frequently seen around red flowers, such as Drummond's Wax-mallow (*Malvaviscus drummondii*), Cardinal Flower (*Lobelia cardinalis*), and various species of *Salvia*.

Both sexes, but especially males, stay around mud puddles for hours. If disturbed, they usually circle around briefly, then return to the moisture. When not nectaring or puddling, males patrol their territory in search of females. Once a receptive female is located, the male attracts her both by ultraviolet light reflected from the outer portion of the dog's face on the forewings and by scent pheromones.

EGG: Elongated, ribbed, yellowish-green, later turning dark red; laid singly on underside of terminal leaf of food plant.

CATERPILLAR: Mostly green but may be striped lengthwise with white bands containing orange dashes or with crossbands of white and black or yellow; makes no shelter.

CHRYSALIS: Green- to bluish-green-striped with white and covered with black dots; held upright from silken mat and strand around body.

FOOD PLANTS: Bearded Dalea (*Dalea pogonanthera*), Black Dalea (*D. frutescens*), Purple

Dalea (*D. purpurea*), Baby Bonnets (*Coursetia axillaris*), False Indigo (*Amorpha fruticosa*), Texas Kidneywood (*Eysenhardtia texana*), and the cultivated Alfalfa (*Medicago sativa*), Soybean (*Glycine max*), and White Clover (*Trifolium repens*).
PARTS EATEN: Foliage.
NOTE: This butterfly is sometimes placed in the genus *Colias*.

White Peacock

(Anartia jatrophae luteipicta)

Family: Brush-footed (Nymphalidae)
Size: 2–2 5/8 inches
Broods: One to several
Flight time: August–December (all year)
Overwinters: Adult
Range: 4, 5 (1, 2, 3, 6, 7)

Although the White Peacock breeds mostly in the Lower Rio Grande Valley, it migrates northward and may possibly be seen anywhere during the summer months, occasionally breeding as far north as the Hill Country. However, it is not a regular migrant to some areas of the state and is never noted for massed groups or noticeable numbers at any one time. It is common within its southern breeding range and can be seen around any patch of wildflowers or in a garden, especially if water or areas of extra moisture are nearby. It is quite common around irrigation canals and dripping faucets or leaking waterlines where its food plants grow.

The White Peacock is closely related to the Buckeyes, with which it often flies, but the upper surface of the White Peacock's wings is washed with a pearly sheen, making the butterfly appear almost white in flight. The white shades into buff, marked with orange crescents forming a border along the wing margins. The hindwings are shortly and bluntly tailed. The undersurface is not as white and has numerous scrawled lines, bands of orange, and various shades of light brown. Both surfaces bear one large black eyespot on the forewing and two smaller, widely separated ones on the hindwing. Summer broods are darker than spring and fall broods.

The White Peacock is not a strong flier and remains low, seeking flowers near the ground. Easily approached if movements are slow, the White Peacock is a fascinating subject to watch while it is feeding. Once it finds a good nectar source, it slowly crawls from flower to flower, seeming to be completely absorbed in nectar gathering. If disturbed, it flutters and glides to the next flower to continue with its feeding. Once really frightened, it flies into nearby grasses or shrubbery, closes its wings, and immediately disappears from view.

In seeking females, the male patrols back and forth in a rather erratic flight interspersed

WHITE PEACOCK

with periods of gliding. Males also perch on low, exposed branches of shrubbery or a blade of grass near the larval food plant to await passing females. To reject a male, the female lands and leans her wings from side to side until the male leaves.

EGG: Pale yellow; laid singly on underside of food plant leaf or on nearby vegetation.

CATERPILLAR: Head with two long, curved, clubbed horns; body dark brown to black, with white or silver spots forming crosswise bands, with four rows of black or orange-red spines; makes no shelter.

CHRYSALIS: Pale green to black, occasionally with small spots; hangs downward from silken mat.

FOOD PLANTS: Texas Frogfruit (*Phyla nodiflora*), Diamond-leaf Frogfruit (*P. strigulosa*), Scented Lippia (*Lippia graveolens*), White-flowered Lippia (*L. alba*), Coastal Water-hyssop (*Bacopa monnieri*), Western Ruellia (*Ruellia occidentalis*), and Violet Ruellia (*Viola nudiflora*). Probably several other species of *Ruellia* are used as well.

PARTS EATEN: Foliage.

Janais Patch (Crimson Patch)

(Chlosyne janais)

Family: Brush-footed (Nymphalidae)
Size: 1 7/8 –2 5/8 inches
Broods: Three or more
Flight time: June–November (all year)
Overwintering: Half-mature caterpillar(adult)
Range: 4, 5, 6 (1, 2, 7)

The Janais Patch is an extraordinarily beautiful butterfly and one not too often seen in the state. Primarily the coloring is soft, velvety black with a sprinkling of white dots on the forewings and very large, prominent patches of burnt-orange to reddish-orange on the upper surface of the hindwings. The lower surface of the forewings is black, heavily dotted in white, while the hindwings are banded in creamy-yellow, reddish-orange, and black. A row of white dots lies within the black band.

This butterfly is usually present in the Rio Grande Valley area, but severe winters can wipe it out. When this happens, it takes only a year or two to become recolonized from Mexico and once again is a welcome sight in gardens and woodlands and along watercourses. Recently it has become an occasional breeding resident as far north as the Austin area, through landscape usage of its larval food plant. During the summer or early fall months, it occasionally strays eastward and as far north as the Rolling Plains area.

This beauty is very specific in its larval food plant, feeding almost entirely on Flame Acanthus (*Aniscanthus quadrifidus* var. *wrightii*). It also uses this plant as a nectar source, along with Buttonbush (*Cephalanthus*

JANAIS PATCH

occidentalis), Beebrush (*Aloysia gratissima*), Annual Sunflower (*Helianthus annuus*), and other good nectar-producing plants near its larval food plant. While breeding, it does not venture far from the Flame Acanthus, preferring to use its flowers for nectaring instead of flying long distances to other sources. Where there is a colony of plants established, there is a great overlap of the breeding cycles, with courtship, egg laying, pupation, and adult emergence all going on at once.

EGG: Cream to pale yellow; laid in clusters on undersides of leaves of food plant.

CATERPILLAR: Head orange-red in upper portion; body white to pale metallic grayish-green with many rows of branched, black spines; makes no shelter.

CHRYSALIS: Yellow, covered in tiny black bumps; hangs downward from silken mat.

FOOD PLANTS: Flame Acanthus and Virginia Frostweed (*Verbesina virginica*).

PARTS EATEN: Buds, flowers, and young leaves.

NOTE: Flame Acanthus can be grown outside its native range and does quite well as far north as the Dallas–Fort Worth area and as far east as Robertson County, but there are no known records of the Janais Patch becoming established that far north or east.

Guava Skipper

(Phocides polybius lilea)

Family: Skipper (Hesperiidae)
Size: 1 5/8–2 ½ inches
Broods: Several
Flight time: (All year)
Overwinters: (Adult)
Range: 5

On wings of brilliant blue, the Guava Skipper flits through gardens and woodlands, stopping frequently at all available flowers. Never far from its larval food plant, this southern straggler to the Valley is becoming more commonly seen because home gardeners are planting guava trees.

No other butterfly of this size or shape in the Valley area has the solid brilliant coloring of the Guava Skipper. The smoky black upper wings are covered with blue scales. The scaling is such that it forms contrasting radiating turquoise streaks along the veins in the wing and on the upper body. There is a vivid, red, two-spotted bar on the upper edge of the wings and a narrow red "collar" behind the head. Both wings are edged with a snowy-white fringe. Underside coloring is the same.

These Skippers seem to be more active in mornings and late evenings—look for them in gardens aroundAnacahuita (*Cordia boissieri*) trees, one of their favorite flowers for nectaring. They will often be found in openings of subtropical woodlands. The Guava Skipper is a fast, erratic flyer but rests often with wings

GUAVA SKIPPER

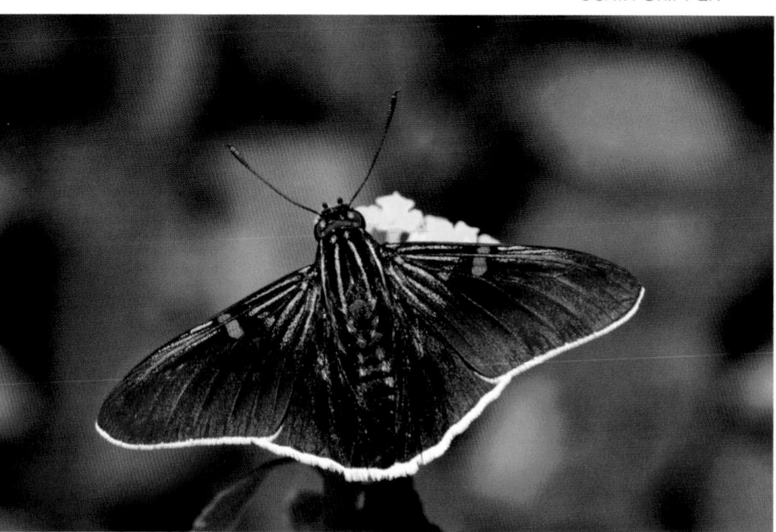

spread wide. During very hot or inclement weather, it takes refuge beneath the foliage of large-leaved plants.

EGG: Bluish-green becoming reddish, laid singly on upperside of young terminal leaf.

CATERPILLAR: Head of mature caterpillar brown with two yellow spots, conspicuously shiny; half-grown caterpillar red with narrow yellow rings around body; body of mature caterpillar chalky-white; eats at night; rests in rolled leaf shelter.

CHRYSALIS: Smooth, dirty-white to greenish, tiny gray and black dots; rests in silk-lined "nest" in rolled-up leaf of food plant.

FOOD PLANTS: The introduced Common Guava (*Psidium guajava*) and Pineapple Guava (*Feijoa sellowiana*).

PARTS EATEN: Foliage.

Red Admiral

(Vanessa atalanta rubria)

Family: Brush-footed (Nymphalidae)
Size: 1 ¾–2 ½ inches
Broods: Two to several
Flight time: March–October (all year)
Overwinters: Chrysalis (adult)
Range: Throughout

First described in Europe in 1758 by Carl Linnaeus, the Red Admiral is one of the best-known and most widespread butterflies. It is one of the most common species in the state, occurring in city parks, gardens, and shrubby fields and along woodland edges and brushy roadsides.

Coloring of the Red Admiral is beautiful in its simplicity. The upper surface of the wings is a soft velvety black, the forewings are crossed with a diagonal band of orange-red or vermillion, and the tips are sprinkled with white dots. A narrower band of the same orange-red color borders the curved hindwings. When the butterfly rests with the wings open, the banding of fore- and hindwings seems connected, forming two bowed or curved lines that almost complete a circle. This banding also appears on the undersurface of the forewing but is much paler in color, appearing more pink than red.

The Red Admiral is similar to the Painted Lady (*V. cardui*) in being widespread along with its larval food plant. Few plants have a wider distribution than nettles (*Urtica* spp.), and where nettles grow, caterpillars of the Red Admiral will almost always be found. Adults are fast fliers and have a tendency to wander away from where they were reared, mainly because the nettle food plants grow in semishaded areas and the adults must seek masses of nectar-producing flowering plants in more sunny areas.

The Red Admiral is particularly fond of salts from human perspiration, often alighting on exposed legs, arms, or hands. It feeds on a variety of substances, including sap, fruit, and dung, as well as regularly visits flowers. On warm winter days it can be seen sipping the sap of trees or the juice of wild fruits that have burst open after freezing, and if the fruit has begun to ferment, the more the butterfly seems to enjoy it. Often it then appears intoxicated and has difficulty flying.

Habitually resting head downward on vertical surfaces and with its wings closed, it is extremely difficult to see. It is one of the first to emerge from hibernation in the spring, spending much time on the ground at moist areas or basking on flat surfaces with opened

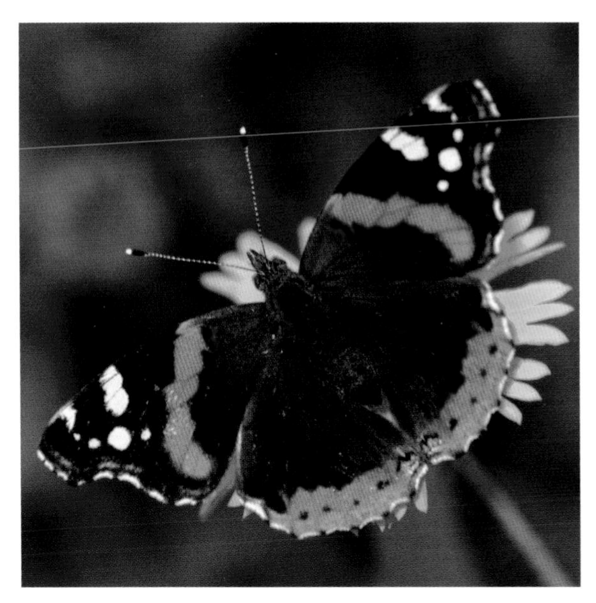

RED ADMIRAL

wings. The male is very territorial and extremely aggressive and will defend his chosen resting site against all intruders.

EGG: Somewhat barrel-shaped, faintly and widely ribbed, pale green; laid singly on upper side of young food plant leaf, but many leaves on the same plant may be used.

CATERPILLAR: Extremely variable but usually blackish; body has many bumps and branching spines; makes silken "nest" or shelter of leaves.

CHRYSALIS: Reddish- to gray-brown with splotches of gold and short bumps; hangs downward from silken mat.

FOOD PLANTS: Heart-leaf Stinging-nettle (*Urtica chamaedryoides*), Florida Pellitory (*Parietaria floridana*), Pennsylvania Pellitory (*P. pensylvanica*), and False Nettle (*Boehmeria cylindrical*).

PARTS EATEN: Leaves, within silken shelter.

NOTE: During summer months the Red Admiral strays as far north as Alaska and is one of the fifteen species of butterflies in Hawaii, where it is now a common resident, having been documented there first about 1882.

Painted Lady

(Vanessa cardui)

Family: Brush-footed (Nymphalidae)
Size: 2 ¼–2 ½ inches
Broods: Two to several
Flight time: March–November (all year)
Overwinters: Hibernating adult (adult)
Range: Throughout

The Painted Lady is the most widespread of all the butterflies in the world. It bears other descriptive common names, such as Cosmopolitan and Thistle Butterfly, the first from its almost worldwide distribution, and the second from its favored food plant. This butterfly is a most familiar sight, being one of the first seen in early spring and one of the last in the fall or seen all year in the southern part of its range. It is a common visitor to the garden during the flowering season but is usually seen in greater numbers in the autumn. In most of its range, it spends the winter months tucked away in some crevice and inactive except on the very warmest days.

The upper wing surface of the Painted Lady is a complex mottling of black and pinkish-orange, with a sprinkling of white dots near tips of the forewings. Patterning on the lower surface is even more complicated, being a mixture of golds, tans, black, and white. Forewings are dominantly rose-pink near the base, patterned with olive, black, and white on the outer margins. A row of small eyespots and a narrow blue band occur near margins of the hindwings. In coloration and markings the Painted Lady is very similar to its near relative, the American Lady (*V. virginiensis*), but differs by having four or five small black marginal spots on the lower surface of the hindwing, whereas the American Lady has only two very large black spots with blue centers.

Painted Ladies occur in almost all environments, as long as they are open, sunny, and filled with flowers. They cannot overwinter in any stage where the temperatures are severe, and because they do not move southward, they perish. But by February or March the overpopulated southern broods begin moving north and east from their warmer wintering grounds, and by late spring the Painted Lady has once again become a common sight throughout North America.

Despite their wide distribution, Painted Ladies do not congregate when feeding. Almost always only one or two are in the garden at a time, unless they are around the larval food source. They are easily attracted to the garden by flowers of the Aster Family (*Asteraceae*), such as Mexican Sunflower (*Tithonia rotundifolia*), Purple Coneflower (*Echinacea purpurea*), various thistles (*Cirsium* spp.), and the single zinnias (*Zinnia* spp.) and marigolds (*Tagetes* spp.).

PAINTED LADY

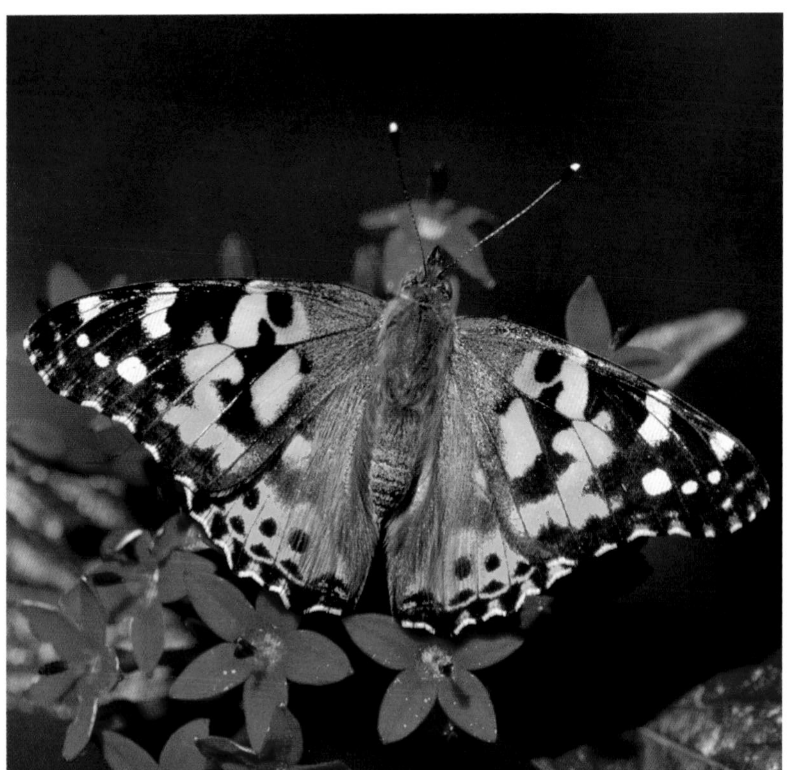

Courtship flights of these butterflies are usually quite elaborate and preferably take place on open hilltops or at least the highest terrain possible. After mating, the female returns to lower ground and seeks out stands of thistle to lay her eggs.

EGG: Elongated, wider in middle, pale green; laid singly on upper surface of food plant leaf.

CATERPILLAR: Grayish-brown or black, with numerous narrow crosswise lines and rows of long, branched spines; larvae live in silk shelter formed by folding leaves of food plant together and binding with silk.

CHRYSALIS: Pale tannish mottled in greens and browns with several gold-colored bumps; hangs downward from silken mat.

FOOD PLANTS: More than one hundred species of plants have been reportedly utilized by this butterfly, but thistles are preferred above all others. Some of the most commonly used are Texas Thistle (*C. texanum*), Horrid Thistle (*C. horridulum*), and Wavy-leaf Thistle (*C. undulatum*). Other species reportedly used include White Sage (*Artemisia ludoviciana* subsp. *mexicana*), Small-flowered Mallow (*Malva parviflora*), Alkali Little-mallow (*Malvella leprosa*), Annual Sunflower (*Helianthus annuus*), American Elm (*Ulmus americana*), and the nonnative and cultivated

White Goosefoot (*Chenopodium album*), Common Mallow (*Malva neglecta*), Milfoil (*Achillea millefolium*), Soybean (*Glycine max*), Alfalfa (*Medicago sativa*), Cotton (*Gossypium hirsutum*), Hollyhock (*Alcea rosea*), and Calendula (*Calendula officinalis*), as well as Borage (*Borage officinalis*), beet (*Beta vulgaris*), and garden bean (*Phaseolus vulgaris*). Many more species of several genera are reportedly used.
PARTS EATEN: Foliage.

AMERICAN LADY

American Lady

(Vanessa virginiensis)

Family: Brush-footed (Nymphalidae)
Size: 1 ¾–2 ½ inches
Broods: Two or more
Flight time: March–November (all year)
Overwinters: Chrysalis (adult)
Range: Throughout

Inhabiting almost all open areas, including gardens, the American Lady is found wherever flowers are abundant. It does not seem particularly attracted to special colors or species and freely visits whatever flowers are available. It occasionally flies along woodland trails or borders because of the flowers found there, such as wild azaleas (*Rhododendron* spp.) or asters (*Symphotrichum* spp.), not because of a preference for shade. When nectaring, it habitually holds its wings open or at least partially so, moving warily about, constantly changing position. It can occasionally be seen basking on bare earth with wings widely spread.

The best time to see the undersurface of the folded wings is to find a butterfly in late evening, after it has gone to roost for the night, in early mornings before it takes flight, or in cool or inclement weather, when it takes shelter beneath a leaf or a blade of grass. At such times the intricate markings, large blue-centered eyespots, and brilliant pink forewing patch can be closely observed and appreciated. The surface of the upper wings is a dark pinkish-orange, with dark markings across the tips of the forewings and sprinkled with a few white dots. A row of small blue eyespots circled in black follows the outer margins of the hindwings; these spots connect or run together to form a band.

If startled, this butterfly takes off in sudden, erratic flight but often returns in a few moments to the former site if there is no further movement. It frequently dashes out at other butterflies that approach the flower it is feeding on, fluttering its wings until the intruder seeks other flowers or leaves the area entirely.

In the more northern regions, this butterfly usually spends winter in the chrysalis stage. In warmer, more southern climes, with the arrival of first really cold nights, it seeks snug, protected places where it will remain mostly inactive during the winter months. With approaching warmer days of early spring, some of them will once again head north, raising two or more broods. The last brood usually emigrates south for the winter.
EGG: Barrel-shaped, pale green or yellowish; laid singly on upper surface of food plant leaf.
CATERPILLAR: Greenish with black marbling or velvety black with groupings of several narrow cross bands of yellow; black, branching spines with red dots at the bases and two large white dots in each black portion; solitary caterpillars live inside nestlike shelter made of the food plant; young spin plant hairs and small bits of the inflorescence into shelters, while older larvae use larger bits of the inflorescence and leaves to make a larger, more compact shelter; caterpillars often pupate inside the shelter.
CHRYSALIS: Variable, ranging from greenish to brownish, mottled in various whites, greens, browns, and blacks, with small bumps and projections; hangs downward from silken mat.
FOOD PLANTS: Fragrant False Cudweed (*Pseudognaphalium obtusifolium*), Gray False Cudweed (*P. canescens*), Purple Cudweed (*Gamochaeta purpurea*), Pennsylvania Cudweed (*G. pensylvanica*), Big-head Evax (*Evax prolifera*), Spring Evax (*E. verna*), Silver Evax (*E. candida*), and Parlin's Everlasting (*Antennaria parlinii*) are the preferred food plants. Many more genera and species are commonly listed as being used.
PARTS EATEN: Buds, flowers, and young leaves.

Common Buckeye
(Junonia coenia)

Family: Brush-footed (Nymphalidae)
Size: 1 5/8–2 ½ inches
Broods: Two or more
Flight time: February–November (all year)
Overwinters: Chrysalis (adult)
Range: Throughout

Wings of the Common Buckeye are generally tawny to dark brown, overlaid with various iridescent colors, and with one white band and two smaller lengthwise orange bars on the forewings. Most noticeable are the large blue and black eyespots set near an orange marginal border on the upper surface of both wings. Beneath, the forewing somewhat resembles the upper surface, while the hindwing is beautifully mottled in soft rose-browns and tans and dotted with much smaller eyespots. The large eyespots on this butterfly give it another common name, Peacock Butterfly.

One of the most frequently seen butterflies, the Common Buckeye ranges almost throughout North America south of the Canadian border. In early spring, some of the butterflies that have overwintered in the adult stage in nonfreezing climes quickly move northward, rearing two or more broods there. Not able to overwinter, in the fall the last brood heads southward in massive movements. For most of the ones in Texas, the winter is spent in the chrysalis stage.

Male Common Buckeyes, not as active as

the females, sit for long periods on the ground or on low shrubbery, basking and waiting for passing females. A male chooses a special perch from which he patrols a territory; and within the chosen boundaries he takes quick flights to intercept passing females or pugnaciously attack other males or any other intruder, no matter its size or description. While most individuals live only an average of ten days, the flight period is long in the southern portion of the Common Buckeyes' range.

Common Buckeyes like to bask with wings spread wide in early mornings or after inclement weather. They are also fond of mud puddles and spend much time there. They are equally fond of flowers and are found in almost all gardens as well as other open areas where flowers are plentiful. A rapid flier, this butterfly tends to be nervous and wary when approached. It usually flies low to the ground, alternately gliding and flapping its wings.

EGG: Flattened on top, ribbed, dark green; laid singly, usually on upper side of a leaf of food plant.

CATERPILLAR: Generally black with lengthwise rows of cream or white and with numerous black, branching spines; rows of spines nearest underside of body conspicuously orange at base; makes no shelter.

CHRYSALIS: Pale cream mottled with reddish-brown to black; hangs downward from silken mat.

FOOD PLANTS: Almost all species of the genera *Agalinis, Castilleja, Linaria, Plantago,* and *Phyla* with the first three most commonly used in Texas. Other food plants include Yellow False Foxglove (*Aureolaria flava*), American Bluehearts (*Buchnera americana*), Violet Ruellia (*Ruellia nudiflora*), Snapdragon Vine (*Maurandya antirrhiniflora*), and the cultivated Snapdragon (*Antirrhinum majas*).

COMMON BUCKEYE

PARTS EATEN: Flower buds, young fruit, and leaves.

Hackberry Emperor

(Asterocampa celtis celtis)

Family: Brush-footed (Nymphalidae)
Size: 2–2 ½ inches
Broods: One to three
Flight time: March–October
Overwinters: Third instar larva
Range: Throughout

The Hackberry Emperor is one of the most common butterflies in the state but is not one to visit flowers very often. On occasion it will be seen gathering fluids from blossoms in the wild, but mostly it feeds on tree sap, rotting fruit, insect honeydew, carrion, and mud.

Even if the Hackberry Emperor is not a great flower visitor, it likes to patrol open areas and bask. If there are any Hackberry (*Celtis laevigata*) trees in or near the garden, the Hackberry Emperor will be seen making frequent forays into the open areas. And if damp sand and rotting fruit are provided, it will be a common visitor to the garden. It is especially fond of fermented dewberries (*Rubus* spp.), mulberries (*Morus* spp.), overripe bananas (*Musa* spp.), peaches (*Prunus* spp.), pears (*Pyrus* spp.), and persimmons (*Diospyrus* spp.). It also loves canned fruit cocktail, especially if a shot of rum or some beer has been added. While feeding on this mixture, the insect becomes so engrossed in sipping the juices it can be approached closely enough to actually be moved about with the fingers without its taking flight.

HACKBERRY EMPEROR

The Hackberry Emperor is a medium-sized butterfly, with the upper wing surface colored a most distinctive orange and olive-brown; the black outer wing tips of the forewing are dotted in white. Patterning and coloring of the lower surface of the wings are complex and variable, with markings of brown, black, and purplish-gray and with eyespots on both wings.

This butterfly is always found in close association with its food plant, Hackberry; the female leaves the trees only while searching for food or basking. Frequently the female will sun with wings spread wide, especially if she is carrying a heavy egg load. The female commonly basks in this position on low vegetation in an open area, where the male "dive-bombs" her in courtship before mating. The male also perches on Hackberry trees, awaiting passing females, or occasionally wanders a short distance, patrolling for feeding females. Some sites are more attractive than others to the male for perching, and an occupant's perching rights are often contested by another male. Male Hackberry Emperors are usually the last butterfly to be seen in the day, continuing to fly until almost dark.

From a favored perching area, males frequently fly out to inspect moving objects, especially if the object is shiny. His main objective is to find a female, of course, but he has a tendency to be curious about any moving object, no matter what its size or color.

This butterfly usually perches on the trunk of the host tree until disturbed, then flies to another perching site, such as a fence post or the handle of a hoe being used. Often it alights on arms or hands, seeking the salts from perspiration.

EGG: Cream to pale green; laid singly or in small groups on underside of leaf or occasionally on stem near leaves.

CATERPILLAR: Head with two small, branched spines; body leaf-green with faint yellow lengthwise stripes, tapered at both ends, forked at rear; each upper straight yellow line ends at a branched spine on the head and a "tail" at the rear; middle instar larvae of last brood hibernate in crevices along a tree trunk or in leaf litter until the following spring; larvae make no "nest."

CHRYSALIS: Yellow to blue-green, yellowish line down back, two short horns on head; hangs downward from silken mat.

FOOD PLANTS: Known to use all species of Hackberry within its range.

PARTS EATEN: Leaves of young trees or new growth on older trees.

Texas Emperor

(Asterocampa clyton texana)

Family: Brush-footed (Nymphalidae)
Size: 2–2 5/8 inches
Broods: One to three
Flight time: March–October (all year)
Overwinters: Third instar caterpillar (adult)
Range: 1, 2, 5, 6, 7

The Texas Emperor is the West Texas form of the Tawny Emperor (*A. clyton clyton*), which ranges more eastward. Both the Texas and Tawny Emperors are very similar to the Hackberry Emperor, but with the orange and black coloring of the wings reversed. The best field marks to look for are the blue eyespots

Texas Emperor

on both the upper and lower surfaces of both the fore- and hindwings of the Hackberry Emperor—the Texas and Tawny Emperors have dots only on the hindwings. Also, the bodies of the Texas and Tawny Emperors are usually orange or reddish-brown, whereas the body of the Hackberry Emperor is much darker and generally gray.

The Texas Emperor has a very swift and powerful flight, but it can and frequently does glide slowly from tree to tree. Generally, it does not fly very far between stops, except the male may fly some distance back and forth from his perching area and the closest larval food plants, where the females usually stay. Mating flights are most often performed from the middle of the day to late evening. The Texas Emperor likes large, mature trees, especially if they bear fruit abundantly. Basking often, it opens its wings wide to fully partake of the sun's warmth. It is commonly seen resting head downward and especially likes to perch on leaves in full sun and on tree trunks, fence posts, rocks, paved roads, and people.

This insect's food preferences are similar to those of the other hackberry-feeding butterflies, and they all are frequently found imbibing liquid from the same source. Sometimes there are so many individuals gathered together, they jostle each other about to try to get the best position. The Texas Emperor also spends a lot of time around mud puddles, either with others of its own kind or with various Sulphurs, Hairstreaks, or Swallowtails.

EGG: Thick sculpturing of usually twenty ridges, cream to greenish; laid in large, moderately tightly packed cluster or layer on underside of leaf or occasionally on bark of host tree.

CATERPILLAR: Pale green with two wide, yellowish lines down back separated by a very narrow blue line; downy, with short hairs, forked at rear, and with two spiny horns on head; larvae gregarious in early instars and commonly found in groups of fifty or more; third instar larvae of the last fall brood hibernate.

CHRYSALIS: Pale green to yellowish-green, raised, sawtooth ridge down portion of back, the ridge narrowly edged in dark yellow with tiny, shiny dot at point of each "tooth," two very short horns on head; hangs downward from silken mat.

FOOD PLANTS: Known to use all species of hackberry (*Celtis* spp.) in the state within the insect's range, except Spiny Hackberry (*C. pallida*).

PARTS EATEN: Mature foliage.

Silver-spotted Skipper

(Epargyreus clarus)

Family: Skipper (Hesperiidae)
Size: 1¾–2½ inches
Broods: Three or more
Flight time: February–November
Overwinters: Chrysalis
Range: 1, 2, 3, 4, 6

One of the largest Skippers, this is a wide-ranging butterfly equally at home in wilderness areas, parks, and suburban gardens or along country roadsides. Its flight is very strong, swift, jerky, and erratic. It is generally very pugnacious in character and will attack just about anything in its range, especially other butterflies, no matter which species they happen to be.

Generally brownish in coloring, the upper

Silver-spotted Skipper

surface of the long forewings has a broad, indistinct band of clearish orange-yellow bars. On the lower surface this band is often hidden by the large hindwings, as the butterfly usually rests with wings folded above the body and covering the forewings. The undersurface of the large hindwings is mostly filled with a large, irregularly shaped patch of silvery-white.

Males often engage in impressive aerial combat flights. From early morning until around noon, they remain on favorite perch sites in open areas to await passing females, and when a female is sighted, the male launches out and tries to persuade her to alight for copulation. Often the female is sighted by more than one male, and the aerial battle is on. Generally the female continues her flight. In the afternoons, males can be seen hanging upside down from beneath leaves when not nectaring.

The Silver-spotted Skipper frequently appears around mud puddles as it sips the moisture there, as well as in open, sunny areas, where it seeks nectar. It visits many species of flowers, briefly visiting one, then quickly flying to another.

EGG: Round, widely ribbed, pale greenish with red band around middle and reddish on top; laid singly on upper side of food plant leaf.

CATERPILLAR: Head dark brownish-red with two large, orange-red oval spots; body yellow or greenish with darker patches or speckles, and with fine, black crosswise lines; larva

builds shelter by binding leaves of food plant together with silken threads.
CHRYSALIS: Pale brown, banded and streaked in darker brown; forms in silken "nest" among foliage or ground litter.
FOOD PLANTS: False Indigo (*Amorpha fruticosa*), Groundnut (*Apios americana*), Kentucky Wisteria (*Wisteria frutescens*), Southern Hog-peanut (*Amphicarpaea bracteata*), Thicket Bean (*Phaseolus polystachios*), Round-head Bush-clover (*Lespedeza capitata*), and Wild Licorice (*Glycyrrhiza lepidota*). Almost all of the tick-clovers (*Desmodium* spp.) are occasionally used as are the nonnative and cultivated Black Locust (*Robinia pseudoacacia*), Bristly Locust (*R. hispida*), Kudzu (*Pueraria montana* var. *lobata*), Japanese Wisteria (*Wisteria floribunda*), and Chinese Wisteria (*W. sinensis*). The genus *Robinia* is reportedly the favored larval food plants, but *Amorpha*, *Apios*, and *Wisteria* are quite readily used, especially in the southeastern portion of the state.
PARTS EATEN: Foliage.

Brazilian Skipper

(Calpodes ethlius)

Family: Skipper (Hesperiidae)
Size: 1 ¾–2 ¼ inches
Broods: Two to several
Flight time: April–December (all year)
Overwinters: (Adult)
Range: 2, 3, 4, 5, 6 (1, 7)

An inspection of any patch of cannas (*Canna* spp.), especially in city gardens, will probably yield this large Skipper in all its stages. Look for rolled leaf edges, large portions of the leaves eaten, or adults zipping about nectaring on the flowers.

The Brazilian Skipper certainly has a personality all its own, and the more you watch this butterfly, the more intriguing it becomes. Quick in flight and almost secretive, it alights, then appears to be watching to see if you have noticed. Then it quickly flies to another perch in plain sight to do the same thing; the game seems endless.

An especially fast flier with strong, powerful wing beats, the Brazilian Skipper often basks with wings spread in the "airplane" pose, at which times the beautiful markings of translucent spots on both wings are very conspicuous against the dark brown scaling. When the wings are folded, the hindwings are marked with three translucent dashes, and the forewings appear to be marked with two smaller ones. Forewings of this butterfly are very long and pointed.

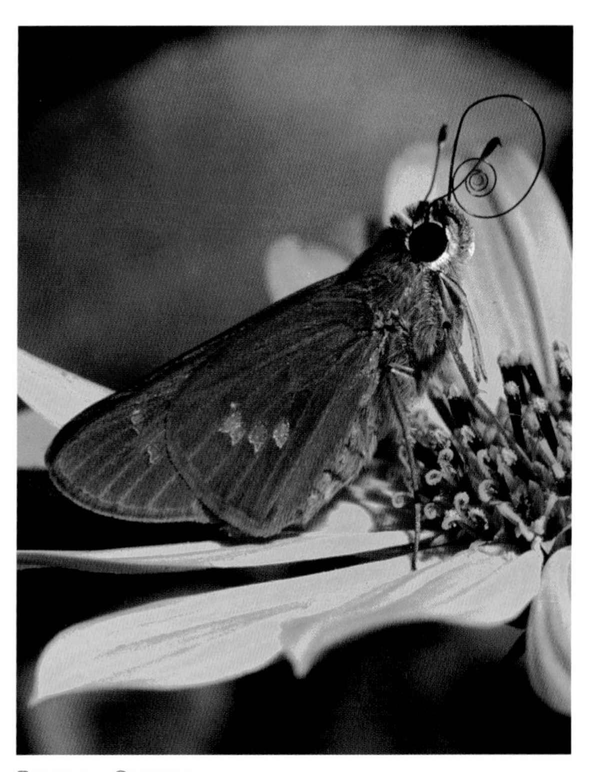

BRAZILIAN SKIPPER

Favorite flowers for nectaring include the Globe Amaranth (*Gomphrena globosa*), Old-fashioned Petunia (*Petunia axillaris*), Summer Phlox (*Phlox paniculata*), and the lantanas (*Lantana* spp.) and cannas.

At times there may be so many larvae on the food plants that practically all of the leaves will be eaten. Often these Skippers disperse to great distances, lay their eggs, and then immediately leave for yet another area. Especially susceptible to viruses, entire colonies are periodically wiped out.

EGG: White to greenish-white becoming reddish; laid singly or in small clusters on upper surface of younger leaves of food plant.

CATERPILLAR: Head of last instar large, orange and black, the body translucent grayish-green with dark line down back and a pale line along each side; rests during day in a leaf stitched together with silk to form tentlike shelter from which it emerges at night to feed.

CHRYSALIS: Head pointed with long, conspicuous proboscis case; body long, slender, pale greenish, pointed at rear, covered with powdery or waxy "bloom"; rests in silken shelter in rolled-up leaf.

FOOD PLANTS: Native Powdery Thalia (*Thalia dealbata*) as well as almost all native and cultivated cannas.

PARTS EATEN: Leaves, usually the midmature ones first.

NOTE: Females reportedly prefer green-leaved cannas with red flowers over the red- or variegated-leaved plants for egg deposition, but some gardeners have found them to show no preference. When a number of caterpillars are feeding, their nighttime chomping can be heard for some distance. If the chrysalis is disturbed, they make a loud "rattling" noise by vibrating rapidly against their leaf shelter. The Brazilian Skipper larvae are sometimes referred to as the "Canna leaf roller" and considered a "pest" on cannas.

Dorantes Longtail (Lilac-banded Longtail)

(Urbanus dorantes dorantes)

Family: Skipper (Hesperiidae)
Size: 1½–2 inches
Broods: One to several
Flight time: April–November (all year)
Overwinters: Chrysalis (adult)
Range: 2, 4, 5, 6

As with other Longtails, the Dorantes is noted for the long, pointed forewings and the hindwings that conspicuously narrow into long, slender tails. The upper surface of the wings is generally dark to grayish-brown, with an

DORANTES LONGTAIL (LILAC-BANDED LONGTAIL)

overall iridescent glimmer of lilac or rosy-lavender covering the hindwings. A grouping of irregularly placed translucent spots occurs on the forewings. The lower surface of the forewings is dark gray; the hindwings are frosty-gray with dark brown spots and banding. Both wings are beautifully tinged purple with iridescent scaling, which is darker and more conspicuous on the hindwings.

This butterfly is found in the coolness of shade more often than in the sun, but it can be found in gardens. It is a great lover of flowers; it feeds for long periods, then darts into the foliage or to a shaded tree trunk to keep cool between feedings. It is a strong flier and rapidly zips back and forth across pathways or small openings before finally settling down on a flower to feed.

It takes nectar from Huisache Daisy (*Amblyolepis setigera*), Missouri Ironweed (*Vernonia missourica*), Narrow-leaf Liatris (*Liatris punctata* var. *mucronata*), Phacelia (*Phacelia congesta*), Summer Phlox (*Phlox paniculata*), Texas Thistle (*Cirsium texanum*), and the lantanas (*Lantana* spp.), verbenas (*Verbena* spp.), mistflowers (*Conoclinium* spp.), and morning glories (*Ipomoea* spp.), as well as many others.

Although the Dorantes Longtail is a common resident only in the southern portion of the state, it emigrates as far north as Kansas and Missouri and eastward to Georgia. It may breed in some areas where the food plant is available. It cannot overwinter in any form in the colder climes, and large numbers perish each year.

EGG: Flattened on both top and bottom, ridged, shiny iridescent green.
CATERPILLAR: Head black; body yellowish-green to reddish-orange, with paler-colored spots, downy with short hairs; forms shelter of silked-together leaves.
CHRYSALIS: Pale brown, has no waxy coating or "bloom" similar to others in this genus; rests in silken "nest" within silked-together leaves of food plant.
FOOD PLANTS: Pigeon-wings (*Clitoria mariana*), Purple Bush-bean (*Macroptilium atropurpureum*), and the cultivated garden bean (*Phaseolus vulgaris*) and lima bean (*P. lunatus*). The native Butterfly-pea (*Centrosema virginianum*) is probably used.
PARTS EATEN: Foliage.

Checkered White

(Pontia protodice)

Family: White/Sulphur (Pieridiae)
Size: 1 ½–2 inches
Broods: Two to several
Flight time: January–November (all year)
Overwinters: Chrysalis (adult)
Range: Throughout

Overall coloration of the Checkered White varies considerably, differing from habitat to habitat and with the seasons; also, the male is less marked than the female. The upper wing surface of spring broods is white with charcoal-gray or brown markings and washes. Veins of the lower surface are lined and speckled in brown or olive-green. The summer male is solid snowy-white, with the exception of a small black dot on the upper surface of the forewing and pale beige or pale brown tracery on the lower surface. Summer females usually have much paler markings than the spring brood. All color gradations of gray, brown, tan, and olive can and do occur in the Checkered White, yet the patterning and flight characteristics are such that identification is not difficult.

Checkered Whites frequent open, sunny spaces and can commonly be found in gardens, fields, and vacant lots and along roadsides. They are great puddlers, and hundreds can be seen at times around small areas of water or temporarily moist areas in a roadway. The Checkered White usually flies in a fast, skipping manner. If disturbed, it flits away into an open area instead of taking refuge among trees or brush. In seeking females for mating, the male flies back and forth near the food plants. Both sexes use ultraviolet reflection instead of scent to identify the opposite sex. The male and the female have different pigments in their wing scales, resulting in ultraviolet light being reflected by the female and absorbed by the male.

These butterflies are known to readily disperse after emerging, traveling as far as the Canadian border and forming new colonies where their weedy food plants are plentiful. Because they cannot overwinter, in the fall there are massive movements back southward.

EGG: Spindle-shaped, yellow, becoming orange; laid singly on bud, flower, or young leaf of food plant.

CATERPILLAR: Blue-green speckled with small black tubercles and with four lengthwise yellow stripes, downy with short, soft, fine hairs; makes no shelter.

CHRYSALIS: Bluish- or grayish-green, marked with black; held upright or horizontally from silken mat and silk strand around body.

FOOD PLANTS: Numerous genera of the Mustard Family (*Brassicaceae*), including native Shepherd's-purse (*Capsella bursa-pastoris*), Virginia Peppergrass (*Lepidium virginicum*), Gregg's Keelpod (*Synthlipsis dissectum*), Clammyweed (*Polanisia dodecandra*), Rocky Mountain Spider-flower (*Cleome serrulata*), and Spectacle-fruit (*Wislizenia refracta*), as well as the cultivated Bird's Rape (*B. rapa*), Field Mustard (*Sinapis arvensis*), Sophia Tansy-mustard (*Descurainia sophia*), Western Tansy-mustard (*D. pinnata*), and broccoli (*Brassica oleraceae* var. *italica*), brussels sprouts (*B. o.* var. *gemmifera*), cabbage (*B. o.* var. *capitata*), and cauliflower (*B. o.* var. *botryris*). Other species reportedly used include Field Pennycress (*Thlaspi arvense*), Tumble-mustard (*Sisymbrium altissimum*), and Sweet Alyssum (*Lobularia maritima*), as well as radish (*Raphanus sativus*).

PARTS EATEN: Flower buds, flowers, fruits, leaves, and tender stems.

NOTE: Both the invasion of the Cabbage

Checkered White

White (*Pieris rapae*) and the agricultural pattern of people are changing the distribution of the Checkered White, continually forcing it to find new areas. With expanding farming practices in the Western states, along with the accompanying introduced or "weedy" members of the Mustard Family, the Checkered White has had no problem becoming established there. Often populations build to serious numbers in some areas, where the Checkered White may be considered a pest. This butterfly was previously placed in the genus *Pieris*.

Bordered Patch

(Chlosyne lacinia adjutrix)

Family: Brush-footed (Nymphalidae)
Size: 1 3/8–2 inches
Broods: Three to several
Flight time: March–December (all year)
Overwinters: Third instar caterpillar (adult)
Range: 1, 2, 4, 5, 6, 7

BORDERED PATCH

One of our most variable butterflies, the Bordered Patch is a little difficult to identify, especially on the wing. However, within its range it is usually abundant and is a frequent visitor to flowers, thus offering many opportunities for closer inspection.

The Bordered Patch is very showy, both with wings open and closed. The upper surface of both wings is primarily black, with a wide band of bright orange. Rows of tiny white dots edge both the band and the wing margins. The lower surface of the wings is black, banded, and dotted in cream or pale yellow and orange.

An avid flower visitor, the Bordered Patch visits almost anything in bloom. It is especially attracted to white and yellow flowers, such as Beebrush (*Aloysia gratissima*), Havana Snakeroot (*Ageratina havanensis*), Annual Sunflower (*Helianthus annuus*), Tooth-leaved Goldeneye (*Viguiera dentata*), Golden Crownbeard (*Verbesina encelioides*), Hispid Wedelia (*Wedelia acapulcensis* var. *hispida*), and dewberries (*Rubus* spp.). When in bloom and available, Beebrush and Golden Crownbeard seem to be top choices for nectaring. It is not uncommon to see several of these butterflies working a large stand of these plants, with a lot of chasing of "intruders" of other species as well as frequent mating pursuits. Males feed on mud, carrion, and dung, as well as flower nectar.

The reproductive cycle of this butterfly is impressive. Females may lay up to five hundred eggs during their lifetime, and the entire life cycle, from egg to adult, is completed in thirty days.

EGG: Pale greenish or yellowish becoming reddish; laid in clusters of more than one hun-

dred on the underside of food plant leaves.
CATERPILLAR: Quite variable, ranging from an all orange to orange-red form, to an all black form with white stripes on the back, to a black form with an orange-red interrupted stripe down the center of the back; younger caterpillars gregarious and usually remain in groups until the fourth or fifth instar, when they begin to disperse; makes no shelter.
CHRYSALIS: Variable, from solid white to white with black markings to solid black; hangs downward from silken mat.
FOOD PLANTS: Annual Sunflower is a major food plant from spring until late summer, with Golden Crownbeard becoming the major choice from late summer until the end of the breeding season in November or December. Giant Ragweed (*Ambrosia trifida*) is used if one of the first- or second-choice plants is nearby as a nectar source. Plants occasionally used include Brown-eyed Susan (*Rudbeckia hirta*), Bush Sunflower (*Simsia calva*), Weak-stem Sunflower (*Helianthus debilis*), Virginia Frostweed (*Verbesina virginica*), Tooth-leaved Goldeneye, and Hispid Wedelia.
PARTS EATEN: Buds, flowers, leaves, and tender stems.

Snout

(Libytheana carinenta)

Family: Snout (Libytheidae)
Size: 1 3/8 –2 inches
Broods: Two or more
Flight time: March–November (all year)
Overwinters: Chrysalis (adult)
Range: Throughout

Snouts are odd-appearing butterflies, still closely resembling their primitive ancestors. Both forewing and hindwings are square-tipped as if deliberately clipped, giving the insect a curiously angular appearance. To make its appearance even more unusual, the two palpi that protect the proboscis are exceptionally long, projecting forward from the head and resembling a "snout" or the beak of a bird. The Snouts are the only butterflies in North America with such long palpi, and the only genus representing the family known as the Long-beaks.

There are several subspecies of this butterfly, with research showing that gradations exist from one subspecies to the next. Coloring varies with the subspecies, but generally the upper surface is blackish-brown with orangish-brown patches and white spots toward the tips of the forewings. The lower surface of forewings is orangish-brown in the basal

SNOUT

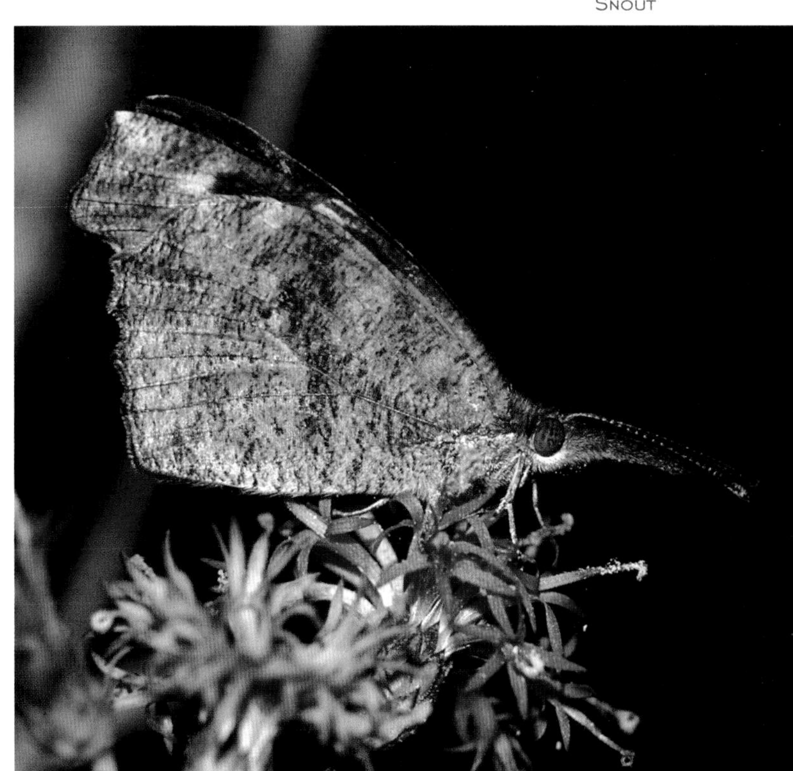

portion, but this area is usually hidden when the insect is at rest. Hindwings are a mottled grayish-brown overlaid with iridescent scales reflecting greens, pinks, and lavenders. In general, females are lighter in hue than the males. The two forelegs of the male are undeveloped, making him appear to have only four legs, but all six legs of the female are well developed.

Snouts do not exhibit any type of mimicry but have evolved an almost perfect leaflike camouflage. When at rest, they fold the wings together and direct the body parallel to a twig, where the wing coloring and shape much resemble a leaf. The forward-projecting palpi and antennae appear as the petiole or "stalk" of a leaf.

Their flight is very swift and rather jerky or fluttery and usually low to the ground, although they will fly several feet high to nectar on flowering shrubs and trees. They are regular visitors to flowers and often gather by the dozens at a flowering plant extra rich in nectar or amino acids, such as Bumelia (*Sideroxylon celastrina*) or plum (*Prunus americana*) and peach (*P. persica*). They are equally attracted to muddy streams and lake margins and are often seen at such sites in the company of various Sulphurs (family Pieridae) and Swallowtails (family Papilionidae). They are rather wary and do not usually allow a close approach.

Snouts are strong voyagers, and massive numbers of them travel northward each year in late summer. These butterflies do not have a return flight southward and perish with the onset of winter.

EGG: Pale green; laid singly on petiole or underside of leaf on young, terminal growth of food plant.

CATERPILLAR: Generally dark green with yellow stripes, the enlarged segment behind head with a pair of black tubercles ringed primarily with yellow; coloring and markings vary greatly according to subspecies; makes no shelter.

CHRYSALIS: Dark green, bluish-green to yellowish-green sprinkled with yellow dots, with whitish or yellowish lines; hangs downward from silken mat but often extends sideways.

FOOD PLANTS: All native hackberries (*Celtis* spp.).

PARTS EATEN: Young foliage.

NOTE: Several subspecies of *L. carinenta* occur in Texas. At this time confusion of names/range occurs in the literature with much more work needed.

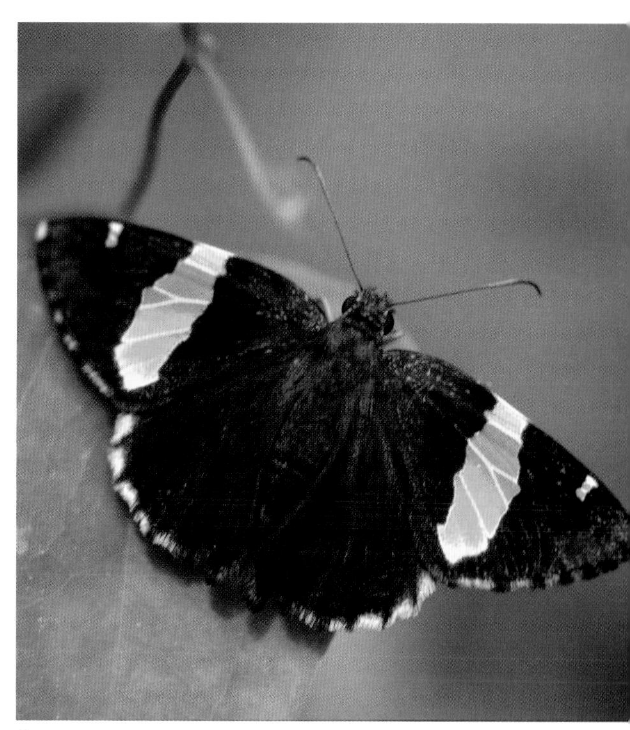

GOLDEN BANDED-SKIPPER

Golden Banded-Skipper

(Autochton cellus)

Family: Skipper (Hesperiidae)
Size: 1 3/8–2 inches
Broods: Two or more
Flight time: April–November
Overwinters: Chrysalis
Range: 2, 3, 6, 7

Not only is the Golden Banded-Skipper one of the less common Skippers in the state but it is also one of the most wary of insects. Even when one flits into view, it is usually difficult to keep it in sight, for it is extremely alert and generally will not tolerate a close approach. When disturbed, it takes off with strong wing beats and in very rapid flight, usually flying quite a distance or even out of sight before alighting again. Even though this Skipper ranges through a large portion of the state, populations are usually small, very local, and often widely separated.

The overall blackish-brown coloring of this Skipper is highlighted by wide, unbroken, golden-yellow bars or bands across both the upper and lower surfaces of the forewings. A small white bar occurs near the tip of each forewing. The fringe of the hindwings is brown-and-white checkered in the upper portion, becoming brownish near the body. The lower surface of the hindwings is gray-frosted with two bands of darker spots near the outer margins.

The Golden Banded-Skipper readily takes nectar, preferring plants in open, moist woodlands near streams, woodland lakes, or humid ravines. It should be looked for on heavy nectar producers such as Buttonbush (*Cephalanthus occidentalis*), False Indigo (*Amorpha fruticosa*), and azaleas (*Rhododendron* spp.), hawthorns (*Crataegus* spp.), ironweeds (*Vernonia* spp.), and milkweeds (*Asclepias* spp.). In the garden it often uses flowering shrubs such as Glossy Abelia (*Abelia* × *grandiflora*) and lilacs (*Syringa* spp.).

EGG: Pale yellow becoming tan or brownish; usually laid in cluster of two or three but occasionally in string of as many as seven or eight, usually at base of food plant leaf.

CATERPILLAR: Head reddish-brown with two eyelike yellow dots; body yellowish-green, yellow dotted or speckled and with broad, clear yellow line along each side; feeds at night, hides during day in rolled or tied leaf shelters; as caterpillars become larger and go to larger leaves, they cut the silken threads that bound the old shelter together, removing signs of their presence.

CHRYSALIS: Dark brown with greenish tint, covered with waxy or powdery “bloom”; remains in silken “nest” in leaf shelter.

FOOD PLANTS: Most commonly Southern Hog-peanut (*Amphicarpaea bracteata*), although Pigeon-wings (*Clitoria mariana*) and Purple Stylisma (*Stylisma aquatica*) are also given in the literature as probable food sources. Wright’s Bean (*Phaseolus filiformis*) is given for the western portion of the state.

PARTS EATEN: Foliage.

Sickle-winged Skipper

Sickle-winged Skipper

(Eantis tamenund)

Family: Skipper (Hesperiidae)
Size: 1 3/8–1 7/8 inches
Broods: One to many
Flight time: May–October (all year)
Overwinters: Chrysalis (adult)
Range: 2, 4, 5, 6

This is one of the largest Skippers, and its size, shape, and coloring are most distinctive. The wings are wide and rounded, with the forewing indented just below the hooked tip, forming a "sickle." The upper surface of this wing is blackish-brown to mahogany-brown, with pale areas of purplish-gray forming bands and irregular groupings of spots. A beautiful iridescent sheen of coppers and lavenders covers both wings. The undersurface of both wings is similar but paler. Females are paler in overall coloration and appear somewhat more mottled, instead of banded as in the males, but the hooked forewing tips and overall violet sheen easily separate this Skipper from all others.

Sickle-winged Skippers visit flowers often, especially those close to shrubby borders or around clusters of trees. They move about with an unusual jerky flight, until alighting on a flower to nectar or on a rock or leaf, where they rest with wings spread tightly against the surface. During midday or periods of extremely high temperatures, it is not uncommon to see them fly to the cooler underside of a leaf after nectaring.

They are especially attracted to Beebrush (*Aloysia gratissima*), Texas Kidneywood (*Eysenhardtia texana*), and lantanas (*Lantana* spp.) and mistflowers (*Conoclinium* spp.).

These Skippers are year-round residents in the Rio Grande Valley area, but in some years they make flights northward as far as Kansas and Arkansas during the summer and fall months.

EGG: Not described; laid singly on upper side of food plant leaf.

CATERPILLAR: Yellowish-green or grayish-blue, with darker stripe down back and broad band of yellow dashes along each side; lives in silk-lined nestlike shelter formed in a leaf.

CHRYSALIS: Green covered with whitish powder or "bloom"; bound by silken thread, resting in "nest" of silked-together leaves of food plant.

FOOD PLANTS: Lime Prickly-ash (*Zanthoxylum fagara*) and possibly some cultivated *Citrus*.

PARTS EATEN: Foliage.

Hoary Edge

(Achalarus lyciades)

Family: Skipper (Hesperiidae)
Size: 1 ½–1 7/8 inches
Broods: Two
Flight time: April–December
Overwinters: Last instar caterpillar/chrysalis
Range: 2, 3, 6

HOARY EDGE

The Hoary Edge is another uncommon and widely distributed Skipper, similar in appearance to both the Golden Banded-Skipper (*Autochton cellus*) and Silver-spotted Skipper (*Epargyreus clarus*). The dark, blackish-brown, triangular-shaped forewings are marked by four or five yellow-orange, squarish bars forming a translucent band above and below. The forewing above is marked by a paler shade of brown between the orange bar and the outer margin. The lower surface of the hindwings is mottled black and dark brown near the body, with the outer half of the hindwing conspicuously frosted in a large, silver-white patch. The irregular shape and amount of "frosting" of this patch give it a "smeared" effect.

While not as fast or strong a flier as its two look-alikes, the Hoary Edge is a mover, and when not basking or perched and waiting for females, the male seems never to be still for long. Its stay at any one flower is brief, but it will continue to nectar in an area at length if not disturbed.

Frequently seen around flowers, these butterflies are continually alert—dipping into the middle of a blossom and then quickly backing out, looking around, and going back in for more feeding. They use many different plants but obviously prefer those in open, sunny spots. The best places to look for these beauties are along the outer edges of wide roadsides where unmowed plants are in flower.

For a courting territory, males usually choose small openings within woodlands or along woodland edges, brushy fencerows, or the sunny edges of shrubby areas in parks or gardens where they perch on outer twigs or leaves and wait for passing females. Females fly by in search of nectar-filled flowers, and the males dash out, circling around them, fluttering their wings, and releasing certain scent pheromones. Often the female is sighted by more than one male, and the males then begin contesting the rights of territory. Courting is forgotten until one of the males is persuaded to leave.

EGG: Whitish or creamy; laid singly beneath leaves of food plant or often on nearby plants.

CATERPILLAR: Head dark reddish-purple to black, no facial markings, covered with very short, stiff hairs; body pale to dark green becoming pinkish in latest instars, blue-green stripe down the back, and a narrow yellow stripe along each side, covered in small yellow dots and short pale hairs.

CHRYSALIS: Pale brown with dark and paler yellowish-tan patches and black dots; rests in silked-together leaves or debris at base of food plant.

FOOD PLANTS: False Indigo (*Amorpha fruticosa*), Downy Bush-clover (*Lespedeza hirta*), Texas Bush-clover (*L. texana*), Little-leaved Tick-clover (*Desmodium ciliare*), Canadian Tick-clover (*D. canadense*), Panicled Tick-clover (*D. paniculatum*), Bare-stem Tick-clover (*D. nudiflorum*), and Trailing Tick-clover (*D. glabellum*).

PARTS EATEN: Foliage.

Amymone (Common Mestra)

(Mestra amymone)

Family: Brush-footed (Nymphalidae)
Size: 1 3/8 – 1 7/8 inches
Broods: One to several
Flight time: March–December (all year)
Overwinters: (Adult)
Range: 4, 5, 6, 7 (1, 2, 3)

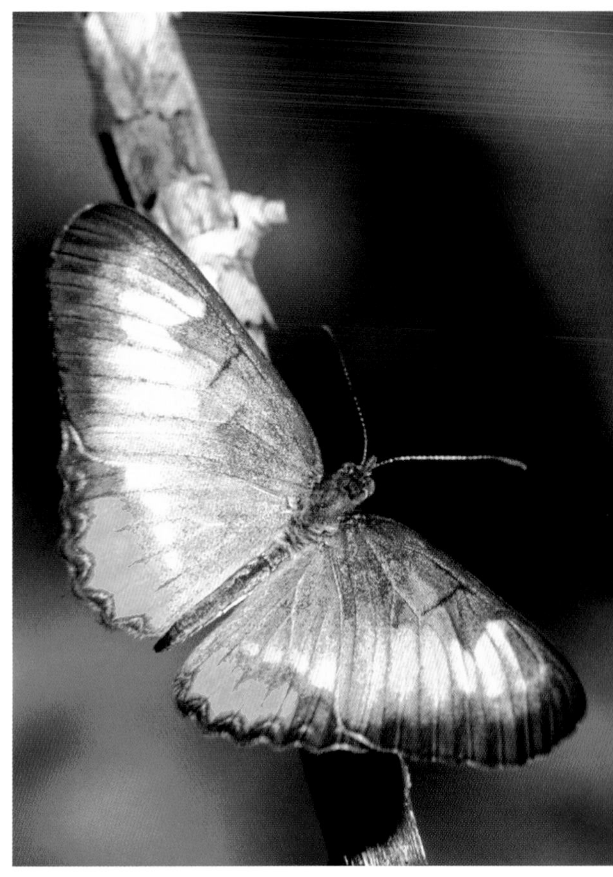

AMYMONE (COMMON MESTRA)

Common in the Rio Grande Valley and occasionally breeding as far north as Austin and possibly Waco in the Hill Country, the Amymone sometimes wanders to Minnesota in small numbers. With colorings and markings quite different from those of most other members of the Brush-foots, the Amymone is easily recognized in the field. The upper surface of both wings ranges in various grays at the base near the body, dusted with a pearly-white sheen in the middle section, and with a wide charcoal band on the tips of the forewings. A bright orange-yellow band edges the hindwings. The lower surface of the wings is brownish-orange, with large, irregular creamy-white spots forming bands. Broods occurring during periods of unusual moisture are reportedly darker in general coloration. The soft, subtle coloring of a freshly emerged adult is truly beautiful, but the colors quickly fade, scales are lost, and the fragile wings easily become tattered and torn.

Flight of the Amymone is slow and sailing, with few wing beats. Never flying far even when disturbed, it stays close to the ground, taking nectar from low-growing flowers and slowly making its way from plant to plant. If really frightened, the Amymone quickly darts into nearby vegetation and disappears. This butterfly is at home wherever the flowers are, whether dense woodlands or the edges of hot, sunny areas.

EGG: Somewhat globe-shaped, ridged to near top, covered in soft hairs, pale yellow.
CATERPILLAR: Head with two longer spines than on body, each of which ends with a knob or crest of smaller spines; body brown with green diamond shapes on back and with eight rows of spines; makes no shelter.
CHRYSALIS: Green or brown; hangs downward from silken mat.
FOOD PLANTS: Catnip Noseburn (*Tragia ramosa*) is the only known food plant.
PARTS EATEN: Foliage.

Fulvia Checkerspot

(Chlosyne fulvia fulvia)

Family: Brush-footed (Nymphalidae)
Size: 1¼–1⅞ inches
Broods: Three
Flight time: April–October
Overwinters: Third instar caterpillar
Range: 1, 6, 7

Seldom wandering far from its food plant, the paintbrushes (*Castilleja* spp.), the Fulvia Checkerspot flies slowly about, nectaring at various flowers and lazily basking in the sun. The general color and pattern of the upper surface of the wings are variable, but the lower surface is distinctive, making this one easy to recognize. With a basic ground color of orange, the upper surface of both wings is bordered and lined with dots and spots of cream, yellow, and black. The veins are black, merging into the checkered marginal fringe. The undersurface of the forewings is orange with black- and cream-colored bands near the outer margins; the hindwings are cream-colored with black veins and a black band enclosing a row of cream-colored spots forming a "chain." Males and females of this species usually look very different from one another, the males appearing smoky or dusky and much darker. Females are usually larger than the males.

While the Fulvia Checkerspots will be found in gardens, where they show a preference for yellow-colored flowers, in the wild

Fulvia Checkerspot

THEONA CHECKERSPOT

they inhabit rocky ridgetops and slopes. They are not widespread but tend to stay in colonies near the food plant. Males gather on hilltops to seek females for mating.
EGG: Cream to pale yellow becoming orange; laid singly or in clusters of ten to thirty on lower surface of leaves near base of host plant.
CATERPILLAR: Yellow to dull orangish with black bands and numerous black spines; gregarious in loose silk web when young, later dispersing.
CHRYSALIS: White, mottled with black stripes and spots and with pale brown or tannish between the black stripes; hangs downward from silken mat.
FOOD PLANTS: The paintbrushes, Tall (*C. integra*), Woolly (*C. lanata*), Purple (*C. purpurea*), Downy (*C. sessiliflora*), and Texas (*C. indivisa*).
PARTS EATEN: Fleshy bracts, tenderest leaves, and buds if foliage not available.

Theona Checkerspot
(Chlosyne theona)

Family: Brush-footed (Nymphalidae)
Size: 1–1 5/8 inches
Broods: Two to several
Flight time: April–October (all year)
Overwinters: Third or fourth instar caterpillar (adult)
Range: 1, 4, 5, 6, 7

Theona Checkerspots are never found very far from their larval food plants. This, their weak flight, and distinctive coloration make for easy identification. Upper wing surfaces of this butterfly are dark grayish or blackish, with orange spots forming bands. A row of pale yellow or cream-colored rectangles forms a wide band across both wings, with another row of orange rectangles just below it, making both wings double-banded. Both wings are bordered in black with a double row of small white dots. Lower wing surfaces are a series of bands of white or cream and reddish-orange highlighted by conspicuous black veins. The abdomen is black with very narrow yellow crosswise bands on the upperside.

This is a butterfly of open country. In its natural habitat it flies slowly among scattered shrubbery of the brushlands or chaparral, stopping to take nectar from blossoms of low-growing herbs. Often after a rain shower it spends much time around mud puddles or concave rocks where the moisture remains standing. Natural seepage areas are an excellent place to look for the adult in the wild. For the home garden within its range, good nectar sources, a group planting of *Leucophyllum* or *Castilleja*, and a constantly moist area readily bring it in.

EGG: Cream-colored; laid in cluster beneath leaf of food plant; often several leaves of the same plant are utilized by the same female.

CATERPILLAR: Velvety brownish-black, dotted and banded with cream, and with many branching spines; young caterpillars tend to stay together in close groups, almost completely defoliating certain portions of the food plant; makes no shelter.

CHRYSALIS: Smooth, white with few black dots and stripes, black stripe down back with orangish spots; hangs downward from silken mat.

FOOD PLANTS: Big Bend Silver-leaf (*L. minus*), Cenizo (*L. frutescens*), Violet Silver-leaf (*L. candidum*), and almost all paintbrushes (*Castilleja* spp.).

PARTS EATEN: Foliage.

NOTE: The Theona Checkerspot enters the United States from Mexico along two different routes, forming two distinct populations. One population enters the United States from western Mexico along the Pacific, with the Texas population entering along the Gulf Coast from eastern Mexico.

Falcate Orangetip

(Anthocharis midea)

Family: White/Sulphur (Pieridae)
Size: 1 3/8–1 3/4 inches
Broods: One
Flight time: February–May
Overwinters: Chrysalis
Range: 1, 2, 3, 4, 5, 6

One of the earliest butterflies to appear each spring, the Falcate Orangetip can be seen flying low to the ground around garden shrubbery and along the edges of open woodlands. It is very local in distribution, but once one is spotted, there are often several in the same area.

Undaunted by unpredictable spring weather, the Falcate Orangetip is on the wing even on very cool or partially cloudy days. The male seems to never perch, continually flying back and forth and often along the same route day after day. Its stops for nectar are numerous but usually very brief. Females are usually seen less often than males but seem to visit flowers more often. They spend a lot of time hovering about, low to the ground, going in and out of brambles, and inspecting numerous plants for the proper ones for egg deposition. Normal flight is composed of a short period of slow sailing or gliding, then a series of quick, jerky wing beats, then more sailing.

For the most part, the upper wing surface of the Falcate Orangetip is a soft, snowy-white. Forewings of the male bear a solitary, elongated black dash about midway in from the margins, and the wings are tipped with a bright orange patch. Forewings of the female bear only the black dash. Lower wing surfaces of both sexes are similar, with the forewings white, black-dotted, and with a patch of greenish-brown mottling near the tip. The hindwings are beautifully marbled in greenish to grayish. Forewings of both sexes are conspicuously hooked (falcate) at the tips. This curving and the orange patch on the wings of the male give this butterfly its common name.

Because it is only a spring visitor to the garden, flower preferences are those that are low to the ground and in full bloom by March and April. Some plants regularly visited are Rose Verbena (*Glandularia canadensis*), Prairie Verbena (*G. bipinnatafida*), False Garlic (*Nothoscordum bivalve*), Spring Beauty (*Claytonia virginica*), Selfheal (*Prunella vulgaris*), and Yellow Star-grass (*Hypoxis hirsuta*), as well as dewberries (*Rubus* spp.), bluets (*Houstonia* spp.), violets (*Viola* spp.), wild onions (*Allium* spp.), and various members of the Mustard Family (*Brassicaceae*).

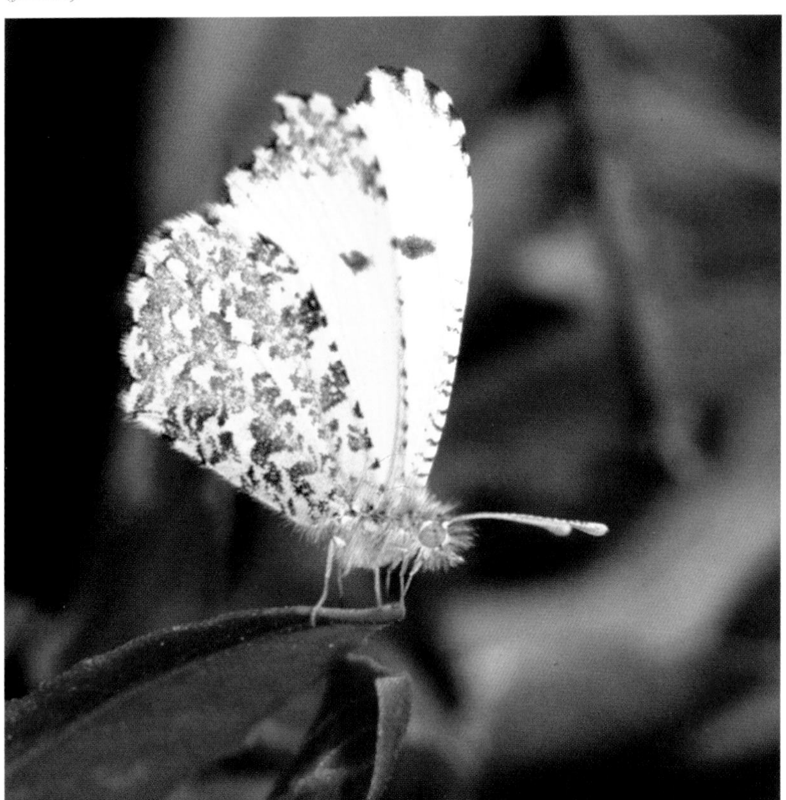
FALCATE ORANGETIP *(female)*

EGG: Elongated, ribbed, yellow-green becoming orange; laid singly, usually at the base of a flower, with rarely more than one egg per plant deposited, but several females may deposit on same plant.

CATERPILLAR: Primarily yellow-green, with conspicuous orange stripe down the center of the back and with blue, white, and yellow stripes along the sides; cannibalistic in early stages; makes no shelter.

CHRYSALIS: Head portion long-pointed, greenish, yellowish becoming tannish-brown with brown and black mottling; slender, crinkled; rests upright from silken mat and silken strand.

FOOD PLANTS: Best-known food plants in Texas are Spring Bittercress (*Cardamine rhomboidea*) in the east and Brazos Rockcress (*Arabis petiolaris*) in the Hill Country and Valley area. Others include Sand Bittercress (*C. parviflora*), Downy Bittercress (*C. hirta*), Virginia Peppergrass (*Lepidium virginicum*), Prairie Peppergrass (*L. densiflorum*), and Woolly-fruit Peppergrass (*L. lasiocarpum*).

PARTS EATEN: Buds, flowers, and seedpods.

NOTE: It usually takes two years for this butterfly to emerge.

Great Purple Hairstreak (Great Blue Hairstreak)

(Atlides halesus halesus)

Family: Gossamer-winged (Lycaenidae)
Size: 1 ¼–1 ¾ inches
Broods: Two to several
Flight time: February–November (all year)
Overwinters: Chrysalis (adult)
Range: Throughout

Wing scaling on the upper wing surface of the Great Purple is one of the most brilliant iridescent blues found among the hairstreaks, making this one of the most unusual and beautiful. The lower wing surface is not quite so lavish in coloring, being more a purplish or bluish sheen over dark charcoal-gray, with spots of bright red on the base near the body and a patch of metallic blue and green spots near the tail. A large patch of brilliant blue covers the base of the forewing but is usually hidden by the hindwing. The upper portion of the abdomen is black with white dots; the rear portion is a spectacular bright reddish-orange. Males are more brilliantly colored than the females but with only one tail on the hindwing, whereas the female has two.

Commonly visiting flowers, the Great Purple Hairstreak can readily be found in gardens if certain conditions are favorable. During most of the year, it does not wander far from trees infested with the semiparasitic mistletoe (*Phoradendron* spp.), its larval food plant. So, if there are trees near your garden with a healthy and thriving growth of this plant, the Great Purple Hairstreak will most likely be a regular visitor to nearby flowers. In some areas greater numbers of these butterflies can be seen in the spring or early summer, but at least a few are almost always around, and anytime this striking beauty is sighted is a special treat.

Some flowers regularly visited for nectar include cultivated fruit tree blossoms such as peach (*Prunus persica*) and plum (*P. americana*). Others include natives such as Mexican Plum (*P. mexicana*), Virginia Frostweed (*Verbesina virginica*), Giant Ironweed (*Vernonia gigantea*), goldenrods (*Solidago* spp.), various wild onions (*Allium* spp.), redbuds (*Cercis* spp.), asters (*Symphotrichum* spp.), and thistles (*Cirsium* spp.).

Males are quite long-lived for a Hairstreak, often surviving three weeks or more. Their chosen area for pursuing mates is usually very local, such as one particular tree and a

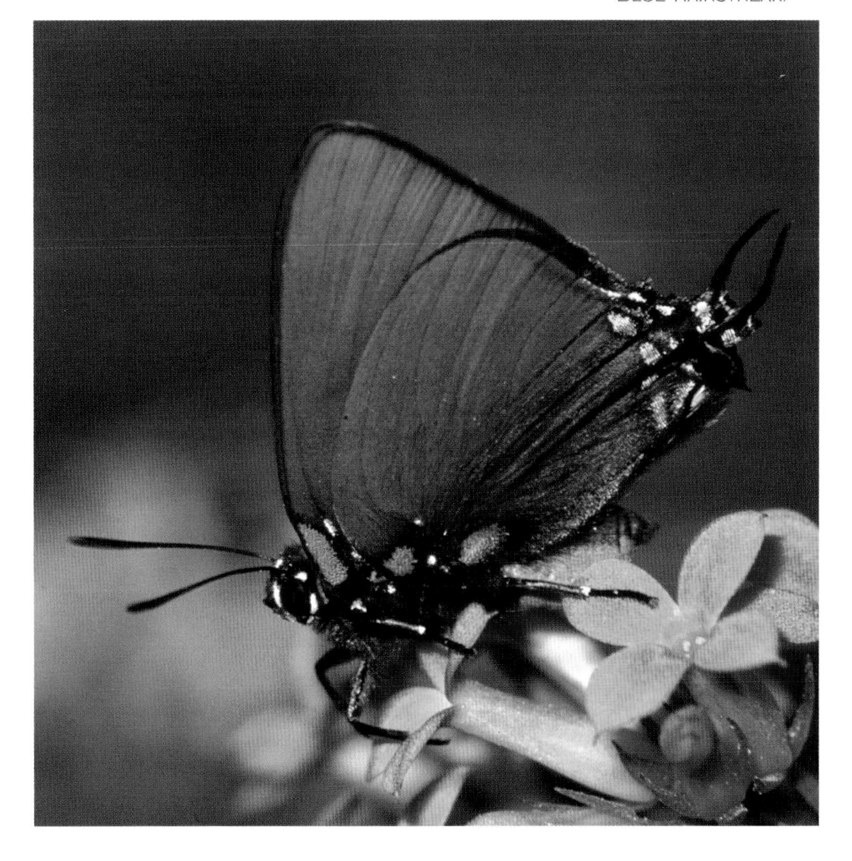

Great Purple Hairstreak (Great Blue Hairstreak)

very small surrounding territory. They usually perch in the open near the top of the tree from midday to dusk.

EGG: Somewhat rounded, flattened on top and bottom with deep depression on top, outer surface covered with tiny bumps arranged in more or less regular horizontal rows, white to creamy, shiny.

CATERPILLAR: Slug-shaped, dull green covered with short hairs and appearing velvety, with tiny, bluish-white diamond behind head; makes no shelter.

CHRYSALIS: Short, stout, rounded, brown, heavily mottled with black; attached only by silken strand beneath bark or among debris at base of food plant tree.

FOOD PLANTS: Christmas Mistletoe (*Phoradendron tomentosum*) growing on Honey Mesquite (*Prosopis glandulosa*) or elms (*Ulmus* spp.) and hackberries (*Celtis* spp.) is preferred. Other host trees include ashes (*Fraxinus* spp.), cottonwoods (*Populus* spp.), oaks (*Quercus* spp.), and willows (*Salix* spp.), as well as Sycamore (*Platanus occidentalis*). Oak Mistletoe (*Phoradendron villosum*) grows on oaks; and Rough Mistletoe (*P. hawksworthii*), on junipers (*Juniperus* spp.). Several other trees are used as well.

PARTS EATEN: Young foliage and occasionally male flowers.

Texan Crescent

(Anthanassa texana)

Family: Brush-footed (Nymphalidae)
Size: 1 ¼–1 ¾ inches
Broods: Three or more
Flight time: March–November (all year)
Overwinters: Caterpillar/chrysalis (adult)
Range: Throughout

With a low, strong, fluttering flight, the Texan Crescent moves from one clump of flowers to another. Once on a good nectar source, such as frogfruit (*Phyla* spp.) or the fall-flowering Joe-Pye Weed (*Eutrochium fistulosum*), however, it remains for long periods, basking in the warm sun while lazily feeding from the plentiful flowers. Many confrontations with bees, wasps, and other butterflies take place, all of them vying for the freshest flowers and most nectar, but the Texan Crescent usually sends them scurrying with strong flicks of its wings.

Primarily a dark brownish-black in color, the upper surface of the Texan Crescent's wings is heavily dotted in white; the large dots, somewhat squarish and variously placed on the forewings, form a horizontal row on the hindwings. Orangish bars and splotches combine to form bright patches at the bases of the wings near the body. The lower surface of the forewings has white markings near the tip, and the hindwings are generally more buff-colored and marked with black lines and dots. A white band crosses the wings about midway. Both body and basal area of the wing near the body are overlaid with iridescent greenish, copper, and purple scaling. Females are usually larger than the males.

Texan Crescents are denizens of low, open, shrubby-type areas, such as in rocky creek bottoms and along edges of thin, rocky woodlands or open trails. It also readily inhabits flower gardens for nectar and will remain in the area if some shrubs or low trees are present, providing it has an escape site. It can be found at various altitudes and is at home on tops of low mountains, along the shrubby edge of a foothill stream, or in flat, coastal grasslands.

In seeking mates, male Texan Crescents usually perch on an exposed twig, rock, or grass blade in a low habitat such as a gully or an open area between hills or along a mountain stream. From his chosen perch within his territory, he flies out to inspect everything that passes by and fiercely chases off other butterflies, especially other male Texan Crescents. Females are an exception, of course; when one flies into a male's territory, he begins an elaborate courtship dance, flying loops behind and above her, attracting her attention, and hopefully persuading her to mate.

Some individuals of the first spring broods of Texan Crescents have a tendency to wander, and specimens are occasionally seen as far north as Minnesota, east to Illinois, and west to California.

EGG: Pale greenish-yellow; laid in clusters on underside of leaves of food plant.

CATERPILLAR: Young larva greenish-brown, with four rows of pale-colored, flattened bumps or tubercles, each bump bearing a hair; mature caterpillar yellow-brown, with the sides striped in black and white; one white band broad and mottled with greenish and brown; spines on lower portion of body greenish-white; all other spines brown; young caterpillars feed and rest in groups in silken shelter on underside of leaves.

CHRYSALIS: Tannish to brown, practically unmarked, short spines down back and on sides; held by silken strand, rests in silken "nest" within folded-over leaf of larval food plant.

FOOD PLANTS: Bracted Dicliptera (*Dicliptera brachiata*), American Water-willow (*Justicia americana*), Lance-leaved Water-willow (*J. ovata* var. *lanceolata*), and the cultivated Flamingo Plant (*Jacobinia carnea*), Mexican Ruellia (*Ruellia brittoniana*), and Green Shrimp Plant (*Blechum Pyramidatum*). In Louisiana, the cultivated King's Crown (*J. suberecta*) and Mexican Honeysuckle (*J. spicigera*) have recently been discovered to be excellent larval food plants.

TEXAN CRESCENT

PARTS EATEN: Foliage.

Little Yellow

(Pyrisitia lisa lisa)

Family: White/Sulphur (Pieridae)
Size: 1 ¼–1 ½ inches
Broods: Three or more
Flight time: February–November (all year)
Overwinters: Caterpillar/chrysalis (adult)
Range: Throughout

Although small in size, the Little Yellow makes up for it in numbers and is probably the state's most plentiful butterfly. It is on the wing year-round in the southern portions of its range and by midsummer is encountered in every field, meadow, garden, and open woodland and along almost every roadside in the rest of its range. It is not a fast flier and stays low to the ground, where it visits flowers readily and also seeks areas of moisture. Inclement weather does not seem to bother the Little Yellow, for it can be seen flying on windy and cloudy days. It is not uncommon to find it in great numbers, along with Hairstreaks and Blues (family Lycaenidae), Clouded Sulphur (*Colias philodice*), or Southern Dogface (*Zerene cesonia*), taking moisture from mud puddles or seepage areas.

The basic color of the male is usually a clear yellow, with a solid black border on the upper surface of the wings. Females are usually a little larger in size, and their wings have spotted borders. Occasionally a female is chalky-white or creamy-white with black markings and is known as forma *alba*. The undersurface of both sexes is yellowish-green with minute dark speckling and brownish blotches and smudges. Females have a large, dark-colored, but indistinct spot near the tip of the hindwing.

A courting male patrols constantly during the day seeking females. The upper surface of the wings of the male reflects ultraviolet light, and he also uses pheromones to attract a mate.

Although Little Yellows migrate north as far as Canada during the summer and even produce two or more broods there, they cannot survive the harsh winters. Each fall the adults either fly southward or perish, with new adults again flying northward the next year.

EGG: Minute, pale green when deposited; laid singly on upper surface of food plant leaf, usually between two leaflets.

CATERPILLAR: Pale green, marked lengthwise with white and green lines, downy with fine, short hair; makes no shelter.

CHRYSALIS: Translucent green tinged bluish, with black dots; held upright from silken mat and with silken strand around middle of body.

FOOD PLANTS: Partridge-pea (*Chamaecrista fasciculata*), Delicate Sensitive-pea (*C. nictitans*), Powderpuff (Mimosa strigillosa), Maryland Senna (*Senna marilandica*), Southern Hog-peanut (*Amphicarpaea bracteata*), Illinois Bundleflower (*Desmanthus illinoensis*), and the introduced Coffee Senna (*S. occidentalis*).

PARTS EATEN: Foliage.

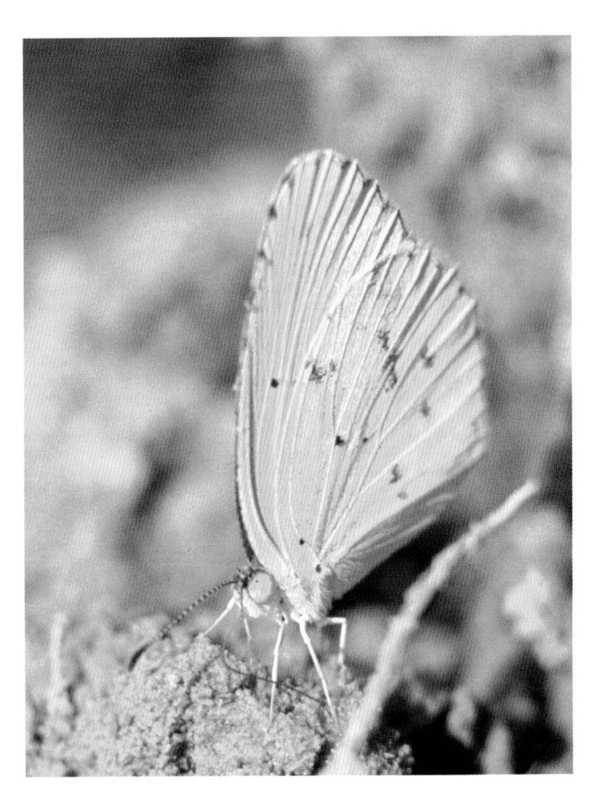

LITTLE YELLOW

Pearl Crescent

(Phyciodes tharos tharos)

Family: Brush-footed (Nymphalidae)
Size: 1 ¼–1 ½ inches
Broods: Two to several
Flight time: February–December (all year)
Overwinters: Third instar caterpillar (adult)
Range: Throughout

Pearl Crescents are one of the most abundant and familiar small butterflies in the state. In almost any open field, meadow, flower garden, or roadside, this little gem can be seen flying low, taking nectar or sipping moisture from wet ground. Males patrol for females with which to mate or perch with closed wings on a bare branch, rock, or grass blade, darting out to inspect everything that passes by. It is especially fond of darting at other butterflies being photographed and sending them off in terrified flight.

The upper surface of the wings is primarily orange, with numerous black blotches, lines, and spots. Bases of both wings are intricately marked with networks of fine curving or scrawly lines. A row of small black dots borders the hindwings near the outer margins. Females have more black than males, and both sexes are darker in the early-spring form. The lower surface of the forewings is pale orange, with black and cream patches vaguely forming a border, while the hindwings are softly mottled in cream and yellows, with fine, brown, curvy lines. Along the darker border of the hindwing is a conspicuous purplish-brown patch surrounding a pearly crescent- or boomerang-shaped mark.

Pearl Crescents often use the same plants for nectaring as ones used for egg deposition and are almost always found around whichever aster (*Symphyotrichum* spp.) is in flower at the time. Other members of the Aster Family (Asteraceae) are preferred to the deeper-throated flowers since the butterfly's proboscis is short, making nectar gathering difficult. While feeding, Pearl Crescents move about with wings opened flat or almost so and characteristically raise them up and down while continually turning the body around and around in slow circles. Flight is usually very low, hardly rising above the grasses or flowers. For protection they usually dart down among plant stems and debris close to the ground.

EGG: Slightly elongated, ribbed, whitish to pale green; laid in a mass, sometimes in a layer, on underside of basal leaves of food plant.

CATERPILLAR: Dark reddish- or chocolate-brown with tiny white dots, lined with

Pearl Crescent

black and cream or yellow, and covered with numerous branching spines; gregarious when young; makes no shelter.
CHRYSALIS: Mottled gray or yellowish to tannish with brownish markings, many bumps along back; hangs downward from silken mat.
FOOD PLANTS: Many species of asters such as Willow-leaf Aster (*S. praealtum*), Bushy Aster (*S. dumosum*), Calico Aster (*S. lateriflorum*), Texas Aster (*S. drummondii* var. *texana*), and the cultivated New England Aster (*S. novae-angliae*). Probably several other species are used, especially in the western and Plains areas.
PARTS EATEN: Foliage; at first only the lower surface but later the entire leaf.

White M Hairstreak

(Parrhasius m-album)

Family: Gossamer-winged (Lycaenidae)
Size: 1–1 ¼ inches
Broods: Two to several
Flight time: February–December
Overwinters: Chrysalis
Range: 2, 3, 4, 6

WHITE M HAIRSTREAK

Like most other Hairstreaks, the White M is an avid feeder on flower nectar and crawls from flower to flower when it finds a good nectar source. While feeding, it is relatively docile but, if startled, takes off in rapid, often erratic flight. Although it is widely distributed and seen in many different habitats, nowhere or at any time is it abundant. Peak emergence occurs usually in the early spring months of March and April. In the fall the butterflies are again frequently seen but are usually very local in distribution and in groups. At this time, they will often be in the company of other Hairstreaks around a good nectar source, and close observation is necessary to separate the different species present.

White M Hairstreaks can be identified quite easily by the large and conspicuous M (or W—depending on which way the wing is viewed) on the hindwings formed by narrow white and black bands. At the outer angle and above the two long, narrow tails are large red and blue areas. The upper sides of the wings are brilliant, iridescent blue with wide black borders along the wing margins. Occasionally this butterfly will partially open the wings while nectaring or basking, but more often even basking is done with the wings closed and turned sideways to the sun. The beauty of the upper sides is generally viewed only briefly as the butterfly alights and takes off in flight.
EGG: Rather flat, pale greenish becoming white; laid singly on buds, young leaves, and twigs of food plant.
CATERPILLAR: Head brownish-black, smooth;

body from dull red to red and green with various markings of olive to light yellowish-green, with darker green stripe along the back and with seven dull, dark green, slanting stripes along each side, downy with soft hairs; makes no shelter.

CHRYSALIS: Brown with darker brown blotches and with black ridge along abdomen segments; when disturbed, the abdomen moves, producing faint "squeaking" sounds; held by silken thread, probably in the leaf litter beneath host oak (*Quercus* spp.) trees.

FOOD PLANTS: Probably any species of oak, especially trees with narrow or very lobed leaves, such as Live Oak (*Q. virginiana*), Water Oak (*Q. nigra*), Post Oak (*Q. stellata*), White Oak (*Q. alba*), and Shumard Red Oak (*Q. shumardii*).

PARTS EATEN: Foliage, especially the reddish to pale green immature leaves or the tenderest portions of mature leaves.

Fiery Skipper

(Hylephila phyleus)

Family: Skipper (Hesperiidae)
Size: 1–1 ¼ inches
Broods: Several
Flight time: March–December (all year)
Overwinters: Chrysalis (adult)
Range: Throughout

A familiar sight in most gardens is this small, rapidly flying wizard among the flowers. With quick, darting take-offs the Fiery Skipper seems to "leap" from one flower to the next and never stays long at any of them. These butterflies are rather pugnacious and will flit at other nectaring butterflies until they leave.

There is a difference in coloration between the male and female of this species. The upperside of the male is bright orange with a long black mark on the forewing and sawtooth marginal border; the female is dark brown with a band of orange spots across the wings. Below, the male is paler orange with a scattering of black spots; the female is pale yellow or somewhat brownish with scattered dark spots and a wide, paler band. The male is smaller than the female.

One of the easiest identifying features of this Skipper is its conspicuously short antennae. The exceedingly long proboscis enables Fiery Skippers to partake of nectar deep within tubular blossoms of such flowers as salvias (*Salvia* spp.), penstemons (*Penstemon* spp.), and morning glories (*Ipomoea* spp.).

EGG: Pale turquoise, shiny; laid singly on food plant leaf.

CATERPILLAR: Gray, pale green to greenish-brown or yellowish-brown, with dark stripes along back and sides; lives in silk-lined shelter at base of food plant.

FIERY SKIPPER

CHRYSALIS: Pale tan, mottled with darker brown dashes; rests horizontally in silken shelter among roots and debris at base of food plant.
FOOD PLANTS: Native grasses such as Southern Crab (*Digitaria ciliaris*), Teal Love (*Eragrostis hypnoides*), the nonnative St. Augustine (*Stenotaphrum secundatum*) and Common Bermuda (*Cynodon dactylon*), and the cultivated Sugar Cane (*Saccharum officinarum*).
PARTS EATEN: Foliage.
NOTE: This is the only North American member of the genus *Hylephila.*

Common Checkered-Skipper

(Pyrgus communis)

Family: Skipper (Hesperiidae)
Size: 1–1 ¼ inches
Broods: Three to several
Flight time: February–November (all year)
Overwinters: Last instar caterpillar (adult)
Range: Throughout

This is possibly the most abundant and most often encountered Skipper in North America. In Texas it can be seen on the wing throughout the year in the southern portion of its range and all through the breeding season in the rest of the state. Common Checkered-Skippers are one of the earliest to fly in spring as well as late fall, visiting whatever is in bloom at the time.

Coloration of this butterfly is quite variable, occasionally causing some confusion in identification. Generally, the upper surface of both wings is dark gray, checkered, and banded with white. The lower surface of both wings is much paler, variously mottled in grays and white. Wing margins are fringed, the fringe checkered dark gray and white. Long, hairlike scales produce a bluish or turquoise sheen at the base of the wings and on the body. Males are somewhat paler in color than females.

Frequently visiting flowers in short, fast, direct flights, this Skipper often stops to bask with wings spread wide. While feeding, it continually turns round and round, and the wings make frequent up-and-down movements. It usually remains low to the ground.

Favorite flowers used for nectaring include New Jersey Tea (*Ceanothus americanus* var. *pitcherii*), Bur-clover (*Medicago polymorpha*), False Garlic (*Nothoscordum bivalve*), Golden Crownbeard (*Verbesina encelioides*), New England Aster (*Symphyotrichum novae-angliae*), Spring Beauty (*Claytonia virginica*), White Sweet-clover (*Melilotus albus*), and various zinnias (*Zinnia* spp.), milkweeds (*Asclepias* spp.), dewberries (*Rubus* spp.), bluets (*Houstonia* spp.), frogfruits (*Phyla* spp.), verbenas (*Verbena* spp.), and violets (*Viola* spp.).

Males are seemingly very territorial and have been described as being pugnacious. It is interesting to watch as a male defends his area, darting out at everything that passes by. He either perches on some exposed twig or branch to await passing females or patrols a regular path, flying slowly back and forth from one boundary to the other. Once he encounters a female, he follows in pursuit until either mating occurs or he has been rejected.
EGG: Bluish-green, changing to cream just before hatching; laid singly on a bud or on upper surface of young leaf of food plant.

CATERPILLAR: Head black, covered in minute tannish hairs; body mostly greenish to pale tan, with darker stripe down back and narrow pale brown and white stripes along sides, downy with short whitish hairs; lives in rolled-up leaf shelter.
CHRYSALIS: Greenish in head portion, brownish toward rear, crossbanded with darker mottling and streaks; attached by silken strand within curled or folded-over leaf.
FOOD PLANTS: Scarlet Globe-mallow (*Sphaeralcea coccinea*), Copper Globe-mallow (*S. angustifolia*), Lindheimer's Globe-mallow (*S. lindheimeri*), Common Mallow (*Malva neglecta*), Three-lobe False-mallow (*Malvastrum coromandelianum*), Alkali Little-mallow (*Malvella leprosa*), Mexican-mallow (*Meximalva filipes*), Carolina Modiola (*Modiola caroliniana*), Tall Poppy-mallow (*Callirhoë leiocarpa*), Crested Anoda (*Anoda cristata*), Arrow-leaf Sida (*Sida rhombifolia*), Prickly Sida (*S. spinosa*), Spreading Sida (*S. abutifolia*), Lindheimer's Sida (*S. lindheimeri*), and the cultivated Hollyhock (*Alcea rosea*).
PARTS EATEN: Foliage.
RELATED SPECIES: Three other very similar Skippers occur in the southern half of the state: White Checkered-Skipper (*Pyrgus albescens*), which can be separated from the Common only by dissection in the lab; Desert Checkered-Skipper (*P. philetas*), in which the dark checks in fringe evenly spaced; and the Tropical Checkered-Skipper (*P. oileus*), in which the dark checks in the fringe of the upper hindwing run together, whereas they are separated in the other two species.

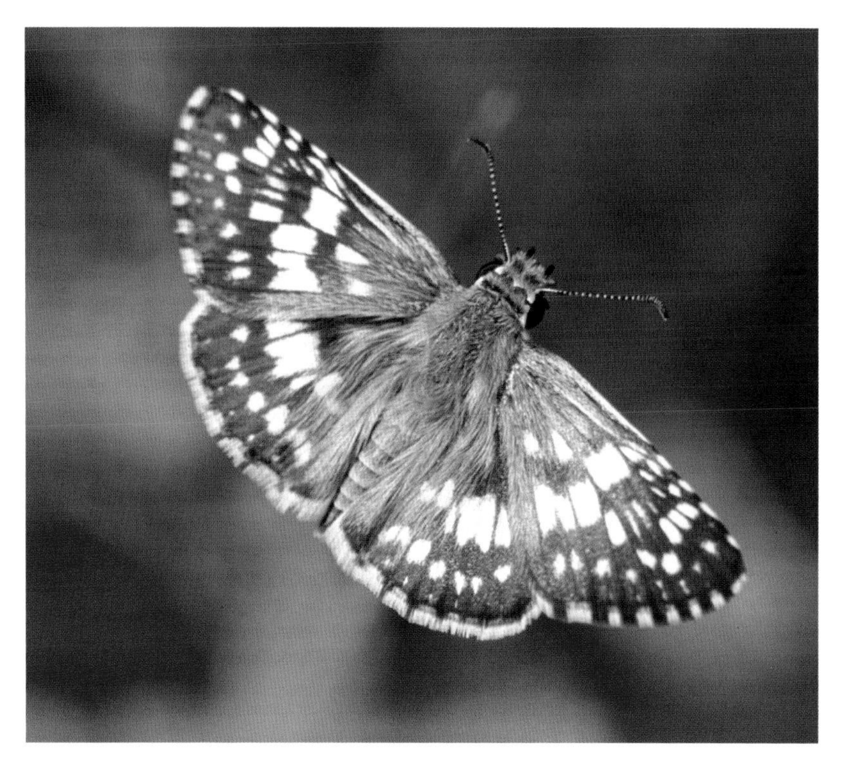

COMMON CHECKERED-SKIPPER

Common Streaky-Skipper

(Celotes nessus)

Family: Skipper (Hesperiidae)
Size: ¾–1 3/16 inches
Broods: Several
Flight time: March–November (all year)
Overwinters: Chrysalis (adult)
Range: 2, 4, 5, 6, 7

One of the state's most interestingly and uniquely marked butterflies, the Common Streaky-Skipper is unmistakable. The upper wing surface is primarily yellowish-brown with various darker lengthwise lines, streakings, spots, and chevrons, creating a folded or "pleated" appearance. The fringe of the wings is conspicuously checkered in dark brown and white, with the hindwings somewhat to prominently indented or scalloped.

The Common Streaky-Skipper is a regular flower visitor and can be readily observed while flying from flower to flower with slow, weak movements. It is a great percher, spending long periods of time just sitting between flights for sips of nectar. In the early morning it spends a lot of time basking in the sun in an open, exposed area, usually on a broad leaf or on a stick or rock low to the ground. Often groups of several individuals are found gathered around areas of moisture. Males can be seen during the day flying back and forth just above the ground, seeking females with which to mate.

Flowers that attract this butterfly to the garden are Bur-clover (*Medicago polymorpha*), Fine-leaf Star-violet (*Stenaria nigricans*), Phacelia (*Phacelia congesta*), Buttonbush (*Cephalanthus occidentalis*), Golden Crownbeard (*Verbesina encelioides*), Huisache Daisy (*Amblyolepis setigera*), and the verbenas (*Verbena* spp.), lantanas (*Lantana* spp.), wild onions (*Allium* spp.), and frogfruits (*Phyla* spp.).

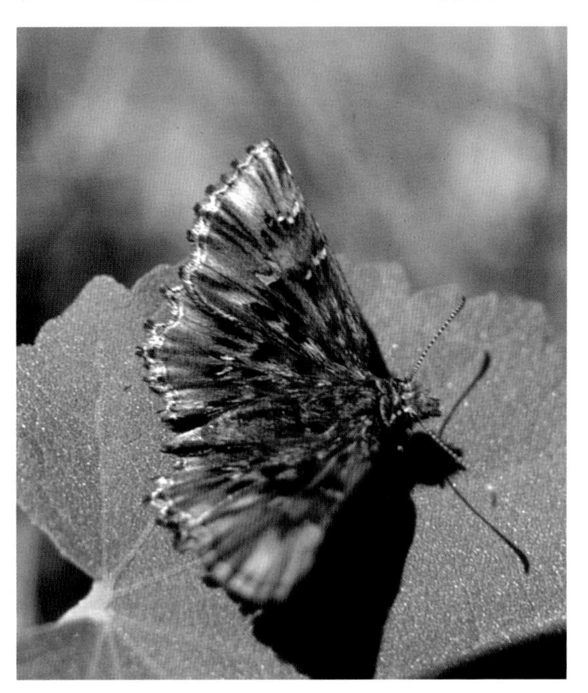

COMMON STREAKY-SKIPPER

EGG: Strongly ridged, white to greenish; laid singly on leaves of food plant.

CATERPILLAR: Pale green with pale yellow stripes along back and sides sometimes present, covered with tiny yellow dots and fine, short hairs.

CHRYSALIS: Not described; rests inside silken shelter on food plant leaf.

FOOD PLANTS: Texas Abutilon (*Abutilon fruticosum*), Copper Globe-mallow (*Sphaeralcea angustifolia*), False Wissadula (*Allowissadula holosericea*), Mexican-mallow (*Meximalva felipes*), and the cultivated Hollyhock (*Alcea rosea*).

PARTS EATEN: Foliage.

NOTE: Although a wide range is given for this butterfly, it is a more "western" species, with its greatest distribution west of a line running from Cook County on the Texas-Oklahoma border to Cameron County in the Lower Rio Grande Valley.

RELATED SPECIES: The Common Streaky-Skipper may possibly be confused with the Chisos Streaky-Skipper (*Celotes limpia*) within their overlapping range in the Big Bend and Guadalupe Mountains area.

Gray Hairstreak

(Strymon melinus)

Family: Gossamer-winged (Lycaenidae)
Size: 7/8 – 1 1/4 inches
Broods: Three to several
Flight time: February–November (all year)
Overwinters: Chrysalis (adult)
Range: Throughout

This is probably the state's most common Hairstreak, found in a wide range of habitats and throughout the season from earliest spring to freezing weather or all year, depending on the region. Anywhere there are flowers, the Gray Hairstreak can be seen flying swiftly about. At times the larvae are so numerous they cause serious damage to cultivated beans, corn, or cotton. Larvae sometimes bore into young flower buds of cotton, seeking the high-protein pollen, and are known as the cotton square borer.

The Gray Hairstreak is colored with soft blackish or bluish-gray scaling on the upper surface of both wings and with an orange spot on each hindwing. Below, the wings are a soft dove-gray, with black and white lines forming narrow bands; there are two large, orange-red and black patches near the tails on the hindwings. The gray coloring is darker in the spring specimens, and males are darker than females. Males also have orange along the abdomen.

As do other Hairstreaks, this one moves the tailed and eyed portion of the wings about while feeding, simulating a head complete with eyes and antennae in case of attack by a predator. In basking, it generally only partially opens the wings and rarely spreads them completely open for any length of time. In the afternoons, males perch on a low tree limb or shrub to await passing females for mating.

This Hairstreak uses many species of flowers for nectaring, especially Chaste Tree (*Vitex agnus-castus*), White Sweet-clover (*Melilotus albus*), and various milkweeds (*Asclepias* spp.), phlox (*Phlox* spp.), and verbenas (*Verbena* spp.). It is an avid partaker of the nectar of Parsley-leaved Hawthorn (*Crataegus marshallii*) and the wild plums (*Prunus* spp.).

Gray Hairstreak

EGG: Pale green; laid singly on a bud or young flower of food plant.

CATERPILLAR: May range from white to various shades of pink, purplish- or reddish-brown, but usually green; white or rose stripes line the side; makes no shelter.

CHRYSALIS: Various shades of brown and cream, mottled with darker brown; attached by silken mat and lies within debris on ground.

FOOD PLANTS: Larval food plants of this butterfly are extensive, with more than ninety species being recorded as utilized. However, members of the Bean Family (*Fabaceae*) and Mallow Family (*Malvaceae*) are preferred. Native species include Downy Bush-clover (*Lespedeza hirta*), Round-head Bush-clover (*L. capitata*), Small-flowered Mallow (*Malva parviflora*), Round-leaf Mallow (*M. rotundifolia*), Scarlet Pea (*Indigofera miniata*), Texas Thistle (*Cirsium texanum*), Texas Lantana (*Lantana

urticoides), Desert Lantana (*L. achyranthifolia*), One-seeded Croton (*Croton monanthogynus*), and Silver-leaf Croton (*C. argyranthemus*). The shrubs False Indigo (*Amorpha fruticosa*) and Parsley-leaved Hawthorn are reported as sometimes used. Some of the more commonly used cultivated species are Alfalfa (*Medicago sativa*), Peanut (*Arachis hypogaea*), and Cotton (*Gossypium hirsutum*). Others include garden bean (*Phaseolus vulgaris*), lima bean (*P. lunatus*), and garden pea (*Pisum sativum*), as well as White Clover (*Trifolium repens*), White Sweet Clover (*Melilotus albus*), Common Mallow (*Malva neglecta*), and the shrub Rose-of-Sharon (*Hibiscus syriacus*).
PARTS EATEN: Flower buds, immature fruits, and occasionally young leaves.

Henry's Elfin
(Callophrys henrici)

Family: Gossamer-winged (Lycaenidae)
Size: 1–1 1/8 inches
Broods: One
Flight time: February–May
Overwinters: Chrysalis
Range: 2, 3, 4, 6, 7

Although it has a wide distribution, Henry's Elfin is very local within its range—even when found, there never seem to be very many of them. They can often be seen taking moisture from the ground but, when disturbed, quickly fly into woodland shrubbery and disappear.

This butterfly will rarely be found anywhere except near its larval food plant. Shrubby or brushy areas along fencerows, woodland edges, and openings in deciduous or pine hardwood forest are much to its liking, but it can also be found in swampy, shady deciduous woodlands.

One of the best ways to find these little beauties is to walk slowly and quietly to one of their larval food plants and gently shake the bush. If a colony of butterflies has become established in the area, more than likely some will fly out, for when not taking nectar or moisture, courting, or laying eggs, they are usually hidden among the foliage. They freely visit the blossoms of any nearby spring flowers, but especially those of its larval food plants.

Coloring of this Elfin varies greatly; the ones in the eastern half of the state are the most darkly colored and marked. These colorings and markings gradually become less conspicuous the farther west this species occurs; the more western ones are rather dull-colored and only faintly marked. The eastern is shown and described here, in which the upper surface of both wings is a dark grayish-brown with reddish scaling. The lower surface of the forewings is a rich, dark brown in the basal two-thirds, with a paler brown covering the marginal third. The two areas are separated by thin, interrupted lines of black and white. Hindwings are dark blackish-brown with brownish scaling, followed by bands of paler brown, then bluish-gray near the wing margin. The two brown areas are separated by thin black and white lines as in the forewings; the lighter brown and grayish area, by a row of dark brown or blackish crescents. Short, stubby, tail-like projections are conspicuous on the hindwings and help in the identification of this species.
EGG: Ribbed, pale green becoming white; laid singly on flowers or near leaf buds.
CATERPILLAR: Green or reddish-maroon

with wide, white slanting dashes along sides; makes no shelter.

CHRYSALIS: Short, stubby, greenish-brown to brown mottled with dark brown to black; attached by silken strand to leaf litter at base of tree.

FOOD PLANTS: In eastern portion of the state Eastern Redbud (*Cercis canadensis canadensis*) is first choice, along with Huckleberry (*Vaccinium arboreum*) and Highbush Blueberry (*V. corymbosum*); Yaupon Holly (*Ilex vomitoria*) and American Holly (*I. opaca*) are sometimes used. In the western portion of the state Texas Redbud (*C. c.* var. *texensis*) and Mexican Redbud (*C. c.* var. *mexicana*) are the preferred choices, but Mexican Buckeye (*Ungnadia speciosa*) and Texas Persimmon (*Diospyros texana*) are also readily used.

PARTS EATEN: Flowers, young fruits, and leaves.

Henry's Elfin

NOTE: This butterfly is sometimes placed in genus *Callophrys*.

Olive Juniper Hairstreak

(Callophrys gryneus gryneus)

Family: Gossamer-winged (Lycaenidae)
Size: 7/8–1 1/8 inches
Broods: Two or more
Flight time: February–December
Overwinters: Chrysalis
Range: 1, 2, 3, 4, 6, 7

The Olive Juniper Hairstreak is the most common green Hairstreak and one not easily mistaken. Although this butterfly is widely spread over the state, it is quite local in some areas in the sense that it remains near good-sized stands of juniper (*Juniperus* spp.), often referred to as cedar, which also aids in identification. An established population remains in a particular area unless some environmental catastrophe forces it to seek a new site. Or, as is often the case, overcrowding will cause a portion of a population to seek another stand of juniper and form a new breeding area. Areas where hardwoods are harvested and not replaced, abandoned fields, and cleared but neglected sites are all beneficial to this Hairstreak. Juniper (or cedar) readily becomes established in such unkempt areas, and the Olive Juniper Hairstreak correspondingly increases in numbers as well as new locations.

The upper wing surface is a dark brown flushed with rust or reddish-brown. The female often has more orange or gold scaling than the male and appears paler. The lower surface of both sexes is a bright green, with a straight line of white dashes set in brownish-copper on the forewings. On the hindwings

Olive Juniper Hairstreak

white and brown dashes form two very irregular marginal bands with two very short lines near the base. A small, indistinct orange spot is set in a patch of blue scaling near the tails on the hindwings. Coloring of the summer broods is usually darker than in the spring and fall broods.

Olive Juniper Hairstreaks visit a large variety of flowers; some of the best liked are Mexican Plum (*Prunus mexicana*), Agarita (*Mahonia trifoliolata*), Eastern Baccharis (*Baccharis halimifolia*), Summer Bluet (*Houstonia purpurea*), and various redbuds (*Cercis* spp.), milkweeds (*Asclepias* spp.), wild onions (*Allium* spp.), and asters (*Symphyotrichum* spp.).

Flight of this butterfly is very swift. When disturbed, it darts about for a few minutes, then frequently returns to the same spot from which it left. Even when disturbed several times, instead of flying off a great distance, it flies to nearby shrubbery (often a juniper tree) and disappears among the branches for a brief period.

To seek mates, males remain around juniper trees, perching on an outermost branch tip near the top of a tree. Mating grounds are more often on hilltops or toward the crests of slopes than in lower gullies or valleys.

EGG: Pale green with white ridges; laid singly on tips of juniper branches.

CATERPILLAR: Dark green, with pale green

or yellowish dashes along sides, forming lines; makes no shelter.
CHRYSALIS: Short, stout, largest in head portion, rich brown with pale greenish sheen, generally covered with black-brown blotches.
FOOD PLANTS: Any native juniper within range, especially Eastern Red Cedar (*J. virginiana*) and Ashe Juniper (*J. ashei*).
PARTS EATEN: Tips of young foliage.
RELATED SPECIES: While the range of this butterfly is given as almost throughout the state, it is far more common in the western half. Another subspecies, Siva Juniper Hairstreak (*Callophrys gryneus siva*), occurs in Region 7 and can be separated from the Olive by having straight marginal lines forming the band on the hindwing and the absence of the two short white dashes near the body.

Fatal Metalmark

(Calephelis nemesis australis)

Family: Metalmark (Riodinidae)
Size: ¾–1 inch
Broods: Three or more
Flight time: March–November (all year)
Overwinters: Chrysalis (adult)
Range: 4, 5, 6, 7

The Fatal Metalmark is probably the most common Metalmark in the state, yet because of its small size, dull coloration, and rather secretive habits, it is often not noticed.

Most Metalmarks are noted for the two wide, shiny, silver bands across the wings, but the Fatal has only thin, inconspicuous lines of silver, a darker brown band near the middle of the wings, and irregularly checkered fringes. The Fatal displays much variation in coloring, and usually the female is much paler and less noticeably marked than the male. The lower surface of the wings of both sexes is a dull yellow-orange with diffuse blackish markings. Males have shorter forelegs, which are less than half the length of the other legs and are not used when walking or perching.

Fatal Metalmarks are most active during the warmest part of the day and at this time seek flowers growing near the ground, where they nectar with wings open or almost so. If heat becomes unbearable, they fly to nearby shrubbery and rest beneath the leaves. They are usually quite calm and do not fly far when disturbed. A local resident, the Fatal rarely flies far from where it was raised and almost never migrates.
EGG: Pitted, reddish with network of white ridges; laid singly on midrib of food plant leaves.

FATAL METALMARK

CATERPILLAR: Dark gray studded with small silvery bumps, a ridge of short, grayish-yellow hairs along back and a "skirtlike" fringe of long hairs low on sides.
CHRYSALIS: Pale dirty-yellow or grayish with few brown spots, sparsely covered with yellow hairs; attached by silken mat and strand to stem or in leaf litter at base of food plant.
FOOD PLANT: Drummond's Virgin's Bower (*Clematis drummondii*).
PARTS EATEN: Foliage.

Western Pygmy-Blue

(Brephidium exilis exilis)

Family: Gossamer-winged (Lycaenidae)
Size: 3/8–3/4 inch
Broods: Two to four
Flight time: March–October (all year)
Overwinters: Chrysalis (adult)
Range: Throughout

Not only is the Western Pygmy-Blue supposedly the smallest butterfly in the state but it is the smallest known in the world at this time, with the Eastern Pygmy-Blue (*B. isophthalma*), Cyna Blue (*Zizula cyna*), and Antillean Blue (*Hemiargus ceraunus*) running a close second. Although often overlooked, it is one of the most beautiful Gossamer-wings. In its range, the Western Pygmy-Blue is very common, flying slow and low to the ground, visiting the smaller flowers. The upper wing surface is a rich reddish-brown, with iridescent blue scaling near the body. The brown undersurface is interrupted with small white striations in the outer portion of the forewings that become mixed with blue-gray near the body. A row of black spots lines the outer margins of the hindwings, the lower spots centered with iridescent blue-green. The wings are delicately fringed in white. Females are generally a little larger than males and somewhat browner.

Despite the fragile appearance of this tiny mite, some of the southern population emigrates northward each year while rearing broods along the way. Since it cannot survive the harsh winters nor does it return south, each year the northward movement is repeated. Its appearance in certain breeding areas is also somewhat sporadic, as in some locales it is abundant in a particular year, then absent the next. Perhaps it depends on food plants of a certain age, but this is not known.

Low-growing or sprawling plants with flowers close to the ground attract this butterfly for nectaring. Also, because its proboscis is very short, the Western Pygmy-Blue uses only flowers with shallow nectaries. Some favorites are Agarita (*Mahonia trifoliolata*), Prairie Verbena (*Glandularia bipinnatifida*), and Phacelia (*Phacelia congesta*), as well as frogfruit (*Phyla* spp.), phlox (*Phlox* spp.), wild onions (*Allium* spp.), bluets (*Houstonia* spp.), zinnias (*Zinnia* spp.), and various asters (*Symphotrichum* spp.), especially New England Aster (*S. novae-angliae*).

Males slowly and continuously fly back and forth over the larval food plants during the day, seeking females with which to mate.
EGG: Pale bluish-green becoming white; laid singly on almost any part of the food plant, particularly on the upper leaf surface.
CATERPILLAR: Variable, most commonly creamy-white to pale green, covered with tiny, white-tipped brown bumps and yellow

WESTERN PYGMY-BLUE

and pinkish stripes on back and sides; when touched by ants, produces a honeydew that is then eaten by the ants; makes no shelter.
CHRYSALIS: Pale yellowish, greenish to pale brown, mottled in browns; attached on top of leaf of food plant by silken strand around body.
FOOD PLANTS: Four-wing Saltbush (*Atriplex canascens*), Slender-leaf Goosefoot (*Chenopodium leptophyllum*), Alkali Seepweed (*Suaeda monquinii*), Utah Swampfire (*Sarcocornia utahensis*), Desert Horse-purslane (*Trianthema portulacastrum*), Winged Sea-purslane (*Sesuvium verrucosum*), and the nonnative Australian Saltbush (*A. semibaccata*), Prickly Russian Thistle (*Salsola tragus*), and White Goosefoot (*C. album*).
PARTS EATEN: Flowers, fruits, leaves, and stems.

Southern Broken-dash *(Wallengrenia otho)* on Texas Thistle *(Cirsium texanum)*

8 Larval Food Plant Profiles

Some of the best larval and nectar food plants that can be used in a butterfly garden are described here and in chapter 9—space prohibits describing all the useful plants to be found in the state. If you use a plant profiled in this chapter and have poor results in attracting butterflies, the best thing to do is consult the more complete additional food plant list that follows, as well as the list given in any good butterfly field guide. If one of the plants listed there is more prolific in your area than the plants shown or listed here, start watching the plants for larvae and larval usage. If locations for obtaining the plants are needed, call members of local garden clubs or butterfly organization. If you live close to a university, visit the botany and entomology departments. Usually someone there is knowledgeable about butterflies and will know the preferred larval food choices for particular species in your area. Some useful information can be obtained from various Web sites.

Become familiar with the "weeds" of your area that are known food plants. Each time you pass through a place where they are growing abundantly, take a close look at them; you will often find larvae. If you are unable to recognize some of the plants of the truly weedy type, visit the herbaria of high schools or universities and study the dried plant specimens.

The profiles have been placed in separate categories—trees, shrubs, vines, and herbs—and placed in alphabetical order by common name.

In the additional larval food plant list at the end of the chapter, the plants have been further separated into native and nonnative/cultivated species. Also on this list are some species of plants that are not especially desirable garden plants. They are included to show that in the wild they are heavily used larval food or nectar sources. If these plants are already present on or near your property, you might want to let them remain.

Each larval and nectar food plant profile includes the following information:

COMMON AND SCIENTIFIC NAMES: *Vascular Plants of Texas: A Comprehensive Checklist including Synonymy, Bibliography, and Index* by Stanley D. Jones, Joseph K. Wipff, and Paul Montgomery (revised, unpublished CD) has been followed for common and scientific names. In some instances, publications by the Botanical Research Institute of Texas were used for common names.

FAMILY: The family is the scientific classification to which each species belongs.

CLASS: This rating has been given the plants in order to better understand the plant and to help in its location. If "native," the plant may have to be obtained from the wild either from seeds or cuttings or from nurseries carrying native plants. If "nonnative," it may possibly be purchased or occasionally found in the wild along with the native species. If "cultivated," then the plant will often be offered by nurseries or can be obtained from a mail-order catalog.

HEIGHT: These measurements are, in most instances, the extremes from lowest to highest as found in nature. Often in a garden situation, growth is faster and a greater height is achieved due to the greater nutrient and moisture availability.

BLOOM PERIOD: Dates given for both the larval and nectar plants are for the flowering period. The information in parentheses indicates the bloom period in the lower southern tip of the state, or the Lower Rio Grande Valley, where the bloom period is much longer. For the larval food plants, in most instances larvae use the foliage, which is available to them over a longer period, usually throughout the entire breeding season.

RANGE: The regions designated here follow the map found on the end sheet of this book and reflect the general ranges of the plants and butterflies. Range of the plants generally follows the *Atlas of the Vascular Plants of Texas* by B. L. Turner. For a larval food plant, only the regions both where the plant can be found and where the butterflies that use the plant as a larval food source are known to breed are given here. Before growing any plant as a larval food source, make sure the butterfly that would use it already occurs in your area.

When "Throughout" is given as the range, the plant either occurs naturally in all the regions or, in the case of a cultivar such as Butterfly Bush (*Buddleja davidii*), may be grown in the garden throughout the state. At least one of the species of the butterflies listed will use the plant in at least one of the regions.

Nurseries now offer many of the natives, which are slowly becoming available over a wider range. For a cultivated species, the ranges given are where the plant is recommended for use in the garden. The first regional number(s) indicates where the plant can most likely be grown and do well; the number(s) in parentheses gives a range where the plant can also be grown, although perhaps not as successfully.

Within each description there is a general, overall view of the plant and its characteristics and then the following information:

CULTIVATION: Generally describes methods for obtaining and growing a plant in the home garden.

USED BY: A list of known butterflies that use this particular plant.

PARTS EATEN: Provided in order for the gardener to better recognize when a butterfly larva (caterpillar) may be using the plant.

NOTE: An additional bit of helpful or interesting information about either the plant or the butterflies that use it.

RELATED SPECIES: Other similar or closely related species that butterflies use or that have some feature of interest.

ASHE JUNIPER

TREES

Ashe Juniper

(Juniperus ashei)

Family: Cypress (Cupressaceae)
Class: Native
Height: To 40 feet
Bloom period: January–March (October)
Range: 1, 2, 6, 7

Basally widely spreading and appearing multitrunked, deeply rooted, evergreen tree usually found growing in colonies. Bark usually blackish to dark gray splotched with paler grays, shredding in long strips. Leaves aromatic, dark green, needlelike, numerous, and forming dense foliage. Flowers tiny, golden-brown, the male and female on separate trees. Male trees (heavy pollen producers) readily distinguished in late winter or very early spring by the overall reddish-brown look of the foliage.

CULTIVATION: Ashe Juniper is rarely if ever offered by nurseries, but plants are so easily obtained from seeds, this is often the first choice for propagation. Collect the berry-like cones in late summer through fall after they have become full and dark blue in color. Gather fruit from several different plants, as viability varies.

Seeds can either be cleaned by soaking in warm water and then rubbed across screen wire, or planted as is, pulp and all. Either plant immediately in fall after fully matured, or cleaned and thoroughly dried seeds can be stored in dry, sealed containers and planted in early spring.

Ashe Juniper grows best in rocky, limestone soils with good drainage. No fertilizing is necessary, but occasional watering until established and growing well should be provided. While the tree is young, a thin autumn mulch of older, shed leaves is beneficial.

USED BY: Olive Juniper Hairstreak (*Callophrys gryneus gryneus*).

PARTS EATEN: Tips of young leaves.

NOTE: May not be a tree you want to plant if property is small, but if already growing, make sure it has plenty of "growing room" to breathe and spread out. The Olive Juniper

Hairstreak prefers the smaller (to fifteen feet) trees for egg laying. It also prefers trees growing in colonies, for the adults never stray far from their larval food plants.
RELATED SPECIES: In the eastern portion of the state, the Olive Juniper Hairstreak uses Eastern Red Cedar (*J. virginiana*), a more slender, single-trunked tree.

Black Locust
(Robinia pseudoacacia)

Family: Bean (Fabaceae)
Class: Nonnative
Height: 40–60 feet
Bloom period: May–June
Range: 1, 2, 3, 6, 7

Medium tall, rather open, spiny, irregularly shaped deciduous tree with black, deeply furrowed bark. Leaves long, bluish-green, divided into numerous leaflets and appearing almost lacy, turn a soft shade of yellow in early fall; each leaf bearing two short spines at the base, with each leaflet folding at night and during inclement weather. Flowers attractive, white, fragrant, bonnet-shaped, numerous, and forming long, pendulant clusters.
CULTIVATION: Not commonly offered in the nursery trade, Black Locust will probably need to be propagated from wild stock. Collect the two- to five-inch-long pods when full and beginning to turn brown. Remove seeds and soak in hot (not boiling) water for several hours or overnight. Have a permanent space prepared where plants are to remain. Black Locust is adaptable to many soil types but grows best in deep, well-drained, calcareous soils.

Black Locust will probably need no added fertilizer other than well-rotted hardwood leaves worked into the growing medium the first few years. Keep plants well watered, especially when young. Once established, Black Locust will need no care except the removal of root sprouts. These sprouts can either be discarded, transplanted, or, if the area is large enough, left to form a natural grove as is normal in the wild. Butterflies choose the younger plants for egg laying, so plants of various ages are most beneficial.
USED BY: Silver-spotted Skipper (*Epargyreus clarus*), Clouded Sulphur (*Colias philodice*), Zarucco Duskywing (*Erynnis zarucco*).
PARTS EATEN: Leaves.

BLACK LOCUST

Black Willow

(Salix nigra)

Family: Willow (Salicaceae)
Class: Native
Height: 40–60 feet
Bloom period: March–April
Range: Throughout

BLACK WILLOW

Sprawling, slender-branched, fast-growing, weak, deciduous tree usually found in more moist sites in the wild; often clump-forming, with several trunks. Bark rough brownish-black with deep, wavy fissures, sometimes loosened in long flakes in lower portion of trunk. Leaves alternate, smooth, thin, long, and slender. Flowers tiny, yellow, forming long clusters, opening before the new leaves, with male and female flowers occurring on separate plants.

CULTIVATION: Black Willow can be easily started either from fresh seeds or cuttings. Seed catkins should be gathered as soon as they begin to turn brown and the "fluff" is first noticed. It is not necessary to extract the tiny seeds; simply scatter the entire catkins where plants are wanted. Keep the area moist. Germination is immediate, usually within twelve to twenty-four hours. After young plants are up and doing well, excess plants can either be pulled and discarded or moved to other sites.

The tree is easily rooted from cuttings; simply take hardwood cuttings in early spring before buds leaf out, insert in prepared beds of sand or sandy loam, and keep moist until well rooted. Under normal conditions, Black Willow can quickly obtain medium tree size and is best suited for a more "natural" situation, such as along the edges of streams or ponds. It can be grown in drier areas, but growth will be slower. It will need quite a bit of space for the gracefully drooping branches to expand.

This may be a tree to be left if already growing near the garden area rather than one to plant, unless property is large and moisture is readily available. Another option would be to either keep it trimmed small or periodically replace when it becomes too large.

USED BY: Red-spotted Purple (*Limenitis arthemis*), Viceroy (*L. archippus*), Mourning Cloak (*Nymphalis antiopa*).

PARTS EATEN: Leaves.

NOTE: The Black Willow is shallow-rooted so is not a tree to be planted near a foundation or any kind of pipes, drains, or buried lines. The wood is soft and weak, with limbs breaking easily in strong winds or ice storms.

RELATED SPECIES: The cultivated Weeping Willow (*S. babylonica*) is often used as a larval food source and may fit into the landscaped garden more appropriately.

Flowering Dogwood
(Cornus florida)

Family: Dogwood (Cornaceae)
Class: Native
Height: To 40 feet
Bloom period: March–May
Range: 2, 3, 4

Small, understory tree with spreading crown and often crooked or leaning trunk; branches thin, delicate, and giving "twiggy" appearance. Bark of older trees dark brown to blackish, shallowly furrowed into small "squared" blocks. Leaves opposite, conspicuously veined, with some remaining green and some becoming brilliant shades of pink to crimson to moody wine in the fall. Flowers very small, greenish-colored, several held in a small cluster surrounded by four large, white (sometimes pink), petal-like notched bracts, the whole appearing as a "flower." Flowering followed by clusters of somewhat elongated berrylike fruits becoming brilliant, shiny red in the fall.

FLOWERING DOGWOOD

CULTIVATION: Easily acquired from nurseries, Flowering Dogwood is one of the most common native trees used for landscaping. Serious thought should be given to its location before planting because specimens more than a year old do not transplant or tolerate moving of any kind very well.

If preferred, new trees can easily be obtained from seeds, cuttings, or "layering." Ripe seeds can be sown immediately after gathering in the fall. Remove pulp from the seeds and cover with one-fourth to one-half inch of fine sand; then mulch over winter. Remove mulch after danger of frost in spring.

Seeds can also be kept over winter for spring planting by removing pulp; placing between layers of moist sand, perlite, or sphagnum moss in a ventilated plastic bag; then storing in refrigerator. Seeds should never be allowed to dry out, so the medium should be misted occasionally.

For softwood or semihardwood cuttings, take the terminal tip of younger growth, cutting just below a node or joint. Dip the cut end in rooting powder, place in sand, and keep cuttings and soil moist and shaded. Leave in rooting bed until the following spring before transplanting. After the plant is well rooted, a light fertilizing occasionally will be beneficial. Hardwood cuttings may be taken and rooted during winter but are often difficult and not especially recommended.

When planting either by seeds or by cuttings, using individual containers might be most beneficial, as the root system would be less disturbed at planting time. If setting permanently in the landscape, place the seedlings in well-drained but moist, sandy, acidic soils. They seem to flower best when in either partial shade or on the eastern side of shaded woodlands where the plants will receive only morning sun.

For landscaping, there can hardly be a better tree than the native Flowering Dogwood. Place this tree alongside Carolina Jasmine (*Gelsemium sempervirens*), or redbud (*Cercis* spp.), native pink-flowered azaleas (*Rhododendron* spp.), and wisteria (*Wisteria* spp.) for a lovely spring show. Use Blue Mistflower (*Conoclinium coelestinum*), Cardinal Flower (*Lobelia cardinalis*), and ironweed (*Vernonia* spp.) at its feet to complement the tree's brilliant fall foliage and clusters of glistening scarlet berries.

In some areas in recent years, Flowering Dogwood has been susceptible to a fungus, *Discula destructiva*, which affects leaves and twigs and, if not treated, eventually kills the tree. Check with the local county agent about using this as a landscape tree in your area and how to treat this disease should it occur. Studies have shown that planting the trees in areas with good air circulation helps in the prevention of this disease.

USED BY: Spring/Summer Azure (*Celastrina ladon*).

PARTS EATEN: Buds, young foliage.

Hercules'-club Prickly-ash

(Zanthoxylum clava-herculis)

Family: Citrus (Rutaceae)
Class: Native
Height: To 30 feet
Bloom period: April–May
Range: 2, 3, 4

Most often remaining a deciduous, short-trunked small tree with broad, rounded crown; large, spine-tipped corky prickles cover trunk and lower branches, upper branches and leaves often bearing reddish-black prickles. Leaves dark green, glossy, divided into several leaflets. Numerous small, greenish-white flowers form large, terminal clusters, followed by clusters of hard fruits resembling unground peppercorns. All parts of tree strongly aromatic when crushed.

CULTIVATION: This is an easily grown tree, not particular about soils. Give adequate moisture, especially during summer, for the healthiest and fastest growth. Habitually a stout or chunky-appearing tree, Hercules'-club rarely needs pruning. It grows relatively fast up to about seven or eight feet, and then growth slows down. This is a plant for the sun, as it will not tolerate shade.

New plants are readily obtained by removing the outer husk from the fruit and planting the inner black, shiny seeds in early fall. This is not a plant usually found in nurseries, so if propagation by seeds seems too slow, try to find a construction site or a kind landowner. Potential plants may be easily found along fencerows, where they have been planted naturally by birds.

Hercules'-club may sometimes be attacked by white flies when in a garden setting, but

HERCULES'-CLUB PRICKLY-ASH

RED BAY

frequent and forceful spraying of the undersides of the leaves with a garden hose usually discourages the pests.
USED BY: Giant Swallowtail (*Papilio cresphontes*).
PARTS EATEN: Young to midmature foliage.
NOTE: Almost all portions of this plant except the fruit produce a numbing effect when chewed; thus, one of its common names is Toothache Tree, as it was used often by earlier settlers to numb the pain of this common ailment. The dried, berrylike fruits can be ground and used as an herbal seasoning instead of black pepper.

Red Bay
(Persea borbonia)

Family: Laurel (Lauraceae)
Class: Native
Bloom period: May–June
Height: To 30 feet
Range: 3, 4, 6

An evergreen or persistent-leaved tree, rarely obtaining maximum height, usually with many trunks from the base and forming large clump. Bark of young tree reddish-brown, relatively smooth, later developing shallow furrows and ridges. Leaves aromatic, thick and leathery, bright green on upper surface and somewhat whitish or silvery underneath. Flowers small, creamy, and borne in small clusters from leaf axils. Small, dark blue to black fruits follow, each containing a solitary seed.
CULTIVATION: Although Red Bay usually grows in naturally moist areas, it can easily be grown in the garden in just about any soil with just a bit of extra care. Use no commer-

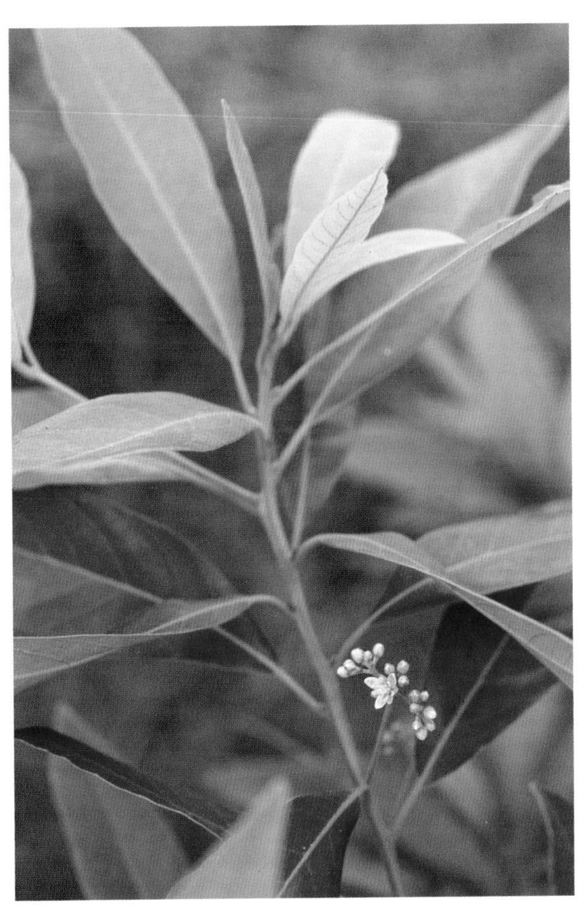
Red Bay

cial fertilizer with this one. Instead, work lots of well-rotted leaves or compost into the soil and add a mulch of shredded leaves or pine straw.

Another requirement for Red Bay is excellent drainage. Unless the tree is planted on a sloping lot, loamy soil alone does not provide enough drainage. The best planting program is to use wide, raised beds at least one foot above the surrounding area and wide enough to give the plant enough area for root growth. A bed six by six feet is large enough for a number of years. Dig the original soil to a depth of two to three feet, and refill with a mixture of sandy loam and rich compost, shredded leaves, or well-rotted sawdust. A mixture of sand and coarse pine bark worked into the soil is excellent, keeping the soil loose, aerated, and well drained. Give Red Bay a little supplemental water now and then, because of the extra drainage required. In the second year after planting, if the shrub is well established and growing nicely, trim a few of the branches back occasionally to promote new growth. The plant will resprout into a clump of new stems with a lot of new growth.

Red Bay is susceptible to a leaf gall that disfigures the leaves, making them unpalatable for butterfly larvae. At the first sign of infestation, spot-spray with a mild Sevin solution. After a day or so, hose the plant down to remove the spray, or the plants will not be used for egg deposition.

USED BY: Palamedes Swallowtail (*Papilio palamedes*), Spicebush Swallowtail (*Pterourus troilus*).

PARTS EATEN: Young foliage.

NOTE: Red Bay is a common component in the moist, humid southeastern portion of the state but is also common in the saline soils of the coast, especially in the Aransas National Wildlife Refuge area. A small, isolated colony thrives in the rocky, caliche soils of the Hill Country. These plants are usually found in colonies, sometimes in dense thickets. The leaves of this plant are very aromatic and when dried, may be used in cooking the same as purchased bay leaves.

Redbud

(Cercis canadensis)

Family: Bean (Fabaceae)
Class: Native
Height: To 40 feet
Bloom period: March–April
Range: 2, 3, 4, 6, 7

One of the most outstanding native small trees for landscaping; beautifully shaped, single or multitrunked, often twisted or gnarled, adding a special character to the garden. Bark on younger trees usually thin and reddish-brown, becoming thicker, scaly, and ridged or furrowed when older. Leaves large, pale green, heart-shaped, deciduous, becoming various shades of yellows and golds in autumn; in the more western regions, somewhat smaller, thicker, glossier, and with a slight waxy surface. Flowers bonnet-shaped, appearing before or with first leaves, producing tight clusters of pale pink to magenta. Flowering followed by clusters of reddish-brown to purplish, flat seedpods, remaining on the tree throughout the winter months.

REDBUD

CULTIVATION: Although Redbuds are great attractants to numerous species of flies, bees, butterflies, and other winged insects, the seeds are notorious for being infertile. To start new plants, gather a generous amount of pods by the end of August or even earlier. Separate seeds from the pod, spread seeds on paper, and allow to air-dry for a few days. Store in a sealed container until ready to plant in spring, or the seeds can be planted directly in the ground after gathering. After planting, place wire "tents" over the seeds for protection from predators.

Redbuds are commonly offered in the nursery trade, which may be the quickest way to obtain plants for the garden. Just be sure of what you are getting—often plants are offered as "natives" yet will be "improved." Also, be sure that the variety is the one from your area.

USED BY: Henry's Elfin (*Callophrys henrici*).

PARTS USED: Buds, flowers, tender foliage, young fruits.

RELATED SPECIES: Three varieties of Redbud occur in the state: Eastern Redbud (*Cercis canadensis canadensis*) in Regions 3 and 4, Texas Redbud (*C. c.* var. *texensis*) in Regions 2 and 6, and Mexican Redbud (*C. c.* var. *mexicana*) in Region 7. A narrow band of hybridization occurs along the western edge of Eastern Redbud and the eastern edge of Texas Redbud. This tree does not occur in the northern Plains area or the southern Rio Grande Valley. Eastern Redbud is shown here, but all varieties make excellent specimen trees for the home garden and grounds within their respective regions. They are used extensively for nectaring.

Sassafras

(Sassafras albidum)

Family: Laurel (Lauraceae)
Class: Native
Height: To 50 feet
Bloom period: March–April
Range: 2, 3, 4, 6

Thinly branched, well-shaped, small deciduous tree rarely attaining its maximum height; twigs bright green, limber, and breaking easily. Bark on younger trees smooth, reddish-brown, becoming grayish, thick and deeply furrowed on more mature trees. Leaves extremely variable in shape, may be entire, one-lobed (mitten-shaped), or distinctly three-lobed. Flowers greenish-yellow, in showy clusters appearing before the leaves or just as leaves begin to unfold; male and female flowers on the same or separate trees. Fruits showy, blue, borne on thickened, cuplike red pedicels, ripening in late fall. Almost all parts of this plant are aromatic.

CULTIVATION: To obtain plants of Sassafras, collect the attractive blue fruits when ripe, anytime between August and October, and sow the cleaned and dried seeds in the garden by late November. Getting a fair-sized sapling from seeds may be a bit slower than desired; on the other hand, well-established plants root-sprout readily and may need thinning or transplanting. Young saplings or root sprouts can more easily be transplanted if moved in early fall. If a sapling is to be moved, it should first be root-pruned to ensure a good root system of its own. To do this, in early spring spade completely around the stem at least a foot away from stem, then leave standing near the mother plant until fall, when it can then be dug and moved to the desired location. Better yet, well-established container-grown plants are frequently offered by nurseries and are the surest way of having adequate food plants.

Sassafras

Although Sassafras can eventually become large, for many years it remains a shrub or small tree. Sassafras grows best and produces the most foliage if given moist, rich, well-drained, sandy loams on the acidic side and plenty of sun.

Under normal conditions the lower branches of Sassafras continually die and fall off as the tree grows taller. Light pruning may be necessary from time to time to ensure the plant retains optimal height for egg laying and to increase foliage.

Very light applications of fertilizer in spring and early summer produce more foliage, but be careful not to overfertilize Sassafras—in this case, a little goes a very long way!

USED BY: Palamedes Swallowtail (*Papilio palamedes*), Eastern Tiger Swallowtail (*P. glaucus*), Spicebush Swallowtail (*Pterourus troilus*).

PARTS EATEN: Foliage, preferably young to midmature.

NOTE: In fall the leaves of Sassafras turn wonderful shades of yellow, orange, rose, pink-purple, and red, so try to plant in groups of three or more in the garden, placing them where they can be easily seen and admired.

SPANISH OAK

Spanish Oak

(Quercus buckleyi)

Family: Beech (Fagaceae)
Class: Native
Height: To 30 feet
Bloom period: February–March
Range: 2, 6

Small, beautifully shaped, often multitrunked deciduous tree, endemic to rocky, limestone soils. Bark either gray and somewhat smooth or brownish to silver-gray, deeply furrowed, and with distinctive platelike scales. Leaves long-petioled, deeply five- to seven- aristate-lobed, turning various shades of apricot, red, and maroon in the fall. Flowers tiny, numerous, of separate sexes with both male and female flowers on same tree, opening before or with first leaves; male flowers arranged in long, slender catkins; female flowers either single or in small clusters.

CULTIVATION: To obtain an oak tree of any size will take awhile, so for the garden, purchasing a large tree may be worthwhile. If time is not a factor, trees can be easily grown from their fruits, or acorns. Gather acorns directly from the tree as soon as they begin turning brown or immediately after dropping. Soak in hot water for about fifteen minutes to kill any infestation by weevils. Discard any seeds that float to the surface during soaking, because they will be sterile.

Immediately after the soaking, plant two or three acorns where the tree is wanted,

spacing the seeds two or three inches apart. Place a chicken-wire tent over seeds for protection from deer or squirrels. Once the seedlings are up and doing well, pull the unwanted ones or move to another area.
USED BY: Horace's Duskywing (*Erynnis horatius*).
PARTS EATEN: Young leaves.
NOTE: Spanish Oak is found in a narrow north/south strip roughly bordering IH-35 from the Red River south to San Antonio. To the west of this band will be found Texas Red Oak (*Q. texana*), and to the east, a much larger area of Shumard Red Oak (*Q. shumardii*). Spanish Oak may one day prove to be a hybrid of these two more dominant species or simply one species with variations because of adaptations to soils and climate.

Western Soapberry
(Sapindus saponaria var. *drummondii)*
Family: Soapberry (Sapindaceae)
Class: Native
Height: To 50 feet
Bloom period: March–June
Range: Throughout

Sturdy, erect-branched, rounded-crowned deciduous tree, usually found in colonies or "groves." Bark tannish-brown to gray, at first divided into narrow plates that eventually begin to flake or peel. Leaves somewhat drooping, consisting of four to eleven pairs of slender, yellowish-green leaflets. Flowers creamy-white, numerous, and forming large clusters, either male or female and borne on separate trees. Fruit berrylike, numerous in showy clusters, each berry large, one-seeded, translucent orange-yellow, shrinking and wrinkling with age but remaining on bare

Western Soapberry

branches through the winter months.

CULTIVATION: Soapberry can be found throughout the state, growing in almost any soil with the exception of wetlands or swamps. Plants for the garden may be obtained by gathering and cleaning the seeds in late fall. Pulp will need to be washed from the seeds, which may be a messy process because the pulp tends to foam and is "gummy," sticking to the hands and utensils, making seed separation difficult. It may be necessary to wash several times until seeds are completely clean.

Another method is to place the seeds on a wire screen off the ground and leave to natural erosion of the pulp. This method usually takes four to six months. Cleaned seeds should be planted in early spring.

Uncleaned seeds can be planted directly in the ground in late fall, but reproduction will not be as reliable. If this method is used, place three or four seeds a few inches apart, removing the unwanted plants after a good root system has developed.

Cuttings of semihardwood can be taken from April to June, treated with a rooting medium, and placed in a sand/perlite mixture under mist and are usually successful.

USED BY: Soapberry Hairstreak (*Phaeostrymon alcestis alcestis*).

PARTS EATEN: Young foliage.

NOTE: This is probably not a tree to plant as a specimen, for it will root-sprout, and no amount of mowing or chopping will prevent new outcroppings. Use this tree around the edges of property or along roadways where a grove or "stand" is needed.

The shiny black, white-spotted soapberry borer (*Agrilus prionurus*) is a recent invader from Mexico, first reported in Austin. Since then it has rapidly spread and killed many Western Soapberry trees in almost fifty counties. Signs of invasion include loose bark, usually at the base of the tree where birds and squirrels have worked it loose while probing for the white larvae feeding beneath the bark. Bayer Advanced Tree and Shrub Insect Control can be used to protect the tree against attack. There are also systemic controls on the market but, if used, will kill any and all butterfly larvae.

Wild Black Cherry
(Prunus serotina)

Family: Rose (Rosaceae)
Class: Native
Height: To 100 feet
Bloom period: March–April
Range: 2, 3, 4, 6, 7

A beautifully shaped deciduous tree with scaly bark on lower portion of trunk. Bark in upper portion of trunk streaked horizontally in shiny, silvery-grays and soft blacks; inner bark green, aromatic, and very bitter to the taste. Leaves bright green, glossy, and bluntly toothed along the edges, turning bright, golden-yellow to reddish in the fall. Flowers small, creamy-white, appearing in long, drooping racemes in early spring, followed by small, round fruits. Fruit edible when fully ripe and dark black, but avoid red half-ripe fruit as it contains hydrocyanic acid, a respiratory poison.

CULTIVATION: Plant this tree in a moist, well-drained situation. It grows in a variety of soils and may be started from either seeds or cuttings. Nurseries dealing in native plants commonly offer these trees, by far the easiest way to get healthy, fast-growing plants.

If starting plants from seeds, gather ripe fruit, remove the pulp, and allow seeds to air-dry for a day or so. Layer dried seeds in moist perlite, and place in the refrigerator until ready to plant. Sow seeds in prepared beds in early fall; then cover the beds with a thin layer of mulch.

Cuttings should be taken from the first year's growth, dipped in a rooting medium, and then placed in a sand/peat mixture. Keep moist and shaded with gauze until well rooted.

Foliage of Black Cherry is sometimes attacked by moth caterpillars. One, commonly called the tent caterpillar (*Malacosoma americanum*), forms large webs or tents of silk in angles of the branches. Close, continual inspection of the trees in early spring and immediate control are the best remedy. When webs are first sighted, squash the mass by running gloved hands up and down the branches. If these caterpillars are not removed, especially from a small tree, the tree can become stressed and sometimes is not able to recover from its loss of foliage.

USED BY: Coral Hairstreak (*Satyrium titus*), Striped Hairstreak (*S. liparops*), Henry's Elfin (*Callophrys henrici*), Red-spotted Purple (*Limenitis arthemis*), Viceroy (*L. archippus*), Spring/Summer Azure (*Celastrina ladon*), Eastern Tiger Swallowtail (*Papilio glaucus*), Two-tailed Tiger Swallowtail (*P. multicaudatus*).

PARTS EATEN: Immature leaves; caterpillars prefer foliage of younger trees.

NOTE: Flowers of Black Cherry are used as a nectar source by many species of butterflies. However, the bloom period is short, lasting only two or three weeks. The ripe fruit is relished by birds, which also consume enormous numbers of butterfly caterpillars.

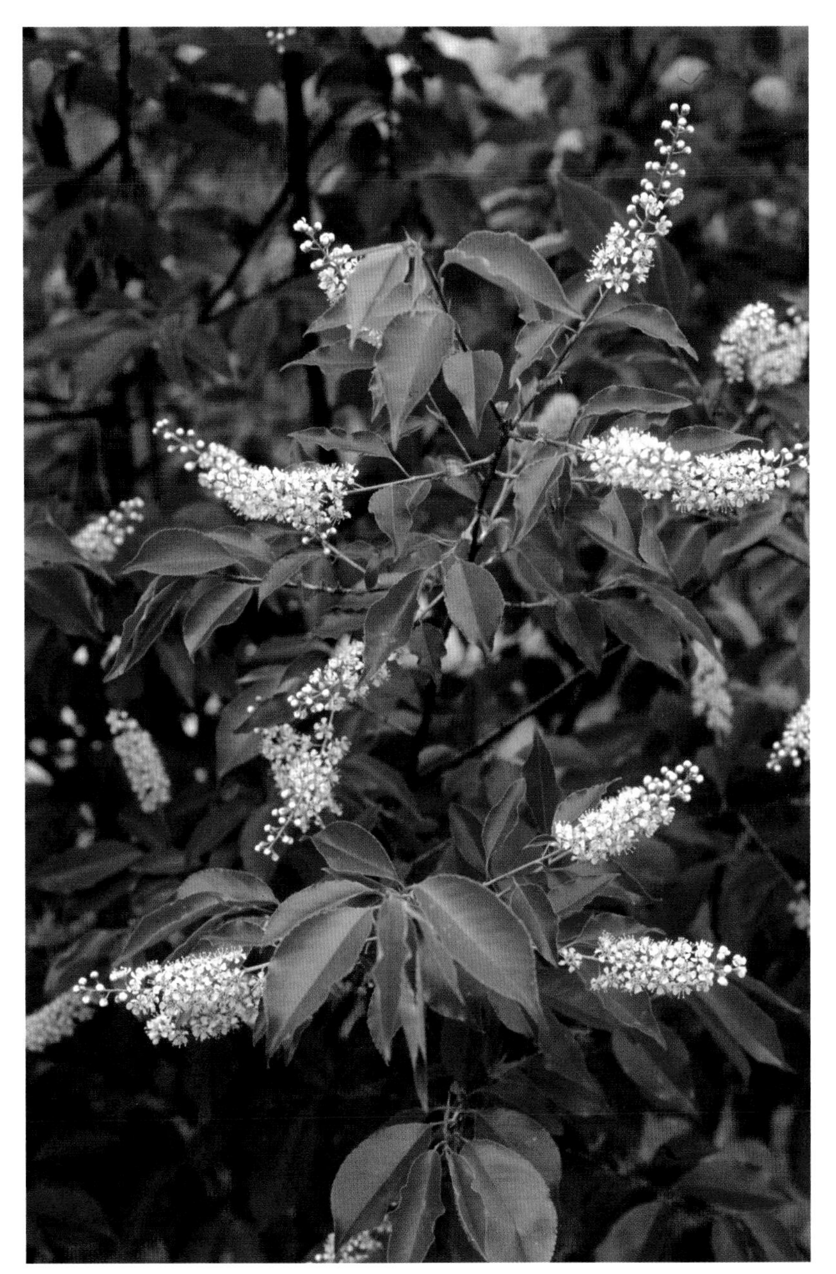

WILD BLACK CHERRY

RELATED SPECIES: There are three varieties in the state: var. *serotina* in Regions 2, 3, and 4; var. *eximia* in Region 6; and var. *virens* in Region 7.

Barbados Cherry

(Malpighia glabra)

Family: Malpighia (Malpighiaceae)
Class: Nonnative
Height: To 9 feet
Bloom period: March–October
Range: 4, 5

An erect, mostly evergreen shrub usually with many slender stems from base. Leaves opposite, extremely variable in shape, somewhat slender, the upper surface smooth and dull green. Flowers delicate, crinkly, pink to reddish-purple, three to seven forming clusters from axils of leaves. Fruit a small, shiny, red, berrylike drupe, similar to a small cherry.
CULTIVATION: This attractive shrub is much used in landscaping in regions as far north as Austin. As an escape, in the wild it can be found in sandy or clayey loams of the Rio Grande Valley. It grows in both full sun or dappled shade.

BARBADOS CHERRY

Propagation can be either by seeds or by cuttings. The three-seeded, glossy drupe can be planted directly in the ground after full maturity or in the spring after all danger of frost is past. Softwood cuttings are best taken in spring, dipped in a rooting medium, planted in a sand/perlite mixture, and kept shaded and under mist until well rooted. Gradually reduce the misting when plants begin to show rooting. If Barbados Cherry is to be moved, move in late winter or early spring. Cut plant back to half its original size, and keep well watered until established.
USED BY: Brown-banded Skipper (*Timochares ruptifasciatus*), Cassius Blue (*Leptotes cassius cassidula*), White-patched Skipper (*Chiomara georgina*).
PARTS EATEN: Foliage.
NOTE: Fruits of Barbados Cherry are edible and may be used to make jellies and preserves. In the wild, it is an important food source for many species of wildlife.

Black Dalea

(Dalea frutescens)

Family: Bean (Fabaceae)
Class: Native
Height: 3–4 feet
Bloom period: June–October
Range: 1, 2, 6, 7

Open, spreading or somewhat rounded, thornless, hard, deciduous shrub. Leaves composed of many tiny, silvery-green leaflets conspicuously dotted on lower surface with minute glands; foliage very aromatic when crushed. Flowers magenta and white, with several compressed into shortened terminal spikes or clusters; flowering usually very prolific, plants often almost solid masses of beautiful color.

CULTIVATION: Black Dalea is a native to dry, rocky, limestone soils and is generally found growing mixed with other brushy-type shrubs.

This is not one to be found in nurseries, so to get plants, you have to plant seeds or take cuttings. By far, the former is the easier method. Collect seedpods after they have become plump and brown. Extract seeds immediately after gathering, and plant either in a seedling bed from which the seedlings can be transplanted or directly into the spot where the plants are to remain. If seeds are to be carried over winter, wait to gather them until as ripe as possible and then air-dry several days after removing from the pods. Fumigate the seeds for a day or so, and then refrigerate for the winter. Sow seeds in spring after danger of frost is past.

Semiwoody tip cuttings can be taken in summer or early fall from the more succulent growth of the current season. Dip cuttings in hormone rooting powder, and place in sand under a plastic frame. Place the rooting container in open, light shade, and keep moist but not soggy.

Black Dalea does quite well even in the poorest of soils and under almost droughty conditions, so do not kill it with kindness in the garden. See that the plants get a good start by placing them in a sunny location with well-drained soils, adding lime if necessary but not fertilizer. Provide the plants with moisture until well established; then water occasionally during the summer if needed, but generally leave them alone. After the plants have become well established, Black Dalea will produce more flowers if trimmed severely each spring.

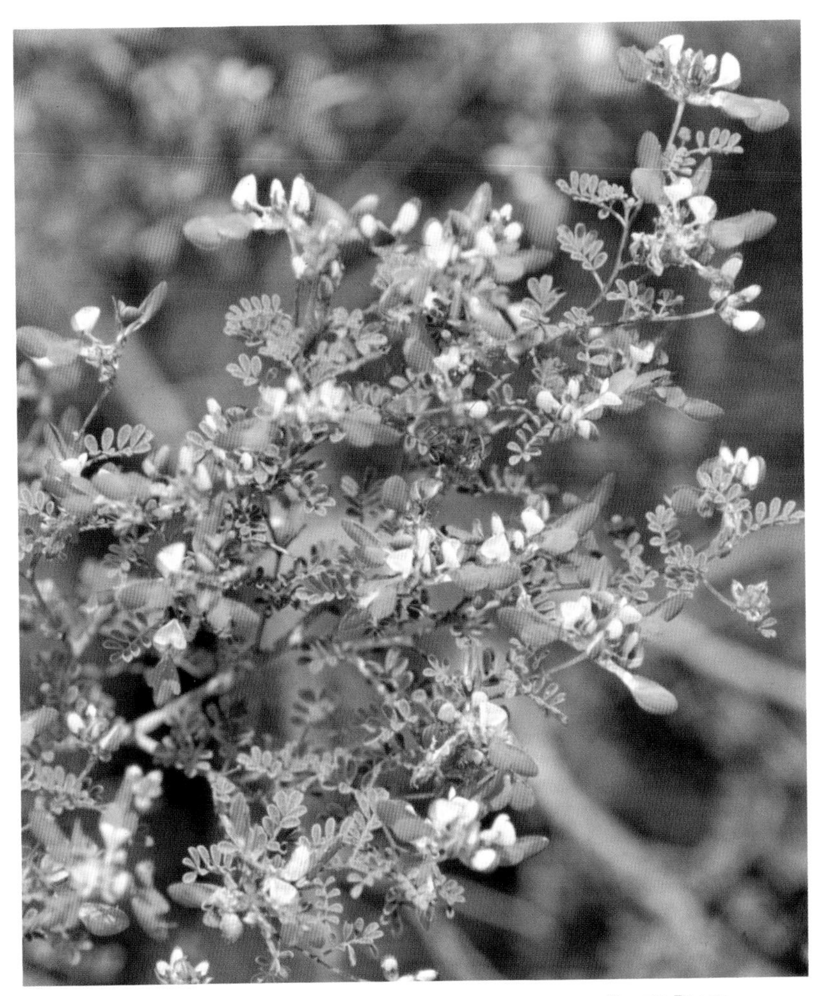

BLACK DALEA

USED BY: Southern Dogface (*Zerene cesonia*).

PARTS EATEN: Leaves.

NOTE: Black Dalea is an excellent nectar source for many species of butterflies.

RELATED SPECIES: Gregg's Dalea (*D. greggii*), a sprawling native perennial naturally occurring in Region 7, can easily be grown as far east as Austin. It is the larval food plant of the Reakirt's Blue (*Echinargus isola*) and an excellent nectar source.

Cenizo

(Leucophyllum frutescens)

Family: Figwort (Scrophulariaceae)
Class: Native
Height: To 8 feet
Bloom period: Summer
Range: 4, 5, 6, 7

A densely branched evergreen shrub; entire plant covered with fine whitish hairs, which make it very soft to the touch and give it its unusual coloration. Leaves pale grayish or "ash-colored," retained during the winter, adding color and interest to the garden. Flowers somewhat bell-shaped, ranging in color from white to various shades of pinks to lavenders and purples.

CENIZO

CULTIVATION: Cenizo does not propagate well from seeds nor does it transplant well. It can easily be raised from cuttings of the current season's growth, and this is the best way to obtain new plants.

Take semihardened cuttings of the current season's growth, usually in April to July, dip cuttings in rooting medium, plant in a peat/perlite mixture, and then place under mist. Once well rooted, cut back to ensure denser growth and a more compact plant.

Container plants are being offered by almost every nursery and are available in a wide variety of leaf and flower colors and in dwarf forms. Not all forms are winter hardy for all areas, so purchase plants recommended by knowledgeable nursery personnel.

Use these plants in a dry, well-drained site, preferably with rocky or gravelly soils. If drainage is a problem, raise the beds. Once the plants are well rooted, a deep, thorough watering during extended droughts is beneficial, but overwatering produces a straggly, sprawling, tender plant that will winter-kill. Contrary to popular belief, it is not rain or watering that induces flowering of Cenizo but high humidity.

All members of *Leucophyllum* need alkaline soils. Add dolomitic limestone to acidic soils, and if more calcium is needed, adding gypsum would be appropriate. Do not use peat moss in the area where Cenizo is growing, as it would make the soil acidic. Cenizo is generally hardy and disease resistant and usually needs nothing but the proper soil for producing spectacular plants. The only serious problem is with cotton root rot, a fungus to which all Cenizos are highly susceptible.

Place several Cenizos together in small

groupings, use as untrimmed hedges or screening, or arrange as the background of a border. These are slow growers, so if you are using them in the last two situations, buy large plants, as small plants may be shaded out by the plants in the front. Do not crowd Cenizo wherever you plant it, for it likes the drying breezes. If the plants need shaping to keep in bounds, trim very sparingly during spring or summer. The plants bloom on old wood as well as new growth, so do not trim to try to force flowering, and take out only what is needed for shaping. Overtrimming ruins the naturally beautiful shape of this shrub.
USED BY: Theona Checkerspot (*Chlosyne theona*).
PARTS EATEN: Foliage preferred; buds and flowers if the larvae run out of leaves.
NOTE: The word *cenizo* means "the color of ashes."

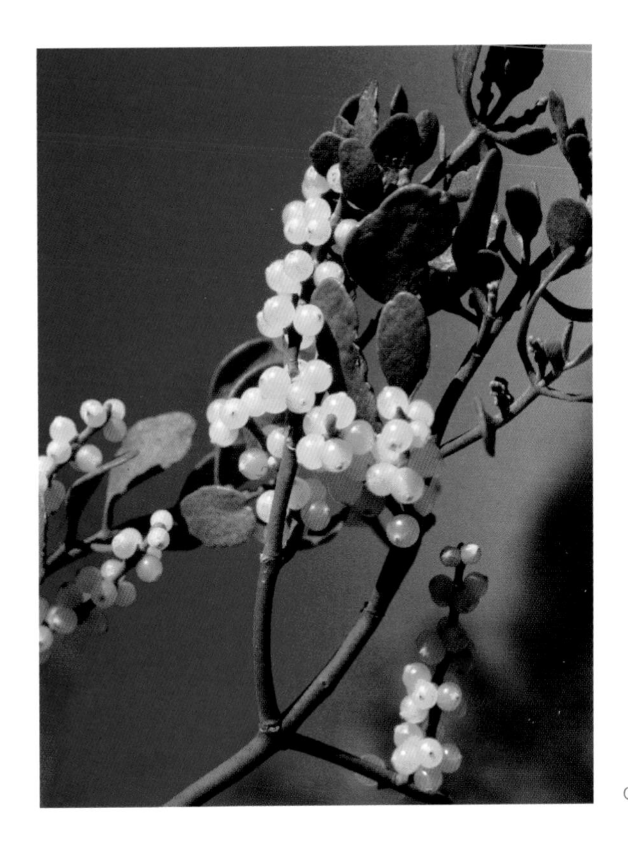

Christmas Mistletoe

Christmas Mistletoe

(Phoradendron tomentosum)

Family: Mistletoe (Viscaceae)
Class: Native
Height: To 3 feet
Bloom period: November–March
Range: Throughout

A parasitic, clumped evergreen shrub growing on both evergreen and deciduous trees. Leaves, small, thick, leathery, either yellowish or dull, dark green. Flowers minute, inconspicuous, male and female occurring on separate plants, forming short, stiff racemes later followed by pearly-white to creamy somewhat translucent berries on female plants.
CULTIVATION: New plants of Christmas Mistletoe are usually started from bird droppings. The fruits or berries consist of a mass of mucilaginous material surrounding the solitary seed. The seeds, even after being eaten, remain whole within some of the glutinous substance and readily become attached to whatever they contact.

Christmas Mistletoe depends on its host tree for certain nutrients. As the seeds sprout, they send minute, hairlike tendrils beneath the bark layer, where they are able to absorb the liquid sap of the tree. If a weak or diseased tree becomes heavily infested with the shrub, the tree can actually be killed. However, one or two clumps on a healthy tree usually will neither weaken nor kill the tree.

These plants use several species of trees as hosts. Some of the common ones are ash (*Fraxinus* spp.), cottonwood (*Populus* spp.), elm (*Ulmus* spp.), hackberry (*Celtis* spp.), willow (*Salix* spp.), almost all of the oaks (*Quercus* spp.), and Honey Mesquite (*Prosopis glandulosa*). Christmas Mistletoe is a plant of open lands and will be found most often on trees in pastures, along roadsides, at woodland edges, or in thin woodlands.

The first requirement when trying to propagate Christmas Mistletoe is to have a healthy host plant that is preferably a few years old. Then, in mid- to late winter, gather the berries and mash them all over the branches and trunk of the tree. Several species of birds savor the berries, in particular, Eastern Bluebird (*Sialia sialis*), Cedar Waxwing (*Bombycilla cedrorum*), and American Robin (*Turdus migratorius*). If a flock of these birds is sighted, gather some of the seed-filled droppings and smear these on the tree where you want mistletoe to grow. Often the percentage rate of germination is greatly increased for seeds that have passed through an animal's digestive tract.

If more than six clumps of plants begin to grow, remove all but three or four after the first year or two. It will do no good to weaken or kill the tree. Each clump will be either male or female, so wait to remove the unwanted clumps until they begin to bear fruit and sex of the plant can be determined.

USED BY: Great Purple Hairstreak (*Atlides halesus halesus*).

PARTS EATEN: Young foliage, occasionally flowers.

NOTE: Berries (from female plants) of all of the mistletoes are often used for decorative purposes but are toxic to humans and should never be eaten.

Dwarf Pawpaw (Small-flowered Pawpaw)
(Asimina parviflora)

Family: Custard-apple (Annonaceae)
Class: Native
Height: 2–8 feet
Bloom period: April
Range: 3

Stout, irregularly branched, tropical-appearing deciduous shrub; stems, branches, and young growth usually covered with soft, reddish-brown hairs. Leaves large, thick, leathery, pale green. Flowers small, purplish or brownish, solitary, appearing in early spring before or with first leaves. Fruits two to three inches, banana-shaped, turning black or dark brown when ripe.

CULTIVATION: Dwarf Pawpaw is an interesting plant for the garden but not one to use for showiness. It is an ideal background plant for the semishaded part of the garden and blends in well with such other natives as Piedmont Azalea (*Rhododendron canescens*), Red Bay (*Persea borbonia*), Sassafras (*Sassafras albidum*), and Spicebush (*Lindera benzoin*).

Dwarf Pawpaw prefers somewhat acidic, moist but well-drained, sandy-loam soils. It grows quite well in the garden under these conditions but cannot tolerate dry, hard soils. It can survive in full sun but does best in open, airy semishade.

Propagation is usually by seeds or cuttings. Both methods are difficult and time-consuming for the home gardener, with very unsure results. It is best to leave propagation to the larger growers who are professionally trained and have the necessary equipment. Native growers sometimes offer Dwarf Pawpaw as

Dwarf Pawpaw

potted plants. This is the quickest way to ensure having a plentiful food supply.

Dwarf Pawpaws are extremely deep rooted and very difficult to impossible to transplant, so place the plants where they are to remain. If transplanting is absolutely necessary, dig deeply and carefully to get as much root as possible. Cut top growth back two-thirds or more, and plant in carefully prepared sites. Dig planting holes much larger than the root balls, and add lots of half-rotted pine needles and sharp sand to the new hole. If the plants were doing well, take some soil from where the plants were growing and add to new hole. After planting, add a deep mulch of pine needles. Keep plants moist but not soggy until new growth begins. Once fully established, Dwarf Pawpaw will root-sprout into small colonies.

USED BY: Zebra Swallowtail (*Eurytides marcellus*).
PARTS EATEN: Foliage.
NOTE: The larvae feed at night or during the day if cloudy; at other times they rest near the base of the plant.

Flame Acanthus

(Anisacanthus quadrifidus var. *wrightii)*

Family: Acanthus (Acanthaceae)
Class: Native
Height: 2–4 feet
Bloom period: June–frost
Range: 5, 6

Many-branched, deciduous shrub usually wider than high; branches slender and brittle. Leaves small, widely spaced, pale green, opposite. Flowers bright firecracker-red or deep orange, numerous, slender, tubular-shaped, forming open, terminal racemes. Fruit a long, club-shaped, flattened capsule containing flat, wrinkled, black seeds.
CULTIVATION: A very drought-tolerant plant and one of the easiest of natives to grow in the home garden if given an exposed, sunny site with dry soils. It is exceptionally disease resistant. Flame Acanthus is one of the last to put forth leaves each spring but one of the last to lose its leaves in the fall. The flowering period is intermittent but continuous, with the heaviest flowering during the hottest, driest periods of late summer and into fall.

New plants can easily be obtained from seeds, seedling transplants, or cuttings. Two to four seeds are formed in each capsule, with seeds maturing throughout the summer and fall after each flowering period. Once they begin ripening, collect the capsules almost daily

Flame Acanthus

until the amount of seeds needed is obtained. Upon drying, the capsule splits, flinging the seeds about the area, sometimes several feet. The seeds should be thoroughly dried after gathering and stored until the following spring. Seeds germinate readily without any pretreatment. After all danger of frost is past, plant in a prepared site in the garden and keep moist but not soggy. Flame Acanthus readily reseeds, and new plants can easily be moved from around the mother plant.

Use this as a specimen planting, placing at least three plants in a prominent area of the lawn or entrance way, or use for an untrimmed, solid hedge. Otherwise, mix with other shrubs in a border. They are very striking when grown in barrels for the porch or patio. When trimmed or shaped, the branches become thicker and denser, but in many instances the more natural look, even if less branched, is more appealing in a garden planting. To obtain more branching and still retain a natural shape, cut the entire plant back to approximately six inches above ground level after a hard freeze or in very early spring.

USED BY: Janais Patch (*Chlosyne janais*), Elada Checkerspot (*Texola elada ulrica*), Texan Crescent (*Anthanassa texana*).

PARTS EATEN: Buds, flowers, young foliage.

Mexican Buckeye

(Ungnadia speciosa)

Family: Soapberry (Sapindaceae)
Class: Native
Height: To 20 feet
Bloom period: March–May
Range: 2, 3, 6, 7

A single or several-stemmed, usually medium-sized deciduous shrub, only occasionally reaching small tree size. Leaves dark green, glossy on upper side, up to a foot long, composed of five to several leaflets. Flowers showy pink to magenta, several and forming large clusters, appearing just before or with first leaves. Fruit capsules decorative, woody, three-lobed, maturing in late fall, hanging from branches on long, slender stems; each lobe or compartment containing a single, shiny, dark brown to blackish round seed.

CULTIVATION: To propagate Mexican Buckeye, gather the capsules when dark brown, beginning to split, and seeds are shiny and very dark, usually from August to November. Remove seeds from pods, and spread on paper to dry for a few days. After they are com-

Mexican Buckeye

pletely dry, store in a sealed container during winter.

Plant seeds in early spring after danger of frost. Studier plants are obtained if seeds are sown directly where plants will remain or in tall pots, allowing the deep taproot to become established. If seeds are planted in pots, keep exposed bottom roots trimmed to force the plant to develop a secondary root system. Potted plants grow fast and can be transplanted from pot to garden by early fall. Plants started from seeds should flower their third year.

In the wild, Mexican Buckeye is most commonly found in rocky soils in the Hill Country westward, usually on slopes, canyon walls, or ridges. Garden soils in these regions should be a little on the lean, dry side.

USED BY: Henry's Elfin (*Callophrys henrici*).

PARTS EATEN: Young fruits.

NOTE: Growth and flower color vary according to range. In the extreme western portion of the state, growth form is smaller, leaves are smaller, and the flowers are very dark in color. In the extreme eastern portion of its range, the growth is more straggly and the flowers much paler in color.

New Jersey Tea

(*Ceanothus americanus* var. *pitcherii*)

Family: Buckthorn (Rhamnaceae)
Class: Native
Height: 1–4 feet
Bloom period: April–June
Range: 2, 3, 6

Small, delicate, deciduous shrub, woody at base, with slender, spreading herbaceous branches; plants often multitrunked from base. Leaves alternate, small, usually covered in short, soft hairs, pointed at tip. Flowers small, white, in frothy terminal racemes.

CULTIVATION: New Jersey Tea is not easily transplanted because of excessively large, burl-like rootstocks, but it can be propagated from seeds or cuttings. Plants are occasionally offered by nurseries, and if found, this is

New Jersey Tea

the best way of obtaining plants. If seeds are to be gathered, watch the capsules carefully. Immediately upon ripening, the capsule splits, throwing seeds quite a distance from the plant and making it practically impossible to find them on the ground. Tie small squares of nylon netting around almost-mature capsules, thus preventing loss of the seeds. Getting these seeds to germinate is sometimes difficult. Soaking seeds in hot water before planting or layering in moist perlite or peat moss for two to three months before planting sometimes promotes better germination. Best results are obtained if seeds are simply cleaned and planted immediately after gathering.

New Jersey Tea will also root from semi-hardwood shoots of the current season's growth. Be careful of the very brittle roots when moving newly rooted cuttings to permanent beds. For better-formed plants, pinch the tops of cuttings when transplanting.

Usually growing in sandy soils, New Jersey Tea is tolerant of a variety of soils in a garden situation if given enough moisture and excellent drainage. It is best used in a lightly shaded portion of a perennial border, along a shady path, or in filtered sun at the edge of a woodland. Plant in groupings of three to five, as butterflies are more attracted to large patches of color and are better able to detect the chemical fragrance of the foliage when in masses. The bloom period lasts only two to three weeks, but seed capsules that form later are a beautiful brown and remain on the plant until the next year. After plants begin showing good growth in early spring, pinch or trim back to make growth bushier and to keep plants lower.

USED BY: Spring/Summer Azure (*Celastrina ladon*), Mottled Duskywing (*Erynnis martialis*).

PARTS EATEN: Foliage.

NOTE: New Jersey Tea is a nectar source for several species of butterflies.

RELATED SPECIES: A very similar-appearing species, Redroot (*Ceanothus herbaceus*) can be found almost throughout the state with the exception of the Rio Grande area and the far western portions. It, too, is probably used as a food plant by the Spring/Summer Azure.

Parsley-leaved Hawthorn

(Crataegus marshallii)

Family: Rose (Rosaceae)
Class: Native
Height: To 25 feet
Bloom period: March–April
Range: 2, 3, 4

Usually low-growing, tough, deciduous shrub with silvery-gray bark peeling off in patches and revealing a reddish-brown inner layer. Leaves distinctively lobed, similar to leaves of parsley (*Petroselinum crispum*). Flowers white, in clusters that cover the branches before or with first leaves; anthers red, conspicuous, making the flowers appear almost pink from a distance. Flowers followed by small, applelike fruits, turning a brilliant scarlet in late fall.

CULTIVATION: Parsley-leaved Hawthorn, as are

PARSLEY-LEAVED HAWTHORN

almost all species of *Crataegus*, is easily grown from stratified seeds; cuttings are not very successful. Seeds should be soaked in water for several days until swollen. Then layer the seeds in moist perlite in an airtight plastic bag, and place in the refrigerator for four to six weeks. Some local nurseries and catalogs offer container plants.

If young plants are to be moved, tag them in the summer or fall when easily identified; then move them after they become dormant in the winter. Parsley-leaved Hawthorn has an extremely long taproot; take care not to break or cut this taproot when transplanting. In the garden, plant in loose, loamy garden soils, fertilize occasionally, and give adequate moisture. Well-established plants can survive brief periods of drought, but prolonged summer dryness can cause the fruit to fall prematurely.

Use this shrub along a naturalized woodland edge or as a specimen planting. It can be left as a single stem by trimming root sprouts that occur. If a bushier plant is desired, cut back to three or four inches from the ground when the plant is two or three years old. New growth results in several stems for a low-branched specimen.

To achieve an immediate clumped effect, place three or four small plants together. For easiest planting, remove approximately one-fourth of the soil from one side of the plants in one-gallon containers. Put all three (or four) of the plants close together in the same hole, and lean the trunks of the plants slightly away from the center. When planted, the trunks should look like a slightly spread, upside-down tripod.

Rarely does Parsley-leaved Hawthorn attain the maximum height given here, perhaps doing so only after many years. It usually remains a slow-growing, low- to medium-height shrub or low tree. With good garden care, there should be no problem growing this shrub well outside its normal range.

USED BY: Gray Hairstreak (*Strymon melinus franki*).

PARTS EATEN: Foliage.

NOTE: When in flower, this small shrub attracts numerous butterflies for nectaring, especially various Hairstreaks (family Lycaenidae), Skippers (family Hesperiidae), and the Swallowtails (family Papilionidae): Zebra (*Eurytides marcellus*), Eastern Tiger (*Papilio glaucus*), and Palamedes (*P. palamedes*).

Southwest Bernardia
(Bernardia myricifolia)

Family: Spurge (Euphorbiaceae)
Class: Native
Height: To 10 feet
Bloom period: March–November
Range: 4, 5, 6

Dense, stout, thickly branched deciduous shrub, usually forming stands or colonies; in

Southwest Bernardia

the wild, usually found on dry, rocky slopes, canyon walls, or caliche ridges. Leaves small, alternate or clustered, dull green, variable in shape, rippled or wavy along edges; all young leaves and branches covered with star-shaped hairs. Flowers small, greenish-white, either male or female, may occur on the same or separate plants; male flowers occurring in short panicles or clusters, the female solitary or in pairs at tips of branches. Fruit a plump, three-lobed, three-seeded capsule.
CULTIVATION: Propagation of Southwest Bernardia is uncertain. Gather seeds from the plant as soon as mature, plant some immediately, place some in a paper bag, and carry through until spring; then plant after all danger of frost is past. Place some seeds in a container, keep in the refrigerator until spring, and then plant.

Take softwood cuttings in midspring, dip in rooting hormone, place in a sand/perlite mixture, and mist or water only when needed.

Native plant nurseries may carry this plant, and if found, may be the best means of obtaining these food plants.
USED BY: Lacey's Scrub-Hairstreak (*Strymon alea*).
PARTS EATEN: Buds, flowers.
NOTE: As this shrub is not a colorfully flowering one, use it at the back of the border or where a thick, warm-season screening is desired.

Spicebush
(Lindera benzoin)

Family: Laurel (Lauraceae)
Class: Native
Height: To 16 feet
Bloom period: March–April (August–September)
Range: 2, 3, 6

Graceful, smooth, aromatic, slender-branched deciduous shrub or small tree, usually with few to several trunks; older bark covered with small, corky bumps and spicy to the taste. Leaves bright green, smooth, glossy, fragrant when crushed. Flowers small, yellow, fragrant, male and female on different plants, forming small clusters close to the branches, opening before new leaves appear. Fruits in clusters, small, bright red, spicy, borne on female plants in late summer or fall.
CULTIVATION: Spicebush can easily be propagated either by seeds or cuttings. Collect seeds in late summer or as soon as fully ripe, and clean immediately. Pulp can be removed by soaking seeds in warm water (a little added detergent will help), or the pulp can be rubbed off by rolling fruits on screen wire.

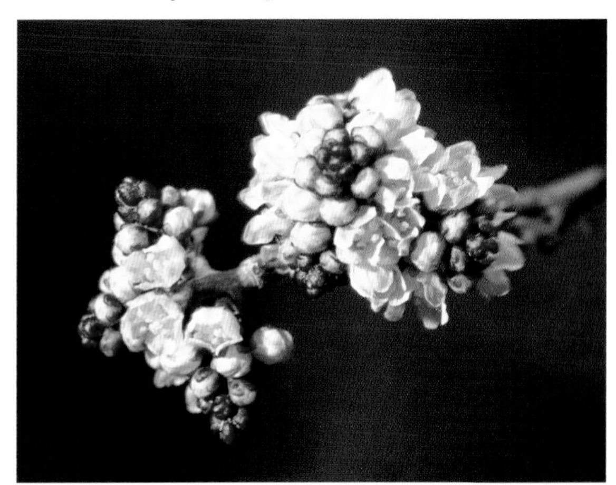

SPICEBUSH

Best germination will be obtained if seeds are sown immediately after gathering and cleaned. If this is not possible, place cleaned seeds between layers of moist sand or peat moss and keep moist and cool. They can be kept for one to two months (although rate of germination will not be as good), then should be planted either in the garden or potted in sandy loam for later planting.

To root cuttings, use two- to three-inch-long mature side shoots of the current season's growth. Cut at a slant, dip cuttings in a rooting hormone, and insert cuttings into a prepared bed of sandy loam. Mist cuttings frequently. Occasional applications of rooting hormone are beneficial. These plants quickly develop a long, deep root system and become difficult to transplant so should be set in a permanent situation as soon as well rooted.

In the eastern garden, place Spicebush in moist sandy or sandy-loam soils on the acidic side in open, thinly shaded areas or where it will receive only morning sun and afternoon shade. For the western regions, it will tolerate dry, thin, alkaline soils as long as partially shaded.

After Spicewood becomes well established, it usually puts out suckers or "sprouts" that can be cut or separated from the mother plant or, if space permits, left to form small colonies.

USED BY: Spicebush Swallowtail (*Pterourus troilus*), Eastern Tiger Swallowtail (*Papilio glaucus*).

PARTS EATEN: Leaves.

NOTE: If purchasing plants, obtain them from native growers in your area. East Texas plants will not survive in the Central Texas soils and climate, or the West Texas ones in East Texas.

Texas Kidneywood

(Eysenhardtia texana)

Family: Bean (Fabaceae)
Class: Native
Height: To 10 feet
Bloom period: April–October
Range: 2, 3, 4, 5, 6, 7

Loosely upright, many-branched, ferny-appearing deciduous shrub with few to several stems from base. Leaves a rich, dark green, divided into numerous tiny leaflets that produce a delightful aromatic fragrance when touched. Flowers numerous, small, white or creamy-colored, fragrant, clustered in slender racemes from the leaf axils or near the branch tips.

CULTIVATION: Texas Kidneywood is easy to propagate from either seeds or softwood cut-

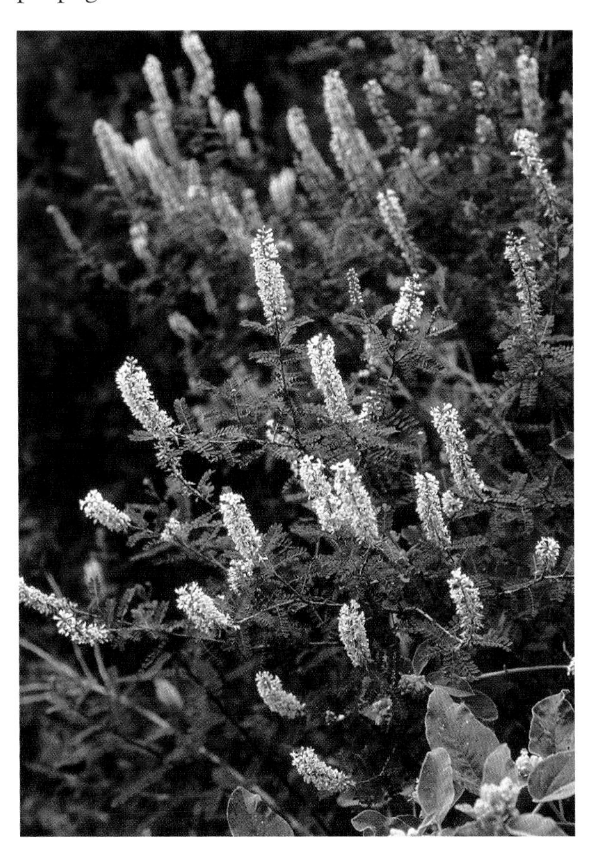

TEXAS KIDNEYWOOD

tings. If using seeds, gather the pods after they have turned brown and are at least partially dry. Place pods in an open container, on a wire screen, or in an open paper bag until completely dry. After drying, remove seeds from the pod or leave the thin, papery pods intact. Place seeds or pods in a paper bag; add a mothball, a few naphtha flakes, or a portion of an insect strip; seal tightly; and store at room temperature until spring.

After danger of frost, plant seeds in the garden where plants are to remain. To better break dormancy, pour boiling water over the seeds and let stand until the water cools, and then plant. Seedling growth is slow at first, but water occasionally with a weak soluble fertilizer. Once the roots become established, tip growth becomes quite rapid.

Softwood or semihardwood cuttings can be taken in summer and early fall. Remove bottom leaves from a four- to six-inch cutting, dip stem in rooting powder, and place either in flats or individual pots. Cuttings should be rooted in three to four weeks, at which time they should be individually potted for the winter. Plant out in the garden the following spring. Young, two-foot plants should be pinched back or trimmed to promote denser growth.

Texas Kidneywood is now available at many nurseries offering native plants. Buying there would be the best and fastest way of obtaining plants already established and large enough to furnish food for larvae.

Use Texas Kidneywood as a specimen plant, as a delicate hedge to line a walk or driveway, or in groupings at the back of a border. It is especially lovely in groups of two or three in a mixed planting with other native shrubs. This is a plant used both as a larval food source and for nectaring, so give it a prominent position in your plantings. Place it in full sun.

In the wild it is more commonly found in calcareous soils but does quite well in sandy or loamy soils. In its native haunts and under normal growing conditions, heaviest flowering is usually in May and again in August and September, but it almost always blooms after any rain. In the garden an occasional heavy soaking during the drier months usually promotes a burst of flowering. An occasional addition of organic fertilizer can help during the growing season but is not mandatory.

USED BY: Southern Dogface (*Zerene cesonia*).
PARTS EATEN: Foliage.

Wafer-ash
(Ptelea trifoliata)

Family: Citrus (Rutaceae)
Class: Native
Height: 2–12 feet
Bloom period: March–July
Range: 1, 2, 3, 4, 6, 7

A large, rounded deciduous shrub with several stems or sometimes a small tree, with aromatic bark and foliage. Leaves alternate, divided into three separate leaflets of various shapes. Flowers small, greenish-white, fragrant, in clusters in spring followed by conspicuously winged, wafer-thin, circular fruits coming to full maturity in late summer or early fall.

CULTIVATION: If not in flower or fruit, Wafer-ash can be mistaken for a small ash (*Fraxinus* spp.). It is easy to grow from either fall- or spring-planted seeds. Propagation by cuttings

WAFER-ASH

is usually less successful and not generally recommended. Gather seeds in late summer or early fall, and plant them in the garden immediately. Mark the planting area well—seedlings do not show until the following spring. Good results can also be obtained by placing the seeds in layers of moist perlite, peat moss, or sand and refrigerating during the winter until ready to plant in the spring. Occasionally, this plant is sold in nurseries, especially those dealing in native species.

Wafer-ash is not especially particular about soil and makes do under near-drought conditions after becoming sufficiently rooted. It tolerates some shade but forms much lusher growth in full sun and with adequate moisture. Trimming or pruning also produces a bushier plant and keeps it to shrub size. If preferred, trim out all the stems except one or two for a small tree form.

USED BY: Giant Swallowtail (*Papilio cresphontes*), Eastern Tiger Swallowtail (*P. glaucus*), Two-tailed Tiger Swallowtail (*P. multicaudata*).

PARTS EATEN: Foliage.

NOTE: Although the flowering period is not long lasting, many species of butterflies use the blossoms for nectaring, especially Hairstreaks (family Lycaenidae) and Skippers (family Hesperiidae). Some gardeners object to the odor of this plant, likening it to the scent of a skunk. If the odor is offensive, place the plant in a part of the garden where it will not be brushed against or bruised.

White-flowered Plumbago (Climbing Plumbago)

(Plumbago scandens)

Family: Plumbago (Plumbaginaceae)
Class: Native
Height: To 4 feet
Bloom period: March–June (all year)
Range: 5, 6, 7

Entire plant sticky with glandular hairs, mound-forming, woody in basal portion, many-branched, the branches long, willowy, and vinelike; plant actually leans or "climbs" to tops of nearby shrubs or clambers into nearby trees. Leaves alternate, pointed at tip, wider at base. Flowers white, the tube long and slender, several in elongating, open cluster.

CULTIVATION: The best method of obtaining new plants of White-flowered Plumbago

White-flowered Plumbago

is by cuttings, since propagation by seeds is not very dependable. Seedlings are easily started by taking semihardwood cuttings in late spring, dipping in rooting medium, and then placing in a sand/perlite mixture. Plant outside when well rooted and showing good leaf growth. Native plant nurseries commonly offer it.

White-flowered Plumbago will grow in almost any soil and prefers bright shade or morning sun. Beneath tall shrubs or small trees or on the east side of house is ideal. Lightly fertilize plants two or three times during the growing season, and water thoroughly when soils become conspicuously dry.

These plants will generally die back to the ground in the more northern portion of its range but will come back quickly in the spring, producing even more growth than in the previous year. Mulching during the colder months will help the deep root system to keep growing.

USED BY: Cassius Blue (*Leptotes cassius cassidula*), Marine Blue (*L. marina*).

PARTS EATEN: Foliage.

NOTE: White-flowered Plumbago is native to the southern and westernmost regions of the state and is also used as a nectar source. It is planted and does well as far north as Austin. A blue-flowered form is also widely planted but is not native and is not used either as a larval food or nectar source.

Yellow Bells (Esperanza)

(Tecoma stans)

Family: Catalpa (Bignoniaceae)
Class: Native
Height: To 8 feet
Bloom period: April–November
Range: 2, 3, 5, 6, 7

Upright, many-branched, deciduous shrub, often with few to several stems and forming a clump. Leaves opposite, divided into five to thirteen leaflets, the margins deeply toothed. Flowers large, yellow, trumpet-shaped, forming showy, elongating terminal clusters. Fruit a long, narrow, flattened capsule filled with numerous flat, two-winged seeds.

CULTIVATION: Yellow Bells is one of the most beautiful native shrubs within its range. In the last few years, it has been offered by many nurseries that sell native plants and has found its way into gardens beyond its native range. This wide distribution demonstrates its ability to readily adapt to a wide range of habitats and various growing conditions.

Plants can be obtained either at nurseries or by starting from seeds or cuttings. Seeds should be gathered when the pods become beige or brownish in color. Separate seeds from the pod, and air-dry on paper for few days. Plant seeds fairly soon after gathering for better germination. Plant where wanted in light, loose soil, and keep moist until sprouted.

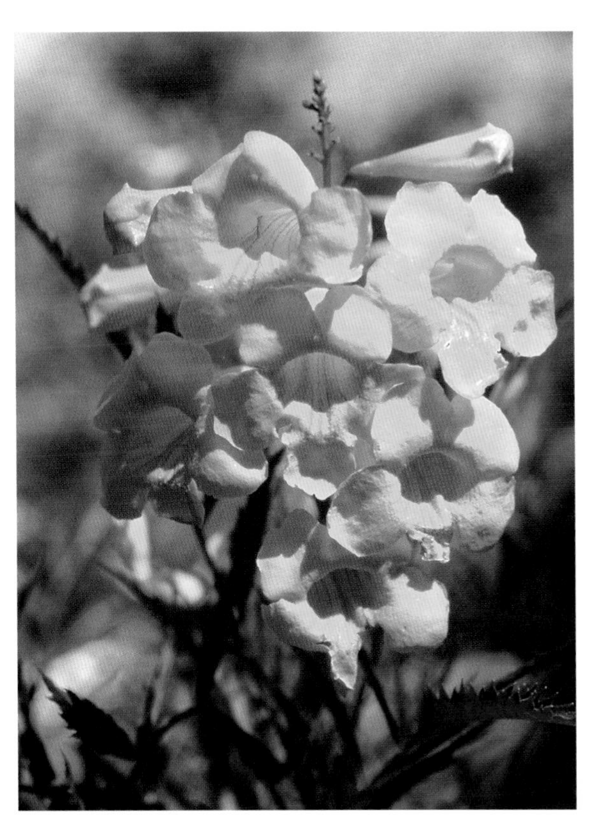

Yellow Bells (Esperanza)

Do not overwater, as seedlings easily root-rot. Good results can be obtained with cuttings by dipping four-inch-long branch sections approximately one-fourth inch across in a rooting hormone powder and potting. Keep potted plants under plastic until well rooted. Cuttings taken in midsummer will be rooted and ready for transplanting to the garden by fall.

Plant in average, well-drained soils, water only in times of drought, and this plant will ask nothing more. This is an airy, graceful shrub in its growth habit, blending beautifully with native and cultivated plants alike. Use it at the back of the border or in a hedge planting along with Beebrush (*Aloysia gratissima*), Barbados Bird-of-Paradise (*Caesalpinia pulcherrima*), Butterfly Bush (*Buddleja davidii*), Desert Willow (*Chilopsis linearis*), Texas Kidneywood (*Eysenhardtia texana*), and any of the lantanas (*Lantana* spp.).

USED BY: Gray Hairstreak (*Strymon melinus franki*).

PARTS EATEN: Flowers and fruits, rarely young foliage.

NOTE: During really hard winters, Yellow Bells will usually freeze back to ground level but, if well established, will resprout from roots, quickly growing to flowering height. Place in beds or borders without soaker hoses. Instead, it needs periodic deep waterings, similar to what occurs in its native desert environment. Removal of spent flowers will ensure a more continuous flowering. These plants are readily used as a nectar source by many species of butterflies, especially Skippers (family Hesperiidae).

Vines

Climbing Milkweed Vine
(Funastrum cynanchoides)

Family: Milkweed (Asclepiaceae)
Class: Native
Height: To 15 feet
Bloom period: April–September
Range: 1, 2, 4, 5, 6, 7

Trailing or climbing, much-branched, mostly smooth, deciduous, unpleasant-scented perennial. Leaves opposite, on long stalks, slender, dull green on upper surface. Flowers numerous, greenish-white to pinkish-purple, crowded into almost round, long-stalked clusters. Fruit a long, slender pod pointed at both ends; seeds numerous, tightly packed in pod, each seed tufted with long, straight, silky hairs; when pod splits, seeds released and carried by

CLIMBING MILKWEED VINE

the wind by the silky hairs, which when dry, become light and fluffy.

CULTIVATION: Climbing Milkweed Vine can be easily propagated by seeds. Watch the pods, and when they turn brown and begin to split, collect and place them in an open paper bag to finish ripening. As soon as the pods are fully open, remove seeds. Place seeds in a container, and lay screen wire or gauze over the top to complete the drying process. Seeds may be directly planted in the garden in a prepared place; mark them well so you can find the seedlings. Be prepared for this plant to climb or sprawl, taking up quite a bit of space.

In the wild this plant will be found in various soils, from sandy to dry, rocky caliche. It will do extremely well in the home garden in soils that are not too rich.

USED BY: Monarch (*Danaus plexippus*), Queen (*D. gilippus thersippus*), Soldier (*D. eresimus*).

PARTS EATEN: Buds, foliage.

NOTE: Due to the unpleasant scent of this plant when disturbed, it might be well to place it where it is not often touched. It is used as a nectar source by the Queen and Soldier as well as numerous other species, especially Skippers (family Hesperiidae) and Hairstreaks (family Lycaenidae).

Drummond's Virgin's Bower (Texas Virgin's Bower)

(Clematis drummondii)

Family: Crowfoot (Ranunculaceae)
Class: Native
Height: 10–15 feet
Bloom period: April–October
Range: Throughout

A climbing or sprawling semiwoody, slender-branched, deciduous, perennial vine. Leaves pale green, usually divided into five to seven leaflets. Flowers small, greenish-white, either solitary or borne in terminal clusters that later form conspicuous and showy feathery seed balls.

CULTIVATION: This is an easily propagated vine, either by seeds or by cuttings. The fluffy seed masses remain on the plant for some time after ripening, so obtaining the brown, mature seeds is usually no problem. Place seeds in a paper bag, and let dry for a few days. *Clematis* seeds require a prechilling (stratification) process to stimulate germination. Place the seeds between layers of moist sand in a container, seal the container in a plastic bag, and leave the bag in the refrigerator for two to three months. The seeds can then be planted directly in the ground in the spring after all danger of frost has passed. Healthy, viable seeds germinate in one to two weeks. If they are planted in a pot, transplant the vine to the garden when its roots have filled the container.

Cuttings from a mature plant can be taken anytime during the growing season. Clip a four- to five-inch section of a branch, strip off the lower leaves, dip the lowest node in rooting hormone, and insert the cutting into sand or vermiculite. Cuttings should be rooted within forty-five days, at which time they may be potted for planting in the garden the following spring.

Drummond's Virgin's Bower is easily divided if dug in early spring or late fall. Dig deeply, and retain a good ball of soil if possible. Clip all aboveground vegetation to about six inches from the crown before digging. Failure to do so often results in the stems being broken off too close to the ground, causing crown damage. Gently separate the root mass, or, if necessary, cut apart with a sharp knife, keeping some roots with each section.

First-year plants are slow to become established and will probably not flower until the second year. Once they are established, however, growth can become lush and rampant and may even require trimming to keep in bounds. These plants thrive in rich garden soils and do not necessarily need fertilizing. Each spring, work organic matter into the soil around the plants, and give an occasional thorough soaking during the growing season, if needed.

If left to climb upward on their own, these plants may become a bit top-heavy by midsummer. To keep them low and full, trim young growth back until quite bushy; then occasionally trim out some of the longer branches. Healthy plants immediately put out new growth just below where they were clipped.

Drummond's Virgin's Bower

This vine is easy to train and is attractive on arbors, trellises, or fences. If clipped for use on a fence, the dense branches make an excellent seasonal screen. But Drummond's Virgin's Bower is perhaps at its loveliest when left to freely clamber over a low split-rail fence or to sprawl across a jumbled pile of rocks.

USED BY: Fatal Metalmark (*Calephelis nemesis australis*).

PARTS EATEN: Foliage.

Groundnut

(Apios americana)

Family: Bean (Fabaceae)
Class: Native
Height: To 12 feet
Bloom period: May–October
Range: 2, 3, 4, 6

Groundnut is a twining, high-climbing, somewhat hairy perennial vine from underground rhizomes producing strings of small, rounded, nutlike tubers; stems usually solitary but

Groundnut

much-branched and often cover nearby vegetation; plants usually form small colonies, with the foliage dying back after frost. Leaves alternate, composed of five to nine leaflets. Flowers bonnet-shaped, brownish-maroon, sweet-scented, numerous, and forming stalked clusters.

CULTIVATION: Groundnut is easily propagated either from seeds or tubers. In obtaining seeds, netting or bagging the long, slender pods before fully mature is recommended, as seeds are flung several feet when mature pods split open. Plant some seeds immediately after they become fully dry, and save some for late summer or fall planting. In the garden, spread seeds where plants are wanted permanently, as moving members of this family is not recommended.

Underground tubers may be dug for transplanting, but if this is done, be very careful, taking only one or two of the topmost in order not to destroy the mother plant. Do not let tubers dry out, so get them into the ground immediately. Place in permanent, previously prepared locations in the garden.

Groundnut is usually found in rich, moist, mostly sandy soils in the wild, but if given sufficient moisture, it will do quite well in most garden settings. As these plants have a tendency to clamber and climb, give them plenty of support, such as a fence, trellis, or brush pile.

If soils are sufficiently rich, no fertilizer other than yearly additions of well-rotted leaves will be needed. Extra moisture will keep plants healthy and the foliage in maximum condition for egg deposition.

USED BY: Silver-spotted Skipper (*Epargyreus clarus*), Southern Cloudywing (*Thorybus bathyllus*), Spring/Summer Azure (*Celastrina ladon*).

PARTS EATEN: Flowers, foliage, fruits.

Kentucky Wisteria
(Wisteria frutescens)

Family: Bean (Fabaceae)
Class: Native
Height: To 40 feet or more
Bloom period: March and April
Range: 3

Twining and high-climbing or sprawling deciduous vine developing a stout, woody, crooked trunk with age. Leaves numerous, smooth, divided into several small leaflets. Flowers fragrant, pale to dark lavender or purple, numerous, bonnet-shaped, thickly clustered in numerous drooping racemes almost covering the vines, appearing after leaves come out. Fruit a long, somewhat flattened, velvety pod containing several large seeds, the pod constricted between each seed.

CULTIVATION: To obtain new plants, planting seeds can be tried. The seedpods of

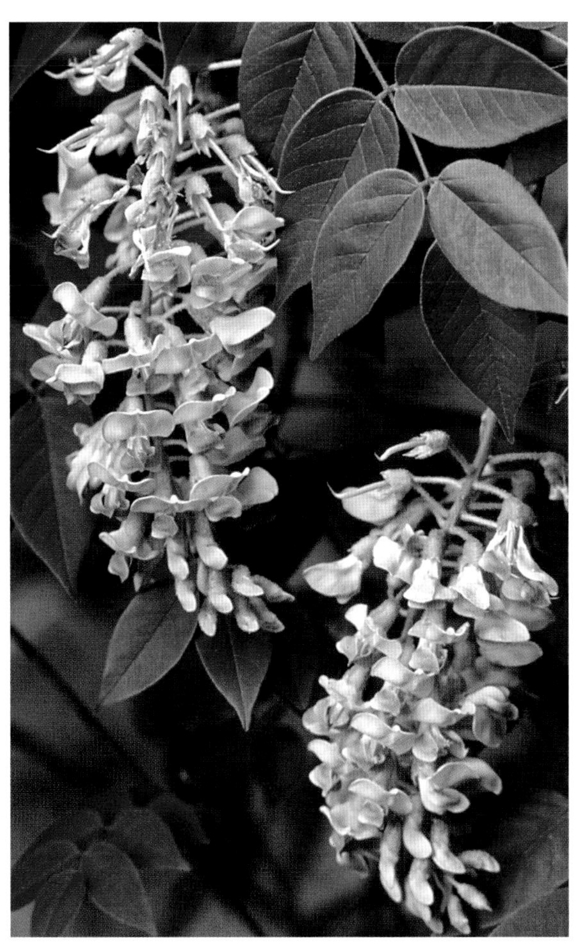

KENTUCKY WISTERIA

Kentucky Wisteria are very hard and tough and, if left to fall naturally, may not open for many months. For starting from seeds, gather pods in midsummer to early fall, let them air-dry, pry open the pods, and remove seeds. Soak the seeds several days; then plant, marking the planting site, for they may not show growth until the following spring. Do the same procedure in fall and again with a spring planting. Seeds may also be soaked in hot water before planting to help sprouting. Plants grown from seeds may not flower for many years.

By far, semihardwood cuttings are the most dependable way of getting new plants. They should be taken in late spring, dipping them first in rooting powder. Place cuttings in a sand/perlite mixture, and place under mist until well rooted. Kentucky Wisteria sparingly spreads by an underground root system, and occasionally young plants can be taken from the mother plant.

USED BY: Silver-spotted Skipper (*Epargyreus clarus*).

PARTS EATEN: Tender to midmature foliage.

Least Snoutbean
(Rhynchosia minima)

Family: Bean (Fabaceae)
Class: Native
Height: To 6 feet
Bloom period: April–December (all year)
Range: 2, 3, 4, 5, 6

A sprawling, trailing or climbing, usually deciduous, perennial vine, often forming dense clumps. Leaves either solitary or most often of three leaflets, dull, dark green on upper surface. Flowers very small, yellow, usually several openly spaced on four-to-five-inch

LEAST SNOUTBEAN

stalk. Fruit a small flattened pod containing one to three seeds.

CULTIVATION: The best and easiest way to obtain plants of Least Snoutbean is from seeds. Once a plant has been found and the pods first begin turning brown, collect the entire pod. Place the pod in a tall paper bag, and let the pod continue to dry. Upon drying, the pods split and the two halves twist and curl, flinging the seeds some distance. Seeds can be planted immediately or saved until fall or the following spring.

Once this plant is started in the garden, there will be an abundance of new plants each spring. Easily transplanted, they can be moved to desired areas. The plants develop extremely deep taproots, so if moving older plants, dig deep.

USED BY: White-striped Longtail (*Chioides albofasciatus*), Ceraunus Blue (*Hemiargus ceraunus astenidas*).

PARTS EATEN: Foliage.

NOTE: Try using these in hanging baskets, placing three to four plants in each container.

Purple Passionflower
(Passiflora incarnata)

Family: Passionflower (Passifloraceae)
Class: Native
Height: to 25 feet
Bloom period: May–frost
Range: 2, 3, 4, 5, 6

An upright, trailing or sprawling, fast-growing herbaceous perennial vine climbing by tendrils from the leaf axils. Stems strong and vigorous, angled when young. Alternate leaves large, dark lustrous green, three-lobed. Two large conspicuous nectar glands on stalk near base of blade. Intricately formed, solitary, short-stemmed, fragrant, three-inch flowers rise from the leaf axils. The flowers are composed of five sepals and five petals, which are quite similar in appearance, subtended by a fringe of wavy or crimped, pale to dark lavender hairlike segments. Rising above are five stamens and a three-parted pistil. The flowers close at night. The fruit is a large, egg-shaped berry, orange-yellow when fully ripe. Containing mostly seed, the scant pulp is edible and the juice makes a tolerable drink.

CULTIVATION: Purple Passionflower is an easily grown vine and can be started either by seed or cuttings. Not particular to soils, it will do well in just about anything except hard clay or overly rich garden soils. One thing it does require is good drainage. It will tolerate long, droughty periods but blooms and fruits best with regular waterings.

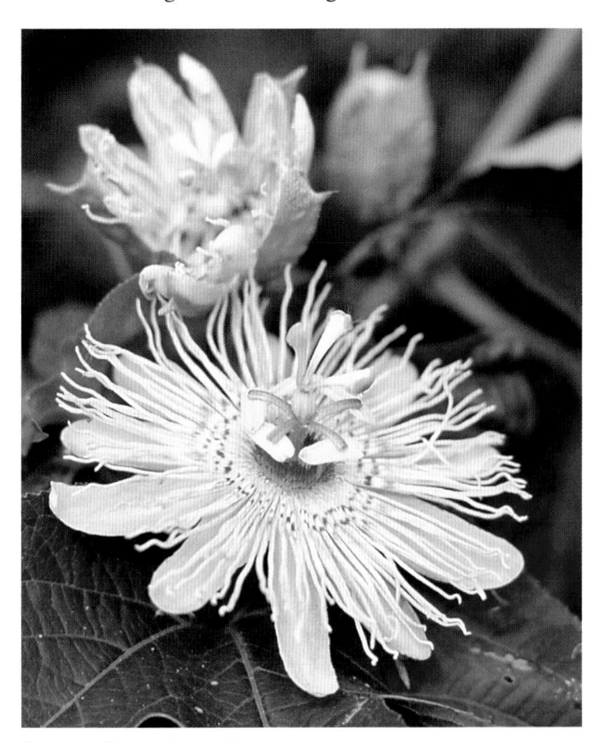

PURPLE PASSIONFLOWER

A light application of 10-5-20 fertilizer two or three times during the growing season may be helpful, but avoid overfertilizing as this can cause serious root damage or even death to the plant. If in doubt, do not fertilize at all. Later in the season if the plant appears to need help, fertilize very rapidly.

In areas where this vine is regularly killed back to the ground each winter, a layer of loose pine straw may be beneficial.

For the most bloom and best fruiting, plant this vine in full sun or in partial shade to protect from hottest afternoon temperatures. If more plants are desired, seed may be taken from the fully ripened and dried seed pods in late fall, stored in a paper bag during winter, and planted in spring after all frost is past. Six-inch tip cuttings can be taken in early summer, dipped in rooting medium, and planted in small pots until rooted. Keep soil of cuttings moist but not soggy until new growth begins. Plant immediately in garden in permanent location.

USED BY: Gulf Fritillary (*Agraulis vanillae incarnata*), Variegated Fritillary (*Euptoieta claudia*), Julia Longwing (*Dryas iulia*), Zebra Longwing (*Heliconius charithonius vazquezae*).

PARTS EATEN: Buds, flowers, tendrils, tender leaves preferred;entire plant if stressed.

NOTE: Purple Passionflower can be used on fences for wonderful warm-weather screening, on decorative trellises, or left to clamber into nearby trees. They are beautiful sprawled across brush piles or low rock walls. Since it is rather a rampant grower, give it plenty of space to spread.

It is not unusual for this vine to remain evergreen in sheltered locations during mild winters.

Red-fruited Passionflower (Tagua Passionflower)

(Passiflora foetida)

Family: Passionflower (Passifloraceae)
Class: Native
Height: To 20 feet
Bloom period: April–November
Range: 4, 5, 6

A strong, vigorous, annual or perennial vine covered throughout with soft grayish or brownish hairs. Leaves light green, three-lobed, with middle lobe the longest. Flowers intricately shaped, pale lavender and purple; subtended by large, beautifully dissected, feathery-appearing bracts; bracts remain after flowering, partially enclosing the brilliant red fruit pods.

CULTIVATION: This hardy vine grows best in sandier soils and delights in adequate moisture and plenty of sun. Light applications of

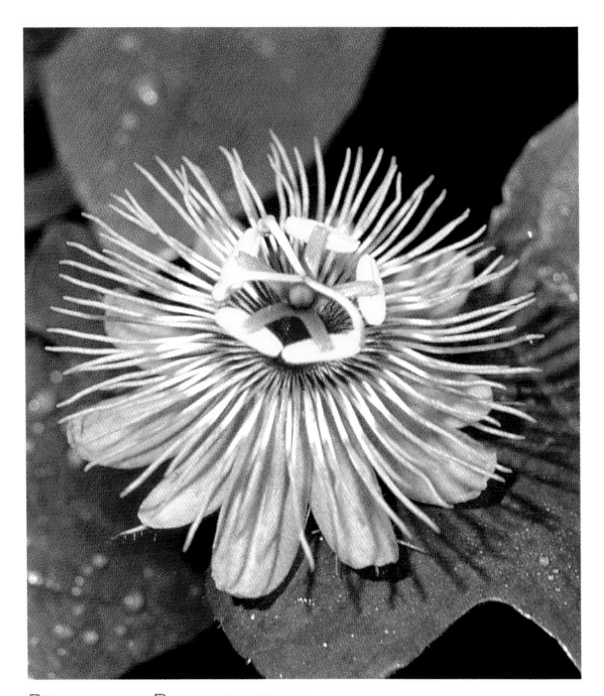

RED-FRUITED PASSIONFLOWER

high-phosphorus fertilizer in spring and again in fall increase the number of flowers. Do not overfertilize, though, as the chemical structure of the foliage would change and not be acceptable as larval food.

Red-fruited Passionflower grows equally well if left to trail on the ground or if given a support to climb. It makes a beautiful "summer cover" for woodpiles or the brush piles often provided by bird-watchers as escape areas for birds. The vines are especially attractive when trained to climb on a low wooden fence at the back of a perennial border.

Seeds may be planted in the fall or spring. After the pods begin losing their red color and become brown, gather and leave them in an open paper bag in a warm area until completely dry and crisp. Crush the pods, and remove seeds. Store seeds in a dark, dry area in a sealed container until spring. Plant them in prepared sites after the ground has become warm. Seeds may also be planted in the fall, with plants appearing in late spring. Mature plants send out numerous sprouts from widely spreading underground roots, and these can easily be dug and replanted where wanted.
USED BY: Crimson-patched Longwing (*Heliconius erato petiverana*), Zebra Longwing (*H. charithonius vazquezae*), Julia Longwing (*Dryas iulia*), Gulf Fritillary (*Agraulis vanillae incarnata*), Mexican Fritillary (*Euptoieta hegesia meridiania*), Variegated Fritillary (*E. claudia*), Mexican Silverspot (*Dione moneta poeyii*).
PARTS EATEN: Buds, foliage, young fruits.
RELATED SPECIES: As far as is known, every native species of *Passiflora* is used as a larval food source by the Longwing butterflies within their breeding range. Choose the ones that grow naturally in your area, or try some that grow just outside your range. Most of the larvae that usually feed on native passionflowers also use many of the cultivated sorts, some of which have gorgeously colored flowers.

Snapdragon Vine
(Maurandya antirrhiniflora)

Family: Figwort (Scrophulariaceae)
Class: Native
Height: 3–8 feet
Bloom period: April–December
Range: 4, 5, 6, 7

A delicate, climbing or trailing, much-twining, herbaceous perennial vine. Leaves very small, smooth, triangular or three-lobed. Flowers beautiful violet or purple, snapdragon-like, solitary from leaf axils.
CULTIVATION: Snapdragon Vine is an easily grown plant that is not particular about soils and does well in whatever is provided. It is drought tolerant and a good choice for the drier, more neglected sites in the garden. It is a prolific grower, producing numerous slender stems that quickly cover its support.

SNAPDRAGON VINE

Snapdragon Vine is not generally offered by nurseries, but it can be raised from seeds. Gather the small, round capsules as they become tannish-colored, and air-dry for a few days. Remove the small, thick, brown, corky-winged seeds by crushing and rubbing the capsule husk away. Plant some seeds in late summer or early autumn and some in early spring. Germination is somewhat unpredictable, so sow plenty of seeds and at various times. Even if spring planted, the vines often bear flowers by midsummer the first year. Not cold tolerant, foliage is killed by the first hard frost but reappears in early spring.

Easily adaptable, this vine does equally well twining through a hedge; trailing over a rock, shrub, or fence; or cascading from a hanging basket, or provide a small, special trellis of its own. If continued bloom is desired, bring potted plants inside in the fall before frost and place them in a sunny window for winter bloom. Gradually move plants outside again in spring as the weather warms. This vine needs only a light support and does not weigh down plants it happens to use in its reach for the sun. It is an excellent choice in any kind of container planting.

USED BY: Common Buckeye (*Junonia coenia*).

PARTS EATEN: Foliage.

NOTE: Damage is usually minimal to these plants, as the Common Buckeye lays eggs singly or few to a plant. These plants are beautiful when used alone or with taller plants in containers or hanging baskets.

Wild Cow-pea

(Vignia luteola)

Family: Bean (Fabaceae)
Class: Native
Height: To 10 feet
Bloom period: All year
Range: 4, 5

Upright, trailing, or sprawling perennial sometimes dying down to ground in winter, with fast-growing annual, vinelike stems and branches quickly forming large mounds or extensive colonies in spring. Leaves alternate, long, on long petioles, each leaf composed of three slender, dark green leaflets. Flowers large, bonnet-shaped, yellow, numerous and clustered in a gradually opening terminal spike, with few to several open at once. Fruit a long, roundish, slender, several-seeded pod.

CULTIVATION: As with most legumes, the pod of Wild Cow-pea opens immediately after becoming fully mature. For propagation, these pods should be gathered when partially brown and placed in a paper bag until completely dry. Remove seeds from the pods, and allow

WILD COW-PEA

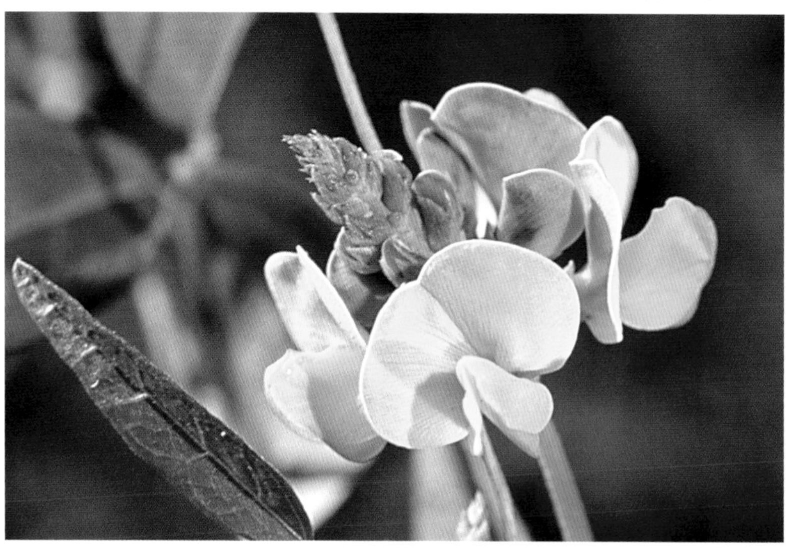

seeds to dry for a few days more. For the best chance of sprouting, plant some seed immediately, and then plant some seeds in the fall (mark the site well) and some the following spring. Propagation by cuttings is not very successful.
USED BY: Mercurial Skipper (*Proteides mercurius*), Proteus Longtail (*Urbanus proteus*).
PARTS EATEN: Foliage.
NOTE: Wild Cow-pea will be found growing naturally in the moister sandy loams and heavy clays in the coastal regions. Due to its spreading habit and tendency to multiply rapidly, this plant may be too large and aggressive for the smaller home garden and may be best used for stabilizing banks or as erosion control.

WOOLLY PIPEVINE

Woolly Pipevine
(Aristolochia tomentosa)

Family: Birthwort (Aristolochiaceae)
Class: Native
Height: To 50 feet
Bloom period: March–June
Range: 2, 3, 4, 6

A high-climbing, deciduous, perennial vine. Leaves large, handsome, heart-shaped, usually woolly with whitish hairs. Flowers elongated, oddly curved, shaped somewhat like a Dutch smoking pipe, an unusual shade of dark purple and greenish-yellow.
CULTIVATION: Woolly Pipevine is not too demanding about soils; it grows naturally in deep sands along upland streams and also in river bottom floodplains. What it does require is good drainage and adequate moisture but with a period of dryness between soakings. Best growth and foliage are produced if planted with the roots in dense shade and where the vines can climb a support to be in bright, filtered sun. It does very well planted at the edge of thick, tall shrubbery or small trees that it can use as a support. Vines die back to the ground each year, but spring growth is rapid and vigorous on healthy, well-established plants. The largest leaves are always the youngest ones, near the tips of the branches.

Propagation is best by seeds or by taking cuttings or runners. The seedpods are large but cannot be found very often in the wild, probably because caterpillars eat them.

If a plant is found in the wild with flowers that look old enough to have been pollinated, then the sleeving process might be tried. Take small pieces of nylon netting, cover each flower, and close the open end around the stem a couple of inches below each flower with a twist tie. When the pods mature, clip and spread them on paper towels until completely dry and beginning to release the seeds. When ready to plant, soak seeds in distilled water for at least twenty-four hours, changing the water every twelve hours. Have the water hand-hot when first adding the seeds. As each seed swells, remove and sow it immediately, before it has time to dry out. Plant the seeds in a professional germinating mix, barely covering the seeds and keeping the soil moist with distilled water. Place a piece of clear plastic over the pot, and put it in an area with lots of light, but not in direct sunlight. Light is very important to the germination of all species of *Aristolochia.*

Mature plants of Woolly Pipevine put out long underground roots or runners with many Ys or Ts, and often a new plant forms at these joints. If well rooted, these runners can be clipped on each side of the newly rooted joint. If propagation is to be tried with these young plants, have a container filled with a commercial potting soil prepared and plant immediately. Moisten with rainwater with a bit of root stimulator added. Place some sticks around the edge of the pot (taller than the plant), and cover with a clear plastic bag. Fasten the bag below the rim of the pot with twine or a rubber band. Do not let the plastic touch the plant. Keep the soil moist. After the plant is firmly established and showing new growth, gradually remove the plastic. Do not let these plants remain too long in containers, as they will grow only roots and make no top growth, so place them in the garden as soon as they are well rooted. Runners can be placed in shallow trenches in the garden, but take extra care to see that they never dry out until well rooted.

Woolly Pipevine is slow to come into flower; it may be two years or more before it blooms. Since these plants are being grown for their foliage, however, the flowers are not important to the caterpillars. Applications of compost or manure from time to time to well-established plants increases foliage.

USED BY: Pipevine Swallowtail (*Battus philenor*).

PARTS EATEN: All aboveground parts of the plant.

Yellow Passionflower

(Passiflora lutea)

Family: Passionflower (Passifloraceae)
Class: Native
Height: To 15 feet
Bloom period: May–September
Range: 2, 3, 4, 6

Usually a thin, delicate, smooth, twining, deciduous vine found in more shaded areas, climbing by solitary tendrils from leaf axils. Leaves solitary, barely three-lobed, dull green on upper surface. Flowers small, pale yellow to greenish, consisting of many hairlike filaments, followed by small, round, globelike fruits turning black at maturity.

CULTIVATION: When fruits have become fully mature and begin to wrinkle, collect and remove seeds. Seeds should be rinsed under running water to remove all gelatinous matter, then air-dried for few days. Sow seeds directly in prepared beds where plants are wanted. Mark the area well. Germination is

YELLOW PASSIONFLOWER

both poor and slow, and it may take two years or more for seeds to sprout.

Six- to eight-inch stem cuttings can be taken early in the season, dipped in rooting medium, and inserted into equal parts sand and perlite. When rooted, they can be transferred directly to the garden.

Once Yellow Passionflower is established, there may be occasional suckers or sprouts from the mother plant that can be taken up and transplanted.

USED BY: Gulf Fritillary (*Agraulis vanillae incarnata*), Mexican Fritillary (*Euptoieta hegesia meridiania*), Variegated Fritillary (*E. claudia*), Julia Longwing (*Dryas iulia*), Zebra Longwing (*Heliconius charithonius vazquezae*).

PARTS EATEN: Entire plant, young foliage preferred.

NOTE: Use these plants in hanging baskets or tall containers, either alone or in combination with other plants.

HERBS

Antelope-horns Milkweed

(Asclepias asperula)

Family: Milkweed (Asclepiadaceae)
Class: Native
Height: 8–24 inches
Bloom period: March–November
Range: Throughout

Low, upright to widely sprawling herbaceous perennial, usually with several unbranched stems from base and forming moundlike clumps. Leaves opposite or almost so, long, slender, the edges often folded together. Flowers greenish, numerous, and crowded into rounded, solitary terminal clusters.

CULTIVATION: Rarely do nurseries offer any species of *Asclepias* other than Butterfly Weed (*A. tuberosa*) and Tropical Milkweed (*A. curassavica*). Antelope-horns are not easily transplanted because of the deep, brittle roots but are easy and fast to grow from seeds. Seeds can be planted immediately after collecting, or hold them over in cold storage until spring. The taproot grows very fast, so seedlings should be transplanted as soon as true leaves appear.

Take three- to four-inch tip cuttings before the plants begin flowering. Remove lower leaves, and insert cuttings into moist, pure sand or a sand and peat mix. Make a tent from clear plastic to cover the cuttings, and keep them moist. Cuttings should be rooted within six weeks.

Antelope-horns are very hardy once established and usually do quite well under normal garden conditions. They grow best and are less prone to stem rot if the soil is a sandy,

Antelope-horns Milkweed

somewhat gravelly type. Good, fast drainage is important. Place plants in full sun, give little or no fertilizer, and water only during periods of drought. Trim back well-established plants to three or four inches above ground level in June, and they will put forth new shoots and bloom again in the fall. If there are several plants in the garden, rotate cutting continually during the entire season, thus providing a constant source of new growth for caterpillars.

USED BY: Monarch (*Danaus plexippus*), Queen (*D. gilippus thersippus*).

PARTS EATEN: Young foliage, occasionally flower buds.

NOTE: When these plants are in flower, they are a much-favored nectar source.

Arrow-leaf Sida (Axocatzin)

(Sida rhombifolia)

Family: Mallow (Malvaceae)
Class: Native
Height: To 2 ½ feet
Bloom period: May–October (all year)
Range: 2, 3

An upright, freely branching, shrubby annual or short-lived perennial with tough, almost woody stems arising from deep, tough roots; plants form large patches or colonies. Leaves alternate, short-stalked, wedge-shaped, deeply toothed on margins and borne on slender branches; lower side of leaves much paler with short, grayish hairs. Flowers solitary, yellow, five-petaled, arranged on long, slender stalks from axils of each leaf.

CULTIVATION: It is not difficult to get Arrow-leaf Sida started growing in the garden. It is as simple as gathering a handful of the small, tannish seed capsules when dried and scattering them where plants are wanted. Plant as soon as gathered; in the warmer portions of its range plants will come up within days and will continue growth through the colder months. In the warmest portions of its range, plants rarely freeze back entirely and will, in most instances, continue blooming. In such conditions, Checkered-Skippers (family Hesperiidae) often continue flying, and egg deposition begins with first new plant growth.

In colder areas, plant some seeds directly in the soil, and then plant more seeds in early

Arrow-leaf Sida

spring after the first frost. This will assure a good stand. These plants are always found in groups or colonies and, for butterfly usage, should be planted in this manner around the home grounds.

This may not be a plant you want to use in regular flower beds and borders. It is a favored and much-used plant, however, so should definitely be considered for planting. Arrow-leaf Sida will self-sow and quickly spread and in some instances may be more aggressive than desired. Use it around outbuildings, toward the back of the property, in "waste" places, at edges of woodlands, or in any open, sunny, neglected part of the home grounds. Almost any kind of soil will do, even hot, dry sand or clay. No fertilizing is necessary and probably should be avoided. Mulching can be tried to help keep down any unwanted reseeding.

USED BY: Common Checkered-Skipper (*Pyrgus communis*), Tropical Checkered-Skipper (*P. oileus*).

PARTS EATEN: Foliage.

NOTE: Once a few inches high and well established, Arrow-leaf Sida is almost impossible to pull up by hand. If eradication is needed, either clip at ground level or hoe unwanted plants.

Coastal Water-hyssop

(Bacopa monnieri)

Family: Figwort (Scrophulariaceae)
Class: Native
Height: To 6 inches
Bloom period: April–December
Range: 2, 3, 4, 5, 6, 7

Low, trailing or sometimes floating, smooth, tender-stemmed perennial with numerous

COASTAL WATER-HYSSOP

stems rooting at the nodes or joints and forming large mats. Leaves opposite, about as wide as long, thick, and firm. Flowers solitary, long-stalked, generally a pale bluish-lavender. Fruit a somewhat longish capsule containing small seeds.

CULTIVATION: Coastal Water-hyssop can occasionally be found offered by aquatic plant nurseries and dealers, and this is the quickest and most assured way of obtaining this plant. Otherwise, gather ripened seed capsules; after seeds are completely dry, plant in a sand/perlite mixture, simply pressing seeds into the soil. Keep soil moist until new plants show. Seeds should be planted at various times in order to get a good stand. If possible, plant seeds outside where plants are to remain.

In placing in the garden, choose a site along the edge of a pond or around a leaky water faucet where a constant moist condition can be provided. Under ordinary conditions these plants will form a solid mat, ground cover, or an extended floating mat on water.

USED BY: Banded Peacock (*Anartia fatima*).

PARTS EATEN: Foliage.

Downy Paintbrush
(Castilleja sessiliflora)

Family: Figwort (Scrophulariaceae)
Class: Native
Height: 4–12 inches
Bloom period: March–May
Range: 1, 2, 6

Downy Paintbrush

Low, upright perennial covered with soft, velvety hairs; often several stems from the base, forming a large, showy clump. Almost all "leaves" somewhat tinged with color, gradually becoming smaller as they extend upward and finally become colorful bracts subtending the flowers in upper portion of stalk; bracts, or modified leaves subtending flowers, long, slender, brightly colored, and often mistaken for "flowers." Flowers long, tubular, yellowish-green, rather inconspicuous, usually extending somewhat beyond the subtending bracts.
CULTIVATION: All of the paintbrushes are almost impossible to transplant, but there should be no problem starting plants in the garden from seeds if some care is taken. Gather seed capsules as soon as they are dry and before they split to release the numerous tiny seeds. Let seeds air-dry for a few days, and then store in the refrigerator for fall sowing. When ready to plant, mix seeds with fine sand for better distribution. All species of paintbrushes are thought to be partially parasitic on roots of grasses during the early stages of growth, so plant the seeds in raked areas next to a clump of native grass for best results.

The percentage of germination will probably be small, but if only one or two plants come up and flower, they usually reseed quite well. A good colony should be established within two or three years. Since little is actually known about the propagation or life cycle of this plant, for best results try planting some seeds as soon as they are ripe, some in early summer, and some in the fall. Also, try them among natural areas of native grasses as well as in perennial beds.

Downy Paintbrush usually grows in dry, rocky, or sandy soils, so in the garden give it the most natural habitat possible. Do not fertilize or overwater these plants.
USED BY: Common Buckeye (*Junonia coenia*), Fulvia Checkerspot (*Chlosyne fulvia fulvia*).
PARTS EATEN: Upper portion of plant.
RELATED SPECIES: The three varieties of Downy Paintbrush each cover a different

local area: var. *citrine* with pale to dark yellow bracts, var. *lindheimeri* with pale to dark salmony-orange bracts, and var. *purpurea* with rose-pink to purple bracts (shown here).

Several other species of *Castilleja* occur in the state, and at least one of them, the annual Texas Paintbrush (*C. indivisa*), is used from Central Texas eastward by the larvae of the Common Buckeye. Many, if not all, of the others are probably used.

Downy Water-willow (Tube-tongue)

(Justicia pilosella)

Family: Acanthus (Acanthaceae)
Class: Native
Height: To 12 inches
Bloom period: April–October
Range: 2, 4, 5, 6, 7

DOWNY WATER-WILLOW

Low, woody-based, multistemmed, deciduous perennial, the stems upright or falling over with plant becoming moundlike. Leaves dark green. Flowers one to several from leaf axils but usually only one flower open at a time. Fruit a four-seeded capsule.

CULTIVATION: Downy Water-willow can sometimes be found at native plant sales or nurseries, and this is by far the easiest way to obtain this plant. To obtain seeds when the plant is found in the wild, watch the flowering closely. When capsules are found, enclose the capsule with netting or watch very closely for ripening. Once capsules become brown, gather, place in a paper bag, and let them dry. Remove seeds, plant some immediately in moist potting soil, and then plant some seeds every month or so. Save some seeds to plant the following spring.

Cuttings can be tried, but success of this method is not known. Try dipping cuttings in rooting powder, and place some in moist potting mix and some in sand.

In the wild, Downy Water-willow can be found in rocky, gravelly soils of the Hill Country to the calcareous silt, clays, and sands of the southern regions. In the home garden, do not make the soil too rich and give good drainage.

USED BY: Tiny Checkerspot (*Dymasia dymas dymas*), Elada Checkerspot (*Texola elada ulrica*), Rosita Patch (*Chlosyne rosita*), Vesta Crescent (*Phyciodes graphica vesta*), Texan Crescent (*Anthanassa texana*).

PARTS EATEN: Foliage.

RELATED SPECIES: A native of the Rio Grande Valley, Runyon's Water-willow (*J. runyonii*) is a larger plant with more flowers open at once. It is used by the Malachite (*Siproeta stelenes biplagiata*), Banded Peacock (*Anartia fatima*), and Texan Crescent.

False Wissadula
(Allowissadula holosericea)

Family: Mallow (Malvaceae)
Class: Native
Height: To 6 feet
Bloom period: April–December
Range: 6, 7

A stout, upright, usually many-branched perennial that becomes semiwoody at base. Leaves to eight inches long and almost as broad, covered in short hairs, making them thick, soft, and velvety to the touch. Flowers numerous, five-petaled, yellow-orange, solitary or in short clusters, with one or two opening at a time in each cluster.

CULTIVATION: Obtain plants of False Wissadula either from seeds or cuttings. Gather seeds in fall immediately upon ripening. Air-dry seeds thoroughly, and place in a paper bag with an insect strip or a mothball for a couple of weeks. Remove, place in a sealed container, and store in the refrigerator until spring. After all danger of frost has passed, plant seeds either directly in the garden or in small pots for later transplanting. Cover seeds with one-fourth to one-half inch of soil. Mark the location well in beds, for sometimes it may take two years for the seeds to germinate.

Six-inch-long cuttings may be taken from strong growing tips of a healthy plant, dipped in a rooting medium, and then planted in well-drained soil. Before dipping them into the rooting medium, cut to just below a node and remove all but the top leaves. Cuttings are best taken in early summer in order for the plants to become established and growing well before fall. The success rate is not very high with cuttings either, so try rooting more than will be needed. Extras will always be welcomed by other gardeners who are trying to attract butterflies.

Cut back well-established, strong-growing plants to ground level during the winter months each year to encourage a robust, bushy habit and more flowering. Do not fertilize, for this would encourage lank, spindly growth with the foliage prone to developing

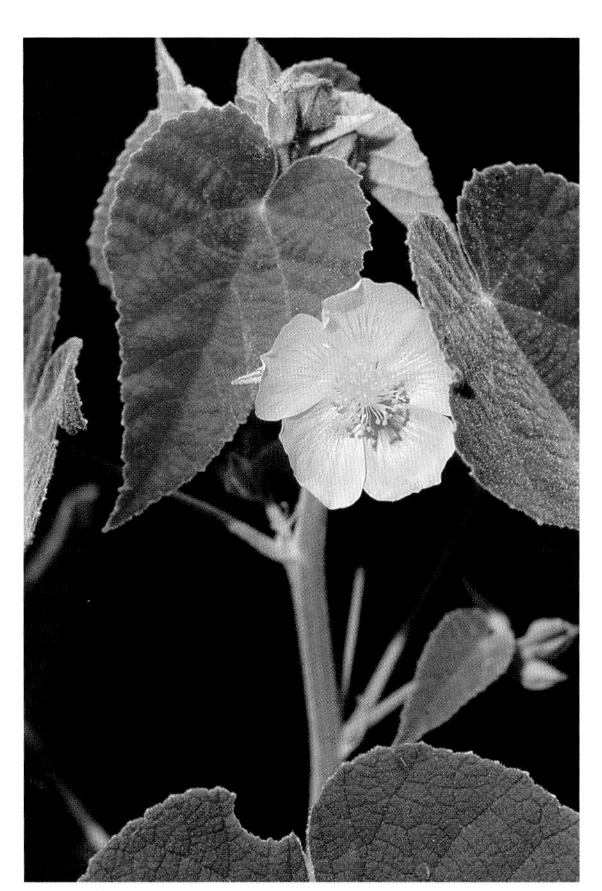

False Wissadula

powdery mildew. Also, keep on the dry side; mature plants cannot tolerate an overabundance of moisture. Butterflies will not use weak, sickly foliage for egg deposition.

Plants may be moved with some success. Before replanting, cut the tips back to about six inches above ground level; keep well watered until new growth appears and the root system is well established. For the first few weeks after transplanting, use a very weak soluble fertilizer about once every two weeks. Provide a bit of shade in the form of leafy branches or tented newspapers the first few days to help reestablish the plants.

USED BY: Common Streaky-Skipper (*Celotes nessus*), Scarce Streaky-Skipper (*C. limpia*), Laviana White-Skipper (*Heliopetes laviana*), Texas Powdered-Skipper (*Systasea pulverulenta*).

PARTS EATEN: Foliage.

Fern Acacia

(Acacia angustissima var. *hirta)*

Family: Bean (Fabaceae)
Class: Native
Height: 2 ½ – 3 feet
Bloom period: June–October
Range: Throughout

An upright to sprawling, deep-rooted, thornless perennial with one to several stems from a woody, persistent base; stems rarely branching, but plants put out numerous underground woody, creeping roots and form small colonies. Leaves alternate, bright green, usually divided into nine to twelve pairs of segments, with each segment bearing eighteen to thirty pairs of tiny leaflets; leaflets sensitive and will fold together during rain, at night, and when touched. Flowers numerous, white to creamy-yellow, in ball-like clusters held on stalks arising from leaf axils in upper portion of stem; flower stalks shorter than subtending leaves, with flower heads remaining intermingled with the fernlike foliage.

CULTIVATION: Fern Acacia is well adapted to a wide variety of soils and climatic conditions, as is shown by its wide distribution. In the wild this plant can be found along roadsides, in open woodlands, in prairie grasslands, or along bluffs, ledges, or outcrops of limestone or shale, often above streams and frequently in shade. In its natural habitat it seems to prefer either tight, heavy, calcareous clay or alkaline soils but can frequently be found in sand. It is a hardy plant that once established will take care of itself and continue to multiply quite readily. It does not multiply at a very rapid rate, however, so the plants do not become obnoxious or hard to control.

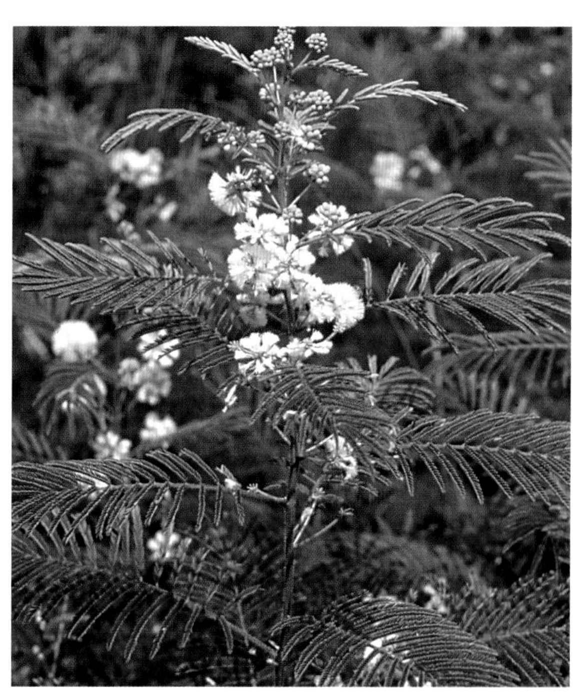

FERN ACACIA

In the northern portion of its range, Fern Acacia will usually be killed to the ground each winter but in the southernmost areas will continue as an evergreen, at most losing only some of its leaves.

Plant Fern Acacia in full sun or dappled shade, giving it good soil and an adequate amount of moisture to get it growing well. Once established, it should not need any special attention. A light application of natural fertilizer or rich compost will enhance both growth and flowering, but do not overdo it. Overfertilizing and too much moisture will cause excessive foliage growth, fewer flowers, and, eventually, crown rot—resulting in death of the plants.

The easiest method to obtain a start of Fern Acacia is by seeds. The brownish seedpods, borne singly in the upper portion of the plant, are from two to three inches long, less than one-half inch wide, and very flat. They open promptly after becoming fully ripe, so it is necessary to observe the wild plants in order to gather seeds at the proper time. Seeds that are at least three-fourths brown in color can be gathered and stored in a cool, dry area for a few days or until completely brown and pods begin to split.

Seeds of Fern Acacia have very hard outer coats, and germination can be very slow. To hasten the process, after gathering the seeds, let them dry for two to four months in a cool, dry area. Refrigeration is not necessary. When ready to plant, pour boiling water over the seeds and soak for at least twenty-four hours. If possible, plant directly in the garden where they are wanted permanently.

USED BY: Acacia Skipper (*Cogia hippalus hippalus*), Outis Skipper (*C. outis*), Ceraunus Blue (*Hemiargus ceraunus astenidas*), Mexican Yellow (*Eurema mexicana*), Reakirt's Blue (*Echinargus isola*).

PARTS EATEN: Buds, flowers, young foliage.

Garden Canna

(Canna generalis 'The President'*)*

Family: Canna (Cannaceae)
Class: Cultivated
Height: To 6 feet
Bloom period: May–December
Range: Throughout

Upright, colony-forming perennial from large bulb. Leaves large, thick, and lush, the foliage quite similar to that of a young banana plant, usually killed to ground by first hard freeze. Flowers large, red, with several forming a terminal spike that extends above the foliage; petals of various widths, with some upright and some recurved, giving the blossom a completely disorganized appearance.

CULTIVATION: Plant Garden Canna bulbs in March or April, spaced twelve to twenty-four inches apart. Divide established clumps at this time. Plants can also be raised from seeds if the seeds are scarified with a file and soaked in warm water overnight. Plant the seeds vertically, about two inches deep. They should sprout in about two weeks.

Garden Cannas grow in almost any soil and habitat, but blooms are more plentiful and colorful if the plants have rich soil and full sun. They demand good drainage; otherwise, the new growth will rot at soil level. When foliage is about eight inches high, apply a complete fertilizer, such as 12-24-12, around the plants at the rate of one to two pounds per one hundred square feet of bed; this amounts

GARDEN CANNA

to a very light sprinkling. For lush, continuous new growth, keep old or damaged stalks cut to ground level. Once well established, healthy clumps need to be dug and divided about every three years. With just a little care as to proper watering and feeding, Garden Canna comes back year after year, the clumps becoming larger and the blossoms lovelier with each season.

USED BY: Brazilian Skipper (*Calpodes ethlius*).

PARTS EATEN: Leaves.

NOTE: Be prepared for plants to look a little ragged at times, for once Brazilian Skipper caterpillars start eating, they consume a surprising amount in a short time. They also cut and roll large portions of the edges of leaves, as if using them for blankets. In the trade, these caterpillars are often referred to as the "canna leaf roller" and exterminated.

RELATED SPECIES: Also used as food plants are the taller-growing, smaller-flowered native species, Golden Garden Canna (*C. flaccida*), Louisiana Canna (*C. glauca*), and Indian-shot (*C. flaccida*).

Globe-mallow

(Sphaeralcea coccinea)

Family: Mallow (Malvaceae)
Class: Native
Height: To 3 feet
Bloom period: May–October
Range: 1, 2, 6, 7

Upright or occasionally sprawling, low-growing herb having a perennial root and one to several woody-based annual stems. Leaves thick, alternate, and usually deeply three-lobed. Flowers mostly orange-pink (rarely scarlet), five-petaled, and loosely arranged in open, branching clusters along stem.

CULTIVATION: This herb is best propagated by seeds. To collect seeds, clip mature seedpods and place in a paper bag. Allow them to dry in a warm area for several days until all moisture is gone and capsules are completely opened. Shake seeds loose or break capsules open, releasing the seeds. If insects are present in the seedpods, add an insect strip to the bag and close tightly for two weeks. Remove cleaned, treated seeds, and store in the refrigerator until spring. Sow seeds thinly in a prepared bed after all danger of frost is passed. Growth will be rapid, and seedlings may be transplanted to permanent places in the garden when they have three to four leaves.

Well suited to the drier areas of the garden, this plant actually grows best and is very long-lived in loose, extremely sandy soils. In tight soils it has a tendency to die out in three or four years. If planted in richer, moister soils than its normal habitat, it will rot at ground level and die.

Use Globe-mallow in scattered groups in beds or borders, along walkways, or in any type of naturalized situation. The unusual coloring of the blossoms is especially lovely when used with Leatherweed Croton (*Croton pottsii*), Prairie Verbena (*Glandularia bipinnatifida*), Western Peppergrass (*Lepidium alyssoides*), White-flowered Plumbago (*Plumbago scandens*), Trailing Lantana (*Lantana montevidensis*), or mistflowers (*Conoclinium* spp.).

USED BY: Common Checkered-Skipper (*Pyrgus communis*), Small Checkered-Skipper (*P. scriptura*).

PARTS EATEN: Foliage.

NOTE: If any of these plants are desired for the garden, propagate by seeds or purchase at a nursery, as it is almost impossible to move established plants without breaking the unusually deep taproot.

RELATED SPECIES: Two varieties of this plant occur in the state: Scarlet Globe-mallow, var. *coccinea*; and Caliche Globe-mallow, var. *elata*—both are used as a larval food plant. Larvae of several species of butterflies use a very similar plant, Copper Globe-mallow (*S. angustifolia*), with narrow, unlobed leaves.

GLOBE-MALLOW

Golden Crownbeard

(Verbesina encelioides)

Family: Aster (Asteraceae)
Class: Native
Height: To 5 feet
Bloom period: February–frost
Range: Throughout

An upright to sprawling, rather coarse, usually much-branching annual. Leaves grayish-green, odoriferous. Flower heads large, yellow, terminal on long, slender stalks.

CULTIVATION: The main problem usually encountered with this plant is that it may be too easily grown. From Austin southward it often does not winter-kill or will freeze only in the upper portion. With the first warm days, new growth shoots forth and the plants may be flowering again by March or April. It also reseeds prolifically, so there is never a shortage of new plants. Seedlings may be transplanted anytime. Golden Crownbeard will survive under the most trying conditions, but for lush, full-flowering plants, give a light fertilizer (any kind will do) occasionally and a bit of extra watering during summer drought periods. Use these plants toward the back of the border—preferably in an out-of-the-way place. The foliage has a rather foul odor when disturbed, which makes for unpleasant working conditions in the garden. Also, even though butterflies nectar on this plant ravenously, they will temporarily leave the area when the plants have been disturbed.

Trimming from time to time may be necessary to keep plants in bounds, but this only makes for a sturdier, bushier plant with even more flowers. Keep as many of the spent flowers clipped as possible, as fruit production will slow down the forming of new flowers. If left undisturbed, Golden Crownbeard will eventually form large colonies for an absolutely spectacular autumn show. Since the plants are so robust, in a garden situation treat them almost as a shrub. They blend beautifully with Beebrush (*Aloysia gratissima*), Butterfly Bush (*Buddleja davidii*), Trailing Lantana (*Lantana montevidensis*), Summer Phlox (*Phlox paniculata*), and bird-of-paradise (*Caesalpinia* spp.) or any of the tall, blue-flowered salvias (*Salvia* spp.). Front them with Tahoka Daisy (*Machaeranthera tanacetifolia*) or mistflowers (*Conoclinium* spp.), liatris (*Liatris* spp.), ironweeds (*Vernonia* spp.), or zinnias (*Zinnia* spp.).

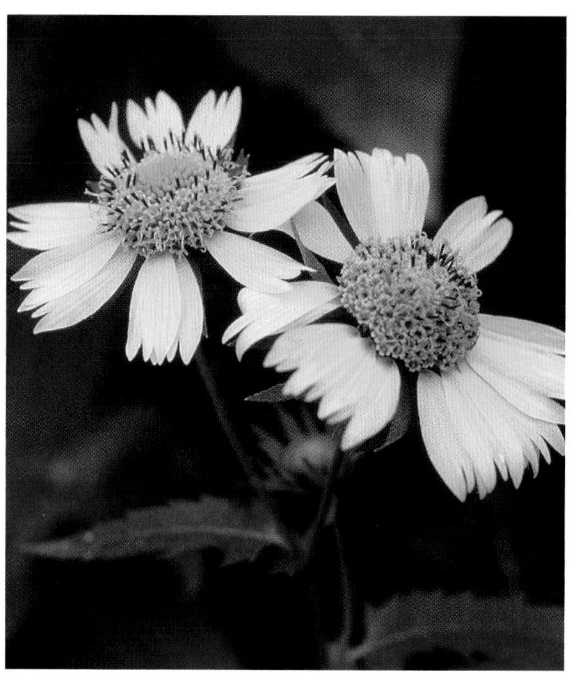
GOLDEN CROWNBEARD

USED BY: Bordered Patch (*Chlosyne lacinia adjutrix*).

PARTS EATEN: Foliage.

NOTE: According to past studies, Golden

Crownbeard is the preferred larval food choice of the Bordered Patch for rearing of the late summer and fall broods, with Annual Sunflower (*Helianthus annuus*) being first choice for the spring and early-summer broods. Giant Ragweed (*Ambrosia trifida*) is frequently used for both spring and summer broods. For best results in attracting this butterfly as a resident, provide a patch of both Golden Crownbeard and Annual Sunflower about the garden, and if present, leave some Giant Ragweed standing if in an out-of-the-way place. Golden Crownbeard is also an excellent nectar plant and is heavily used by many species of butterflies.

Heart-leaf Stinging-nettle
(Urtica chamaedryoides)

Family: Nettle (Urticaceae)
Class: Native
Height: 2 – 2 ½ feet feet
Bloom period: February–July
Range: 2, 3, 4, 5, 6

An upright or somewhat lax, slender annual, usually branching near base and from main stem; almost entire plant covered with short, stiff, stinging bristles. Leaves thin, opposite, long-stalked along lower portion of stem, sometimes purplish on lower surface. Flowers tiny, greenish-white, in clusters from axils of upper leaves; either male or female, but both on same plant.

CULTIVATION: This is not a plant offered by the nursery trade and not one to be planted for any kind of showiness. But a nettle is almost imperative if the Red Admiral (*Vanessa atalanta rubria*) is to be attracted as a resident. The best method of propagation is by digging young plants. Since this is one of the more "weedy" species and one most folks are trying to get rid of, anyone who has a colony growing will probably be quite happy to share a few—or a lot. When handling the plant, wear thick gloves and a long-sleeved shirt, for any contact between this plant and bare skin results in a most painful sting that lasts for hours.

Ideally, the Heart-leaf Stinging-nettle should be given a place of its own. A semi-shaded area beneath high-branched shrubbery or small trees, or in afternoon shade provided by a fence or building, would work quite well. If it has to be planted in a bed or border, place it where it can be left alone and where you will not have to work around it. It likes fairly

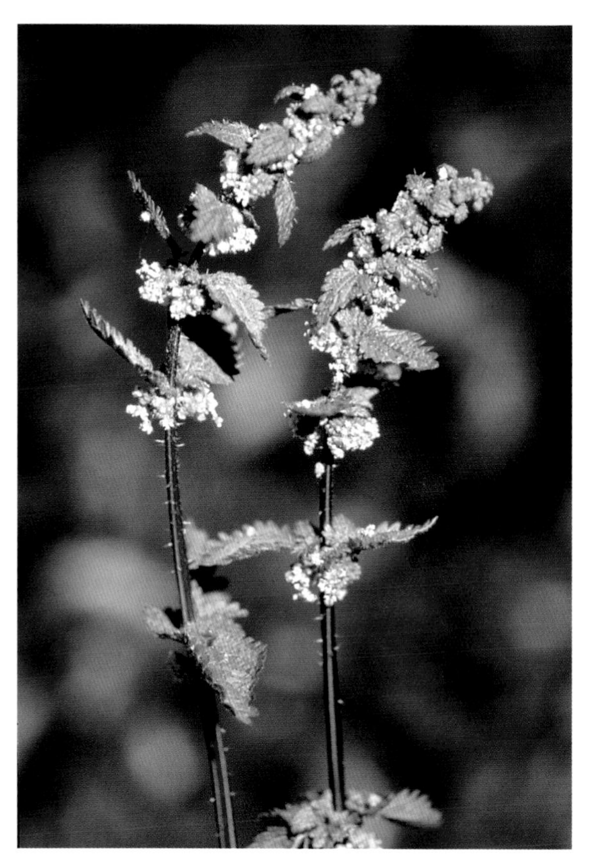

Heart-Leaf Stinging-nettle

rich, well-drained loamy soils but can tolerate sand or clay as long as shade and moisture are provided. To promote new growth and provide a longer period of usable larval food, trim back occasionally.

If growing conditions are suitable, Heart-leaf Stinging-nettle will reseed itself each year. If it begins to spread beyond its designated area, remove young plants as they come up. Thin to eight to twelve inches apart, if needed, for healthier plants with more useful foliage. The plants are easy to contain by planting in a twenty-gallon plastic pot or an old galvanized tub with the bottom removed. Sink the container to ground level before filling with soil and planting.

USED BY: Red Admiral.

PARTS EATEN: Young foliage.

Heath Aster

(Symphyotrichum ericoides)

Family: Aster (Asteraceae)
Class: Native
Height: To 3 feet
Bloom period: October–December
Range: Throughout

Basically upright, many-branched perennial with numerous arching or reclining branches, spreads from underground stems, forming colonies. Leaves along branchlets very numerous and crowded, short, stiff, very narrow, and heathlike. Flower heads of yellow disk flowers and white ray flowers; heads numerous and almost cover plant during flowering.

CULTIVATION: Heath Aster is a strong, hardy plant, well adapted to many soil types and amounts of drought or moisture. When planted in full sun, it makes an exceptional autumn-flowering plant. It performs beautifully under almost any garden condition but grows taller and lusher when given fairly good soil and a bit of moisture.

Heath Aster multiplies readily and is easy to propagate from root sprouts. When a colony of plants is found, there almost always are numerous young plants around the mother clump. These can be loosened by inserting a spade and gently lifting. Clip the young plant from the main plant, leaving a plentiful cluster of rootlets on the lifted plant.

To start plants from seeds, gather the dried seed heads after first frost. Plant the

HEATH ASTER

seeds immediately in prepared beds where the plants are to remain, or transplant them in late spring. Sow the seeds thickly, as the viability of aster seeds is never very good. If plants come up too close together, thin by transplanting some to other areas. Seedlings benefit from weekly applications of a weak liquid fertilizer solution until they become established and are growing well; usually, no further fertilizing is required. This aster can also be propagated by taking tip cuttings of young growth in early spring. Dip cuttings in a rooting hormone, and then insert in a mixture of equal parts peat and sand. Cover cuttings and container with plastic to maintain high humidity until rooted.

Old clumps of Heath Aster may decline in vigor if allowed to become excessively crowded. For healthier, more robust plants, dig and divide the clumps every three or four years.

Give Heath Aster plenty of room in the border, where the graceful "weeping" branches can droop and spread. When in flower, the plants appear as large masses of snowy-white, forming a wonderful backdrop for various purple and yellow fall-flowering species. Try Heath Aster backed with the purple-flowered New England Aster (*S. novae-angliae*) and fronted with the native yellow-flowered Huisache Daisy (*Amblyolepis setigera*), Woolly Paper-flower (*Psilostrophe tagetina*), or Hispid Wedelia (*Wedelia acapulcensis* var. *hispida*).

USED BY: Pearl Crescent (*Phyciodes tharos tharos*).

PARTS EATEN: Leaves.

Lindheimer's Senna

(Senna lindheimeriana)

Family: Bean (Fabaceae)
Class: Native
Height: To 5 feet
Bloom period: September–November
Range: 5, 6, 7

Upright, strong, many-stemmed perennial from a woody base; almost entire plant covered in shiny, velvety hairs. Leaves divided into four to eight pairs of small leaflets; when disturbed, foliage strongly and, to some, unpleasantly scented. Flowers golden-yellow, several borne in terminal clusters held well above foliage.

CULTIVATION: Lindheimer's Senna produces an abundance of long pods filled with small "beans." Gather these seeds when mature, and either plant in late fall or store in the refrigerator and plant the following spring. As this is a perennial, sow where it is to remain in the garden, especially if planting seeds in

Lindheimer's Senna

the fall. If planting in spring, plant either directly in the garden or in pots to be transferred to the garden later. Germination is best if seeds are planted immediately after harvesting and before the pods are totally dry.

When plants are up and showing true leaves, use a weak soluble fertilizer about every two weeks until plants are well established. Discontinue all fertilizing after plants are showing a lot of healthy growth. Lindheimer's Senna does not need an overabundance of moisture, but an occasional deep soaking during July and August ensures good growth and nice flowering for the fall.

Because of its height and growth habit of producing several stems from the base, this plant is ideal at the back of a border or for an annual hedge. Frost kills the plants back somewhat, but new growth begins again in early spring. Trimming well-established plants back even further, to approximately six or eight inches above the ground in late fall after a frost, promotes even more growth the following year.

This plant may do well just outside its natural range and is certainly worth a try. If being transplanted to an entirely different habitat, gather some small limestone rocks along with the seeds. In the garden, place the plants among a grouping of the rocks and even work some large limestone gravel into the planting hole. This provides aeration to the roots along with needed nutrients.

USED BY: Sleepy Orange (*Abaeis nicippe nicippe*), Orange-barred Sulphur (*Phoebis philea*), Funereal Duskywing (*Erynnis funeralis*), Boisduval's Yellow (*Eurema boisduvaliana*).

PARTS EATEN: Buds, young foliage.

Missouri Violet

(Violaceae sororia var. *missouriensis)*

Family: Violet (Viola)
Class: Native
Height: To 6 inches
Bloom period: February–April
Range: 1, 2, 3, 4, 6

Small, low perennial herbs forming tight clumps deeply rooted from rhizomes. Leaves extremely variably shaped, long-stalked, deeply notched at base, lowest sometimes gently lobed near base, toothed along edges. Flowers purple, five-petaled, the lower petal with nectar-bearing spur at base, several extended on long, slender stems from rhizome. Earliest fruit a purple to brown-mottled, slightly elongated capsule on long stems, later fruit formed from nonopening (self-fertilizing) flowers on short stems near base of plant.

CULTIVATION: Collecting seeds for propagation of the Missouri Violet will take a bit of care. Plants in the wild should be watched closely; when pods first start showing above or near the top of leaves, gather the pods. Place in a closed paper bag, and once all pods have expelled their seeds, plant some seeds immediately, then again in fall and the following spring.

Once plants are up and well established (two to three years), each plant will have formed more or less a "clump" by extending the underground rhizome. For new plants these rhizomes can be cut apart, keeping a leaf node and roots to each section and replanting each section.

USED BY: Variegated Fritillary (*Euptoieta claudia*).

PARTS EATEN: All parts of plant.

NOTE: The variety *sororia*, bearing whitish to

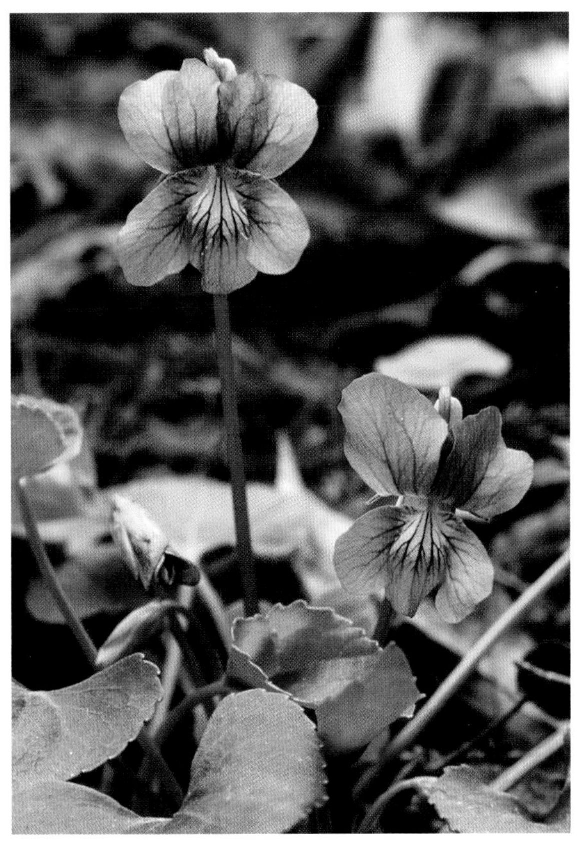

Missouri Violet

grayish petals veined or streaked with blue or purple are known as the "Confederate" Violet. All violets serve as a much-needed nectar source for the early-appearing butterflies, especially the Falcate Orangetip (*Anthocharis midea*).

RELATED SPECIES: Almost any native smooth, large-leaved violet and most cultivated violets and pansies (*Viola* spp.) are used as larval food plants. The related native, tall-growing Green Nod-violet (*Hybanthus verticillatus*) is also used.

Nuttall's Prairie Parsley
(Polytaenia nuttallii)

Family: Parsley (Apiaceae)
Class: Native
Height: To 3 feet
Bloom period: April–June
Range: 2, 3, 4, 6

Upright, stout, usually several-branched biennial. Leaves mostly near the base, with each leaf divided into several broad segments. Flowers numerous, tiny, yellow, grouped into half-round, terminal clusters.

CULTIVATION: The best way to obtain plants of Nuttall's Prairie Parsley is by sowing seeds. Since this is a biennial, which means the plants come up one year and bloom the next, care must be taken not to destroy tiny seedlings while working the beds.

Seeds should be mature about a month after flowering, so keep a close watch on the plants to get good, healthy, mature seeds. Collect the entire head when seeds are light tan or yellowish. Thrash the head over an open paper bag or large tray, and then blow or pick as much debris out as possible. Store seeds in a dry, sealed, labeled container, and place in the refrigerator. The following spring, sow seeds in prepared beds where plants will remain or in individual pots for transplanting later. Seedlings should be up within a few days. Water occasionally with a weak soluble fertilizer. Transplant a potted plant to its permanent location when roots have filled the pot.

Once started, growth is usually rapid, forming large rosettes of leaves aboveground and sending the long taproot deep into the soil. Plants overwinter as rosettes and flower

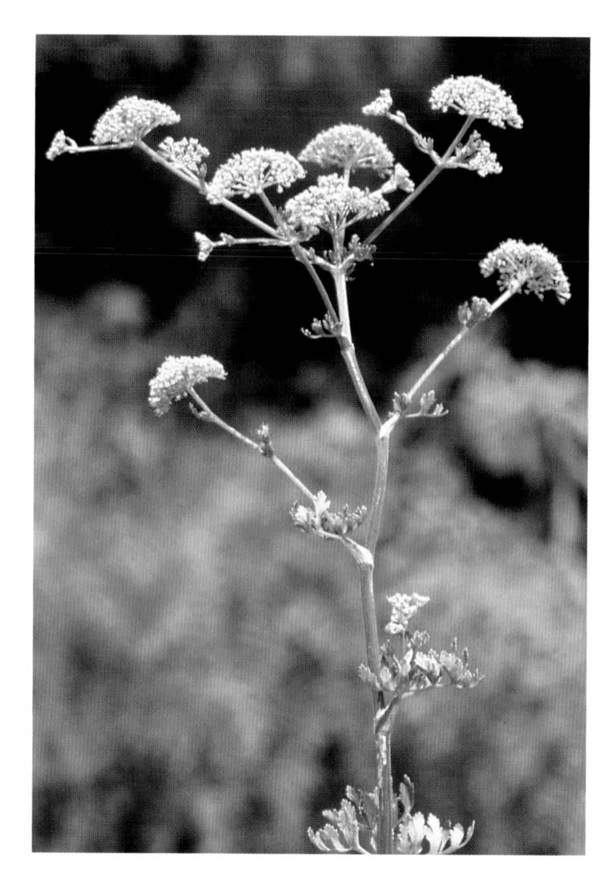

NUTTALL'S PRAIRIE PARSLEY

the following spring. If seeds are broadcast in a natural area instead of regular beds, make sure seeds are well covered with soil.

These plants grow in a good garden loam or even in clay but do not do well in extreme limestone, caliche, or acidic soils. A light application of fertilizer now and then makes for studier plants, as does a little extra water. Since plants may become lusher under garden conditions, they may need staking. These are sun-loving plants, so place them in full sun and give them plenty of room to breathe and spread out. Almost all of the foliage is near the ground, and even this may be disappearing at bloom time, so use this plant toward the middle or back of a border and plant something in front of it.

Nuttall's Prairie Parsley readily reseeds itself, so once it is established in an area, there will be an abundance of new plants appearing in late summer. These seedlings may be moved while still very young; because the taproot is long, however, moving is not advisable if seedlings are more than two months old.

USED BY: Eastern Black Swallowtail (*Papilio polyxenes asterius*).

PARTS EATEN: Foliage.

NOTE: In order for the butterfly to lay her eggs on the leaves, this plant needs to be in a fairly open area, so do not crowd the plants. Place them in small groups in order for there to be a large selection of leaves. Females usually deposit only one egg to a leaf but will leave two or three eggs per plant and use several plants in a group.

RELATED SPECIES: The similar Texas Prairie Parsley (*Polytaenia texana*) is common on Blackland and Coastal Prairies.

Partridge-pea
(Chamaecrista fasciculata)

Family: Bean (Fabaceae)
Class: Native
Height: 1–5 feet
Bloom period: May–frost
Range: Throughout

Upright, smooth to hairy, usually many-branched annual herb. Leaves composed of many small leaflets and appearing almost fernlike; foliage partially closes when touched. Flowers yellow, with five petals of unequal size, opening in short clusters along the branches; only one flower within each cluster opening at a time.

CULTIVATION: Partridge-pea is easily established in sandy or loamy soils anywhere in its range by simply scattering a few seeds. It is rather late to come up in the spring but grows rapidly once sprouted. This plant does not transplant well, so sow seeds where plants are desired. This may cause a bit of a problem if plants are to be in a border since the seeds need to be planted in the fall but the plants do not come up until late spring. The best way is to choose the desired site for the plants and scatter seeds generously in the bed in September. Mark the area, using metal tags or metal or wooden stakes. In the spring after the seeds have sprouted and plants are about four inches high, pull out or thin plants into the configuration desired. For a solid mass of mature plants (or a continuous border effect), leave the plants standing approximately two feet apart.

Under normal garden conditions of richer soil and more water than normal, these plants may easily reach their maximum size. They also grow larger and are much healthier if there is not severe competition with other plants, so give them plenty of space. If plants are still crowded later in the season, simply pull out a few. The remaining ones will spread out and remain lower and stockier instead of growing tall and becoming spindly.

This is an excellent plant to use in "natural" or untended areas, as well as in the border. Simply scatter a handful of seeds along a fence; around mulch, compost, or brush piles; next to the garage; or at the back of the lawn. Rake seeds in well, and gently tamp the surface.

Gathering ripe seeds of Partridge-pea may be a little tricky. Upon ripening, the long, flat pods split, with each side making a couple of twists and flinging the seeds for several feet. Since it does not take but a few seeds to get a start of this plant for the garden, gathering good seeds is worth a little extra effort. The best way is to enclose the almost mature pod in a little bag made of a square of cheesecloth or nylon net wrapped around the seedpod and closed at the base with a cord or band. Do not gather until pods have split and loosened the seeds. Another method is to gather seedpods after they have turned dark brown but before they have split. The ratio of plants obtained per seed gathered will not be nearly as great with this method since some seeds may still be immature, so be sure to plant plenty.

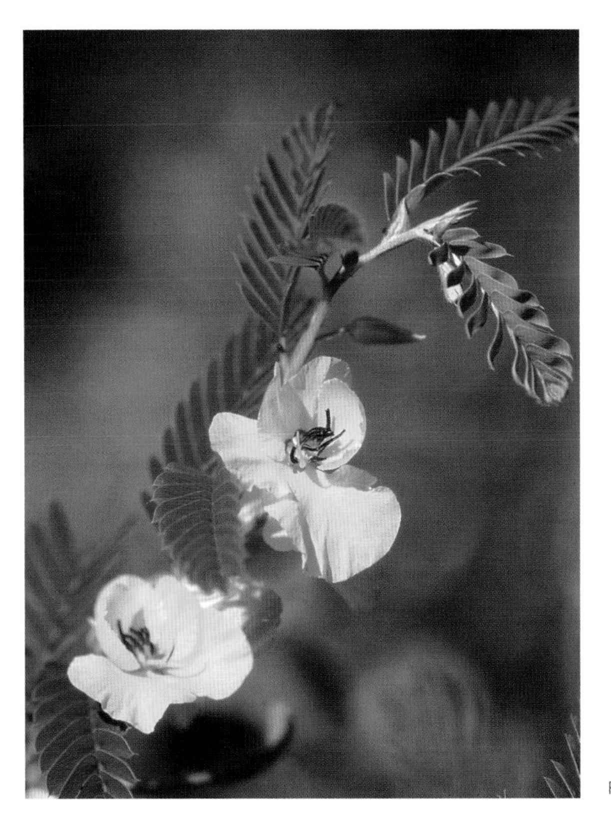

Partridge-pea

At the end of the growing season, if Partridge-pea has been allowed to mature its seeds, there should be an abundance of new plants the following year. However, due to the method of dispersal, the next season's plants may not be exactly where you would like them. To be sure plants are in desired areas, it is best to plant them there.

Partridge-pea is quite attractive when used in small groups or as a somewhat continuous row in a long border. Some studies have shown that butterflies have a tendency to choose the more isolated plants for egg deposition, so plants should be tried in various situations in the garden to determine best usage.

USED BY: Cloudless Sulphur (*Phoebis sennae*), Clouded Sulphur (*Colias philodice*), Little Yellow (*Pyrisitia lisa lisa*), Sleepy Orange (*Abaeis nicippe nicippe*).

PARTS EATEN: Buds, flowers, leaves.

NOTE: This plant produces abundant "extra-floral nectar" along the stem and at leaf nodes and is much used by several species of butterflies.

Pennsylvania Pellitory

(Parietaria pensylvanica)

Family: Nettle (Urticaceae)
Class: Native
Height: To 8 inches
Bloom period: March–June
Range: Throughout

Tender, weak-stemmed annual or biennial, stems slender and leaning or often many-branched at base and forming spreading clump. Leaves thin, flimsy, green to reddish or bronzy. Flowers tiny, inconspicuous, in small clusters.

CULTIVATION: One of the more "weedy" species, this is not a plant for prominence in the flower bed, but instead seek an out-of-the-way, semishaded location. Or it will make a most attractive early ground cover beneath shrubs.

To get it started around your garden, when wild stands are found, mark the location. Late in the flowering season, gently clip the top portion of plants and place them into paper bags. Do not clip all the plants, leaving some for natural propagation the following year.

In the garden spread the clipped plant portions into various sites. In most portions of its range, this plant will be up and growing in late fall and a couple of inches high by January. If things go well, there should be a good stand of plants. Remember that this plant will look like a "weed" when first coming up, so mark locations well. By June, plants of Pennsylvania Pellitory will have become withered and straggly. At this time, clip, shallowly bury plants, and fill space with summer annuals

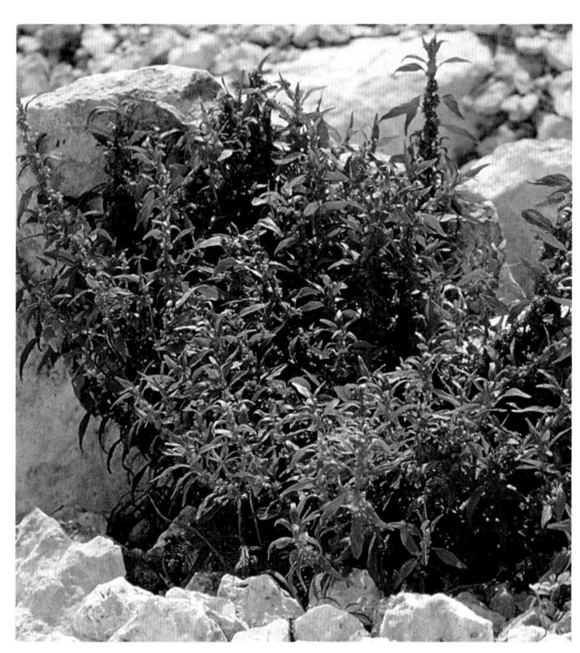

PENNSYLVANIA PELLITORY

such as single zinnias (*Zinnia* spp.) and marigolds (*Tagetes* spp.) or Old-fashioned Petunia (*Petunias axillaris*). Pennsylvania Pellitory readily reseeds, so once they are started, the only thing necessary will be to start colonies where wanted.

USED BY: Red Admiral (*Vanessa atalanta rubria*).

PARTS EATEN: Foliage.

NOTE: Although not long-lived, this plant is a good nonabrasive substitute for Heart-leaf Stinging-nettle and may be more welcome in the garden. The Red Admiral will begin laying eggs on these plants by late February or early March.

Powderpuff

(Mimosa strigillosa)

Family: Bean (Fabaceae)
Class: Native
Height: To 6 inches
Bloom period: April–frost
Range: 2, 3, 4, 5

Low, sprawling or trailing perennial, usually forming solid mats with annual stems to six feet in length, stems bearing bristles but are not stiff and prickly. Leaves appearing ferny, composed of numerous tiny leaflets that close at night, during cloudy weather, or when touched. Flowers numerous and forming roundish ball, consisting mostly of long, pink-colored hairlike stamens conspicuously tipped with yellow pollen. Fruit a decorative flattened, jointed, oblong pod containing several seeds; when mature, pod breaks into sections.

CULTIVATION: Powderpuff is probably not a plant for the flower beds, as it can eventually cover quite a large area. Instead, use it as a ground cover for open, sunny slopes or along a driveway, meadow path, or pond edge. It will grow in just about any soil but will do best in loamy or sandy soils.

POWDERPUFF

To propagate this plant, gather the flattened pods when they are brown and feel stiff. Let pods air-dry for several days. Remove seeds, and plant some immediately, some in fall, and some the following spring. Try soaking some seeds in hot water before planting. Not much is known about the propagation of this plant, so seed planting should be varied. Powderpuff wilts immediately after being cut, so tip-rooting is not generally satisfactory.

USED BY: Little Yellow (*Pyrisitia lisa lisa*), Mimosa Yellow (*P. nise nelphe*), Reakirt's Blue (*Echinargus isola*).

PARTS EATEN: Buds, foliage.

NOTE: Once this plant is established, it requires little maintenance except maybe an occasional watering, depending on its location.

Purple Agalinis

(Agalinis purpurea)

Family: Figwort (Scrophulariaceae)
Class: Native
Height: 2–4 feet
Bloom period: August–November
Range: 3, 4

An erect to somewhat sprawling, many-branched annual; branches slender, wiry, and widely spreading. Leaves few, small, very narrow. Flowers pale pink to dark lavender, somewhat tubular, opening one or a few at a time from upper leaf axils.

CULTIVATION: Purple Agalinis is another of the beautiful natives that is not offered by nurseries or through catalogs. The plants are supposedly partially parasitic on the roots of grasses and perhaps other herbs, but they may need host plants only for the first weeks of growth. Little is actually known about the cultivation of this native in the home garden.

For new plants, gather seed capsules as soon as they start to become brownish in color and before they split and release the numerous tiny seeds inside. Finish air-drying in an open paper bag or on paper towels. Since so little is known about its growth requirements, stagger the planting of seeds to be on the safe side. Plant some immediately after they are completely dry, plant some in late fall, and save some seeds in cold storage for planting in early spring. Try stratifying some seeds by placing in layers in moist sand or between moist paper towels. Put them in a resealable plastic bag in the refrigerator for two or three weeks before planting. When sowing, try some in flower beds and others in a more naturalized area where native grass

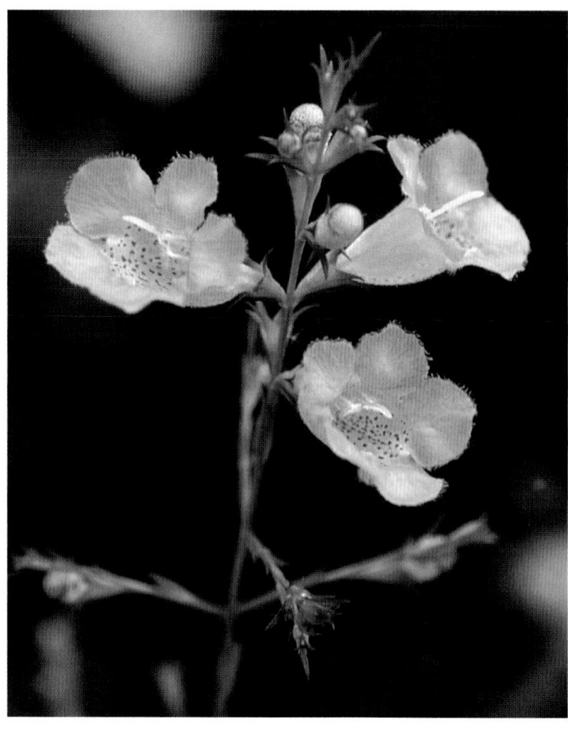

PURPLE AGALINIS

roots are available. If they are in pots, once the seedlings are up and showing leaves, do not try transplanting until their roots are well developed.

Withhold fertilizer until plants are well up and growing; then use sparingly if at all. Keep soil moist, especially during droughty summer months. Purple Agalinis grows in a wide diversity of soils and habitats so should do well in just about any good garden situation. Plant in full sun for best flowering.

Purple Agalinis will not be showy in the border until late summer or early fall, so plant with summer-flowering species, such as Brown-eyed Susan (*Rudbeckia hirta*), Kansas Liatris (*Liatris pycnostachya*), Tropical Milkweed (*Asclepias curassavica*), Pentas (*Pentas lanceolata*), and single zinnias (*Zinnia* spp.) or any of the low-growing lantanas (*Lantana*

spp.) or verbenas (*Verbena* spp.). Some good companions for fall would be Blue Lobelia (*Lobelia siphilitia*) and Golden Crownbeard (*Verbesina encelioides*), as well as goldenrod (*Solidago* spp.), mistflower (*Conoclinium* spp.), any of the asters (*Symphyotrichum* spp.), or blue-colored salvias (*Salvia* spp.).
USED BY: Common Buckeye (*Junonia coenia*).
PARTS EATEN: Buds, foliage, young fruits.
NOTE: Caterpillars feed during the day. They are extremely well camouflaged and make no effort to hide.
RELATED SPECIES: There are many species of *Agalinis* in the state, and the Common Buckeye uses almost all of them.

Purple Cudweed
(Gamochaeta purpurea)

Family: Aster (Asteraceae)
Class: Native
Height: 4–16 inches
Bloom period: February–May
Range: 2, 3, 4, 6

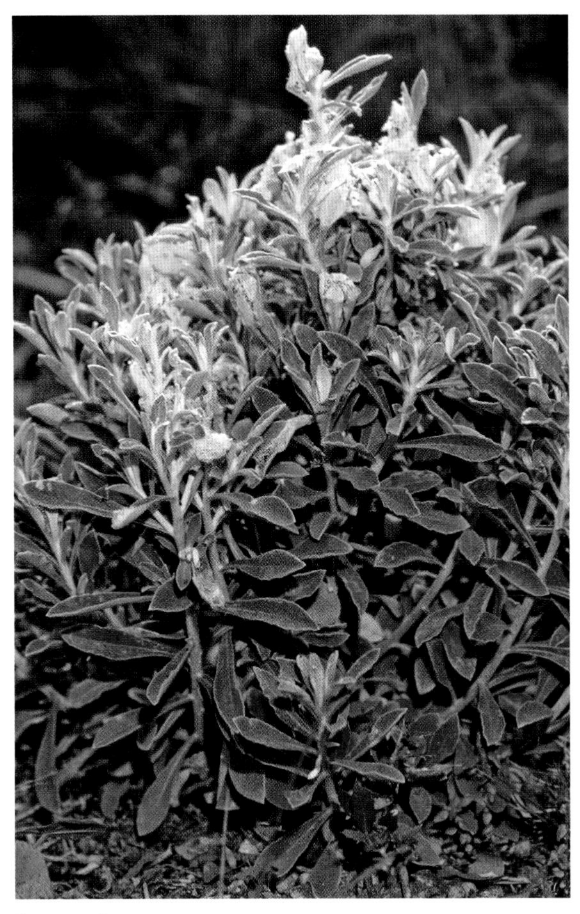

Purple Cudweed

Low annual, biennial, or perennial with usually several stems from base and forming rounded mounds. Leaves alternate, densely silvery-woolly on lower surface, a brighter green and sparsely woolly on upper surface; fresh foliage fragrant when crushed. Disk flowers minute, whitish or yellowish, few, in small clusters, with several clusters grouped in short, spikelike arrangements in axils of upper leaves.
CULTIVATION: Purple Cudweed is not a showy plant with brilliant displays of flowers, but its green- and silver-colored foliage makes it a beautifully subtle companion to brighter-colored flowers such as Purple Coneflower (*Echinacea purpurea*), Old-fashioned Petunia (*Petunia axillaris*), and various zinnias (*Zinnia* spp.). Purple Cudweed is a strong, hardy plant and one easy to grow from seeds. When seeds begin to ripen, the entire tip of the plant will look cottony or fluffy. Seeds should be ripe and ready for harvesting a few weeks after flowering. When ready to gather, grasp the entire fluffy mass on each stem and strip it gently but firmly from the plant. Otherwise, clip the entire seed-bearing portion of the plant. Place seeds and the fluff in a paper bag, and store in a cool, dry place until ready to plant.

Sow seeds in early fall either where they

are to grow or in a bed for transplanting later. Plants will be ready for transplanting by late fall or can be moved as late as the end of January. Moving after this would not give the plants time to become fully established and would usually result in small, nonmounding plants. The healthiest and best-formed specimens are those planted in early fall where they are to stay; as the seeds come up, thin the plants to approximately a foot apart. Try saving some seeds and planting them around the first of May. Since larvae use the young, tender plant growth, the availability of young plants later in the season could possibly extend the length of breeding and rearing of butterflies in the garden.

Purple Cudweed is a drought-hardy plant that grows its finest in full sun in well-drained soils. If it is growing in ordinary garden soil, no fertilizer should be necessary. In fact, overfertilizing and overwatering will cause excessive growth, resulting in weak, sprawling, rather unattractive plants. When growing under such conditions, the plants will not be as readily used by female butterflies for egg deposition, if at all. Purple Cudweed will also grow in open, semishaded conditions, but again, plants generally will not produce as many stems, and clumps will not be as full or mound-forming.

Use Purple Cudweed toward the front of beds or borders where it can be readily found by searching females. Do not crowd the plants, but group three to five together, with several groupings scattered about the garden.

USED BY: American Lady (*Vanessa virginiensis*).

PARTS EATEN: Buds, flowers, young foliage.

NOTE: The female American Lady lays eggs on younger foliage, with the larvae making silk "nests" in the upper leaf clusters so the plants appear "matted" at the top.

Silver-leaf Croton

(Croton argyranthemus)

Family: Spurge (Euphorbiaceae)
Class: Native
Height: To 2 feet
Bloom period: April–September
Range: 2, 3, 4, 5

Low, bushy, somewhat reddish-stemmed perennial. Leaves slender, conspicuously silver on lower surface. Flowers small, white, forming feathery terminal clusters; male flowers appearing near tip of cluster, with female flowers below.

CULTIVATION: Silver-leaf Croton does not transplant well even under the most ideal circumstances. The best way to obtain plants is by seeds. Capsules of Silver-leaf Croton split open when ripe, flinging the seeds for several feet. Thus, to get a good supply of seeds, collect large, firm, but unopened capsules, and

SILVER-LEAF CROTON

let them finish maturing in a paper bag. When gathered before fully ripe, some may not mature properly or become viable, so gather more seeds than the number of plants you need. Plant directly in the garden immediately. Keep the planting medium moist but not saturated, as excessive moisture will cause the seeds to rot.

Silver-leaf Croton has a very deep root system that needs plenty of aeration. Its permanent growing soil should be as sandy as possible. If your garden is not naturally sandy, then mix a large quantity of sand in the area where the plants are to grow. They do best when there is plenty of room to breathe, yet they like the company of their own kind, so always plant in groups of at least three to five and space the plants so they will be barely touching when mature. After they have become root-established, there is no need to fertilize and water unless the plants look wilted.

In most of its range, Silver-leaf Croton dies down to ground level each fall but puts out new growth in the spring. It readily reseeds if the soil and other growing factors are favorable, so there will be plenty of new plants each year. Unwanted plants will need to be removed to prevent overcrowding.

USED BY: Goatweed Leafwing (*Anaea andria*), Gray Hairstreak (*Strymon melinus franki*).

PARTS EATEN: Flower buds, foliage.

RELATED SPECIES: There are many species of *Croton* in the state, with almost all of them used as larval food plants.

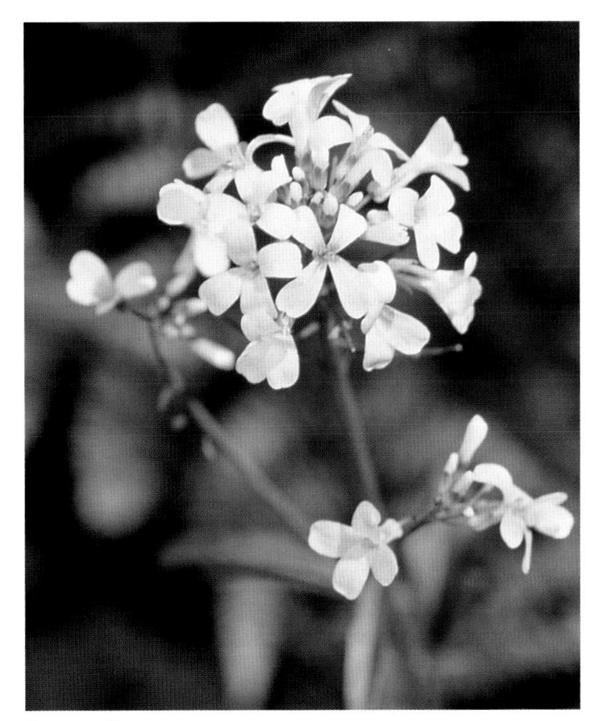

SPRING CRESS

Spring Cress

(Cardamine rhomboidea)

Family: Mustard (Brassicaceae)
Class: Native
Height: To 2 feet
Bloom period: February–May
Range: 3, 4

Upright, unbranched, smooth perennial from short, thick tubers, the stems several and forming a clump. Leaves mostly basal, dark green, shiny, lobed, and long-stalked. Flowers white to rarely pink, forming loose, elongating, terminal cluster; plants usually start flowering when stems very short, with stems becoming longer as flowering season advances.

CULTIVATION: Spring Cress is another of the wildings not generally offered by nurseries, so seeds must be gathered from the wild. To collect, it is best to “bag” the long, slender pods with a finely woven material such as cheesecloth. The seeds are very small and will slip through most materials.

As this is not a commonly used garden plant, some experimenting may be necessary to obtain the best germination in your area. To start, choose a place in the garden with soils as closely resembling the native soil as possible. Immediately after collecting, plant some seeds where plants are to remain and then try planting some seeds in pots or flats of river sand or light loam. Keep soil moist but well drained at all times. Later, well-established clumps can be lifted and carefully separated for more plants.

These plants will need more moisture and shade than average garden plants so should be placed in a special bed along with Cardinal Flower (*Lobelia cardinalis*), Joe-Pye Weed (*Eutrochium fistulosum*), Pickerel Weed (*Pontederia cordata*), and the milkweeds—Red-flowered Milkweed (*Asclepias rubra*), Swamp Milkweed (*A. incarnata*), and Lance-leaf Milkweed (*A. lanceolata*)—where they can be given a soaker hose all their own. If placed in a humus-rich soil, no fertilizing should be needed other than yearly additions of well-rotted hardwood leaves.

Ideally, Spring Cress should be permanently placed at outer edges of deciduous shrubs or hardwood trees or along a woodland trail. To flower and produce the healthiest foliage, they like the early-spring light before trees fully leaf out, then later need the shade to protect the tender foliage until it withers away.

USED BY: Falcate Orangetip (*Anthocharis midea*).

PARTS EATEN: Buds, flowers, but mainly young fruits.

Stiff-stem Flax

(Linum berlandieri)

Family: Flax (Linaceae)
Class: Native
Height: 8–14 inches
Bloom period: February–September
Range: Throughout

Upright or somewhat sprawling smooth annual; stems usually several, many-branched, and forming rounded clumps. Leaves very small, stiffly upright. Flowers few to numerous, forming open, terminal clusters, with one flower opening at a time in each cluster; petals golden-yellow or copper-colored, splashed with reddish near base.

CULTIVATION: Nice, healthy clumps of this plant appear to be "all flowers," for the blossoms are quite large in comparison to the tiny leaves and delicate stems. Flowers of many species of *Linum* have a tendency to fall early in the day or shatter when touched or blown by the wind. The flowers of Stiff-stem Flax remain open and windproof until at least

STIFF-STEM FLAX

midafternoon, making this plant a good subject for the border as well as caterpillar food.

Stiff-stem Flax is not usually offered by nurseries, nor are the seeds readily available. Transplanting these plants is just about impossible, so to obtain a start, you must gather seeds. Gather seedpods as soon as they turn a light brown or beige color. Place in an open paper bag until the capsules are completely dry and easy to open. Plant seeds as soon as they are dry, scatter them about in the area where the plants are wanted, and rake seeds in thoroughly. Lightly tamp the soil, and sprinkle gently until the ground is thoroughly moistened. Otherwise, wait until a week of fall rains is forecast and plant the seeds before the rains begin. Seedlings should appear shortly, live through winter, and flower the following season.

Although tolerant of a fairly wide range of growing conditions, Stiff-stem Flax thrives in sandy, well-drained soils kept on the dry side. An occasional soaking during the summer, if conditions are dry, keeps them producing more flowers. Grow the plants in an open, sunny exposure, and they develop a tight, rounded habit. A lightly shaded setting will do, but clumps will be looser, more spreading, and with fewer flowers. Also, they will be less likely to attract butterflies for egg laying.

Stiff-stem Flax is most attractive when used toward the front of a bed or border along with Prairie Verbena (*Glandularia bipinnatifida*), Dwarf Crownbeard (*Verbesina nana*), Western Peppergrass (*Lepidium alyssoides*), and Woolly Paper-flower (*Psilostrophe tagetina*), as well as frogfruit (*Phyla* spp.).

USED BY: Variegated Fritillary (*Euptoieta claudia*).

PARTS EATEN: Almost all aboveground parts of the plant.

Texas Frogfruit

(Phyla nodiflora)

Family: Vervain (Verbenaceae)
Class: Native
Height: To 10 inches
Bloom period: March–November
Range: Throughout

Low, creeping or trailing, mat-forming perennial rooting at the nodes. Leaves long, slender, conspicuously toothed. Flowers numerous, tiny, white, forming a small terminal cluster, with flowers opening in a circle; new circles continuing to open during the season, with the flowering portion eventually becoming elongated.

CULTIVATION: Wherever this plant is encountered, there is usually a large patch with the creeping branches rooting at each node. The best method of obtaining plants is simply to clip one of the branches near the main root and gently loosen the roots along the branch.

Texas Frogfruit

When replanting, in order to obtain the most coverage for the largest area, cut the branch between each node and plant each rooted section separately.

Texas Frogfruit is not particular about soils but does exceptionally well in rich garden loam with good drainage. A little extra moisture and occasional fertilizing make for faster and more luxurious growth.

Once the plants have spread and have sufficiently covered the ground, take branches to start plants in new areas. Although this plant spreads and multiplies rapidly and quite readily, it never becomes obnoxious or uncontrollable. Cutting back to the desired area with either clippers or a hoe is all that is needed to keep it in bounds.

Virtually disease resistant, Texas Frogfruit is an ideal plant for edgings, as a ground cover, or as a replacement for problem spots in lawns. It can also be used in beds that have been mulched. The branches will not root as readily since they will not be in contact with the soil but will instead spread out into most attractive designs and patterns.

Texas Frogfruit grows in semishade but produces thicker, more compact foliage if planted in full sun. Also, butterflies more readily use plants in an open, sunny area for egg deposition.

USED BY: White Peacock (*Anartia jatrophae luteipicta*), Common Buckeye (*Junonia coenia*), Phaon Crescent (*Phyciodes phaon phaon*).

PARTS EATEN: Foliage.

RELATED SPECIES: Several species of frogfruit are found in the state. All of them are used as a larval food source and are usually sought for the plentiful nectar produced.

Texas Thistle

(Cirsium texanum)

Family: Aster (Asteraceae)
Class: Native
Height: To 4 feet
Bloom period: May–July
Range: Throughout

Upright, bristly-spiny biennial or perennial, with usually only one many-branched stem. Leaves long, slender, covered with fine, woolly hairs on lower surface and appearing silver—a beautiful contrast to the bright green, smooth upper surface. Flowers numerous, small, pink to rose-purple, clustered in showy, terminal, somewhat rounded heads.

CULTIVATION: This is an easy plant to grow from seeds. Gather seeds when fully ripe and "fluffed up" on the plant. Sow immediately or at least by early September. Plants will come up in the fall and form rosettes that live over the winter. Texas Thistle readily reseeds, so a good colony of plants form the following year. Remove unwanted plants as soon as they come up in the fall. In flower beds, thin the plants, leaving from two to three feet between each plant for sufficient growing space. A small colony is a better attractant than solitary plants scattered through a border or bed.

Plant in full sun, and keep somewhat on the dry side. They become scraggly and have a tendency to fall down, with the lower foliage developing mildew or turning brown if plants are placed in shade or receive too much moisture. Plant toward the middle or back of the border or away from foot traffic, as the plants are rather prickly. Often these plants are not impressive in the wild, but under garden conditions of rich soil and adequate

TEXAS THISTLE

moisture, they become magnificent specimens. They are prolific bloomers, and the season can be lengthened by keeping faded blossoms trimmed off. Cut the long flower stalk just above the leaf node below the flower, and new flowering stalks will form. This trimming back to the node keeps the plant lower and bushier and prevents an invasion of plants the following season.

USED BY: Painted Lady (*Vanessa cardui*).

PARTS EATEN: Foliage. The caterpillars also make a shelter of several leaves, gathering them together with strands of silk.

NOTE: Texas Thistle (as are most other thistles) is an excellent nectar source for many species of butterflies, especially Swallowtails (family Papilionidae).

Texas Toadflax

(Nuttallanthus texanus)

Family: Figwort (Scrophulariaceae)
Class: Native
Height: 28 inches
Bloom period: February–May
Range: Throughout

Upright, slender, delicate, smooth annual or biennial; stems usually solitary, almost hairlike, rising from small, flat rosette of short, trailing, sterile branches at base; plants usually forming colonies. Leaves small, narrow, few, and scattered along stem and basal branches. Flowers numerous, bluish-violet, fragrant, forming showy terminal clusters; lower lip of flower ending in conspicuously curved spur.

CULTIVATION: Texas Toadflax is not offered by nurseries or seed catalogs, so propagation is best done by gathering seeds from the wild. As seedpods are often mature at the base of a cluster while flowers are still opening in the top portion, several "gatherings" may be necessary. The clusters can simply be held over a paper bag and gently tapped or shaken to release the shiny, black seeds.

If plants are wanted in a space that is relatively undisturbed during the summer season, then plant seeds immediately after gathering. Otherwise, wait until late summer, and then scatter the seeds where they are wanted, barely covering with a very thin layer of sand. Sprinkle gently but thoroughly, and keep sand barely moist until plants are up and well established.

If using in a "wild" or untended situation, scratch various bare spots with a hand digger, scatter the seeds, cover very thinly with soil,

TEXAS TOADFLAX

buildings, at front edges of open borders, or thickly scattered throughout a border or, ideally, in open, sparsely vegetated meadowlike situations.

USED BY: Spring brood of Common Buckeye (*Junonia coenia*).

PARTS EATEN: Mainly foliage, sometimes buds or young fruits. Often the entire young plant will be consumed.

NOTE: This plant was previously placed in the genus *Linaria*.

RELATED SPECIES: A similar but smaller-flowered species, Old-field Toadflax (*N. canadensis*), is also readily used as a larval food plant.

and pack down lightly. Sprinkle gently but thoroughly if possible. Once established, the plants readily self-sow.

Texas Toadflax prefers soils on the drier, sandier side but will tolerate heavier soils if dry. Since this is one of the earliest-flowering plants, usually spring rains provide all the moisture required. It does not require nor will it tolerate heavy fertilizing.

Texas Toadflax is not a long-lasting plant, and once the last seedpods have matured, the tiny basal foliage yellows and shrivels away, leaving space for summer annuals.

Since these plants are so delicate, they are not of great use as "show" flowers in a border and are best used in an open, out-of-the-way space. Also, this is a plant to definitely use in masses. One butterfly caterpillar will consume several plants during its lifetime, so many plants will be needed for a female to be enticed to lay eggs. In a garden situation, try to use Texas Toadflax along open fences, against

Violet Ruellia (Wild Petunia)

(Ruellia nudiflora)

Family: Acanthus (Acanthaceae)
Class: Native
Height: To 2 feet
Bloom period: March–December
Range: 2, 3, 4, 5, 6

Upright to semisprawling plant from woody base; stem usually solitary or may be several and clump-forming as plant ages. Leaves an attractive dark green. Flowers lavender to purple, trumpet-shaped, several forming a showy terminal cluster, with few to several clusters per plant; open about sunrise, then fall from plant in afternoon heat.

CULTIVATION: Depending upon which portion of this plant's range it is in, it may be a deciduous perennial or an evergreen. In the northern part of its range, it dies back to the ground each year, while in the Rio Grande Valley area, it may remain green and flowering all year if the winter is mild. Violet

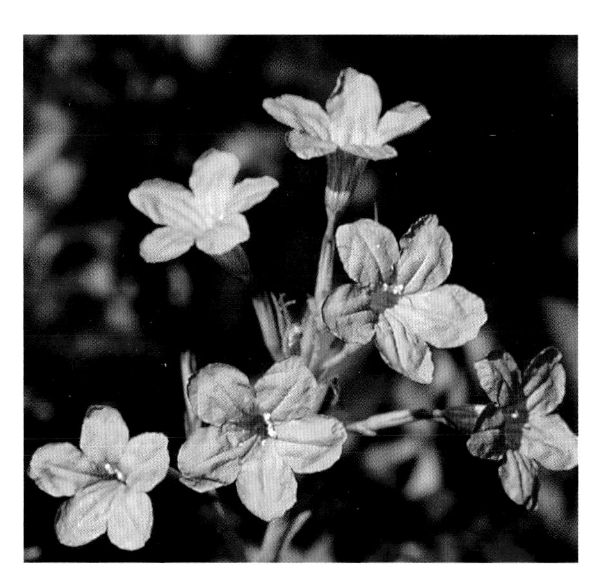

Violet Ruellia (Wild Petunia)

Ruellia grows in most types of soil except dry, sandy ones, but soil supplemented with organic matter promotes larger, more floriferous specimens.

To start plants, collect dried seed capsules as soon as they become a pale to medium brown and before they begin to split open. Sow in prepared areas of the garden immediately. Germination is slow, and seedlings will probably not show until the following spring. Once up, the seedlings are fast growing. Weekly applications of an all-purpose fertilizer enhance their growth.

Some flowering can be expected the first year, but more abundant flowering comes from fully mature plants of the second season. Give mature plants adequate moisture and light dressings of fertilizer twice a year for really spectacular flowering. Clip off spent blossoms until you want the seeds to set, usually in late summer or early fall.

An open, barely semishaded setting is recommended for Violet Ruellia, such as along woodland edges or in borders beneath taller plants. Violet Ruellia is especially lovely when combined with Cherry Sage (*Salvia greggii*), Flame Acanthus (*Anisacanthus quadrifidus* var. *wrightii*), White-flowered Plumbago (*Plumbago scandens*), or Barbados Cherry (*Malpighia glabra*).

USED BY: Common Buckeye (*Junonia coenia*), Tropical Buckeye (*J. evarete*), Malachite (*Siproeta stelenes biplagiata*), Banded Peacock (*Anartia fatima*), White Peacock (*A. jatrophae luteipicta*), Texan Crescent (*Anthanassa texana*).

PARTS EATEN: Foliage.

Western Peppergrass

(Lepidium alyssoides)

Family: Mustard (Brassicaceae)
Class: Native
Height: To 28 inches
Bloom period: February–August
Range: 1, 6, 7

Low, upright, or somewhat sprawling perennial, woody at the base, usually with several stems, and forming large, rounded clump. Leaves in basal area long, divided or lobed, being shorter and narrow along stem. Flowers numerous, white, in elongating terminal racemes, making plants appear as mounds of snow.

CULTIVATION: Plants of Western Peppergrass do not transplant well except in a very young stage, and then a large amount of soil should be taken with the roots. For a large number of plants, it is best to sow seeds in late fall where plants are wanted and then thin to desired stands. This plant is not particularly soil selective, but soil must be very well drained and plants kept on the dry side after they are up and growing well. Plant in full sun, as flowering will be drastically reduced in shade and plants will become leggy and sprawl excessively.

Because of the beautiful mounding effect when these plants are grown in full sun, groupings of five or seven plants throughout the beds or as a continual irregular border make a most dramatic effect. Or they can be used as a higher-than-average edging in front of taller flowering perennials and shrubs, such as Purple Coneflower (*Echinacea purpurea*) and Trailing Lantana (*Lantana montevidensis*). Shear off spent flowers for continual bloom. Also, as flowers begin to fade, cut some plants back about one-half to one-third to induce new growth, providing plenty of tender larval food. Plants remain green throughout the winter if cut back after the first frost.

USED BY: Checkered White (*Pontia protodice*).

PARTS EATEN: Buds, flowers, young seedpods, young leaves.

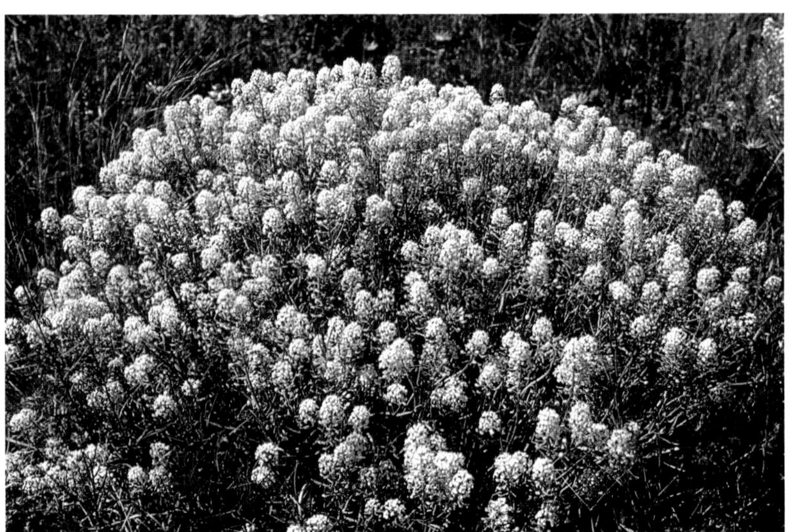

WESTERN PEPPERGRASS

White Clover

(Trifolium repens)

Family: Bean (Fabaceae)
Class: Nonnative
Height: To 8 inches
Bloom period: March–November
Range: 1, 2, 3, 4, 6

Low, creeping, shallow-rooted perennial forming large mats or patches. Leaves long-stalked, numerous, and parted into three separate leaflets. Flowers numerous, small, pale pink or white, tightly congested into long-stalked terminal clusters.

CULTIVATION: White Clover is not a difficult plant to get started, and one some folks would like to get rid of. In some areas it is an unwanted "weed" in the lawn or garden, but when growing along roadsides and in abandoned areas, it is quite lovely. Plants of White Clover start easily from seeds. Gather the seed heads after they have turned brown, and place them in an open paper bag for a few days until completely dry. Shake the seeds loose, and store in the refrigerator until spring.

Plants are also easy to obtain from a rooted branch from a mature plant. Gently dig the roots loose at each node along the branch, and then clip it loose from the mother plant. When replanting in the garden, clip the branch between each node and plant each rooted section. Even if no roots are showing at the node, make a shallow depression, lay the branch in the depression, and cover it with soil. It usually quickly roots and forms a new plant.

White Clover is best adapted to loamy soils a little on the heavy side with clays and silt. However, do not let the soil type be detrimen-

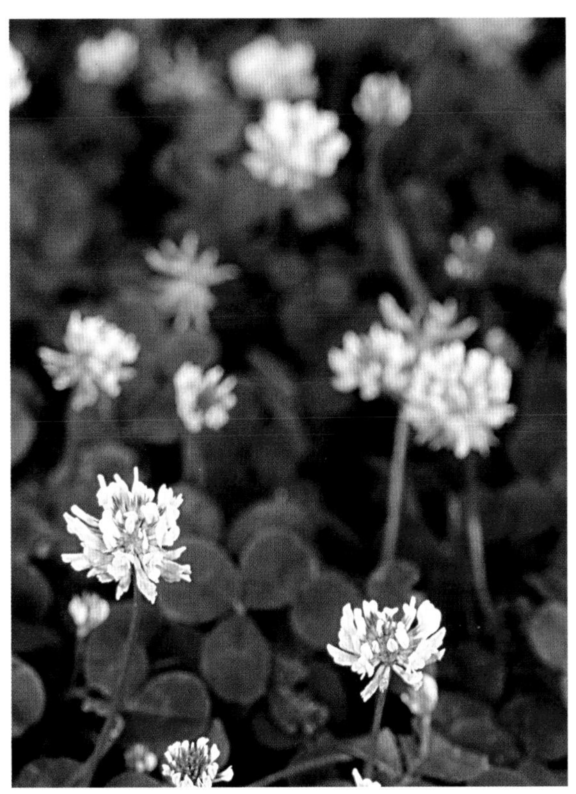
WHITE CLOVER

tal in trying this one, for it grows on caliche, limestone, sand, or salt. What it does like is a little extra moisture and good drainage. Occasional fertilizing through the growing season promotes lusher vegetation.

Once established, White Clover readily reseeds and needs very little care. To keep it within desired bounds, simply hoe back each spring. In the flower garden it should not become obnoxious or uncontrollable. White Clover can be most effective in problem areas where grass or other ground covers do not grow well. Use it in low areas where moisture has a tendency to stand, on badly eroding slopes, or in just about any area where nothing else grows. This plant dies out during the summer but returns with cooler fall weather and in some areas remains green all winter.

USED BY: Clouded Sulphur (*Colias philodice*), Orange Sulphur (*C. eurytheme*), Southern Dogface (*Zerene cesonia*), Eastern Tailed-Blue (*Cupido comyntas texana*), Reakirt's Blue (*Echinargus isola*), Barred Yellow (*Eurema daira*), Gray Hairstreak (*Strymon melinus franki*), Northern Cloudywing (*Thorybes pylades*), and probably others.

PARTS EATEN: Leaves, flowers, young fruits.

Woolly Stemodia

(Stemodia lanata)

Family: Figwort (Scrophulariaceae)
Class: Native
Height: To 5 inches
Bloom period: April–August
Range: 4, 5

Low, trailing, white-woolly, widely branching perennial forming large mats, the branches to four feet or more long, rooting at nodes. Leaves small, almost lost in the white hairs covering plant. Flowers small, usually solitary, bluish to dark purple.

WOOLLY STEMODIA

CULTIVATION: Seeds of Woolly Stemodia are very small, black, and contained in a two-part capsule. Ripened capsules can be gathered and the seeds extracted and air-dried for several days before planting. As little is known about propagating this plant by seeds, try planting at various times, both in indoor pots and outside where plants are wanted.

Clipped, rooted sections can be taken from established plants. Native plant nurseries often offer this plant and are by far the best possible source.

USED BY: Tropical Buckeye (*Junonia evarete*).
PARTS EATEN: Foliage.
NOTE: In the garden, this plant makes a beautiful ground cover beneath other plants and is especially impressive when allowed to trail over a rock edging or when used between the stones in a dirt-laid patio.

Yellow False Foxglove (Oak-leech)

(Aureolaria flava)

Family: Figwort (Scrophulariaceae)
Class: Native
Height: To 5 feet
Bloom period: May–October
Range: 3

Upright, robust, mostly smooth, shrublike perennial with many stems from base and forming large clump. Leaves opposite, the lower ones usually deeply cut with upper ones gradually becoming smaller, irregularly alternate and mostly unlobed or cut. Flowers large, showy, tubular, yellow, forming slender terminal spikes, with few flowers opening at a time.

CULTIVATION: False Foxglove is noted for being semiparasitic on the roots of various species of trees from the white oak group (family Fagaceae) but occasionally shrub species from the Heath Family (Ericaceae) will be used. Obtaining a start of these plants may require a little more work than for some others but will be well worth the effort. Not only is this a preferred late-season food plant for the Common Buckeye (*Junonia coenia*) but it also makes a large, beautiful shrubby plant in one season. Usually it blooms in early summer, but with first freezes it withers away, leaving sunny winter openings for other plants to obtain needed light and moisture.

This is definitely not a plant found in nurseries or catalogs. Nor can it *ever* be transplanted. Do not even think about it. It will have to be started from seeds, and several tries may be necessary before success is achieved. It is best to drive through the countryside in early summer and find and mark some flowering plants. Before the large seedpods start ripening, select some pods, tie small bags of nylon net over the pods, and wait until the pods begin to split open. At this time, cut only completely ripened pods from the spike, leaving others to complete ripening.

As the viability of seed and the existence of appropriate microbes in the soil for "sprouting" will be so uncertain, it would be best to plant some seeds immediately after gathering, then make more plantings spaced two weeks apart. Save some seeds over the winter in paper bags in a cool, dark place and some in paper bags in the refrigerator. Plant these in the spring after all chance of a freeze is past. With *every planting* place seeds near a thriving oak tree in the white oak group (*Quercus* spp.) or a species of *Vaccinium*, such as Huckleberry (*V. arboreum*), Highbush

Yellow False Foxglove

Blueberry (*V. corymbosum*), or Deerberry (*V. stamineum*).

Once started and thriving well, the plants continue to form ever-larger clumps and become truly shrublike for a season. At this time, make a flower bed around them, instead of trying to get them started in a flower bed—usually an impossible task.

Do not use commercial fertilizer with these plants. Instead, use dry to moist well-drained sandy soils and enrich only with a sparse layering of leaf mulch in the fall. After plants are established and growing well, Yellow False Foxglove is spectacular in combination with Mountain Mint (*Pycnanthemum incanum*), various blue-flowered salvias (*Salvia* spp.), any of the monardas (*Monarda* spp.), and mistflower (*Conoclinium* spp.). In the wild they are often found in combination with another semiparasitic member of the same family, agalinis (*Agalinis* spp.).

USED BY: Common Buckeye (*Junonia coenia*).

PARTS EATEN: Leaves, buds, young fruits.

NOTE: In late autumn after seeds have fallen, the empty seedpods may be gathered to be used in dried floral arrangements.

Yellow Prairie Grass (Indian Grass)

(Sorghastrum nutans)

Family: Grass (Poaceae)
Class: Native
Height: To 7 ½ feet feet
Bloom period: June–August
Range: 1, 2, 3, 4, 6, 7

Tall, robust perennial grass usually forming small to large colonies or stands. Leaves flat, long, and very slender, often somewhat bluish-green. Flowers not very showy; long, terminal, golden to tawny-colored panicles of ripening seeds in late summer through late fall most striking.

CULTIVATION: Gather the seeds of Yellow Prairie Grass in early fall or before frost. Separate seeds by working them between your fingers. Plant some of the seeds immediately; then plant more in early spring. These can simply be broadcast and left on top of the ground, or for best results, scratch the soil and barely cover seeds. When possible, plant-

Yellow Prairie Grass

ing just before a rain is ideal. If a small patch is wanted, lay the entire seed spike on the ground, cover, and water in thoroughly.

Once plants are well established, a "clump" or portion of the colony can be dug and divided. Dig deep when transplanting because prairie grasses form deep, extensive root systems. Make sure the new holes are deep enough to accommodate the roots and that roots are not bent but pointing straight downward in the hole. This grass not only readily self-seeds but naturally spreads occasionally by rhizomes, so be careful when lifting plants.

USED BY: Pepper and Salt Skipper (*Amblyscirtes hegon*).

PARTS EATEN: Leaves.

NOTE: Yellow Prairie Grass is one of the four dominant grasses found in well-established prairies and is adapted to fire. Since the use of fire will probably not be an option for smaller properties, trim or mow back every third year to ensure better clumping and deeper root growth.

Additional Information and More Larval Food Plants

The following list includes more larval food plants than shown and described in the text in order to broaden the scope of exciting possibilities for attracting butterflies to the garden. Plants have been placed on each list within the region(s) where the butterflies that use them usually occur for breeding, not necessarily where the plants naturally occur within the state. Even though a plant may grow throughout the state, there is no need to plant it in the garden as a larval food source if the butterfly that uses it does not breed there.

As this is a food plant list and not a complete listing of the butterflies that occur in the state, many species of butterflies that may be found in Texas are not given because their food plants are unknown or because the literature researched gave only the plant genus and no species.

Sometimes a plant listed, such as the elms, oaks, ashes, and some of the grasses, will already be present in a particular habitat. Such plants are listed not to be considered first for planting but as ones not to destroy if they already exist in or around the garden. In some instances, the butterflies that use such plants require a number of plants for breeding, and one or two plants in the garden would not be adequate. If the plants are available in sufficient numbers nearby to support a colony, you will most likely get the butterflies around the nectar plants in your garden.

As with the nectar and larval food plants shown and described elsewhere, the plant species are placed here in separate categories of trees, shrubs, vines, and herbs and listed in alphabetical order by common name. In this list, they have been further separated into native and nonnative plants. Within the listing, common and scientific names of the plant are given, followed by the region(s) in which both the plant and butterfly that uses it can be found.

Native Trees

Tree	***Range***	***Butterfly***
Ash, Berlandier's (*Fraxinus berlandieriana*)	5 6 7	Two-tailed Tiger Swallowtail
Ash, Green (*Fraxinus pensylvanica*)	2 3 4 6	Eastern Tiger Swallowtail
	6	Two-tailed Tiger Swallowtail
Ash, Velvet (*Fraxinus velutina*)	7	Two-tailed Tiger Swallowtail
Ash, White (*Fraxinus americana*)	2 3 6 7	Mourning Cloak
	2 3 6	Eastern Tiger Swallowtail
Aspen, Quaking (*Populus tremuloides*)	7	Mourning Cloak Red-spotted Purple Viceroy
Basswood, Carolina (*Tilia americana* var. *caroliniana*)	2 3 6	Mourning Cloak Red-spotted Purple White M Hairstreak
Bay, Red (*Persea borbonia*)	2 3 4 6	Spicebush Swallowtail
	3 4 6	Palamedes Swallowtail
Boxelder (*Acer negundo*)	2 3 4 6	Banded Hairstreak
Catalpa, Northern (*Catalpa speciosa*)	2 3 4 6	Eastern Tiger Swallowtail
Cherry, Choke (*Prunus virginiana*)	1	Coral Hairstreak
	7	Red-spotted Purple Spring/Summer Azure
Cherry, Wild Black (*Prunus serotina*)	2 3 4 6 7	Red-spotted Purple Spring/Summer Azure Viceroy
	2 3 4 6	Eastern Tiger Swallowtail
	6 7	Two-tailed Tiger Swallowtail
	2 3 4	Striped Hairstreak
	2	Coral Hairstreak
Cottonwood (*Populus deltoides*)	1 2 3 4 5 6 7	Mourning Cloak Viceroy
	2 3 4 5 6	Eastern Tiger Swallowtail
	2 3 4 6 7	Red-spotted Purple
Dogwood, Flowering (*Cornus florida*)	2 3 4	Spring/Summer Azure
Ebony, Texas (*Ebenopsis ebano*)	4 5	Cassius Blue Large Orange Sulphur

Tree	***Range***	***Butterfly***
Elm, American (*Ulmus americana*)	1 2 3 4 5 6 7	Mourning Cloak Painted Lady Question Mark
	2 3 6	Eastern Comma
Elm, Cedar (*Ulmus crassifolia*)	2 3 4 5 6	Mourning Cloak Question Mark
Elm, Slippery (*Ulmus rubra*)	2 3 6	Mourning Cloak Question Mark
Elm, Winged (*Ulmus alata*)	2 3 4	Question Mark
Hackberry (*Celtis laevigata*)	1 2 3 4 5 6 7	Hackberry Emperor Mourning Cloak Question Mark Snout
	1 2 3 4 5 6	Tawny Emperor
Hackberry, Western (*Celtis occidentalis*)	1	Hackberry Emperor Mourning Cloak Question Mark Snout Tawny Emperor
Hickory, Black (*Carya texana*)	2 3 6	Banded Hairstreak
Hickory, Shag-bark (*Carya ovata* var. *ovata*)	2 3	Banded Hairstreak
Holly, American (*Ilex opaca*)	2 3 4	Henry's Elfin
Hop-hornbeam, Eastern (*Ostrya virginiana*)	2 3 4	Mourning Cloak Red-spotted Purple
Hornbeam, Carolina (*Carpinus caroliniana*)	3 4	Eastern Tiger Swallowtail Red-spotted Purple Striped Hairstreak
Huisache (*Acacia farnesiana* var. *farnesiana*)	2 3 4 5 6 7	Marine Blue
Juniper, Ashe (*Juniperus ashei*)	1 2 6 7	Olive Juniper Hairstreak
	7	Siva Juniper Hairstreak
Mesquite, Honey (*Prosopis glandulosa*)	1 2 3 4 5 6 7	Marine Blue Reakirt's Blue
	2 3 4 5 6 7	Ceraunus Blue
	2 3 4 5	Proteus Longtail
	7	Leda Ministreak Palmer's Metalmark

Tree	Range	Butterfly
Mulberry, Little-leaf (*Morus microphylla*)	1 2 6 7	Mourning Cloak
Mulberry, Red (*Morus rubra*)	1 2 3 4 5 6	Mourning Cloak
Oak, Black (*Quercus velutina*)	2 3	Horace's Duskywing Juvenal's Duskywing
Oak, Black-jack (*Quercus marilandica*)	1 2 3 4 5 6	Horace's Duskywing
	1 2 3 4 6	Juvenal's Duskywing
	2 3 4 6	White M Hairstreak
Oak, Bur (*Quercus macrocarpa*)	2 3 4 6	Banded Hairstreak Juvenal's Duskywing
	2 3 6	Sleepy Duskywing
	2 6	Arizona Sister
	2	Edwards's Hairstreak
Oak, Chinquapin (*Quercus muhlenbergii*)	2 6	Horace's Duskywing
Oak, Emory (*Quercus emoryi*)	7	Arizona Sister Juvenal's Duskywing Poling's Hairstreak
Oak, Gambel (*Quercus gambelii*)	7	Arizona Sister
Oak, Laurel-leaf (*Quercus laurifolia*)	3 4	Horace's Duskywing Northern Oak Hairstreak White M Hairstreak
Oak, Live (*Quercus virginiana*)	2 3 4 5 6	Gray Hairstreak Horace's Duskywing
	2 3 4 6	Juvenal's Duskywing Northern Oak Hairstreak White M Hairstreak
	2 6	Arizona Sister
Oak, Post (*Quercus stellata*)	1 2 3 4 5	Horace's Duskywing
	1 2 3 4 6	Northern Oak Hairstreak
	1 2 3 4	Juvenal's Duskywing
	2 3 4 6	White M Hairstreak
Oak, Southern Red (*Quercus falcata*)	2 3 4	Banded Hairstreak Horace's Duskywing Juvenal's Duskywing White M Hairstreak
Oak, Spanish (*Quercus buckleyi*)	2 6	Horace's Duskywing White M Hairstreak
Oak, Water (*Quercus nigra*)	2 3 4	Horace's Duskywing Juvenal's Duskywing Northern Oak Hairstreak White M Hairstreak

Tree	*Range*	*Butterfly*
Oak, White (*Quercus alba*)	2 3	Banded Hairstreak Juvenal's Duskywing White M Hairstreak
Oak, Willow (*Quercus phellos*)	2 3 4	Horace's Duskywing White M Hairstreak
Pecan (*Carya illinoinensis*)	1 2 3 4 5 6 7	Gray Hairstreak
Pine, Loblolly (*Pinus taeda*)	3	Eastern Pine Elfin
Pine, Long-leaf (*Pinus palustris*)	3	Eastern Pine Elfin
Pine, Short-leaf (*Pinus echinata*)	3	Eastern Pine Elfin
Pine, Slash (*Pinus elliottii* var. *elliottii*)	3	Eastern Pine Elfin
Plum, Mexican (*Prunus mexicana*)	2 3 4 6	Eastern Tiger Swallowtail Viceroy
Prickly-ash, Hercules'-club (*Zanthoxylum clava-herculis*)	2 3 4	Giant Swallowtail
Redbud (*Cercis canadensis*)	2 3 4 6 7	Henry's Elfin
Red Cedar, Eastern (*Juniperus virginiana*)	1 2 3 4	Olive Juniper Hairstreak
Sassafras (*Sassafras albidum*)	2 3 4 6	Spicebush Swallowtail
	2 3 4	Eastern Tiger Swallowtail
	3 4 6	Palamedes Swallowtail
Soapberry, Western (*Sapindus saponaria* var. *drummondii*)	1 2 4 6 7	Soapberry Hairstreak
Sweet-bay (*Magnolia virginiana*)	3	Eastern Tiger Swallowtail Palamedes Swallowtail Spicebush Swallowtail
Sweetleaf (*Symplocos tinctoria*)	3	King's Hairstreak
Tree, Tulip (*Liriodendron tulipifera*)	3	Eastern Tiger Swallowtail Spicebush Swallowtail
Walnut, Black (*Juglans nigra*)	1 2 3 4 6	Banded Hairstreak
Walnut, Little (*Juglans microcarpa*)	1 2 6	Banded Hairstreak
Willow, Black (*Salix nigra*)	1 2 3 4 5 6 7	Mourning Cloak Viceroy
	2 3 4 6 7	Red-spotted Purple
Willow, Peach-leaved (*Salix amygdaloides*)	1 6 7	Mourning Cloak
Willow, Sandbar (*Salix exigua*)	1 2 4 6 7	Mourning Cloak Viceroy
	2 4 6 7	Red-spotted Purple

Nonnative/Cultivated Trees

Tree	Range	Butterfly
Apple (*Malus pumila*)	1 2 3 4 5 6	Gray Hairstreak Viceroy
	2 3 4 5 6	Eastern Tiger Swallowtail
	2 3 4 6	Red-spotted Purple Spring/Summer Azure
	2 3 4	Striped Hairstreak
Avocado (*Persea americana*)	3 4	Palamedes Swallowtail
Camphor Tree (*Cinnamomum camphora*)	2 3 4	Eastern Tiger Swallowtail Spicebush Swallowtail
Grapefruit (*Citrus* × *paradisi*)	3 4 5	Giant Swallowtail
Lemon (*Citrus limon*)	3 4 5 6	Giant Swallowtail Gray Hairstreak
Lime (*Citrus aurantifolia*)	3 4 5 6	Giant Swallowtail
Locust, Black (*Robinia pseudoacacia*)	1 2 3 6 7	Clouded Sulphur
	1 2 3 4 6	Silver-spotted Skipper
	2 3 4	Zarucco Duskywing
Mimosa, Flowering (*Albizia julibrissin*)	1 2 3 4 6	Reakirt's Blue
Orange, Sour (*Citrus aurantium*)	5	Giant Swallowtail
Orange, Sweet (*Citrus sinensis*)	3 4 5 6	Giant Swallowtail
Peach (*Prunus persica*)	2 3 6	Eastern Tiger Swallowtail
Pear (*Pyrus communis*)	2 3 6	Mourning Cloak Red-spotted Purple Viceroy
Plum, American (*Prunus americana*)	3 4	Striped Hairstreak
Poplar, Silver-leaf (*Populus alba*)	2 3	Red-spotted Purple Viceroy
Willow, Weeping (*Salix babylonica*)	2 4 6	Mourning Cloak Viceroy

Shrub	*Range*	*Butterfly*
Abutilon, Shrubby (*Abutilon abutiloides*)	4 5	Laviana White-Skipper Texas Powdered-Skipper
Acacia, Blackbrush (*Acacia rigidula*)	5 6	Coyote Cloudywing
Acacia, Cat-claw (*Acacia greggii*)	1 4 5 6 7	Marine Blue
Acacia, Roemer's (*Acacia roemeriana*)	1 5 6 7	Reakirt's Blue
Acanthus, Flame (*Anisacanthus quadrifidus* var. *wrightii*)	5 6	Elada Checkerspot Janais Patch Texan Crescent
Agave, New Mexico (*Agave parri*)	7	Orange Giant-Skipper
Agave, Slim-footed (*Agave gracilipes*)	7	Orange Giant-Skipper
Baccharis, Willow-leaf (*Baccharis salicifolia*)	5 6 7	Elada Checkerspot
Bean, Coral (*Erythrina herbacea*)	2 3 4	Spring/Summer Azure
Bear-grass, Foot-hill (*Nolina erumpens*)	7	Sandia Hairstreak
Bernardia, Southwest (*Bernardia myricifolia*)	4 5 6	Lacey's Scrub-Hairstreak
Blueberry, Highbush (*Vaccinium corymbosum*)	3	Henry's Elfin Spring/Summer Azure Striped Hairstreak
Buckeye, Mexican (*Ungnadia speciosa*)	2 3 4 6 7	Henry's Elfin
Buckwheat, Wright's Wild (*Eriogonum wrightii*)	7	Acmon Blue
Bundleflower, Illinois (*Desmanthus illinoensis*)	1 2 3 4 6 7	Reakirt's Blue
Carlowrightia, Small-flowered (*Carlowrightia parviflora*)	4 5	Banded Patch Janais Patch
Carlowrightia, Texas (*Carlowrightia texana*)	5 6	Banded Patch
Cenizo (Texas Silver-leaf) (*Leucophyllum frutescens*)	4 5 6 7	Theona Checkerspot
Cherry, Barbados (*Malpighia glabra*)	4 5	Cassius Blue
	5	Brown-banded Skipper White-patched Skipper

Shrub	***Range***	***Butterfly***
Crucita (*Chromolaena odorata*)	4 5	Rounded Metalmark
Dalea, Bearded (*Dalea pogonanthera*)	4 5 6 7	Reakirt's Blue Southern Dogface
Dalea, Black (*Dalea frutescens*)	1 2 6 7	Reakirt's Blue Southern Dogface
Deerberry (*Vaccinium stamineum*)	3	Red-spotted Purple
Guayacan (*Guajacum angustifolium*)	4 5 6 7	Gray Hairstreak Lyside Sulphur
Hackberry, Dwarf (*Celtis tenuifolia*)	2 3 4	Hackberry Emperor Snout Tawny Emperor
Hackberry, Spiny (*Celtis pallida*)	4 5 6 7	Empress Leilia Hackberry Emperor Snout
	4 5 6	Red-bordered Metalmark
Hawthorn, Parsley-leaved (*Crataegus marshallii*)	2 3 4	Gray Hairstreak
Holly, Yaupon (*Ilex vomitoria*)	2 3 4	Henry's Elfin
Huckleberry (*Vaccinium arboreum*)	2 3 4	Henry's Elfin Striped Hairstreak
Indigo, False (*Amorpha fruticosa*)	1 2 3 4 6	Eastern Tailed-Blue Gray Hairstreak Silver-spotted Skipper Southern Dogface
	2 3 6	Hoary Edge
Kidneywood, Texas (*Eysenhardtia texana*)	2 4 5 6 7	Southern Dogface
Lantana, Desert (*Lantana achyranthifolia*)	4 5 6 7	Gray Hairstreak Painted Lady
Lantana, Texas (*Lantana urticoides*)	2 3 4 5 6	Gray Hairstreak Painted Lady
Lechuguilla (*Agave lechuguilla*)	6 7	Mary's Giant-Skipper
	6	Coahuila Giant-Skipper
Locust, New Mexico (*Robinia neomexicana*)	7	Mexican Yellow
Mistletoe, Christmas (*Phorandendron tomentosum*)	1 2 3 4 5 6 7	Great Purple Hairstreak
Pawpaw, Dwarf (*Asimina parviflora*)	3	Zebra Swallowtail
Pawpaw, Tall (*Asimina triloba*)	3	Zebra Swallowtail

Shrub	*Range*	*Butterfly*
Persimmon, Texas (*Diospyros texana*)	2 4 5 6 7	Gray Hairstreak
	2 4 6 7	Henry's Elfin
Plum, Chickasaw (*Prunus angustifolia*)	1 2	Coral Hairstreak
Possumhaw (*Viburnum nudum*)	2 3 4	Spring/Summer Azure
Prickly-ash, Downy (*Zanthoxylum hirsutum*)	2 4 5 6 7	Giant Swallowtail
	2 4 5 6	Sickle-winged Skipper
Prickly-ash, Lime (*Zanthoxylum fagara*)	2 4 5 6	Giant Swallowtail Sickle-winged Skipper
Redroot (*Ceanothus herbaceus*)	2 3 6	Mottled Duskywing Spring/Summer Azure
Rock-spirea (*Petrophyton caespitosum*)	6 7	Spring/Summer Azure
Screwbean, Dwarf (*Prosopis reptans*)	4 5	Clytie Ministreak Tailed Orange
Silver-leaf, Big Bend (*Leucophyllum minus*)	7	Chinati Checkerspot
Snakeroot, Havana (*Ageratina havanensis*)	6	Bordered Patch Rawson's Metalmark
Snowbells, American (*Styrax americanus*)	3	Eastern Tiger Swallowtail
Spicebush (*Lindera benzoin*)	3 6	Eastern Tiger Swallowtail Spicebush Swallowtail
Sumac, Flame-leaf (*Rhus copallina*)	2 3 4	Red-banded Hairstreak
Sumac, Fragrant (*Rhus aromatica*)	2 3	Red-banded Hairstreak
Tea, New Jersey (*Ceanothus americanus* var. *pitcherii*)	2 3 6	Mottled Duskywing Spring/Summer Azure
Wafer-ash (*Ptelea trifoliata*)	1 2 3 4 6 7	Giant Swallowtail
	1 6 7	Two-tailed Tiger Swallowtail
	2 3 4 6	Eastern Tiger Swallowtail
Wax-mallow, Drummond's (*Malvaviscus drummondii*)	4 5 6	Mallow Scrub-Hairstreak
	4 5	Turk's-cap White-Skipper
	5	Glassy-winged Skipper
Wax-myrtle, Southern (*Morella cerifera*)	2 3 4	Red-banded Hairstreak
Yucca, Arkansas (*Yucca arkansana*)	2 3 5 6	Yucca Giant-Skipper

Shrub	Range	Butterfly
Yucca, Banana (*Yucca baccata*)	7	Yucca Giant-Skipper
Yucca, Buckley's (*Yucca constricta*)	2 5 6	Yucca Giant-Skipper
	6	Strecker's Giant-Skipper
Yucca, Louisiana (*Yucca flaccida*)	2 3	Yucca Giant-Skipper
Yucca, Smooth (*Yucca glauca*)	1 6 7	Yucca Giant-Skipper
	1 6	Strecker's Giant-Skipper

Nonnative/Cultivated Shrubs

Shrub	Range	Butterfly
Flamingo Plant (*Jacobinia carnea*)	4 5 6	Texan Crescent
Lantana, Trailing (*Lantana montevidensis*)	2 3 4 6	Painted Lady
Lantana, West Indian (*Lantana camara*)	2 3 4 5 6	Gray Hairstreak Painted Lady
Lilac, Fragrant (*Syringa vulgaris*)	2 3 6	Eastern Tiger Swallowtail
Locust, Bristly (*Robinia hispida*)	2 3	Silver-spotted Skipper
Orange, Trifoliate (*Poncirus trifoliata*)	2 3 4 6	Giant Swallowtail
Rattle-bush, Showy (*Sesbania punicea*)	3 4	Zarucco Duskywing
Senna, Argentina (*Senna corymbosa*)	2 3 4 5 6	Cloudless Sulphur Sleepy Orange
	4 5 6 7	Orange-barred Sulphur
Senna, Candle-stick (*Senna alata*)	2 3 4 5 6	Cloudless Sulphur Gray Hairstreak
	4 5 6	Orange-barred Sulphur
Senna, Coffee (*Senna occidentalis*)	3 4 5 6	Cloudless Sulphur Little Yellow
	4 5 6	Orange-barred Sulphur

Vine	Range	Butterfly
Balloon-vine, Common (*Cardiospermum halicacabum*)	4 5 6	Mallow Scrub-Hairstreak Silver-banded Hairstreak
Balloon-vine, Tropical (*Cardiospermum corundum*)	5	Silver-banded Hairstreak
Bean, Slick-seed Wild (*Strophostyles leiosperma*)	2 3 4	Southern Cloudywing
Bean, Trailing Wild (*Strophostyles helvula*)	2 3 4 6	Proteus Longtail
	2 3 4	Southern Cloudywing
Bindweed, Field (*Convolvulus arvensis*)	1 7	Painted Crescent
Butterfly-pea, Virginia (*Centrosema virginianum*)	2 3 4 5 6	Funereal Duskywing
	2 3 4	Southern Cloudywing Zarucco Duskywing
	2 4 5 6	Dorantes Longtail
Cow-pea, Wild (*Vigna luteola*)	4 5	Cassius Blue Dorantes Longtail Gray Hairstreak Proteus Longtail
Groundnut (*Apios americana*)	2 3 4 6	Silver-spotted Skipper Spring/Summer Azure
	2 3 4	Southern Cloudywing
Hog-peanut, Southern (*Amphicarpaea bracteata*)	2 3	Golden Banded-Skipper Gray Hairstreak Silver-spotted Skipper
Honeysuckle, Coral (*Lonicera sempervirens*)	2 3 4 6	Spring/Summer Azure
Milk-pea, Downy (*Galactia regularis*)	2 3 4	Cassius Blue
Milkweed Vine, Climbing (*Funastrum cynanchoides*)	1 2 4 5 6 7	Monarch Queen
	4 5	Soldier
Milkweed Vine, Wavy-leaf (*Funastrum crispum*)	1 2 4 5 6 7	Queen
Passionflower, Bracted (*Passiflora affinis*)	6	Zebra Longwing
Passionflower, Purple (*Passiflora incarnata*)	2 3 4 5 6	Gulf Fritillary Julia Longwing Variegated Fritillary Zebra Longwing
Passionflower, Red-fruited (*Passiflora foetida*)	4 5 6	Gulf Fritillary Julia Longwing Variegated Fritillary Zebra Longwing
	5	Mexican Fritillary Mexican Silverspot

Vine	Range	Butterfly
Passionflower, Slender-lobed (*Passiflora tenuiloba*)	5 6 7	Gulf Fritillary Julia Longwing
Passionflower, Yellow (*Passiflora lutea*)	2 3 4 6	Gulf Fritillary Julia Longwing Variegated Fritillary Zebra Longwing
Pipevine, Woolly (*Aristolochia tomentosa*)	2 3 4 6	Pipevine Swallowtail
	4 6	Polydamas Swallowtail
Snapdragon Vine (*Maurandya antirrhiniflora*)	4 5 6 7	Common Buckeye
Snoutbean, Least (*Rhynchosia minima*)	2 3 4 5 6	Ceraunus Blue White-striped Longtail
	2 4 5 6	Dorantes Longtail
Snoutbean, Texas (*Rhynchosia senna* var. *texana*)	2 4 6 7	Northern Cloudywing
	2 4 5 6	White-striped Longtail
Snoutbean, Woolly (*Rhynchosia tomentosa*)	3	Southern Cloudywing
Virgin's Bower, Drummond's (*Clematis drummondii*)	4 5 6 7	Fatal Metalmark
Wisteria, Kentucky (*Wisteria frutescens*)	3	Proteus Longtail Silver-spotted Skipper Zarucco Duskywing

Nonnative/Cultivated Vines

Vine	Range	Butterfly
Kudzu (*Pueraria montana* var. *lobata*)	2 3 4	Silver-spotted Skipper Spring/Summer Azure
Passionflower, Blue (*Passiflora caerulea*)	3 5 6	Gulf Fritillary Julia Longwing Variegated Fritillary
Passionflower, Scarlet (*Passiflora coccinea*)	3 4 5 6	Zebra Longwing
Pea, Annual Sweet (*Lathyrus odoratus*)	3 4 6	Marine Blue
Pigeon-wings, Asian (*Clitoria ternata*)	2 3 4 5 6	Proteus Longtail
Pipevine (*Aristolochia macrophylla*)	4 5 6	Pipevine Swallowtail Polydamas Swallowtail
Wisteria, Chinese (*Wisteria sinensis*)	2 3 4 6	Marine Blue Proteus Longtail Silver-spotted Skipper Zarucco Duskywing
Wisteria, Japanese (*Wisteria floribunda*)	2 3 4 6	Silver-spotted Skipper

Herb	*Range*	*Butterfly*
Abutilon, Texas (*Abutilon fruticosum*)	2 4 5 6 7	Common Streaky-Skipper
	2 4 6 7	Mallow Scrub-Hairstreak
	4 5 6 7	Texas Powdered-Skipper
	4 5 6	Laviana White-Skipper
Abutilon, Wright's (*Abutilon wrightii*)	4 5 6	Laviana White-Skipper
Abutilon, Yellow (*Abutilon malacum*)	7	Scarce Streaky-Skipper
Acacia, Fern (*Acacia angustissima* var. *hirta*)	1 2 3 4 5 7	Ceraunus Blue Mexican Yellow Reakirt's Blue
	1 2 5 6	Outis Skipper
	6 7	Acacia Skipper
Agalinis, Bunch-leaved (*Agalinis fasciculata*)	2 3 4	Common Buckeye
Agalinis, Plateau (*Agalinis edwardsiana*)	6	Common Buckeye
Agalinis, Prairie (*Agalinis heterophylla*)	2 3 4 6	Common Buckeye
Agalinis, Purple (*Agalinis purpurea*)	2 3 4	Common Buckeye
Agalinis, Slender-leaved (*Agalinis tenuifolia* var. *leucanthera*)	2 3	Common Buckeye
Amaranth, Red-root (*Amaranthus retroflexus*)	1 2 4 5 6 7	Common Sootywing
	2 4 5 6	Mazans Scallopwing
	7	Mexican Sootywing
Amaranth, Slender (*Amaranthus hybridus*)	2 4	Common Sootywing
Anise-root (*Osmorhiza longistylis*)	2	Black Swallowtail
Anoda, Crested (*Anoda cristata*)	6 7	Common Checkered-Skipper
Aster, Flat-top (*Doellingeria sericocarpoides*)	2 3	Silvery Checkerspot
Aster, Heath (*Symphyotrichum ericoides*)	1 2 3 4 5 6 7	Pearl Crescent
Aster, Texas (*Symphyotrichum drummondii* var. *texana*)	2 3 4 6	Pearl Crescent
Aster, Willow-leaf (*Symphyotrichum praealtum*)	2 3 4 5 6	Pearl Crescent
Bagpod (*Glottidium vesicarium*)	2 3 4 6	Funereal Duskywing
	2 3 4	Zarucco Duskywing

Herb	***Range***	***Butterfly***
Beard-tongue, Cobaea (*Penstemon cobaea*)	1 7	Dotted Checkerspot
Beard-tongue, White-flowered (*Penstemon albidus*)	1	Dotted Checkerspot
Berula, Stalked (*Berula erecta*)	1 6 7	Black Swallowtail
Bittercress, Spring (*Cardamine rhomboidea*)	2 3 4	Falcate Orangetip
Bladder-pod, Slender (*Lesquerella gracilis*)	2 6	Olympia Marble
Bluebonnet, Big Bend (*Lupinus havardii*)	7	Little Yellow
Bluebonnet, Texas (*Lupinus texensis*)	2 3 4 5 6 7	Eastern Tailed-Blue Gray Hairstreak Little Yellow
	2 3 4 6 7	Henry's Elfin
Bluehearts, American (*Buchnera americana*)	2 3 4 5 6	Common Buckeye
Bluestem, Big (*Andropogon gerardii*)	1 2 3 6 7	Delaware Skipper
	1 2 3 6	Large Wood-Nymph
	1 2 3	Dusted Skipper
	2 3	Cobweb Skipper
	2 6	Arogos Skipper
	1	Ottoe Skipper
	7	Orange-headed Roadside-Skipper
Bluestem, Little (*Schizachyrium scoparium*)	1 2 3 6 7	Delaware Skipper
	1 2 3	Dusted Skipper
	1	Ottoe Skipper
	2 3 4 5 6 7	Celia's Roadside-Skipper
	2 3 4	Swarthy Skipper
	2 3	Cobweb Skipper Crossline Skipper Meske's Skipper
	2 6	Arogos Skipper
Broom-sedge (*Andropogon virginicus*)	2 3	Cobweb Skipper
Brown-eyed Susan (*Rudbeckia hirta*)	2 3 4 6	Silvery Checkerspot
	2 4 5 6	Bordered Patch
Bulrush, Woolly-grass (*Scirpus cyperinus*)	2 3	Dion Skipper
Bush-bean, Purple (*Macroptilium atropurpureum*)	4 5	Dorantes Longtail White-striped Longtail

Herb	*Range*	*Butterfly*
Bush-clover, Downy (*Lespedeza hirta*)	2 3	Eastern Tailed-Blue Gray Hairstreak Hoary Edge Northern Cloudywing Southern Cloudywing
Bush-clover, Round-head (*Lespedeza capitata*)	2 3 4	Eastern Tailed-Blue Gray Hairstreak Northern Cloudywing Silver-spotted Skipper Southern Cloudywing
Butter-and-eggs (*Linaria vulgaris*)	2 6	Common Buckeye
Butterfly Weed (*Asclepias tuberosa*)	1 2 3 4 6 7	Gray Hairstreak Monarch Queen
Cane, Giant (*Arundinaria gigantea*)	3	Creole Pearly-eye Lace-winged Roadside-Skipper Southern Pearly-eye
Caric-sedge, Cedar (*Carex planostachys*)	6 7	Apache Skipper
Caric-sedge, Thin-scale (*Carex hyalinolepis*)	2 3	Dion Skipper
	3	Duke's Skipper
Clammyweed (*Polanisia dodecandra*)	1 2 3 4 5 6 7	Checkered White
Clover, Buffalo (*Trifolium reflexum*)	3	Orange Sulphur
Crotolaria, Arrow-leaf (*Crotolaria sagittalis*)	2 3 4	Wild Indigo Duskywing
	2 3	Frosted Elfin
Crotolaria, Woolly (*Crotolaria incana*)	4 5	Cassius Blue
Croton, Leather-weed (*Croton pottsii*)	1 5 6 7	Goatweed Leafwing
Croton, One-seeded (*Croton monanthogynus*)	1 2 3 4 5 6 7	Goatweed Leafwing Gray Hairstreak
Croton, Silver-leaf (*Croton argyranthemus*)	2 3 4 5	Goatweed Leafwing Gray Hairstreak
Croton, Woolly (*Croton capitatus*)	2 3 4 5 6	Goatweed Leafwing Gray Hairstreak
Crownbeard, Golden (*Verbesina encelioides*)	1 2 3 4 5 6 7	Bordered Patch
Crownbeard, Gravel-weed (*Verbesina helianthoides*)	2 3	Silvery Checkerspot Spring/Summer Azure
Cudweed, Fragrant False (*Pseudognaphalium obtusifolium*)	2 3 4 6	American Lady
Cudweed, Pennsylvania (*Gamochaeta pensylvanica*)	2 3 4 5 6	American Lady

Herb	***Range***	***Butterfly***
Cudweed, Purple (*Gamochaeta purpurea*)	2 3 4 6	American Lady
Dalea, Gregg's (*Dalea greggii*)	6 7	Reakirt's Blue
Dalea, Purple (*Dalea purpurea*)	2 6 7	Marine Blue Reakirt's Blue Southern Dogface
Deer-vetch, Single-leaf (*Lotus unifoliatus*)	2 3 4 6	Eastern Tailed-Blue
Dicliptera, Bracted (*Dicliptera brachiata*)	2 3 4 5 6	Texan Crescent
	5	Rosita Patch
Dock, Bitter (*Rumex obtusifolius*)	1	Gray Copper
Dock, Curly (*Rumex crispus*)	1	Gray Copper
Dock, Mexican (*Rumex triangulivalvis*)	1	Gray Hairstreak
Dogbane, Spreading (*Apocynum androsaemifolium*)	7	Monarch
Dogweed, Bristle-leaf (*Thymophylla tenuiloba*)	3 4 5 6	Dainty Sulphur
Dogweed, Common (*Thymophylla pentachaeta*)	1 4 5 6 7	Dainty Sulphur
Dutchman's-breeches, Texas (*Thamnosma texana*)	1 2 5 6 7	Black Swallowtail
Dyssodia, Fetid (*Dyssodia papposa*)	1 7	Dainty Sulphur
Eupatorium, Late-flowering (*Eupatorium serotinum*)	2 4 5 6	Rounded Metalmark
Evax, Big-head (*Evax prolifera*)	1 2 6 7	American Lady
Evax, Silver (*Evax candida*)	2 3 4 6	American Lady
Evax, Spring (*Evax verna*)	1 2 3 4 5 6 7	American Lady
Everlasting, Parlin's (*Antennaria parlinii*)	2 3	American Lady
Flax, Berlandier's (*Linum berlandieri*)	1 2 3 4 5 6 7	Variegated Fritillary
Flax, Blue-flowered (*Linum lewisii*)	1 2 6 7	Variegated Fritillary
Flax, Grooved (*Linum sulcatum*)	2 3 4	Variegated Fritillary
Flax, Prairie (*Linum pratense*)	1 2 6 7	Variegated Fritillary

Herb	Range	Butterfly
Flax, Rock (*Linum rupestre*)	2 6 7	Variegated Fritillary
Flax, Stiff-stem (*Linum rigidum*)	1	Variegated Fritillary
Flax, Texas (*Linum medium* var. *texana*)	2 3 4	Variegated Fritillary
Frogfruit, Diamond-leaf (*Phyla strigulosa*)	2 4 5 6 7	Common Buckeye Phaon Crescent
	4 5	White Peacock
Frogfruit, Lance-leaved (*Phyla lanceolata*)	1 2 3 4 6	Common Buckeye Phaon Crescent
Frogfruit, Texas (*Phyla nodiflora*)	1 2 3 4 5 6 7	Common Buckeye
	2 3 4 5 6 7	Phaon Crescent
	4 5	White Peacock
Frostweed, Virginia (*Verbesina virginica*)	2 4 6	Bordered Patch Silvery Checkerspot Spring/Summer Azure
Glasswort, Virginia (*Salicornia virginica*)	4	Eastern Pygmy-Blue
Globe-mallow, Copper (*Sphaeralcea angustifolia*)	1 2 6 7	Common Checkered-Skipper
	6 7	Common Streaky-Skipper Texas Powdered-Skipper
	7	Scarce Streaky-Skipper
Globe-mallow, Scarlet (*Sphaeralcea coccinea*)	1 2 6 7	Common Checkered-Skipper
Goldeneye, Tooth-leaved (*Viguiera dentata*)	2 6 7	Bordered Patch
Goosefoot, Berlandier's (*Chenopodium berlandieri*)	1 2 4 5 6 7	Common Sootywing Western Pygmy-Blue
Goosefoot, Slender-leaf (*Chenopodium leptophyllum*)	1 6 7	Western Pygmy-Blue
Grass, Basket (*Oplismenus hirtellus*)	2 3 4 5	Carolina Satyr
Grass, Buffalo (*Buchloe dactyloides*)	1 2 6 7	Green Skipper
Grass, Common Witch (*Panicum capillare*)	1 2 3 4 6 7	Clouded Skipper
Grass, Downy Woolly (*Erioneuron pilosum*)	1 2 6 7	Green Skipper
Grass, Dwarf Bristle (*Setaria pumila*)	1 2 4 6 7	Nysa Roadside-Skipper
Grass, Fall Witch (*Digitaria cognata*)	2 6	Dotted Skipper
	1	Ottoe Skipper

Herb	*Range*	*Butterfly*
Grass, Fluff (*Dasyochloa pulchella*)	6 7	Green Skipper Pahaska Skipper
Grass, Forked Panic (*Panicum dichotomum*)	3	Northern Broken-Dash
Grass, Rice Cut (*Leersia oryzoides*)	1 2 3 4 6	Least Skipper
Grass, Sideoats Grama (*Bouteloua curtipendula*)	1 2 3 4 5 6 7	Orange Skipperling
	1 2 6 7	Green Skipper
	2 6	Dotted Skipper
	1	Ottoe Skipper
Grass, Slender Crab (*Digitaria filiformis*)	2 3	Tawny-edged Skipper
Grass, Slender Grama (*Bouteloua gracilis*)	1 2 6 7	Green Skipper
	1 7	Pahaska Skipper Uncas Skipper
	1	Ottoe Skipper
	7	Simius Roadside-Skipper
Grass, Slender Prairie (*Sorghastrum elliottii*)	3	Pepper and Salt Skipper
Grass, Southern Crab (*Digitaria ciliaris*)	1 2 3 4 5 6 7	Fiery Skipper
	2 3 4 5 6	Southern Broken-Dash
Grass, Switch (*Panicum virgatum*)	1 2 3 6 7	Delaware Skipper
	2 6	Dotted Skipper
Grass, Teal Love (*Eragrostis hypnoides*)	2 3 4 5	Fiery Skipper
Grass, Yellow Prairie (*Sorghastrum nutans*)	3	Pepper and Salt Skipper
Greenthread, Fine-leaf (*Thelesperma filifolium*)	1 2 3 4 5 6 7	Dainty Sulphur
Greenthread, Rayless (*Thelesperma megapotamicum*)	1 4 5 6 7	Dainty Sulphur
Ground-plum (*Astragulus crassicarpus*)	1 2 3 6 7	Clouded Sulphur Gray Hairstreak Orange Sulphur
Horse-purslane, Desert (*Trianthema portulacastrum*)	4 5 7	Western Pygmy-Blue
Indigo, Blue Wild (*Baptisia australis*)	2	Frosted Elfin Wild Indigo Duskywing
Indigo, Lindheimer's (*Indigofera lindheimeriana*)	6 7	Reakirt's Blue
	6	False Duskywing
Indigo, Nuttall's Wild (*Baptisia nuttalliana*)	2 3	Frosted Elfin

Herb	***Range***	***Butterfly***
Indigo, White Wild (*Baptisia alba*)	2 3 4	Wild Indigo Duskywing
Indigo, Wild (*Baptisia bracteata*)	2 3 4	Wild Indigo Duskywing
Iresine, Rootstock (*Iresine rhizomatosa*)	2 3 6	Hayhurst's Scallopwing
Licorice, Wild (*Glycyrrihiza lepidota*)	1 7	Gray Hairstreak Marine Blue Melissa Blue Orange Sulphur Reakirt's Blue
	1	Silver-spotted Skipper
Little-mallow, Alkali (*Malvella leprosa*)	7	Mallow Scrub-Hairstreak
Mallow, Rio Grande False (*Malvastrum americanum*)	4 5	Laviana White-Skipper
Mallow, Three-lobe False (*Malvastrum coromandelianum*)	2 4 5 6	Common Checkered-Skipper
	4 5 6	Tropical Checkered-Skipper
Mallow, Velvet-leaf (*Wissadula amplissima*)	4 5 7	Common Streaky-Skipper Laviana White-Skipper Texas Powdered-Skipper
	7	Scarce Streaky-Skipper
Mallow, Wright's False (*Malvastrum aurantiacum*)	2 4 5 6	Tropical Checkered-Skipper
Mexican-mallow, Slender-stalked (*Meximalva filipes*)	2 4 5 6	Common Streaky-Skipper
	4 5 6	Laviana White-Skipper
Milk-vetch, Bent-pod (*Astragalus distortus*)	2 3 4	Southern Cloudywing Wild Indigo Duskywing
Milk-vetch, Canadian (*Astragalus canadensis*)	2 3 4	Wild Indigo Duskywing
Milk-vetch, Lotus (*Astragalus lotiflorus*)	1 2 6 7	Ceraunus Blue Eastern Tailed-Blue Marine Blue
Milk-vetch, Nuttall's (*Astragalus nuttallianus*)	2 3 4 6	Northern Cloudywing
Milk-vetch, Woolly (*Astragalus mollissimus*)	1 6 7	Gray Hairstreak
	1 7	Melissa Blue
Milk-vetch, Wooton's (*Astragalus wootonii*)	7	Acmon Blue
Milkweed, Antelope-horns (*Asclepias asperula*)	1 2 3 4 5 6 7	Monarch Queen
Milkweed, Blunt-leaf (*Asclepias amplexicaulis*)	2 3	Monarch Queen
Milkweed, Broad-leaf (*Asclepias latifolia*)	1 2 6 7	Monarch

Herb	*Range*	*Butterfly*
Milkweed, Green (*Asclepias viridis*)	2 3 4 6	Monarch Queen
Milkweed, Prairie (*Asclepias oenotheroides*)	1 2 4 5 6 7	Monarch Queen
Milkweed, Swamp (*Asclepias incarnata*)	6	Monarch Queen
Milkweed, Texas (*Asclepias texana*)	6 7	Monarch Queen
Mistflower, Gregg's (*Conoclinium dissectum*)	5 6 7	Rawson's Metalmark
Mock Bishop's-weed, Nuttall's (*Ptilimnium nuttallii*)	2 3 4 6	Black Swallowtail
Mock Bishop's-weed, Thread-leaf (*Ptilimnium capillaceum*)	2 3 4	Black Swallowtail
Nettle, False (*Boehmeria cylindrical*)	1 2 3 4 6 7	Question Mark Red Admiral
	2 3 6	Eastern Comma
Noseburn, Catnip (*Tragia ramosa*)	4 5 6 7	Amymone
Paintbrush, Downy (*Castilleja sessiliflora*)	1 6 7	Common Buckeye Fulvia Checkerspot
Paintbrush, Purple (*Castilleja purpurea*)	1 2 6	Common Buckeye
	1 6	Fulvia Checkerspot
Paintbrush, Texas (*Castilleja indivisa*)	2 3 4 5 6	Common Buckeye
	6	Fulvia Checkerspot
Pansy, Field (*Viola bicolor*)	1 2 3	Variegated Fritillary
Parsley, Nuttall's Prairie (*Polytaenia nuttallii*)	2 3 4 6	Black Swallowtail
Partridge-pea (*Chamaecrista fasciculata*)	1 2 3 4 6	Cloudless Sulphur Little Yellow Sleepy Orange
	1 2 3 6	Clouded Sulphur
Paspalum, Thin (*Paspalum setaceum*)	1 2 3 4 5 6 7	Clouded Skipper
	2 3 4 5 6	Whirlabout
Pea, Scarlet (*Indigofera miniata*)	1 2 3 4 5 6	Funereal Duskywing Gray Hairstreak Reakirt's Blue
	2 3 4 5 6	Cassius Blue
Pellitory, Florida (*Parietaria floridana*)	4 5	Red Admiral
Pellitory, Pennsylvania (*Parietaria pensylvanica*)	1 2 3 4 5 6 7	Red Admiral

Herb	Range	Butterfly
Peppergrass, Prairie (*Lepidium densiflorum*)	1 2 3 4 5 6 7	Checkered White
	1 2 3 4 5 6	Falcate Orangetip
	1 2 3 4 6 7	Cabbage White
	4 5	Great Southern White
Peppergrass, Virginia (*Lepidium virginicum*)	1 2 3 4 5 6 7	Checkered White
	1 2 3 4 5 6	Falcate Orangetip
	1 2 3 4 6 7	Cabbage White
	4 5	Great Southern White
Peppergrass, Western (*Lepidium alyssoides*)	1 6 7	Checkered White
Pigeon-wings (*Clitoria mariana*)	2 3 4 6	Funereal Duskywing Proteus Longtail
	2 3 4	Southern Cloudywing Zarucco Duskywing
	2 3 6	Golden Banded-Skipper
	2 4 6	Dorantes Longtail
Pipevine, Net-leaved (*Aristolochia reticulata*)	2 3	Pipevine Swallowtail
Pipevine, Upright (*Aristolochia erecta*)	2 3 4 5 6	Pipevine Swallowtail
Pipevine, Virginia (*Aristolochia serpentaria*)	3 6	Pipevine Swallowtail
Plantain, Narrow-leaved (*Plantago lanceolata*)	2 3 4 6 7	Common Buckeye Painted Lady
Plumbago, White-flowered (*Plumbago scandens*)	5 7	Marine Blue
	5	Cassius Blue
Powderpuff (*Mimosa strigillosa*)	2 3 4 5 6	Little Yellow Reakirt's Blue
	4 5 6	Mimosa Yellow
Parsley, Nuttall's Prairie (*Polytaenia nuttallii*)	2 3 4 6	Black Swallowtail
Rabbit-tobacco, Silver (*Diapera candida*)	2 3 4 6	American Lady
Ragweed, Giant (*Ambrosia trifida*)	1 2 4 5 6 7	Bordered Patch
	1 2 3 6	Gorgone Checkerspot
Rattle-bush, Drummond's (*Sesbania drummondii*)	2 3 4 5 6	Funereal Duskywing
Rice, Southern Wild (*Zizanopsis miliacea*)	2 3 4 6	Broad-winged Skipper

Herb	***Range***	***Butterfly***
Rockcress, Balcones (*Arabis petiolaris*)	2 6	Falcate Orangetip
Rockcress, Sickle-pod (*Arabis canadensis*)	1 2 4 6 7	Falcate Orangetip Olympia Marble
Rocket-mustard (*Sisymbrium irio*)	1 2 4 5 6	Falcate Orangetip
	1 2 4 6 7	Cabbage White
Ruellia, Violet (*Ruellia nudiflora*)	2 4 5 6	Common Buckeye Texan Crescent
	4 5	Tropical Buckeye White Peacock
	5	Banded Peacock Malachite
Ruellia, Western (*Ruellia occidentalis*)	4 5 6	Texan Crescent
	4 5	White Peacock
Rye, Canada Wild (*Elymus canadensis*)	2 3	Zabulon Skipper
Saltbush, Four-wing (*Atriplex canescens*)	1 2 4 5 6 7	Western Pygmy-Blue
	1 7	Saltbush Sootywing
Scaleseed, Spreading (*Spermolepis inermis*)	1 2 3 4 5 6 7	Black Swallowtail
Sea-purslane, Winged (*Sesuvium verrucosum*)	1 4 5 7	Western Pygmy-Blue
Senna, Lindheimer's (*Senna lindheimeriana*)	5 6 7	Funereal Duskywing Orange-barred Sulphur Sleepy Orange
	5 7	Boisduval's Yellow
Senna, Maryland (*Senna marilandica*)	2 3 4	Cloudless Sulphur Little Yellow Sleepy Orange
Senna, Two-leaved (*Senna roemeriana*)	1 2 6 7	Cloudless Sulphur Sleepy Orange
Sensitive-pea, Delicate (*Chamaecrista nictitans*)	2 3 4	Cloudless Sulphur Little Yellow Sleepy Orange
Sensitive-pea, Texas (*Chamaecrista flexuosa* var. *texana*)	4 5	Tailed Orange
Sesbania, Drummond's (*Sesbania drummondii*)	2 3 4 5 6	Funereal Duskywing Gray Hairstreak
Shepherd's-needle (*Bidens alba*)	3 4 5	Checkered white Dainty Sulphur
Shepherd's-purse (*Capsella bursa-pastoris*)	1 2 3 4 5 6 7	Checkered White
	4 5	Great Southern White
Sida, Arrow-leaf (*Sida rhombifolia*)	2 3 4 5	Common Checkered-Skipper Tropical Checkered-Skipper

Herb	*Range*	*Butterfly*
Sida, Lindheimer's (*Sida lindheimeri*)	2 3 4 5 6	Common Checkered-Skipper
Sida, Prickly (*Sida spinosa*)	2 3 4 5	Common Checkered-Skipper
Sida, Spreading (*Sida abutifolia*)	2 3 4 5 6 7	Common Checkered-Skipper
	5 6 7	Desert Checkered-Skipper
Sneezeweed, Autumn (*Helenium autumnale*)	1 2 6	Dainty Sulphur
Sneezeweed, Purple-headed (*Helenium flexuosum*)	2 3 4	Dainty Sulphur
Speedwell, Water (*Veronica anagallis-aquatica*)	2 6	Common Buckeye
Spiderling, Erect (*Boerhaavia erecta*)	1 2 3 4 5 6 7	Variegated Fritillary
Spiderling, Scarlet (*Boerhaavia coccinea*)	1 2 3 4 5 6 7	Variegated Fritillary
Sprangletop (*Leptochloa dubia*)	1 2 4 5 6 7	Orange Skipperling
Stemodia, Woolly (*Stemodia lanata*)	4 5	Tropical Buckeye
Stenandrium, Shaggy (*Stenandrium barbatum*)	7	Definite Patch
Stinging-nettle, Dwarf (*Urtica urens*)	2 3 4	Red Admiral
Stinging-nettle, Heart-leaf (*Urtica chamaedryoides*)	2 3 4 5 6	Red Admiral
Stinging-nettle, Slim (*Urtica dioica*)	1	Red Admiral
Stylisma, Purple (*Stylisma aquatica*)	2 3	Golden Banded-Skipper
Sunflower, Annual (*Helianthus annuus*)	1 2 3 4 5 6 7	Bordered Patch Painted Lady
	1 2 3 6	Gorgone Checkerspot
	2 3 4 6	Silvery Checkerspot
Sunflower, Plains (*Helianthus petiolaris*)	1 2	Gorgone Checkerspot
Tansy-mustard, Western (*Descurainia pinnata*)	1 2 3 4 5 6 7	Checkered White
	1 2 6	Olympia Marble
Tephrosia, Lindheimer's (*Tephrosia lindheimeri*)	2 4 5 6	White-striped Longtail
Thalia, Powdery (*Thalia dealbata*)	2 3 4	Brazilian Skipper
Thistle, Horrid (*Cirsium horridulum*)	2 3 4	Painted Lady
	3	Little Metalmark

Herb	***Range***	***Butterfly***
Thistle, Texas (*Cirsium texanum*)	1 2 3 4 5 6 7	Painted Lady
Three-awn, Purplish (*Aristida purpurascens*)	2 3	Meske's Skipper
	2 6	Dotted Skipper
Tick-clover, Little-leaved (*Desmodium ciliare*)	2 3 4	Hoary Edge Southern Cloudywing
Tick-clover, Maryland (*Desmodium marilandicum*)	2 3	Eastern Tailed-Blue Silver-spotted Skipper
Tick-clover, Panicled (*Desmodium paniculatum*)	1 2 3 4 6	Eastern Tailed-Blue Proteus Longtail
	2 3 4 6	Northern Cloudywing White-striped Longtail
	2 4 6	Dorantes Longtail
Tick-clover, Round-leaved (*Desmodium rotundifolium*)	3	Confused Cloudywing Hoary Edge Northern Cloudywing Silver-spotted Skipper Southern Cloudywing
Tick-clover, Trailing (*Desmodium glabellum*)	2 3 4	Northern Cloudywing Silver-spotted Skipper
	2 3	Hoary Edge
Tick-clover, Woolly (*Desmodium canescens*)	2 3	Eastern Tailed-Blue Hoary Edge Proteus Longtail
Toadflax, Texas (*Nuttallanthus texanus*)	1 2 3 4 5 6 7	Common Buckeye
Tridens, Purpletop (*Tridens flavus*)	1 2 3 4 6	Dun Skipper
	1 2 3 6	Large Wood-Nymph
	2 3	Crossline Skipper
	3 4	Little Glassywing
Tridens, Slim (*Tridens muticus*)	1 2 6 7	Green Skipper
Tuberose, Spotted (*Manfreda maculosa*)	5	Manfreda Giant-Skipper
Tuberose, Variegated (*Manfreda variegata*)	5	Manfreda Giant-Skipper
Vetch, Common (*Vicia sativa*)	2 3 6	Eastern Tailed-Blue Orange Sulphur
Vetch, Louisiana (*Vicia ludoviciana*)	1 2 3 4 5 6 7	Eastern Tailed-Blue Funereal Duskywing Little Yellow Southern Dogface

Herb	*Range*	*Butterfly*
Vetch, Woolly-pod (*Vicia villosa*)	1 2 3 6	Eastern Tailed-Blue
Violet, Bayou (*Viola sororia*)	1 2 3 4 6	Variegated Fritillary
Violet, Lance-leaved (*Viola lanceolata*)	3 4	Variegated Fritillary
Violet, Primrose-leaved (*Viola primulifolia*)	2 3	Variegated Fritillary
Water-cress (*Rorippa nasturtium-aquaticum*)	1 2 3 6 7	Cabbage White
Water-hemlock, Spotted (*Cicuta maculata*)	1 2 3 4 6	Black Swallowtail
Water-hyssop, Coastal (*Bacopa monnieri*)	4 5	White Peacock
Water-willow, American (*Justicia americana*)	1 2 4 6	Texan Crescent
Water-willow, Downy (*Justicia pilosella*)	1 2 4 5 6 7	Texan Crescent Vesta Crescent
	2 4 5 6 7	Elada Checkerspot
	1 7	Painted Crescent
	5 6 7	Tiny Checkerspot
	5	Rosita Patch
Wedelia, Hispid (*Wedelia acapulcensis* var. *hispida*)	2 4 5 6 7	Bordered Patch
Wissadula, False (*Allowissadula holosericea*)	6 7	Common Streaky-Skipper Texas Powdered-Skipper
	7	Scarce Streaky-Skipper
Wood-oats, Broad-leaf (*Chasmanthium latifolium*)	2 3 4 5 6	Celia's Roadside-Skipper
	2 6 7	Bronze Roadside-Skipper
	2 3	Bell's Roadside-Skipper Common Roadside-Skipper
	3	Pepper and Salt Skipper
Wood-oats, Slender (*Chasmanthium laxum*)	2 3 4	Gemmed Satyr

Nonnative/Cultivated Shrubs

Herb	***Range***	***Butterfly***
Alfalfa (*Medicago sativa*)	1 2 3 6 7	Clouded Sulphur
	1 7	Melissa Blue
	2 3 4 5 6 7	Ceraunus Blue Funereal Duskywing Gray Hairstreak Marine Blue Orange Sulphur Painted Lady Reakirt's Blue Southern Dogface
	2 3 4 6 7	Northern Cloudywing
Alyssum, Sweet (*Lobularia maritima*)	1 2 3 4 5 6 7	Checkered White
	1 2 3 4 6 7	Cabbage White
Amaranth, Princess Feather (*Celosia argentea*)	1 2 4 5 6 7	Common Sootywing
Bean, Garden (*Phaseolus vulgaris*)	1 2 3 4 5 6 7	Gray Hairstreak Painted Lady
	1 2 3 4 6	Silver-spotted Skipper
	2 3 4 5 6	Proteus Longtail Cassius Blue
	2 4 5 6	Dorantes Longtail
Bean, Lima (*Phaseolus lunatus*)	1 2 3 4 5 6 7	Gray Hairstreak
	2 3 4 5 6	Proteus Longtail
	2 3 4 5	Cassius Blue
	2 4 5 6	Dorantes Longtail
Beet, Garden (*Beta vulgaris*)	1 2 3 4 5 6 7	Painted Lady
Borage (*Borage officinalis*)	1 2 3 4 5 6 7	Painted Lady
Broccoli (*Brassica oleraceae* var. *italica*)	1 2 3 4 5 6 7	Checkered White
	1 2 3 4 6 7	Cabbage White
	4 5	Great Southern White
Brussels Sprouts (*Brassica oleraceae* var. *gemmifera*)	1 2 3 4 6 7	Cabbage White
	4	Great Southern White
Bur-clover, Toothed (*Medicago polymorpha*)	2 3 4 5	Funereal Duskywing
Cabbage (*Brassica oleraceae* var. *capitata*)	1 2 3 4 6 7	Cabbage White Checkered White
Cabbage, Bastard (*Rapistrum rugosum*)	2 6 7	Checkered White

Herb	Range							Butterfly
Calendula (*Calendula officinalis*)	1	2	3	4	5	6	7	Painted Lady
Cane, Sugar (*Saccharum officinarum*)					5			Eufala Skipper Fiery Skipper Obscure Skipper Proteus Longtail
Canna (*Canna* × *generalis*)		2	3	4	5	6		Brazilian Skipper
Canna, Common (*Canna indica*)		2	3	4	5	6		Brazilian Skipper
Carrot, Garden (*Daucus hybrida*)	1	2	3	4	5	6	7	Black Swallowtail
Cauliflower (*Brassica oleraceae* var. *botrytis*)	1	2	3	4	5	6	7	Checkered White
	1	2	3	4		6	7	Cabbage White
				4	5			Great Southern White
Clover, White (*Trifolium repens*)	1	2	3	4		6		Eastern Tailed-Blue Gray Hairstreak Orange Sulphur Reakirt's Blue Southern Dogface
	1	2	3			6		Clouded Sulphur
		2	3	4		6		Northern Cloudywing
Corn (*Zea mays*)	1	2	3	4	5	6	7	Eufala Skipper
	1	2	3	4	5	6		Least Skipper
	1	2	3			6	7	Clouded Skipper
Cotton (*Gossypium hirsutum*)	1	2	3		5	6		Gray Hairstreak Painted Lady
Crown-vetch, Purple (*Coronilla varia*)				4				Orange Sulphur Wild Indigo Duskywing
Dill (*Anethum graveolens*)	1	2	3	4	5	6	7	Black Swallowtail
Fennel, Common (*Foeniculum vulgare*)	1	2	3	4	5	6	7	Black Swallowtail
Flamingo Plant (*Jacobinia carnea*)	1	2	3	4	5	6	7	Texan Crescent
Flower-of-an-hour (*Hibiscus trionum*)	1	2				6		Common Checkered-Skipper
Goosefoot, White (*Chenopodium album*)	1	2		4	5	6	7	Common Sootywing Painted Lady Western Pygmy-Blue
		2		4	5	6		Mazans Scallopwing
		2	3			6		Hayhurst's Scallopwing

Herb	*Range*	*Butterfly*
Grass, Buffel (*Pennisetum ciliare*)	2 4 5 6 7	Clouded Skipper Eufala Skipper
Grass, Common Bermuda (*Cynodon dactylon*)	1 2 3 4 5 6 7	Eufala Skipper Fiery Skipper Orange Skipper Sachem
	1 2 6 7	Green Skipper Red Satyr
	1 2 3	Common Roadside-Skipper
	2 3 4 5 6 7	Julia's Skipper
	2 3 4 5 6	Gemmed Satyr Southern Skipperling Whirlabout
	2 3 4	Carolina Satyr
	4 5	Obscure Skipper
	7	Umber Skipper
Grass, Dallis (*Paspalum dilatatum*)	2 3 4 5 6 7	Julia's Skipper
Grass, Downy Crab (*Digitaria sanguinalis*)	1 2 3 4 5 6 7	Sachem
	1 2 4 5 6 7	Nysa Roadside-Skipper
	2 3 4 5 6	Southern Broken-Dash
	3	Northern Broken-Dash
Grass, Goose (*Eleusine indica*)	2 3 4 5 6	Sachem
Grass, Johnson (*Sorghum halepense*)	1 2 3 4 5 6 7	Eufala Skipper
	2 3 4 5 6 7	Clouded Skipper Julia's Skipper
	2 3	Bell's Roadside-Skipper
Grass, Large Barnyard (*Echinochloa crus-galli*)	1 2 3 4 5 6 7	Eufala Skipper
	1 2 4 5 6 7	Nysa Roadside-Skipper
Grass, Perennial Rye (*Lolium perenne*)	1 2 6 7	Green Skipper
Grass, St. Augustine (*Stenotaphrum secundatum*)	2 3 4 5 6	Carolina Satyr Celia's Roadside-Skipper Clouded Skipper Eufala Skipper Fiery Skipper Julia's Skipper Sachem Southern Broken-Dash Whirlabout
	2 3 4 6	Little Wood-Satyr
	2 4 6	Nysa Roadside-Skipper
	2 6	Red Satyr

Herb	*Range*	*Butterfly*
Grass, Vasey (*Paspalum urvillei*)	2 3 4 5 6	Whirlabout
Hedge-mustard (*Sisymbrium officinale*)	2 3 6	Cabbage White Falcate Orangetip
	2 6	Olympia Marble
Hollyhock (*Alcea rosea*)	1 2 3 4 5 6 7	Common Checkered-Skipper Painted Lady
	2 4 5 6 7	Common Streaky-Skipper
	3 4 5 6	Tropical Checkered-Skipper
Indigo, Anil (*Indigofera suffruticosa*)	2 3 4 5 6	False Duskywing
Mallow, Common (*Malva neglecta*)	1 2 3 6 7	Common Checkered-Skipper Gray Hairstreak Painted Lady
	7	West Coast Lady
Mallow, High (*Malva sylvestris*)	2 6	Common Checkered-Skipper Painted Lady
Mallow, Small-flowered (*Malva parviflora*)	2 4 5 6 7	Common Checkered-Skipper Gray Hairstreak Painted Lady
Medic, Black (*Medicago lupulina*)	1 2 3 4 5 6 7	Eastern Tailed-Blue Orange Sulphur
Medic, Toothed (*Medicago polymorpha*)	2 3 4 5 6	Orange Sulphur
Milkweed, Tropical (*Asclepias curassavica*)	2 3 4 5 6	Monarch Queen
Nasturtium (*Tropaeolum majus*)	1 2 3 4 6 7	Cabbage White
	4	Great Southern White
Nut-grass, Yellow (*Cyperus esculentus*)	1 2 3 4 5 6 7	Dun Skipper
Oats (*Avena fatua*)	1 2 3 6	Large Wood-Nymph
Pansy, Garden (*Viola × wittrockiana*)	1 2 3 4 5 6 7	Variegated Fritillary
Parsley (*Petroselinum crispum*)	1 2 3 4 5 6 7	Black Swallowtail
Pea, Garden (*Pisum sativum*)	1 2 3 4 5 6 7	Gray Hairstreak Orange Sulphur Painted Lady Sleepy Orange
	1 2 3 6 7	Clouded Sulphur
	2 3 4 5	Proteus Longtail
Pipevine, Fringed (*Aristolochia fimbriata*)	1 2 3 4 5 6 7	Pipevine Swallowtail
	4 5 6	Polydamas Swallowtail

Herb	***Range***							***Butterfly***
Rice, Cultivated (*Oryza sativa*)			3	4				Ocola Skipper
Rue, Fringed (*Ruta chalapensis*)	1	2	3	4	5	6	7	Black Swallowtail Giant Swallowtail
Rue, Garden (*Ruta graveolens*)	1	2	3	4	5	6	7	Black Swallowtail Giant Swallowtail
Ruellia, Mexican (*Ruellia brittoniana*)		2	3	4	5	6		Texan Crescent
			3	4	5	6		Common Buckeye
Senna, Coffee (*Senna occidentalis*)		2	3	4	5			Cloudless Sulphur Little Yellow Sleepy Orange
Senna, Sickle-pod (*Senna obtusifolia*)		2	3	4				Cloudless Sulphur Sleepy Orange
Shrimp Plant (*Justicia brandegeana*)			3	4	5	6		Common Buckeye Texan Crescent
Shrimp Plant, Green (*Blechum pyramidatum*)					5			Banded Peacock Malachite
				4	5			White Peacock
Sour-clover, Yellow (*Melilotus indicus*)		2	3	4	5	6	7	Eastern Tailed-Blue Reakirt's Blue
Soybean (*Glycine max*)	1	2	3		5			Orange Sulphur Painted Lady Southern Dogface
Sunflower, Mexican (*Tithonia rotundifolia*)	1	2		4	5	6	7	Bordered Patch
Sweet-clover, White (*Melilotus albus*)	1	2	3	4	5	6	7	Eastern Tailed-Blue Gray Hairstreak Orange Sulphur Reakirt's Blue
		2	3	4	5	6	7	Clouded Sulphur
	1						7	Acmon Blue
Sweet-clover, Yellow (*Melilotus officinalis*)	1	2	3	4		6		Eastern Tailed-Blue Orange Sulphur Reakirt's Blue
		2	3	4		6		Spring/Summer Azure
Vetch, Crown (*Coronilla varia*)			3	4		6		Orange Sulphur
Wormseed (*Chenopodium ambrosioides*)	1	2		4	5	6		Common Sootywing
		2		4	5	6		Mazans Scallopwing
Wormwood, Common (*Artemisia absinthium*)	1	2	3	4	5	6	7	American Lady

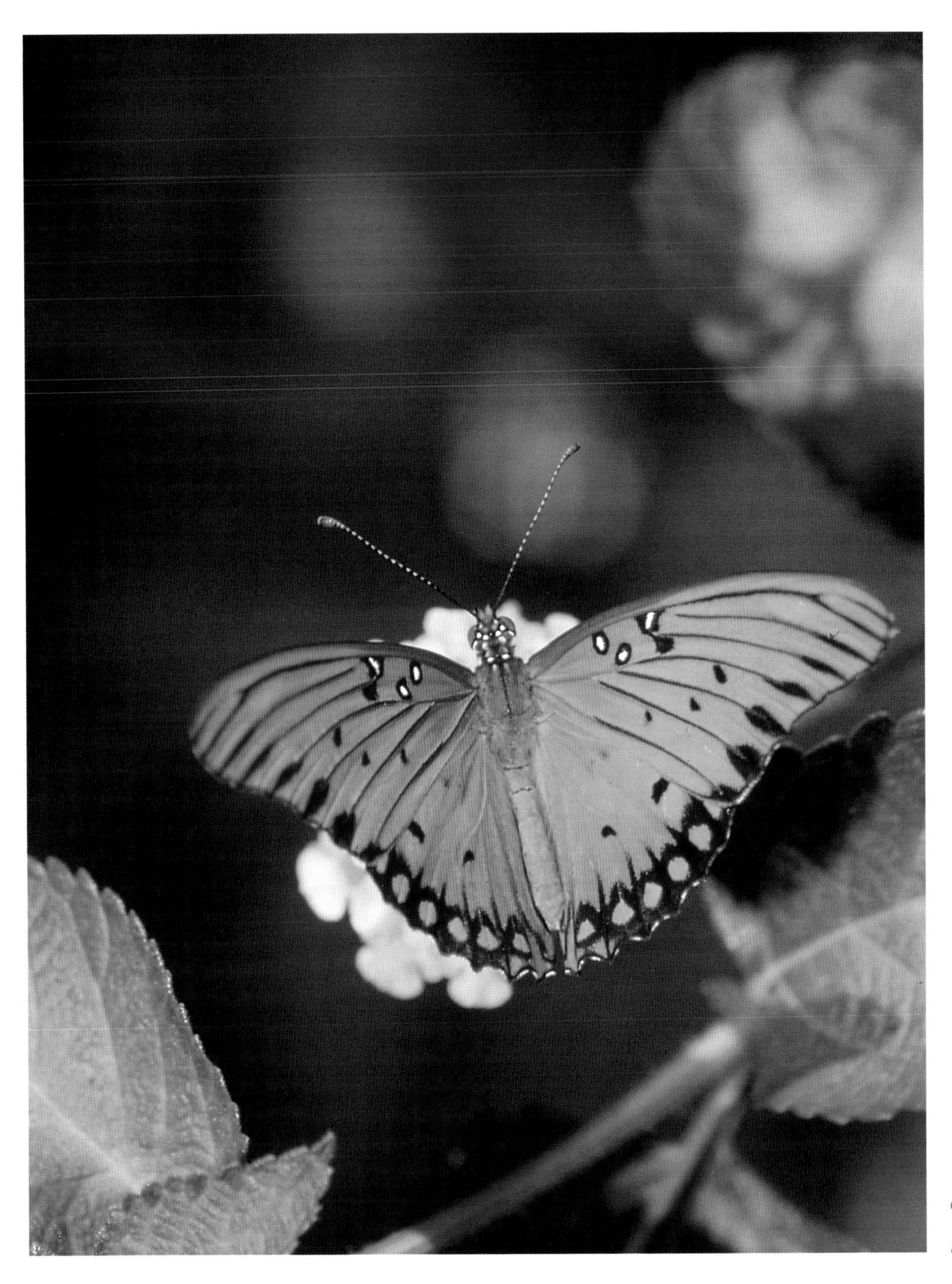

Gulf Fritillary
(Agraulis vanillae incarnata)

9 Nectar Plant Profiles

Nectar and the plants that produce it are the material that forms the very foundation of butterfly attracting. A knowledge of good nectar-producing plants is most important, and a few familiar plants known to be good nectar producers are shown and described here.

See chapter 8 for general explanations of the categories used in each profile. In this chapter, under "Range," numbers in parentheses indicate the region or regions where the plant does not naturally occur but can be grown, although it may be short-lived due to various climatic factors. The category "Related species" is occasionally added when there are other species that may be of interest to the gardener.

In the list of nectar plants following the descriptions, notice that some of the most common plants, such as the Showy Evening Primrose (*Oenothera speciosa*), Mexican-hat (*Ratibida columnifera*), and Black-foot Daisy (*Melampodium leucanthum*), along with such cultivars as double zinnias (*Zinnia* spp.) and ruffled petunias (*Petunia* spp.), are not included. These plants, although they seem to have the characteristics a butterfly wants, evidently lack the thing most important—plentiful nectar. Remember that butterflies have to use what is there. So look around before gathering seeds or purchasing plants—even though butterflies may be nectaring on a plant (or trying to), it may not be a preferred nectar source. However, if several other plants are also in flower and butterflies are still coming repeatedly to a particular species or if several species of butterflies are coming to the plant, the plant is most likely a good nectar producer and preferred choice.

Red Admiral *(Vanessa atalanta rubria)* on Purple Coneflower *(Echinacea purpurea)*

Anacahuita (Mexican Olive)

(Cordia boissieri)

Family: Borage (Boraginaceae)
Class: Native
Height: To 24 feet
Bloom period: Spring–frost (all year)
Range: 5 (6)

Spreading, round-topped tree often obtaining only large shrub size. Leaves large, thick, fuzzy, alternate along stout branches. Flowers showy, yellow-throated white, numerous, in large, loose clusters at tips of branches.

CULTIVATION: Propagation by seeds of this plant is usually uncertain and often difficult. Planting seeds may be tried by collecting after they are fully mature, usually from July through September. If some can be found on the ground with the pulpy flesh naturally disintegrated, so much the better. Collect these to try; such seeds sprout much more readily than freshly cleaned seeds. If seeds are fresh, remove the fleshy pulp, air-dry, and then store in a cool, dry area. However seeds are cleaned, sprouting will vary from year to year and from site to site from where collected. However, once germinated, the seedlings are fast growing, often reaching eight to ten feet in two years.

Propagating plants from soft or semihardwood cuttings taken in summer, dipped in rooting hormone powder, and placed in a moist rooting medium is usually successful. Anacahuita is offered by most nurseries in its growing range, which may be the best and most assured method of obtaining this plant.

Not too demanding about permanent planting soils, this tree survives in sand, loam, clay, or caliche and in the wild is found along stream banks, on gravelly slopes, or in sandy thickets. What it does demand is good drainage. Occasionally water young plants the first year or two, and then under ordinary circumstances, they can make it on their own.

NOTE: The white to yellow-green pulp within the fruit is edible and sweet but if consumed in excessive amounts causes dizziness and intoxication.

ANACAHUITA (MEXICAN OLIVE)

Chaste Tree

(Vitex agnus-castus)

Family: Vervain (Verbenaceae)
Class: Nonnative
Height: 10–20 feet
Bloom period: June–September
Range: 2, 3, 4, 5, 6, 7

Deciduous, long-lived, large shrub or small tree usually with more than one trunk from the base and appearing broad, spreading, or loose in general outline. Leaves divided fan-like into three to seven narrow leaflets, dark green on upper surface, a cool gray beneath, both twigs and leaves strongly but pleasantly aromatic when crushed. Flowers numerous, small, lilac to lavender, fragrant, forming slender spikes terminally and from upper leaf axils.

CULTIVATION: Although not native, Chaste Tree is widely planted and in many regions has escaped and may be found in various "wild" situations. It may even reseed under such situations but never becomes a problem.

Chaste Tree tolerates a wide variety of soils but requires plenty of summer heat for the most richly colored and profuse bloom, so plant this one in an open spot in full sun. Give it a well-drained site with rather poor soil, and do not overwater. Fertilizing sparingly in the spring with a high-phosphorus fertilizer increases bloom production. Extra-rich soil and too much moisture result in lush foliage but only a few pale-colored flowers low in nectar. New plants may be started from tip cuttings or easily obtained from nurseries or catalogs.

Chaste Tree is late to put out foliage in the spring and in some areas may be killed to ground level during extremely cold winters. Severe pruning only enhances its shape; if cut to within six inches of the ground in late winter or early spring, it quickly resprouts and forms new blossoming stems. New stems easily attain three to five feet the first year before flowering. In this respect Chaste Tree is quite similar to Butterfly Bush (*Buddleja davidii*) and can be treated more as an herbaceous perennial. In this form it makes a beautiful hedge or "living fence." It also works well in borders when kept small, and the lower branches offer closer butterfly viewing. If not pruned and in areas where it does not freeze back each year, it may obtain small tree size. The lower branches can be trimmed into a tree form to make a lovely specimen planting. When trimmed this way, Chaste Tree is especially nice used for shade in a small yard.

NOTE: A large, healthy potted plant from a nursery should flower the first or second year. In making a selection, start with only one or two plants to be sure they are good nectar producers. Blue- or purple-flowered plants will probably attract more butterflies than white- or pink-flowered ones.

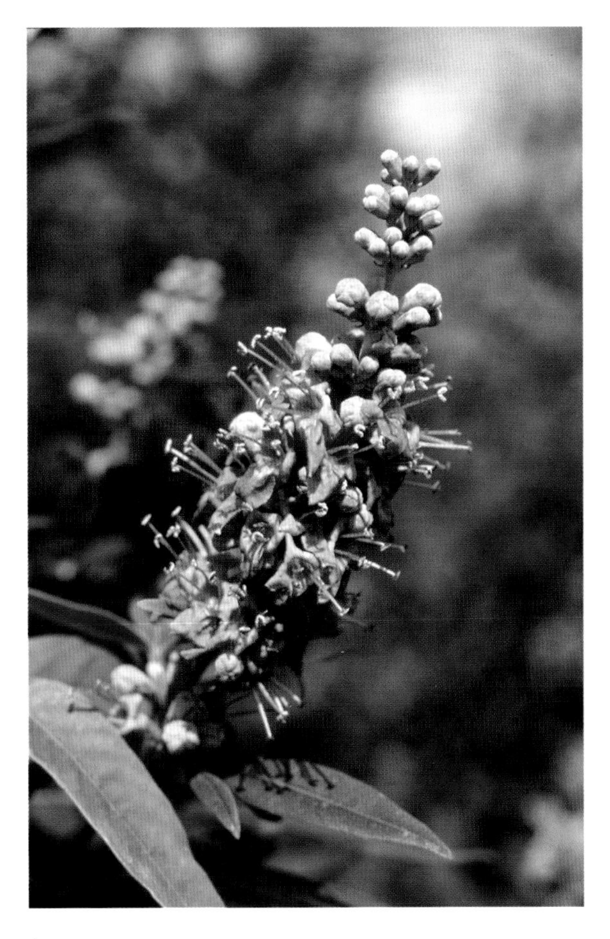

Chaste Tree

Desert Willow
(Chilopsis linearis)

Family: Trumpet-creeper (Bignoniaceae)
Class: Native
Height: 6–30 feet
Bloom period: April–June
Range: 1, 2, 5, 7

Native shrub or small deciduous tree, usually with several trunks from the base. Leaves long, slender, and willowlike. Flowers large, orchidlike, sweetly fragrant, loosely clustered in large, terminal panicles; do not last long but appear after each rain. Flowers replaced by long, slender seedpods, which remain dangling from branches and aid in identifying this tree long after flowers are gone.
CULTIVATION: Desert Willow can be grown from seeds, cuttings, or nursery transplants. If starting from seeds, gather the seedpods as soon as they are dry and brown-colored. Remove seeds from pods and, when completely dry, store in the refrigerator until spring.

DESERT WILLOW

Before planting, soak the seeds in water for a few hours. Semihardwood cuttings can be taken in late summer and should root in two or three weeks. Transplant the rooted cuttings to the garden in late fall, or carry over until spring.

Whether from seeds or cuttings, plants grow rapidly and will produce flowers even when very young. This is another shrub that tolerates drought conditions but responds to an occasional watering and a light application of fertilizer. Planted in full sun, Desert Willow tolerates various soils but prefers limestone soils. It demands good drainage. In the western and northern portions of its range, it may occasionally be broken or damaged by snow or wind, so in these areas it is best to use it in protected places, such as the south side of buildings, fences, or taller-growing trees or shrubs.

With its open, somewhat sprawling growth habit, Desert Willow is at its best when used at the end of a border or as a specimen planting. In areas where it is not bothered by freezing, trim the lower branches to eventually form an airy, graceful tree shape. If a more shrubby shape is desired, prune back the plant severely; it actually produces more flowers when so treated.

Flowering Mimosa
(Albizia julibrissin)

Family: Bean (Fabaceae)
Class: Nonnative
Height: 15–25 feet
Bloom period: April–August
Range: 1, 2, 3, 4, 5, 6

Hardy to semihardy, tropical-appearing deciduous tree with drooping, slender branches. Leaves large, twice divided into numerous tiny leaflets and appearing fernlike, folding together at night. Flowers delicate, fragrant, fluffy pink, in large clusters at tips of branches. Leaves and flower clusters occur on upper sides of branches, giving this tree an exotic look.

CULTIVATION: This native of Asia is not particularly choosy about soils but does like adequate moisture. An addition of iron to regular feeding produces more and darker-colored flowers. It is at its very best in areas of high summer heat and humidity, so it usually blooms more profusely in its more southern range. It is one of the few cultivated trees that can tolerate both drought and air pollution quite well.

Flowering Mimosa is easily started from seeds and grows very fast, often flowering the second year. It also reseeds easily and can often be seen growing as escapes along roadsides or around trash dumps where trimmed limbs and branches have been discarded. Young saplings may be dug either in the spring or fall for transplanting in the home garden, or potted plants are usually available at local nurseries.

Branches of Flowering Mimosa have a tendency to droop somewhat, so trim off lower limbs at the trunk when planted near a traffic area. The open branching and lacy leaves of this tree allow filtered sunlight through and do not kill lawn grasses as most shade trees do. If the branches are trimmed high, this can be used as a specimen tree in a bed with low shrubs and perennials planted beneath. Unfortunately, Flowering Mimosa is relatively short-lived and the wood is brittle; sometimes large limbs are lost to storms or high winds. It is also often attacked by a borer that kills the tree if not treated.

Limbs of Flowering Mimosa are sometimes cut about halfway back, but this totally destroys the natural beauty of this tree and also makes it more susceptible to attack by insects and fungus diseases. If Flowering Mimosa is never trimmed, growth of the branches naturally slows down, and the tree forms an absolutely beautiful sight with the branches arching almost to the ground and smothered in feathery pink blossoms. The unique, rather flat-topped or gently mounded shape of its branches makes a perfect canopy for a patio. The undulating form and almost solid mat of flowers held above the foliage make this tree especially nice when viewed from a second-story window, balcony, or deck.

NOTE: When Flowering Mimosa starts opening its fragrant pink flowers in early spring, butterflies flock to its readily available nectar and continue to work the tree throughout the day. The nectar is so plentiful at times that even hummingbirds (family Trochilidae) and

FLOWERING MIMOSA

orchard orioles (*Icterus spurius*) can be seen dipping their beaks into the flowers. In late afternoon the fragrance becomes stronger, making this a desirable tree for placing near a porch or patio.

Due to its ability to readily reseed, in some areas this tree is considered an invasive species. But in the home garden, unwanted plants are easily removed when small by simply pulling or clipping at ground level.

Mexican Orchid Tree
(Bauhinia mexicana)

Family: Bean (Fabaceae)
Class: Nonnative
Height: To 18 feet
Bloom period: June–frost (all year)
Range: 3, 4, 5, 6

Widely spreading, usually multitrunked small tree rounded in top portion, prolifically covered with clusters of delicate, spidery flowers. Leaves dull green, deciduous or semievergreen, alternate, oddly shaped, doubly pointed at tip, appearing as if two leaves have grown together. Flowers numerous, white-colored, five-petaled, forming clusters at tips of slender branches, with the stamens and pistil extending far beyond the petals, giving the flowers an airy, exotic appearance.

Mexican Orchid Tree

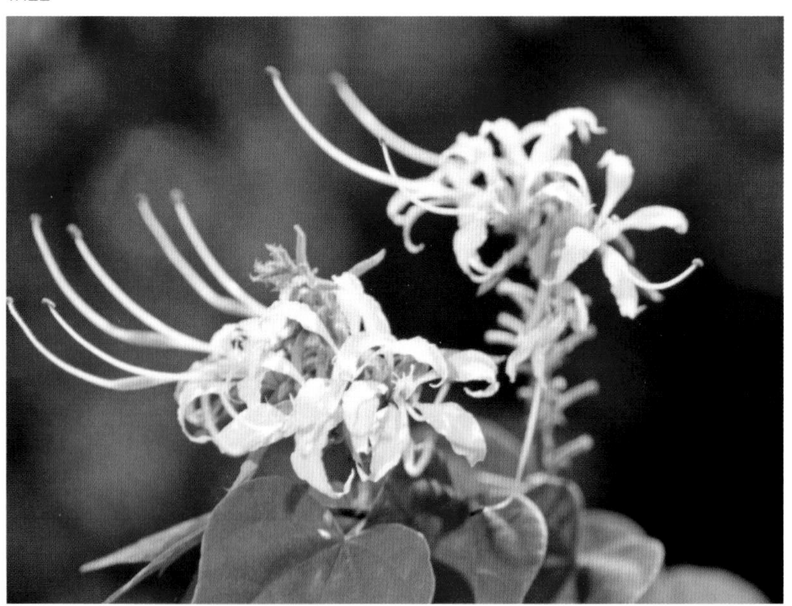

CULTIVATION: Although this is classified as a tree and does reach tree height in the southernmost portion of its range, it is easily raised from seeds, which is the easiest method of obtaining plants. The one to four smooth, flattened, brown seeds are contained in a flattened pod. When the pod is completely dry, it twists open, quickly dispersing the seeds. To obtain these seeds, before pods are totally dry, cover them with netting or cheesecloth, allowing the pod to dry completely on the plant. When the pod has opened, remove seeds, cleaning all pod debris from seeds. Place seeds in a wet paper towel, place the towel in a plastic bag, but do not seal all the way. Keep the towel moist until seeds sprout. Plant immediately upon sprouting. Or dried seeds can be stored in a sealed container for later planting. If seeds are stored, acid scarification may be necessary to break dormancy.

Mexican Orchid Tree can withstand temperatures as low as ten degrees but likes all the protection it can get. Plant it on the south side of a building or in the most protected place possible by other trees or shrubs, preferably evergreen. Plant in full sun, morning sun, or dappled shade. It demands good drainage. Small, occasional applications of organic fertilizer the first year or two are beneficial; then only occasional waterings during extreme dryness will be necessary. Mulch lightly during winter. Usually the tree will grow four or five feet and bloom the first or second year.

If plants are frozen back to the ground, they will almost always resprout; and if the roots are well established and healthy, the trees will bloom the first year.

If desired, Mexican Orchid Tree can be pruned to retain a more shrublike habit or trimmed to a single-trunked tree.

Mexican Orchid Tree is now being used in landscaping in the more eastern regions of Houston and Beaumont. It does well as far north as Austin. Use it with Drummond's Wax-mallow (*Malvaviscus drummondii*), Mexican Oregano (*Poliomintha longiflora*), Shrimp Plant (*Justicia brandegeana*), Cigar Plant (*Cuphea micropetala*), Mexican Sunflower (*Tithonia rotundifolia*), Summer Phlox (*Phlox paniculata*), the vines Queen's Wreath (*Antigon leptopus*) and Mexican Flame Vine (*Pseudogynoxys chenopodioides*), and various lantanas (*Lantana* spp.).

NOTE: In the Rio Grande Valley area, the Mexican Orchid Tree is the larval food plant of the Cassius Blue (*Leptotes cassius cassidula*), Gold-spotted Aguna (*Aguna asander*), and Tailed Aguna (*A. metophis*).

Mexican Plum

(Prunus mexicana)

Family: Rose (Rosaceae)
Class: Native
Height: To 25 feet
Bloom period: March
Range: 2, 3, 4, 6

Large, beautifully shaped, single-trunked, deciduous tree with spreading branches. Bark distinctively banded, furrowed, and peeling. Leaves thick, prominently veined, becoming yellow in fall. Flowers, white, sweetly fragrant, covering tree in clusters before leaves unfold in spring. Fruit a delicious plum, covered with powdery "bloom," changing from greenish-yellow to mauve to purplish as it ripens in fall.

CULTIVATION: Nice-sized plants of Mexican Plum can be purchased from native plant dealers or, if time to blooming is not a factor, may be started from seeds or cuttings. Collect the plums when ripe, and clean all pulp from the seeds by cutting or scraping, then by washing. Fill a bucket with one gallon of warm water, mix in two packets of yeast and two tablespoons of sugar, add seeds, and let soak three or four weeks. Rinse seeds, and plant where tree is to remain. Mark well, as germination will not take place until spring.

Six- to eight-inch semihardwood tip cuttings can be taken in the summer. Strip off lower leaves, dip in rooting medium, place in a sand/perlite mixture, and keep under occasional mist until well rooted.

In a planting, Mexican Plum is drought

MEXICAN PLUM

resistant but will show faster growth with frequent waterings. It can be found in almost solid limestone rock to rich bottomland soils but does require good drainage. This is a tough tree and can tolerate full sun but does its best in dappled shade, making an excellent understory tree.

NOTE: Mexican Plum is one of the loveliest of spring-flowering trees and is especially outstanding when growing close to Texas Redbud (*Cercis canadensis* var. *texensis*) and Mexican Buckeye (*Ungnadia speciosa*) in the west or Eastern Redbud (*C. c.* var. *canadensis*) and the vines Carolina Jasmine (*Gelsimium sempervirens*) and wisteria (*Wisteria* spp.) in the eastern portion of the state. The nectar-rich flowers attract various insects, especially bees and numerous species of butterflies. It is occasionally used as a larval food source for the Eastern Tiger Swallowtail (*Papilio glaucus*) and the Viceroy (*Limenitis archippus*). The attractive fruit is edible and can be made into delicious jams and jellies.

Shrubs

Barbados Bird-of-Paradise

(Caesalpinia pulcherrima)

Family: Bean (Fabaceae)
Class: Nonnative
Height: To 15 feet
Bloom period: May–October (all year)
Range: Throughout

Most impressive, fast-growing, dense, deciduous shrub. Leaves ferny, dark green, twice divided into numerous leaflets. Flowers interestingly shaped, numerous, in large terminal clusters; long, red stamens protruding conspicuously, curving upward from base of petals. Foliage and flowers rather unpleasantly scented. Fruit an elongated, flattened pod containing numerous seeds.

CULTIVATION: An almost perfect shrub for the garden, Barbados Bird-of-Paradise should be one of the first considered for butterfly plantings. A native of the West Indies, it has long been in cultivation, and the new forms are more frequently being used in landscaping. It is a very common plant in South Texas and Mexico.

Easy to grow and hardy, it is now being offered by most nurseries that sell native plants. It is also easy to propagate from seeds. Seeds are usually produced in abundance, with pods maturing in the lower portion of the raceme

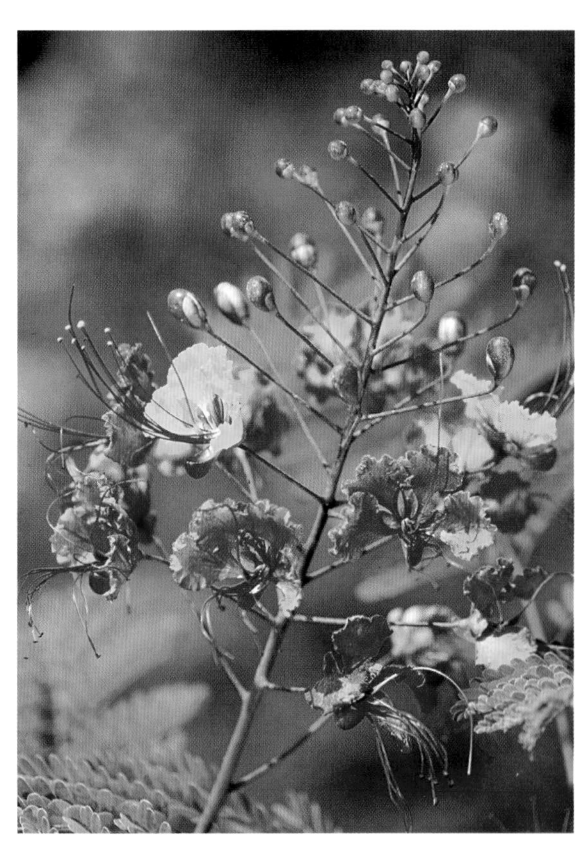

Barbados Bird-of-Paradise

as flowers continue to open at the tip. Fully mature seeds can be planted either in fall or spring. With a sharp knife, nick the seeds, then soak several hours or overnight in warm water. If seeds are planted indoors in early spring and seedlings transplanted to the garden after the last frost, it would not be unusual to have flowers by fall. Place young plants in full sun in a well-drained location, keeping the soil moist until roots are well established. Add a small amount of fertilizer containing phosphorus at planting time, then sparingly a couple of times during the growing season for added growth and bloom. Many of the seeds that fall to the ground will germinate, and there will usually be a number of young seedlings around the mother plant. By the second year plants should produce a wealth of exotic blooms, with flowering increasing each year.

In the more northern portion of its range, reduce watering toward the end of the flowering season. After the first frost, trim stems back to three or four inches from the ground and cover with loose mulch of dried leaves, straw, or hay. In spring, do not water until the leaf buds begin to swell; then deeply soak plants every two weeks to once a month, depending on natural rainfall.

Barbados Bird-of-Paradise is very hardy and not generally bothered by insects or diseases. Its worst enemy is too much moisture or shade, which can cause it to become weak and spindly and with few or no flowers. Under such adverse conditions, it may also become infested with red spider mites. The best solution is to move the plant to an open, sunny, well-drained spot in the garden.

NOTE: This plant readily reseeds, and seedlings may require frequent removal. For stability, this plant develops large roots, so do not plant near a patio, walkways, or foundations of buildings.

RELATED SPECIES: The species *C. pulcherrima* has yellow flowers blotched in red. There are now several forms to be found in the trade; the red-flowered form commonly known as Red Bird-of-Paradise is shown here. Mexican Bird-of-Paradise (*C. mexicana*) is a shrub with strongly fragrant yellow and red flowers cultivated in the Lower Rio Grande Valley. Desert Bird-of-Paradise (*C. gilliesii*) is cultivated and occasionally escapes to be found in the wild. The latter two start flowering when young shrubs, and both can easily be raised from seeds.

Beebrush

(Aloysia gratissima)

Family: Verbena (Verbenaceae)
Class: Native
Height: To 9 feet
Bloom period: March–December
Range: 2, 5, 6, 7

Slender, much-branched, usually deciduous shrub, the branches gray, almost square and noticeably brittle. Leaves small, opposite, wonderfully aromatic when touched, often with a bundle of smaller ones in the axils. Flowers fragrant, widely spaced, yellow-throated white, borne from axils; racemes several and forming bouquetlike panicles.

CULTIVATION: This plant is easily started from either seeds or cuttings. Within a few years Beebrush usually forms a nice colony or a small thicket if left unattended. New, unwanted plants can easily be uprooted. In the wild state the shrubs may become a little

Beebrush

unruly, but this should present no problem in the garden. If they begin to get out of bounds, simply cut them to the ground in late winter, and in the spring, again cut the resprouted growth back to what is desired. The remaining plants will respond by putting out a multitude of new stems and growing to a nice shrub size the first year. Trimming periodically produces best flowering.

Plant in full sun and poor soil, keep it on the dry side, and watch the profusion of flowers that appear after each rain. This is a shrub to use in the driest and poorest soil of the garden; if the soil is gravelly or rocky, the plants will be even happier.

NOTE: Flowers of Beebrush have a strong vanilla-like fragrance and are a great attraction to many species of butterflies. Bees also find it highly desirable, and in its natural range it is considered one of the best nectar plants for the production of honey.

Butterfly Bush
(Buddleja davidii)

Family: Vervain (Verbenaceae)
Class: Cultivated
Height: To 12 feet
Bloom period: May–frost
Range: Throughout

Deciduous or semievergreen, widely spreading shrub with slender, arching branches. Leaves thick, felty, dull green on upper surface, dull silvery beneath. Flowers numerous, small, fragrant, densely compacted into large, thick, terminal, often drooping racemes.

CULTIVATION: A native of China, Butterfly Bush is an exceptionally hardy, easily cultivated shrub blooming the first year after planting. It requires little care after becoming established. Plant in soil liberally enriched with peat moss, leaf mold, or compost. The plants do not like to be crowded, so place them at least six to ten feet apart to allow plenty of room to spread. Plants will remain healthy and blooming with a light fertilizing (5-10-5) in early spring to encourage blooms. Adequate moisture during the hot, dry summer months is necessary for continued nectar production.

Butterfly Bush blooms best on new wood and should be cut back to three or four buds from the base of the old wood (approximately ten inches above the ground) in early February or at least before spring growth begins. It will not put forth new growth until late spring, but once started, it grows rapidly. Pruning keeps the plant at a lower height and spread of around six feet. However, unpruned shrubs begin flowering earlier in the spring, so pruning of different plants should be rotated each year. To keep plants flowering all summer, keep all spent racemes clipped.

This plant can be grown in almost any soil but needs good drainage and full sun. A regular addition of iron may be necessary to prevent iron chlorosis if the available water is alkaline. In hotter, drier regions, Butterfly Bush may occasionally be bothered by red spider mites. Treat with one of the organic methods discussed elsewhere in this book.

New plants are easy to obtain from semi-hardwood cuttings taken anytime from late summer to late fall using normal rooting methods. Root the cuttings directly in the garden in a sheltered area or in a cold frame.

NOTE: There is probably no better cultivated shrub for attracting butterflies to the home garden than Butterfly Bush. They are usually prolific bloomers, producing masses of flowers throughout the season. Butterflies will not leave them. There will be from one to several species feeding almost constantly. The racemes are sturdy enough to hold the weight of larger butterflies such as Fritillaries (family Nymphalidae) and Swallowtails (family Papilionidae), which often perch, nectaring at length. At times there will be so many butterflies feeding around these plants that the butterflies actually jostle one another for the best nectar spots. The wonderful honeylike fragrance is very strong, especially in late evening and early night, so Butterfly Bush is especially nice planted near a window, porch, or patio.

There are many cultivars of *B. davidii*, with flower colors ranging from white to yellow through pinks, blues, and purples. Butterflies generally seem to like the lavender or purple shades the best. To be sure of purchasing a good nectar producer, spend some time at the nursery watching to see which ones attract the butterflies.

BUTTERFLYBUSH

Buttonbush

(Cephalanthus occidentalis)

Family: Madder (Rubiaceae)
Class: Native
Height: 3–15 feet
Bloom period: June–September
Range: Throughout

Medium to large, rather coarse deciduous shrub, often with few to several trunks from the base. Leaves long, shiny, dark green, solitary, opposite, or in whorls of three or four along slender branches. Flowers numerous, fragrant, creamy-white to pinkish with conspicuously protruding stamens forming perfect spheres.

CULTIVATION: In the wild, Buttonbush is always found in areas of plentiful moisture or even actually growing in shallow water, but it

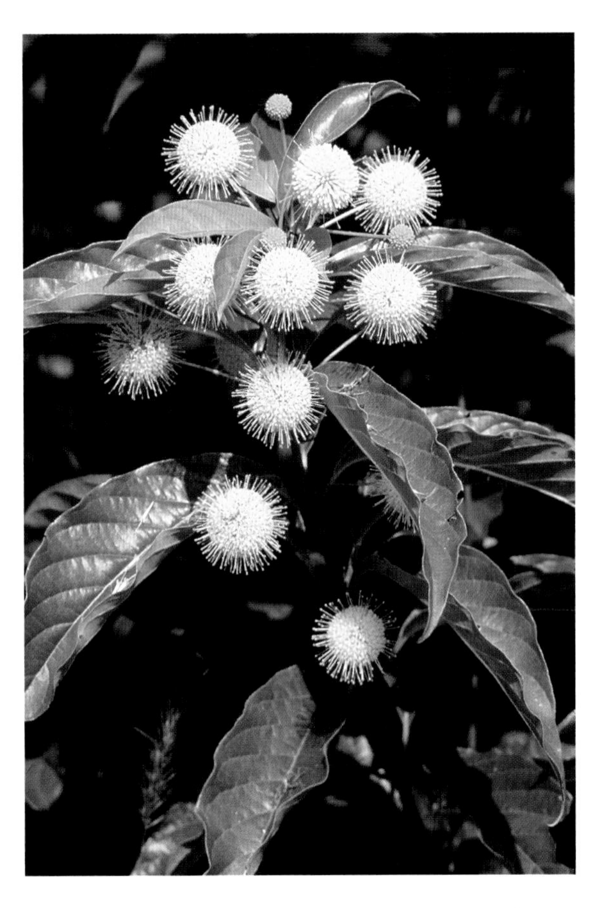

Buttonbush

tolerates much drier conditions under cultivation. For best flower production, it still needs more water than other shrubs in the garden, so plant it near a dripping water faucet or where extra water can be provided with no problem. It is a hardy shrub and grows in almost any soil. Buttonbush can be planted in either sun or semishade but produces more flowers with richer nectar in the sun. It is ideal when used in front of taller shrubs or trees to form an "edge" effect.

Plants may be obtained from seeds, or cuttings or may occasionally be offered by nurseries. Extract ripe seeds from the dried balls in late fall, and sow immediately in a moist, sandy site. Young seedlings can later be transferred to permanent locations in the garden. Unless a very large space is available, only one or two plants are needed because these shrubs become quite wide and require a lot of growing space. Also, they eventually root-sprout, providing even more plants. Under normal growing conditions, Buttonbush remains a large shrub six to eight feet high and wide. As it gets older, the trunks become dark, gnarled, and very picturesque.

If a small tree form is desired, select one, two, or three of the strongest, straightest trunks; trim all the other trunks to ground level; and then trim the lower branches to the height wanted. This will never become a real tree, so do not trim the branches too severely or so high the feeding butterflies cannot be enjoyed.

NOTE: When this plant is in flower, the fragrant balls are completely covered with butterflies and other insects. Butterflies become so engrossed in feeding that they can sometimes be picked up by their wings. Buttonbush seems to be a favorite with all species, but especially Hairstreaks (family Lycaenidae), Skippers (family Hesperdiidae), and Swallowtails (family Papilionidae).

Crucita

(Chromolaena odorata)

Family: Aster (Asteraceae)
Class: Native
Height: To 6 feet
Bloom period: September–frost (all year)
Range: 4, 5

Robust, sprawling to somewhat viny or scandent shrub, woody in basal portion, many-branched, and forming large mounds. Leaves opposite, rather triangular, usually toothed along edges. Flowers lilac to pale purplish-blue, very small, in terminal clusters.

CULTIVATION: In the wild, this plant is usually found in more rocky or clayey soils around edges of shrub thickets, oak mottes, open woodlands, and palm groves. In the home garden almost any soil will do as long as it is not too rich and has good drainage.

Crucita is offered by almost all nurseries carrying native plants but can easily be started either from seeds or cuttings. In late fall after the flowers have become a dark tannish, fluffy mass, clip the entire head. Seeds can be extracted, or plant the entire mass in flats containing loose, well-drained soil. Barely cover with a layer of fine sand, and keep the soil moist. New growth should be showing in two to three weeks.

Growing from cuttings is the fastest and easiest method for obtaining new plants. Take three- to six-inch softwood or semihardwood cuttings in summer or fall. Remove leaves from the lower nodes, dip in rooting powder, and place the cuttings in potting soil under mist. Once a root system is established, place summer-rooted plants directly in the garden. Fall-rooted plants may need to be carried over with protection until spring.

Old clumps of Crucita can easily be divided by trimming back to four to six inches, then digging, separating into rooted sections, and replanting. Keep them watered the first few days until established. Crucita will grow in semishade, but for most flowering and richest nectar production, place in full sun with a lot of room to spread.

Once plants are well established in the garden, an early spring trimming will make foliage thicker and lusher, but generally leave them alone. An occasional watering in the driest months may be needed, but fertilizing is not recommended and the plants are usually insect- and disease-free.

NOTE: Crucita is one of the larval food plants for the Rounded Metalmark (*Calephelis perditalis perditalis*). Blooming over a long period and rich in nectar, this plant is an absolute magnet, drawing in numerous species such as Malachite (*Siproeta stelenes biplagiata*), White Peacock (*Anartia jatrophae luteipicta*), Sickle-winged Skipper (*Eantis tamenund*), Soldier (*Danaus eresimus*), various Longtails (family Hesperiidae), and Metalmarks (family Riodinidae).

RELATED SPECIES: The lower-growing Betony-leaf Mistflower (*Conoclinium betonici-*

CRUCITA

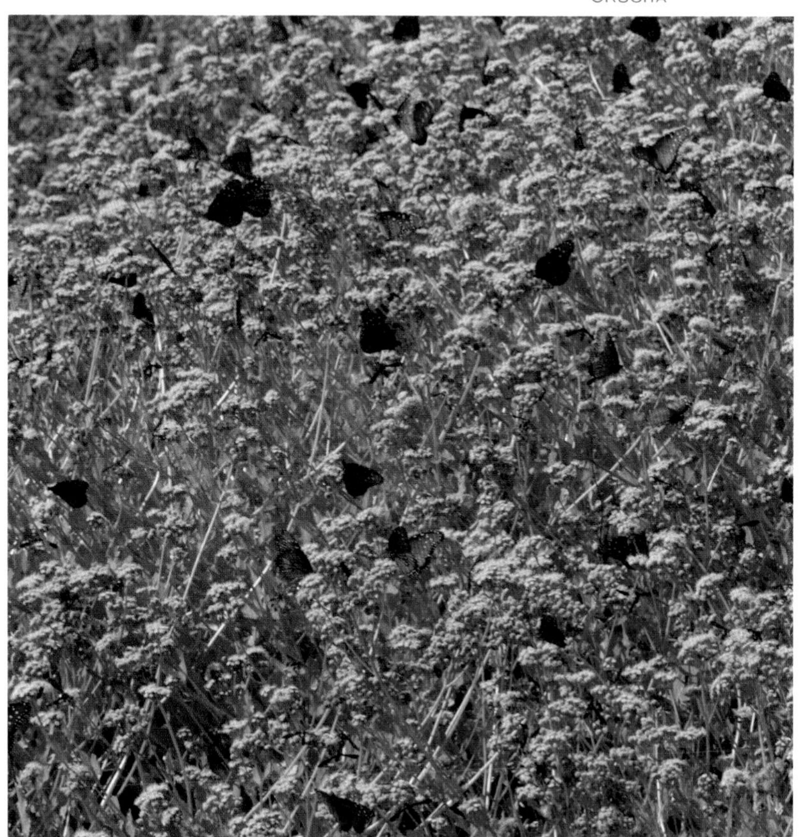

folium), with a natural range along the coast, does exceptionally well throughout the Valley area. It, too, is used as a larval food source by the Rounded Metalmark and is also heavily used as a nectar source by many species of butterflies.

Drummond's Wax-mallow (Turk's Cap)
(Malvaviscus drummondii)

Family: Mallow (Malvaceae)
Class: Native
Height: 2–10 feet
Bloom period: March–frost (all year)
Range: 2, 3, 4, 5, 6 (1, 7)

Upright to sprawling, many-branched, tropical-appearing herbaceous shrub, woody or semiwoody at base. Leaves large, dark green, stalked, and three- to five-lobed, soft and velvety on lower surface. Flowers small, numerous, scarlet-colored, formed from five erect, whorled petals; the stamen conspicuously protruding an inch or more beyond petals. Fruit small, berrylike, tubular-shaped, red when ripe, edible, tasting similar to an apple, although more seeds than pulp.

CULTIVATION: Buying from a native plant nursery is the easiest and quickest method of getting this plant started in the garden. After it has become well established and growing, it should produce an abundance of small fruits. When fruits begin to dry and shrivel, collect the fruit and spread on wire or paper until crispy dry. When they are papery, rub to remove any remaining matter and loosen seeds. Plant seeds in spring after soil is thoroughly warm.

Well-established plants can be lifted in the spring and divided. Four- to six-inch tip cuttings can be taken in late summer to fall, dipped in rooting powder, and the potted cutting placed under a plastic bag until rooted. Preferring native to poor soils, once established, this native requires nothing more than occasional trimming and watering during the driest of seasons.

Severe pruning or trimming of seedlings, rooted plants, and well-established garden plants should be done in winter to encourage bushier growth and better flowering. When trimmed in this manner, it is an excellent plant when used at the edge of or beneath trees or large shrubs. It can be used alone in large colonies or combined with shrubby perennials such as Agarita (*Mahonia trifoliolata*), Tropical Sage (*Salvia coccinea*), Broadleaf Wood-oats (*Chasmanthium latifolium*),

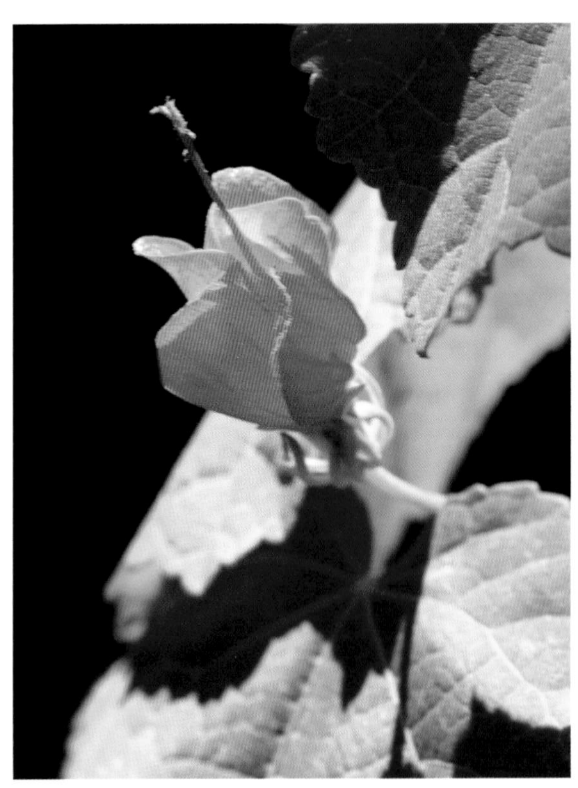

DRUMMOND'S WAX-MALLOW (TURK'S CAP)

various lantanas (*Lantana* spp.), and the nonnative Mexican Oregano (*Poliomintha longifolia*).

Under ideal growing conditions in the southern portion of its range, it reaches its maximum height, but for most of its growing range, without trimming, it will sprawl and never reach much higher than three or four feet. Depending on the region, it will remain semievergreen or deciduous, coming back from the roots each year.

NOTE: This plant is also a larval food source for the Turk's-cap White-Skipper (*Heliopetes macaira*), Mallow Scrub-Hairstreak (*Strymon istapa istapa*), and Glassy-winged Skipper (*Xenophanes tryxus*).

False Indigo

(Amorpha fruticosa)

Family: Bean (Fabaceae)
Class: Native
Height: 2–15 feet
Bloom period: April–June
Range: 1, 2, 3, 4, 6 (5, 7)

Upright, slender-stemmed, deciduous shrub or small tree, often with several stems and forming a clump. Leaves interesting, ferny, up to nine inches long, formed from many small, opposite leaflets. Flowers numerous, small, dark purple, with protruding, bright orange anthers; densely compacted into long, gracefully tapering racemes. Fruit a short, somewhat flattened, curved pod containing one or two seeds.

CULTIVATION: False Indigo may be easily propagated by either seeds or cuttings. Obtain the dark brown, shiny seeds by gathering pods in late summer or early fall or as soon as they turn yellowish-brown. Remove seeds, and place them on an open tray or in an open paper bag for a few days. Store in the refrigerator until ready to plant.

Plant the seeds in either fall or early spring. For spring planting, first soak seeds in hot water for ten or fifteen minutes. Sow where plants are to remain, covering the seeds to a depth of one-half inch. After seedlings are up and growing well, transplant if necessary, but do so as soon as possible, for this plant quickly develops a very long taproot. Seeds germinate better if sown after the ground has become thoroughly warm.

Take softwood cuttings anytime from late spring to early fall. Make the cut just below a node, and root in sand. Plants from cuttings grow rapidly, often attaining a height of six to eight feet the first year from fall-rooted cuttings.

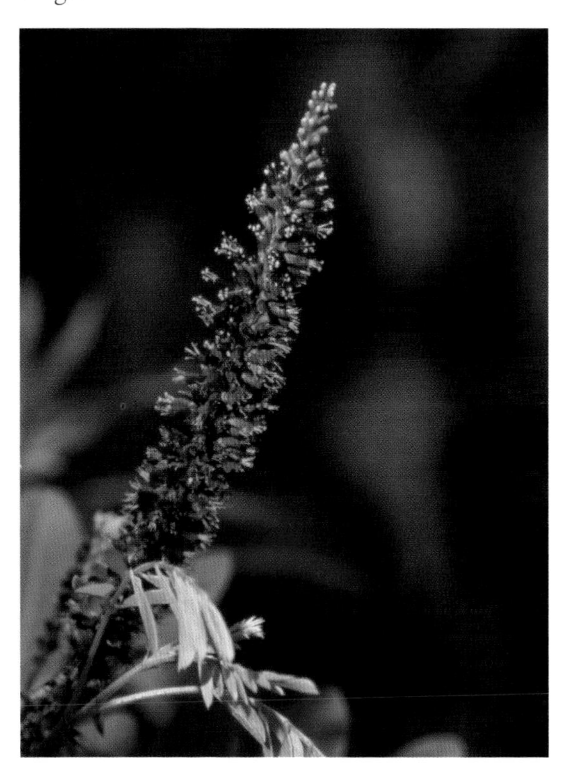

FALSE INDIGO

Mature stands of False Indigo often have numerous stems, which can usually be separated for transplanting in spring, fall, or winter. After planting, trim the stems back two-thirds and water in with a root stimulator. Keep moist but not soggy until new growth appears, and then give them a mild solution of soluble fertilizer. Once the plants start growing well, thin the clumps by trimming, leaving five or six stems for a more shapely grouping.

Found naturally in very mesic habitats, such as along stream banks, at the edges of wet meadows or marshes, or along roadside ditches, False Indigo does quite well in a garden if given just a bit of extra moisture. It thrives better and requires less care and attention if used with other moisture-demanding plants in a special area or bed of its own. It is not too choosy about soil, but good garden loam produces more and better growth and flowering. False Indigo also produces more and darker-colored flowers in full sun, so use this plant in open, exposed areas, such as at the back of perennial borders, along moist woodland edges, or as a specimen planting. False Indigo should be placed somewhat away from perennials, as it is a rather aggressive surface feeder.

NOTE: False Indigo is also the larval food plant for the Southern Dogface (*Zerene cesonia*), Hoary Edge (*Achalarus lyciades*), Silver-spotted Skipper (*Epargyreus clarus*), Gray Hairstreak (*Strymon melinus franki*), and Eastern Tailed-Blue (*Cupido comyntas texana*). It has become such a favorite for the home garden that it is now often available as a container-grown plant at nurseries offering native stock.

Fragrant Lilac
(Syringa vulgaris)

Family: Olive (Oleaceae)
Class: Cultivated
Height: To 12 feet
Bloom period: March and April
Range: Throughout

Well-shaped, many-branched deciduous shrub. Leaves smooth, numerous, pale green. Flowers numerous, small, four-petaled, fragrant, various shades of lavender, forming long, showy racemes.

CULTIVATION: Since Fragrant Lilacs do not readily produce seeds, propagation must be either by purchasing plants or by planting semihardwood cuttings. *Syringa* is an old genus and can be readily found in the trade. Six- to eight-inch tip cuttings from the branches should be made after flowering; remove lower leaves, and cut the base of the stem at a slant. Dip in rooting hormone, and then place some cuttings in sand, some in potting soil. Keep

FRAGRANT LILAC

soil moist until there is good root growth. Set out in a permanent location in early fall, disturbing roots as little as possible. Cover with wire for protection against predators until in full summer leaf.

Lilacs demand full sun and slightly acidic to neutral, moist, well-drained soils. Generous additions of organic matter from time to time are beneficial to established plants.

If pruning becomes necessary on fully mature plants, do so immediately after flowering, as plants flower on old wood.

RELATED SPECIES: Numerous varieties and cultivars of lilac are offered, with flowers ranging from white to yellow to darkest purple. Many of them have been bred for various attributes, but none for better nectar production. Better nectar production will be found in the older species, especially the "original" shown here. This is the old-fashioned lilac so often seen in Southern gardens and is almost always covered in butterflies, especially Swallowtails (family Papilionidae). If purchasing plants, make sure the plant is mildew resistant, fragrant, and noted as being a butterfly attractant.

Glossy Abelia
(Abelia × grandiflora)

Family: Honeysuckle (Caprifoliaceae)
Class: Cultivated
Height: To 8 feet
Bloom period: May–October
Range: Throughout

Hardy, evergreen to semideciduous shrub with graceful, arching branches. Leaves small, conspicuously shiny, bronze when young and again in the fall. Flowers fragrant, white to pale pink, tubular or bell-shaped, occurring in terminal clusters and from leaf axils along the branches.

GLOSSY ABELIA

CULTIVATION: The shrub known today as Glossy Abelia is a cross between two species, *chinensis* and *uniflora*. It is a very hardy, fast-growing shrub with no significant insect or disease problems. It tolerates neglect and grows in shade, but for the healthiest plant producing the most flowers, give it a moist but well-drained, humus-enriched soil in full sun. It takes some drought but grows much more vigorously with adequate but normal watering. This shrub does not like wind; in such areas it grows and blooms better if planted where protection is offered by a building or a fence of some solid material. Flowers appear on both old and new wood, so prune at any time, although early spring is generally best.

New plants may be started from the trimmed branches. Strip leaves from the last two nodes of a six- to eight-inch tip cutting, dip in rooting powder, and then insert in a potting medium. Keep soil moist until roots are well developed, and then place in a permanent location.

This is a plant commonly offered by nurseries and mail-order catalogs, so obtaining nice, well-developed plants should be no problem.

Glossy Abelia is often used as hedges and screens, where it is severely trimmed, but it is much more impressive as a specimen plant. When used as such and left unpruned, it forms a beautiful, large mound, with slender drooping or "weeping" branches covered with dense, lush foliage and a wealth of flowers. A long hedge of unpruned or moderately trimmed plants being avidly used by butterflies is truly a wondrous sight.

NOTE: The nectar-rich flowers of Glossy Abelia are especially liked by Swallowtails (family Papilionidae), who spend much time flying lazily about the bush, then periodically sailing in to nectar long and contentedly on some choice cluster.

Havana Snakeroot
(Ageratina havanensis)

Family: Aster (Asteraceae)
Class: Native
Height: 1–10 feet
Bloom period: September–November
Range: 2, 6

An open, semiwoody, many-stemmed shrub forming a dense, rounded clump or becoming scandent and clambering into nearby shrubs and trees. Leaves opposite, broadest at base, toothed along margins. Flowers numerous, scented, white, forming showy terminal clusters; stamens protruding far beyond the corolla giving the flowers an airy, frothy look.

CULTIVATION: New plants of Havana Snakeroot can be easily obtained by seeds or cuttings. For seeds, gather the entire flower cluster after it has become brown, dry, and fluffy. Place clusters in an open paper bag to air-dry for a few days, and then store in the refrigerator. In late winter, plant seeds in flats in loose, well-drained soil and at medium temperatures until up and growing well. Transplant to the garden in early spring for autumn flowering.

To start plants by cuttings, take a four- to six-inch portion of a well-developed branch showing new growth, remove the bottom leaves, dip the branch in rooting powder, and place in flats or individual pots. Once they are rooted, plant in a permanent position in the garden.

Havana Snakeroot can be easily divided in either spring or fall. Trim foliage to a few inches above the ground before digging; then separate the clump and replant where wanted.

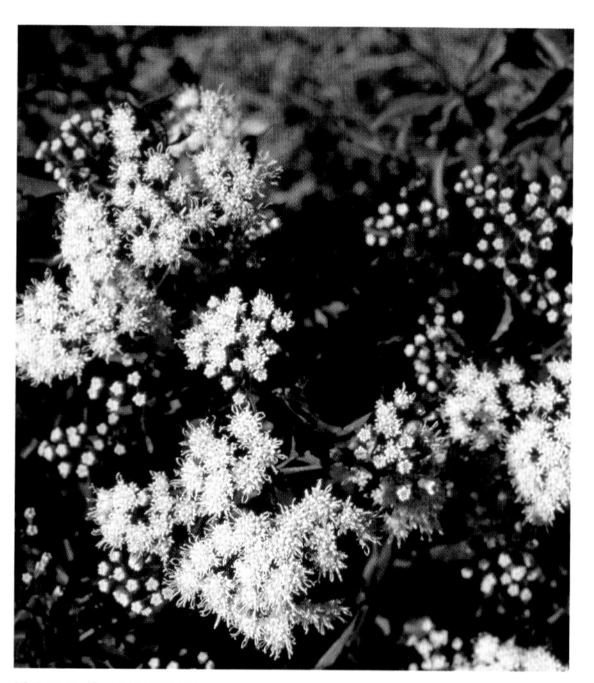

HAVANA SNAKEROOT

Keep new plantings watered thoroughly until well established. Nurseries commonly offer container plants that are usually strong and healthy stock, ensuring much larger and bushier plants for first-year flowering.

This is a plant for the less fertile and more arid sites in the garden, performing beautifully with practically no attention. Fertilizing is beneficial if done sparingly, and an occasional extra watering during the hottest months ensures better growth and more flowers in the fall. For really bushy plants with the most flowers, trim them severely in late winter. If not trimmed, plants eventually climb or lean on nearby shrubbery or low trees. This plant is very hardy and not usually bothered by pests or diseases.

For spectacular fall displays in the garden, combine Havana Snakeroot with other species such as purple-flowered Trailing Lantana (*Lantana montevidensis*), Lindheimer's Morning Glory (*Ipomoea lindheimeri*), Sharp-pod Morning Glory (*I. cordatotriloba*), Bush Sunflower (*Simsia calva*), Maximilian Sunflower (*Helianthus maximiliani*), Cardinal Flower (*Lobelia cardinalis*), or any of the asters (*Symphyotrichum* spp.), agalinis (*Agalinis* spp.), or verbenas (*Verbena* spp.).

NOTE: One of the few white-flowered plants used extensively by butterflies, Havana Snakeroot is one of the best natives for drawing them into the garden. The fragrance is very strong and may not be pleasing to some, but butterflies apparently love it.

RELATED SPECIES: Wright's Snakeroot (*Ageratina wrightii*) is a native to the far west (Region 7) but grows beautifully as far east as Austin and as far north as the Dallas–Fort Worth area.

Piedmont Azalea
(Rhododendron canescens)

Family: Heath (Ericaceae)
Class: Native
Height: To 10 feet
Bloom period: March–May
Range: 3

Upright, many-branched but open, deciduous perennial shrub, flowering before or just as first leaves put forth. Leaves alternate, much longer than wide, covered with soft hairs on both surfaces. Flowers pale pink to dark rose, trumpet-shaped, growing in loose clusters at ends of branches, often in great profusion; the large flowers richly fragrant, the scent from a few plants in bloom almost intoxicating.

CULTIVATION: Piedmont Azalea may be started from seeds but is a little difficult and requires more care than most plants. The easiest and fastest method of propagation is by taking four- to six-inch cuttings from new growth in the spring. Crush cut ends of the cuttings, dip in rooting hormone, and pot in a

Piedmont Azalea

peat and perlite mixture. Place cuttings under a glass jar or a plastic tent, and mist occasionally until rooted. Remove the covering often to give the cuttings fresh air.

Simple rooting of the cuttings is not sufficient for survival. The cuttings would die the following spring unless new top growth has been forced by frequent additions of a weak-strength fertilizer throughout the past season. Healthy, well-established plants of Piedmont Azalea are offered by some nurseries, and this is by far the best means of obtaining plants.

To keep plants happy and growing well, place them in open semishade—ideally beneath tall pines and high-branched hardwoods. Place them in a raised bed to ensure the good drainage they require, and work plenty of humus-rich organic matter into the soil. Half-rotted pine needles are excellent for this, for these plants thrive best in an acidic soil. If pine needles are not available, use leaf mold, rotten sawdust (from pines if possible), or peat moss. Peat moss is the best to use in a new planting because it absorbs and retains the moisture so essential to azaleas. Add plenty of sharp sand if the soil is not naturally sandy. Another method is to spread a layer of sand one or two inches deep on top of a prepared bed, place the plants on top of the sand, and fill around the roots with a mixture of sand, bark, and peat. Azaleas are very shallow rooted, so good mulching is mandatory for their survival. After planting, spread a four- or five-inch layer of pine needles, shredded pine bark, or coarsely shredded cedar bark around the shrub and well beyond its branches. A mixture of materials is preferable since the mulch naturally deteriorates at different times.

If using Piedmont Azaleas against a house or other building with a brick or concrete foundation, take care to prevent lime seepage from the concrete. Line the back of the bed and at least partway under the bed with plastic before adding soil. Do not use bonemeal or wood ashes near azaleas, both of which contain lime. To retain the high acidity necessary for azaleas, add a new layer of mulch each year. Azaleas also benefit from specially formulated azalea food, which should be applied each spring after flowering, at midsummer, and again in early fall. Manure is not recommended for azaleas because of the high alkaline reaction.

Azaleas require consistent moisture year-round, with extra watering during the hot summer months. It is especially critical the first two or three years after planting. Keep a close watch, and do not let the soil become so dry that the plants begin to wilt. After severe wilting, no amount of water will save them, so frequently check the soil to be sure it is moist at least two inches deep.

Once an azalea is planted and thriving, do not attempt to move it—any kind of transplanting almost always results in death of the plant.

In the garden, Piedmont Azaleas are especially lovely when used in a shrubby border or at the edge of woodlands and mingled with such other natives as Tall Pawpaw (*Asimina triloba*), Flowering Dogwood (*Cornus florida*), and Parsley-leaved Hawthorn (*Crataegus marshallii*). In the foreground use great drifts of some of the lower-growing perennials such as Moss Phlox (*Phlox subulata*), Prairie Phlox (*P. pilosa*), Rose Verbena (*Glandularia canadensis*), and the vine Carolina Jasmine (*Gelsimium sempervirens*). All of these bloom simultaneously with azaleas and in the wild form spectacular displays.

NOTE: Almost all azaleas, both native and cultivated, are much liked by butterflies, especially the large ones such as Swallowtails (family Papilionidae), Gulf Fritillary (*Agraulis vanillae incarnata*), and spring-migrating Monarch (*Danaus plexippus*). Smaller Hairstreaks (family Lycaenidae) and Skippers (family Hesperiidae) crawl into the large blossoms in order to reach the nectar with their shorter proboscises.

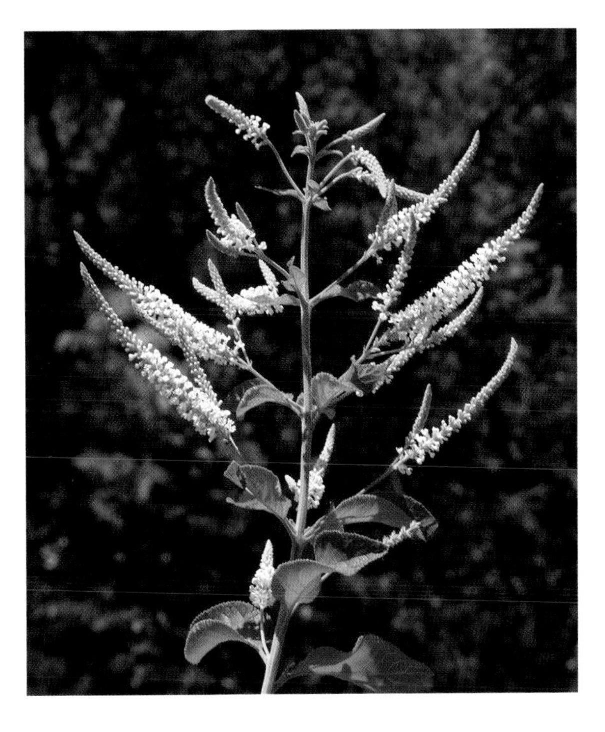

SWEET ALMOND VERBENA

Sweet Almond Verbena
(Aloysia virgata)

Family: Verbena (Verbenaceae)
Class: Nonnative
Height: To 15 feet
Bloom period: May–frost
Range: 3, 4, 5 (2, 6)

Upright, widely branched, delicate-appearing perennial woody shrub, flowering well after leaves have put out. Leaves opposite, dark green on upper side, pale and softly fuzzy beneath. Flowers small, white, intensely fragrant, in long, slender, upright clusters.

CULTIVATION: Sweet Almond Verbena does not readily bear seeds but is now offered by most nurseries, which may be the best way to obtain a mother plant. After the plant has become fully established and is growing well, cuttings can be taken for more plants.

Before beginning propagation, prepare the needed containers. For each cutting, fill a four-inch pot with potting soil, thoroughly soak the soil, and then press down soil to extract all extra moisture. Cuttings should be taken after leaves have appeared in early spring but before any flower buds are showing. Cut four- to six-inch lengths of tip growth with three or four sets of leaves, cutting stem at a slant or forty-five-degree angle. Remove leaves from approximately two inches of the lower portion of the stem, and dip in rooting powder, thoroughly coating these last two inches. Punch a hole two inches deep in prepared moist soil, place the cutting in the hole, and firm soil around the cutting. Barely moisten the soil just around the cutting. Place the potted cutting in a large, clear plastic bag; almost close the bag but leave a small "breathing" space. Place bag near a window or door but never in the sun. Check daily, making sure soil is still wet but the bag is not dripping moisture. Soil must be moist at all times.

Plants should be rooted in approximately two weeks. If a cutting looks healthy and appears to be growing, remove the bag, and place the pot outside in dappled shade. Water plants carefully, and do not let them dry to the point of wilting. After plants have added three to four inches of new growth, they can be planted in a sunny, wind- and freeze-protected area in the garden. Under normal growing conditions, they will bloom the second season. In the most northern portion of their range (Region 2) they like protection from northern winds and freezes, or to be really safe, overwinter them in a greenhouse. In the Hill Country area they need protection, but if winter-killed, they will usually resprout in spring. In the southern coastal regions, they often flower most of the year.

Eventually becoming taller than most flowering shrubs, Sweet Almond Verbena is excellent when used at the back of wide borders, as a hedge, or as a single plant standing alone. It can be fronted with lower-growing shrubs such as Butterfly Bush (*Buddleja davidii*), Salt-marsh Mallow (*Kosteletzkya virginica*), Havana Snakeroot (*Ageratina havanensis*) or Wright's Snakeroot (*A. wrightii*), or False Indigo (*Amorpha fruticosa*). It is lovely mixed with the taller-growing perennials Joe-Pye Weed (*Eutrochium fistulosum*), Swamp Milkweed (*Asclepias incarnata*), Tooth-leaved Goldeneye (*Viguiera dentata*), and Argentina Verbena (*Verbena bonariensis*) bordered with lower-growing annuals and perennials. Wherever the plants are placed, butterflies of many species will be drawn in by the sweet fragrance and abundant nectar.

NOTE: A native of Argentina, this plant grows in the wild as far north as southern Mexico. It is closely related and similar to the South Texas and Hill Country native shrub Beebrush (*Aloysia gratissima*).

Texas Colubrina (Hog-plum)
(Colubrina texensis)

Family: Buckthorn (Rhamnaceae)
Class: Native
Height: 3–6 feet
Bloom period: March–May
Range: 4, 5, 6, 7

Densely branched, low-growing deciduous shrub, with numerous silvery-barked, zigzag, twiggy branches. Leaves small, solitary or clustered, grayish-green. Flowers small, greenish-yellow, oddly shaped, borne in clusters in abundance in early spring. If more than one shrub planted to provide cross-fertilization, flowering followed by numerous hard, dry fruits.

CULTIVATION: In its natural habitat Texas Colubrina is usually common but not intrusive. Since it does not bear thorns, it is one of the more desirable native western shrubs. Whether plants are started from seeds or by cuttings, Texas Colubrina is one of the more difficult ones to propagate. Unless eaten by

TEXAS COLUBRINA (HOG-PLUM)

wildlife, the hard, dry seed capsules remain on the plants for several months after ripening and may be gathered at any time. Plant mature fruit either in fall or spring, although fall planting seems to result in better germination. Try soaking some seeds overnight, soaking some in hot water, or placing some between layers of moist sand for several days. Plant seeds outdoors in permanent locations.

Make cuttings by taking six-inch-long semihardwood tips, dipping in a rooting powder, and potting individually in a good rooting medium.

If it is necessary to transplant established plants, move when the plants are young and retain as much surrounding soil as possible. Cut branches back at least two-thirds, and water thoroughly after planting. Fertilize two or three times a year with a very mild water-soluble fertilizer the first year or two. This helps plants become established but will not be necessary later on. After plants are two or three years old, occasional fall trimming produces thicker, more compact growth.

This is a wonderful plant to use in the driest, rockiest portion of the garden and worth the trouble of establishing. If your soil is not already sandy or rocky, it is beneficial to buy gravel at a business that mixes cement and combine it with soil in the planting hole.

Use Texas Colubrina at the back of a flower border, or mix it with other native shrubs to form a woodland edge. It also makes striking specimen plants or solid borders.

NOTE: Scientific studies have shown that Texas Colubrina is avidly visited by several species of butterflies, especially the Texas Emperor (*Asterocampa clyton texana*) and Empress Antonia (*A. celtis antonia*). The majority are females whose objective, however, is not a sugar-rich nectar. Instead, the butterflies work the central disks of the flowers, possibly obtaining nitrogen to aid in better egg quality.

Trailing Lantana (Weeping Lantana)
(Lantana montevidensis)

Family: Vervain (Verbenaceae)
Class: Nonnative
Height: To 10 feet
Bloom period: February–September
Range: Throughout

Woody, deciduous perennial forming large, many-stemmed clumps with spreading or trailing, vinelike branches; entire plant very aromatic. Leaves opposite, coarsely toothed. Flowers pale lilac to magenta or purple, numerous, and forming rather flat or somewhat rounded showy clusters.

CULTIVATION: Trailing Lantana is a hardy shrub that does best in the garden if cut back to about four to six inches from the ground in late fall after flowering or after the first hard

TRAILING LANTANA (WEEPING LANTANA)

freeze. The roots are strong, and many shoots put forth each spring from healthy plants. When planting, give Trailing Lantana plenty of room to spread, placing the plants at least eight feet apart and in full sun. By the middle of the growing season, they will have formed lush, solid masses, spreading equally as far forward and backward as well as upward. As new growth starts, the long, slender branches begin extending upward; if there are nearby shrubs or trees, Trailing Lantana will lean on or climb into the branches. If no support is near, the branches will eventually fall back to the ground in a weeping effect, sprawling gracefully into large, beautiful clumps or mounds. These plants are spectacular cascading down a bank or stone wall.

Trailing Lantana makes almost unbelievable growth if a little extra fertilizing and moisture are available. Light topdressings of horse manure in the spring and fall are excellent. Lantanas are usually exceptionally drought tolerant, but adequate moisture ensures the most bloom over the longest period. Good drainage is necessary; otherwise, mildew will be a problem. When the plant is overwatered, blooming slows down until the soil moisture has become balanced enough to produce new flowering shoots. In the garden, lantanas are sometimes infested with white flies (especially if crowded), but frequent and forceful spraying of the underside of the leaves with a water hose in the early morning usually keeps pests under control.

A corner or bed filled with several plants of Trailing Lantana is well worth considering if space allows. Butterflies will readily repay you. During the middle of the day at the height of nectar production, there will often be dozens of butterflies of various species around the plants. Certainly the number of butterflies will be in direct proportion to the number of plants in the garden, so try to include as many as the garden will bear. Even if there is room for only one, plant it. To prolong the flowering period and ensure the greatest flower productivity, keep the clusters of violet-black berries trimmed off.

Trailing Lantana is propagated by seeds, cuttings, or divisions. The berrylike drupes may be gathered when fully mature and either planted immediately or any time up until fall. Four- to eight-inch cuttings—taken in late spring, dipped in a rooting hormone, and inserted in a peat and perlite mixture—should root in three to four weeks. Place a clear plastic bag over the rooting container, and keep the plants and rooting mixture moist but not soggy. Trailing Lantana also roots readily at each node, and new plants are easily obtained by cutting on each side of the rooted node and gently lifting from the soil. Replant immediately.

NOTE: Foliage of all species of *Lantana* has a strong, pungent scent when disturbed, which some people find unpleasant. Butterflies also have a dislike for the fragrance and usually fly to another part of the garden until the scent has dissipated.

RELATED SPECIES: Many color and growth forms of *Lantana* species are available at nurseries as cultivars (usually of more dwarf form and with smoother foliage), but these are hybrids bred for showy flowers and contain little if any nectar. Butterflies totally ignore most of them.

Climbing Hempweed
(Mikania scandens)

Family: Aster (Asteraceae)
Class: Native
Height: To 20 feet
Bloom period: June–November
Range: 2, 3, 4, 5, 6

Under good growing conditions, slender branches extend several feet, forming thick, dense masses and often covering nearby vegetation; in the wild, usually a sprawling, twining, luxuriant, low-climbing herbaceous perennial; in the garden can be tamed to a wonderful background vine to trail on a fence or trellis. Stems dark maroon to almost black, in stark contrast to the dull green foliage. Leaves opposite, often wavy or toothed along margins, giving the plant an unusual "lacy" appearance. Flowers small, whitish (some plants with flowers whiter than others), disk-shaped, forming numerous, showy clusters from leaf axils, resembling the blue-flowered mistflowers.

CLIMBING HEMPWEED

CULTIVATION: Not often offered in the trade, this plant is best started from seeds or tip cuttings taken in the wild. Seeds are tiny and black with a tuft of whitish bristles. They may be shaken from the dried, brown bloom cluster. Plant immediately where the plants are wanted, or store in a closed container until spring. Semihardwood tip cuttings can be taken in mid- to late summer, dipped in rooting powder, and potted in a sand/perlite mixture.

In the garden, place with species that require more moisture, such as Joe-Pye Weed (*Eutrochium fistulosum*), Cardinal Flower (*Lobelia cardinalis*), Purple Marsh-fleabane (*Pluchea odorata*), and Pickerel Weed (*Pontederia cordata*). For more continuous flowering, keep all dead flower heads trimmed off.

NOTE: Climbing Hempweed is an important late-fall nectar source for many species of butterflies and is heavily utilized by various bees.

Mexican Flame Vine
(Pseudogynoxys chenopodioides)

Family: Aster (Asteraceae)
Class: Nonnative
Height: 8–10 feet (30 feet)
Bloom period: May–frost (all year)
Range: 2, 5, 6

Trailing, sprawling, or climbing, fast-growing annual or perennial. Leaves thick, coarsely toothed, dark green, making lush green background. Flowers daisylike, orange-red outer petals with yellow center, occurring in small, numerous clusters; become darker with age, almost red in coloring.

CULTIVATION: Finding this plant in any form is very uncertain, so get it however it is

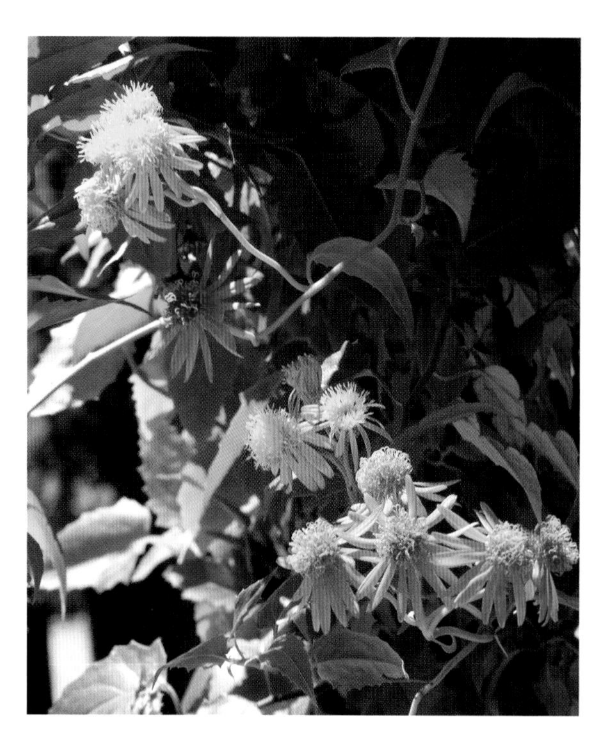
Mexican Flame Vine

encountered—buy plants or beg for seeds or cuttings. For attracting butterflies, it is one of the very best.

Seeds must have light to germinate, so spread thinly over moist soil, gently pressing into soil. Do not cover. Keep the area moist by misting daily or as needed. If seeds will be planted outside, wait until the soil has thoroughly warmed (70–80 degrees). Germination of seeds is usually erratic, so plant plenty. If seeds are gathered from a growing plant, take only fully mature, dry seed heads.

Success is much greater if a four- to six-inch tip cutting is taken, dipped in rooting powder, and then potted. Some instructions suggest placing the cutting in a small square of wet floral foam. Both foam and soil should be tried. Whatever the method, when roots form, plant in the ground where wanted. If the foam has been used, plant foam and all in the ground. This plant also roots at nodes whenever branches touch the ground, so layering should produce needed new plants. These plants will root-sprout, but any unwanted plants can easily be pulled out.

Place plants in light to medium rich garden soil and water regularly, but do not overwater. Foliage is fairly sensitive and is a good indicator of water needs. Fertilizing can be done lightly but regularly. Overfertilizing will result in a lot of lush foliage but fewer blooms.

For best flowering, plant in full sun. Plants will produce some blooms in just afternoon sun or dappled shade, but butterfly usage may not be as good.

In the extreme tip of the Valley, this plant is usually evergreen and blooms all year. In the Austin area it will be killed to the ground each winter but will root-sprout each spring. North of the Austin area plants will need to be trimmed, dug, potted, and brought inside for the winter. Most gardeners in this area plant them in large pots, put the pots in the ground each year, and then take up the potted plant just before the first freeze. If this is done, and the potted plant is placed near a sunny window, the plants will grow through the winter, with nice plants ready to start rapid growth when reset in the yard in spring.

If a climbing vine is wanted, this plant may need some help in the beginning. Tie to a support with soft cloth strips or twist ties. These can be removed once the plant has twined throughout the support. It makes an ideal screen for porches, railings, or chain-link fences. If growth becomes too rampant, trim back as much as needed; the plant will not be harmed.

The startling orange-colored flowers may be a bit much for some gardeners, but if plants in

a soft pink color are used as companion plants, such as some Summer Phlox (*Phlox paniculata*) and West Indian Lantana (*Lantana camara*), the effect will be softened considerably.
NOTE: Previously known under the name *Senecio confusus*, this vine is a native of Central America and grows wild from Honduras to Mexico.

Queen's Wreath (Coral Vine)
(Antigonon leptopus)

Family: Knotweed (Polygonaceae)
Class: Nonnative
Height: To 40 feet
Bloom period: August–December
Range: 2, 3, 4, 5, 6

Rampantly climbing or sprawling deciduous, perennial vine from large, tuberlike root, climbing by tendrils. Leaves large, thin, pale green, deeply indented at base, conspicuously veined. Flowers white to rose-pink, hanging from slender, wiry stems in branched, terminal clusters.
CULTIVATION: A native of Mexico and Central America, where it is known as Chain-of-love, Queen's Wreath makes a spectacular late summer and autumn display. This vine loses its leaves each winter in the northernmost limits of its range, and in cold winters most top growth may die to the ground, but it recovers quickly. Where the temperature drops below 25 degrees, protect roots with a deep, loose, straw mulch. In the Rio Grande Valley, this vine often remains evergreen and hardly stops flowering.

Queen's Wreath is offered in pots by many nurseries, rooted sections can be taken from the mother plant, or it can be started from seed. Whichever way is used, plant in loose, well-drained soil, and keep moist until new growth begins to show. Fertilize twice yearly with a good, all-around fertilizer such as 20-20-20. Give this plant the hottest spot in the garden and full sun. It revels in high summer heat but appreciates regular and thorough waterings. If allowed to become too dry, the plant will shed its leaves.

Queen's Wreath is at its very best when shading a patio or terrace or draping its foliage and long, trailing sprays of blossoms over a fence or garden wall. If using it as a climber, at its base use White-flowered Plumbago (*Plumbago scandens*), Golden Dew-drop Duranta (*Duranta erecta*), Argentina Verbena (*Verbena bonariensis*), Gregg's Mistflower (*Conoclinium dissectum*), or New England Aster (*Symphyotrichum novae-angliae*), as well as liatris (*Liatris* spp.).
NOTE: In its native areas, the huge nutty-flavored tuber of this plant is eaten.

QUEEN'S WREATH (CORAL VINE)

Amelia's Sand-verbena

(Abronia ameliae)

Family: Four-o'clock (Nyctaginaceae)
Class: Native
Height: To 24 inches
Bloom period: December–July
Range: 5

Upright to widely sprawling, usually many-branched perennial; almost the entire plant covered with glandular hairs, making it sticky to the touch. Leaves opposite, thick. Flowers trumpet-shaped, fragrant, lavender to violet-purple, numerous, forming an almost spherical terminal cluster.

CULTIVATION: Throughout its range, Amelia's Sand-verbena is an absolutely fabulous plant to use in the garden for attracting butterflies. It is extremely rich in nectar, and butterflies of many species find it much to their liking. Beginning to flower as early as December in South Texas, it continues blooming until the extreme heat of summer does it in. About the end of June or July, it stops flowering, and the aboveground portion of the plant just withers away.

AMELIA'S SAND-VERBENA

Members of *Abronia* are not easy to grow, but some, such as this one, are so well liked by butterflies that they are worth the special effort necessary to get them to grow. They demand deep, well-drained sand for growing, which means almost dunelike. Unless you live where these plants grow naturally, prepare a bed especially for them where they will have the best chance of thriving and surviving. Either build a raised bed and fill it with loose sand, or dig out the soil to a depth of one and one-half feet and fill the bedding space with the sand. In most cases, the raised bed will probably work better and be easier to prepare.

The foliage of Amelia's Sand-verbena is very tender and fragile and breaks easily, especially when young, so try not to work the beds any more than absolutely necessary. Once the plants have started growing, they require very little water and usually do best with no fertilizer. Do not mulch, for this will cause plants to rot at ground level and die.

It is virtually impossible to transplant any of the sand-verbenas, but they usually reseed themselves, producing young seedlings around the mother plant. However, there are always adverse years when natural conditions are not the best for germination, so it does not hurt to lend a hand. Gather seeds as soon as ripe, let them air-dry for a few days, and store in the refrigerator until time to plant. Planting time is very critical for best germination; plant seeds during October. For best results, make two to four plantings during the month.

Scatter seeds about on an area that has been lightly raked, and then tamp the soil down lightly with a board or the back of a hoe. Sprinkle the planted area until thoroughly moist but not soggy.

After the first flush of flowering, trim back any plants that have started to sprawl, and they will respond with another burst of bloom.

Because these plants demand deep, dry, sandy soil, companion plants for the garden are rather limited. Use them with Huisache Daisy (*Amblyolepis setigera*), Rose Palafoxia (*Palafoxia rosea*), or Western Peppergrass (*Lepidium alyssoides*), or fill the bed solid with Amelia's Sand-verbenas for a spectacular sight and an area that will be much appreciated by butterflies. After the plants have faded in midsummer, intermingle with nursery-purchased Moss Rose (*Portulaca grandiflora*) and Globe Amaranth (*Gomphrena globosa*), which are shallow-rooted and will not disturb Amelia's Sand-verbena. Do not do any unnecessary digging or cultivating in this area if young plants of Amelia's Sand-verbena are needed to come up the next year.

Argentina Verbena
(Verbena bonariensis)

Family: Verbena (Verbenaceae)
Class: Nonnative
Height: 2–4 feet
Bloom period: April–Frost
Range: Throughout

Upright, widely spreading annual or perennial, the stems several, four-angled, slender, occasionally branching, giving plant an open, airy appearance. Leaves few, opposite, slender, dark green, rough, toothed on margins. Flowers blue-purple, small, but formed in terminal showy compact clusters.

CULTIVATION: Argentine Verbena often self-sows, so obtaining new plants is usually no problem. Simply transplant new plants where wanted.

Six-inch tip cuttings can be taken, cutting above a node. Trim the upper tip growth from the cutting. Dip the lower end of the cutting in rooting powder (up to the second node), and plant in a four-inch pot that has been filled with soil-less seed mix. Press soil around cutting, water the soil, and spray or mist the cutting. Place a plastic bag over the cutting and the pot. Keep under a bag (checking soil often for moisture content) until new growth is observed. Then, transplant either into a larger pot or a permanent position in the flower bed.

These plants prefer a good garden soil, not overly rich or overly moist. Good drain-

ARGENTINA VERBENA

age is most important. Plant in full sun, and give them room to breathe. If crowded, they sometimes develop mildew, which can usually be controlled by trimming nearby vegetation.

These easy-to-grow plants are definitely low maintenance, generally just needing an occasional watering and periodic trimming of old flowering clusters to ensure continuous bloom during the entire season. As with most verbenas, they are rich in nectar and readily attract numerous species of butterflies.

Depending on how wide the flower bed is, these plants can be used either at the back of a border or near the middle.

Brown-eyed Susan
(Rudbeckia hirta)

Family: Aster (Asteraceae)
Class: Native
Height: 1–3 feet
Bloom period: April–frost
Range: 2, 3, 4, 5, 6

Annual or short-lived somewhat bristly perennial, forming stiffly upright to somewhat sprawling, coarse mounds; many slender stems form nice mounded clumps of pure color. Leaves alternate, mostly in lower portion of plant. Flowers yellow, formed at tips of long, slender stems; the slender ray flowers petal-like and with tight clusters of tiny chocolate-brown disk flowers forming a flattened conelike cluster in the center; "petals" often blotched with brown or maroon in basal portion.

Brown-eyed Susan

CULTIVATION: Brown-eyed Susan readily reseeds, and simply replanting the young seedlings in desired locations in the beds is easy. Seeds are tiny, dry, and hard to obtain. Young seedlings usually appear in fall and should be moved to permanent locations at this time. The roots become well established over the winter. Once growing well in the garden, some of the seedlings from the first flowering may bloom in the late summer or fall.

Plant Brown-eyed Susan in full sun in fertile, well-drained soils. If necessary, amend tight soils with organic matter. Adding a small amount of slow-release fertilizer every four or five weeks will be very beneficial. After the plants are placed in beds, add a layer of mulch to keep the soil cool. Afternoon dappled shade will promote more profuse flowering and a longer bloom period.

To keep the plants flowering longer, keep them watered, the foliage green and growing well, and all spent flowers trimmed off. Do not let them go to seed until late in summer. After they have made seeds, flowering will stop and foliage will wither.

Butterfly Weed
(Asclepias tuberosa)

Family: Milkweed (Asclepiadaceae)
Class: Native
Height: 1–3 feet
Bloom period: June–August
Range: 1, 2, 3, 4, 6, 7

A hardy, long-lived perennial from a deep, stout root, sending up many flowering stems from a central axis; the milky sap typical of other members of this genus absent in this species. Leaves alternate, numerous, somewhat crowded, margins wavy or sometimes rolled toward lower surface. Flowers large, flat clusters of small, uniquely shaped blossoms, ranging from yellow to red, but most commonly a brilliant orange, borne in the terminal portions of slender stems.

CULTIVATION: Butterfly Weed is easily grown from seeds. Gather pods six to eight weeks after flowering or just as they begin to split open. Strip the silken down from seeds, and sow outside immediately, preferably where the plants are to remain. Seed germination is rather erratic, but it is possible for seeds sown in late summer or fall to produce blooming plants the following season. Flowering is most abundant if plants receive full sun.

Butterfly Weed can be moved about the garden, but the taproot is both brittle and extremely deep so must not be broken or cut off in the moving. Transplanting young, smaller-sized plants gives best results. Nurseries and mail-order catalogs commonly offer healthy, well-established plants; this is the best method to obtain starter plants. Because these are slow to show growth in the spring, permanently mark these plants in the garden to prevent their being stepped on or dug up by mistake.

In the wild, Butterfly Weed commonly grows in thin, dry, usually sandy soils. In the garden with somewhat richer soil and more moisture, plants form lush clumps and produce an abundance of flower clusters. Take caution, however, for in a too-rich, moisture-retaining soil, the roots will rot. Good drainage is absolutely essential for this plant. If your soil is not naturally sandy, work in a couple of shovelfuls of sand or small pea gravel per plant to guarantee good drainage. This plant is perfectly hardy once established and within two or three years will have formed a large, spectacular clump under good garden conditions. To get the most bloom from Butterfly Weed, after the first flowering is over, trim back a healthy thriving clump to about four inches from the ground. The plant will resprout and bloom again two or three months later.

NOTE: Well named, Butterfly Weed is an

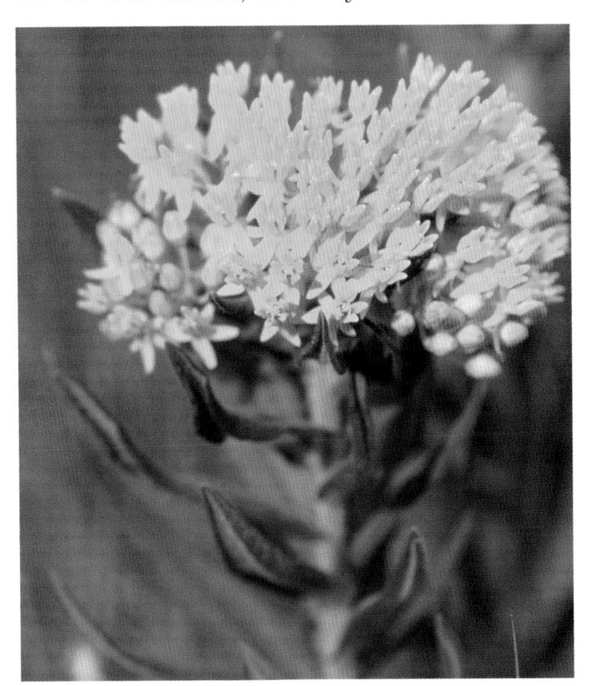

BUTTERFLY WEED

excellent nectar producer and attracts multitudes of butterflies of many different species throughout the flowering period. Butterfly Weed may occasionally be used as a larval food plant by the Monarch (*Danaus plexippus*) and the Queen (*D. gilippus thersippus*) but is not preferred because of the tough, stiff-haired leaves and its low concentration of poisonous chemicals.

RELATED SPECIES: Almost all species of *Asclepias* produce showy clusters of flowers and are attractive to butterflies. Some others that are excellent for use in the garden as nectar sources include the pale pink to dark rose-colored Common Milkweed (*A. syriaca*), Purple Milkweed (*A. purpurascens*), Swamp Milkweed (*A. incarnata*), and the snowy-blossomed White-flowered Milkweed (*A. variegata*). The cultivated Tropical Milkweed (*A. curassavica*), with two-toned red and orange flowers, is easy to raise from seeds, forming nice plants and producing flowers about five months after early-spring sowing. It is root hardy only in the Rio Grande Valley area but is easy to grow as an annual or in pots or tubs to be brought inside during the freezing months elsewhere. If kept over winter, it will begin producing flowers earlier in the year (or, in many instances, bloom all year). It is an especially good nectar producer as well as a favored larval food plant.

Cardinal Flower
(Lobelia cardinalis)

Family: Bluebell (Campanulaceae)
Class: Native
Height: 6 inches–4 feet
Bloom period: July–frost
Range: Throughout

CARDINAL FLOWER

A tall, stately, short-lived perennial, one of the most striking wildflowers, demanding center stage wherever planted. Leaves a deep, dark green, massed along the stem beneath a long spike of brilliant scarlet, two-lipped flowers.

CULTIVATION: Seeds of Cardinal Flower are extremely small, and viability is not very dependable, although new plants can be expected to crop up around the garden. The best method for increasing plants of this beauty is by layering. Choose an outer stem, gently bend to the ground, scratch the soil beneath a node, and then lay a rock across the stem at the node. It will root at the nodes along the stem. When well rooted, clip from the mother plant between each rooted node, and plant each section in the garden where wanted.

Every two or three years, in early spring or late fall, the old clump can be dug and the vigorous outer portions sectioned off and replanted.

Cardinal Flower is generally a flower of marshes, bogs, floodplains, and stream banks. It likes moisture, but it also adapts. It can be found growing from between layers of limestone along Hill Country streams, and with a bit of extra moisture it does quite well in the home garden.

The growth of this plant is directly related to its habitat. If it is in a very shaded location, it will need less moisture but will remain low. It grows taller in rich soil and with abundant moisture. And in this moister environment, if it receives a half a day or more of sun, it will produce the lushest foliage, the greatest abundance of flowers, and a lot of nectar. If given afternoon or dappled shade, the flowers will be more brilliant in coloring.

Each plant of Cardinal Flower will produce only one stalk of flowers, so for best effect in the garden, group three to five plants together. Mulch around plants and surrounding area to retain moisture.

The flower color of this plant is so dramatic that its companion plants should be subtle and softly hued. Gregg's Mistflower (*Conoclinium dissectum*); late-flowering lavender, pink, or white Summer Phlox (*Phlox paniculata*); Pickerel Weed (*Pontederia cordata*); Swamp Milkweed (*Asclepias incarnata*); and a pale yellow-flowered goldenrod (*Solidago* spp.) would blend beautifully.

Globe Amaranth
(Gomphrena globosa)

Family: Amaranth (Amaranthaceae)
Class: Cultivated
Height: To 3 feet
Bloom period: Spring–frost
Range: Throughout

One of the "everlasting" plants, an annual favored by many species of butterflies; stems simple, purplish, usually much-branched, swollen at the nodes. Leaves alternate, bright green, thick, sometimes red-veined, mostly in lower portion of plant. Flowering head consisting of numerous tiny purple flowers backed by stiff scales or bracts, clustered in showy, rounded, or globelike heads, the heads terminal on long, slender stalks.

CULTIVATION: As Globe Amaranth is a common annual, the best method to obtain plants is by purchasing seeds from a favorite nursery or catalog. If at all possible, purchase the older purplish-flowered form.

GLOBE AMARANTH

Most nurseries offer this plant in the spring, but often they are "improved" forms; although the newer colors may be unique or "prettier" and the plants shorter, the older, taller form offers the most nectar.

Seeds should be planted around mid-March, and as soon as true leaves show, pot singly. After last frost, place in a permanent position in a bed or border in good garden soil. Or seeds can be planted directly in the garden where wanted after the last frost. Lightly sprinkle a little liquid fertilizer in the planting hole, and then water thoroughly after planting to settle soil around roots. Keep plants moderately moist during the growing season. For longer flowering, keep spent flower heads removed. The old-fashioned form of this plant usually readily reseeds, so new plants are available the following spring.
NOTE: When flowers are at their peak of color, the stems can be clipped and hung upside down to dry in a cool, dark place. The flowers retain their color after drying (thus the "everlasting" name) and are most attractive in dried arrangements.

Gregg's Dalea
(Dalea greggii)

Family: Bean (Fabaceae)
Class: Native
Height: To 9 inches
Bloom period: May–September
Range: 6, 7

The perfect perennial ground cover within its range, with long, wiry, trailing stems rooting at the nodes and forming solid mats. Leaves small, compound, blue-green, thickly covered with silvery-gray hairs giving the plant an airy, lacy look. Flowers numerous, small, reddish-lavender to purple, nestling throughout foliage.

GREGG'S DALEA

CULTIVATION: Gregg's Dalea is easily rooted from softwood cuttings taken in midsummer. Dip four- to six-inch cuttings in a rooting hormone, spray a growing medium until moist, and place the pot in an open plastic bag. Keep soil moist until cuttings are rooted. When placed in the garden, shade plants a few days until well established.

This dalea is not too particular about soil; it grows and thrives in sand, loam, or rocky limestone. Soils must be well drained, and at no time should these plants be overwatered. In fall after the first frost or in spring before the first new growth, shear plants to a couple of inches above ground level. Shearing back in summer if foliage becomes too leggy or leafless will result in denser, lush foliage. Keep this one dry during the winter or plants may rot.
NOTE: Over time, healthy specimens of Gregg's Dalea form a thick, woody base, becoming almost a subshrub. This is one of the larval food plants of the Reakirt's Blue (*Echinargus isola*).

Gregg's Mistflower (Palmleaf Mistflower)
(Conoclinium dissectum)

Family: Aster (Asteraceae)
Class: Native
Height: To 2 feet
Bloom period: April–frost
Range: 5, 6, 7

A perennial from creeping roots, eventually forming extensive colonies but easily controlled. Leaves opposite, pale green, finely cut or lobed into fanlike sections, becoming smaller toward top of stem. Flowers pale lavender-blue, congested in small head, the heads several and forming terminal cluster on long, leafless stalks.

CULTIVATION: Obtaining new additions of Gregg's Mistflower is no problem. It continually sends out underground runners, and one plant quickly becomes a colony. If new colonies are wanted, simply dig or pull rooted sections and plant. The rooted section will bloom the first year, and a nice colony will have become established by the second year.

It also reseeds profusely, and new plants will appear throughout the garden. Even with all this abundant reproduction, these plants are easily contained by simply pulling out the outside or unwanted shoots.

Gregg's Mistflower can grow in native soils but will be much lusher and produce more flowers in average garden soils and with a bit of added moisture.

The soft colors of these flowers will blend with just about anything, but especially soft pinks to dark rose, purples, and pale yellows.

NOTE: Not noticeably fragrant, these blossoms do attract every species of butterfly around, especially the Queen (*Danaus gilippus thersippus*). At times the flowers themselves can hardly be seen for the many Queens flying about.

GREGG'S MISTFLOWER

Hispid Wedelia
(Wedelia acapulcensis var. *hispida)*

Family: Aster (Asteraceae)
Class: Native
Height: To 3 feet
Bloom period: May–frost
Range: 2, 4, 5, 6, 7

Upright to spreading or sprawling, woody-based perennial or low shrub; usually deciduous in upper portion of range and dies to the ground each fall, but remains evergreen and flowering in more southern regions. Leaves rough, prickly, densely covering lower portions of stems, upper stem portion leafless. Flowers orange-yellow, daisylike, tipping upper stems, extending well above foliage.

CULTIVATION: Hispid Wedelia is usually offered at native plant sales and by nurseries carrying native plants and is the best way to get a "start" of this plant. Once growing and flowering well, it usually reseeds, although sparingly, and these seedlings can be transplanted when small. Older plants do not transplant well because the root is too deep.

Seeds can be gathered from the wild by collecting thoroughly dry seed heads in late summer before they disintegrate. Break heads apart after gathering, and let the pieces dry for few more days; then either plant immediately or store in a sealed container until the following spring.

Three- to four-inch semihardwood tip cuttings can be taken from large, healthy plants in midsummer to fall. Dip in rooting powder, and place cuttings in a peat/perlite mixture. Keep moist but not soaking wet until rooted. Pot or permanently plant immediately after roots appear. Fall-rooted cuttings will need to be carried over winter in a protected area. If larger, established plants must be moved, do so in winter after cutting back. Water well after moving.

These plants are generally long-lived and, other than mildly reseeding, are no problem in a garden bed or border. With trimming, they work well as edgings. Easily cultivated, they grow equally well in extremely dry or moist situations but produce more flowers with adequate watering.

NOTE: Hispid Wedelia is one of the larval food plants for the Bordered Patch (*Chlosyne lacinia adjutrix*).

HISPID WEDELIA

Huisache Daisy

(Amblyolepis setigera)

Family: Aster (Asteraceae)
Class: Native
Height: 4–20 inches
Bloom period: March–July
Range: 1, 4, 5, 6, 7

Low, mounded, or sprawling annual. Leaves mostly in basal portion of plant, becoming smaller and clasping in upper portion of stem. Flower heads fragrant, yellow, solitary, and terminal on long, slender stalks; tips of ray flowers conspicuously squared and notched.

CULTIVATION: Huisache Daisy is an excellent plant for drier sites and grows best in sandy, limestone, or chalky soil. It grows very well in rocky soil and disturbed habitats. This plant will not be found in nurseries but readily reproduces from seeds. For a naturalized area, gather seeds from a choice plant, and sow in raked soil in late fall for flowering plants the following season. For use in the beds, seeds can be planted where wanted or young plants can be transplanted one to two feet apart. These plants need no fertilizer. Keep moist until they are growing well, and then water only occasionally for best growth and bloom. When so treated, the plants stay mounded

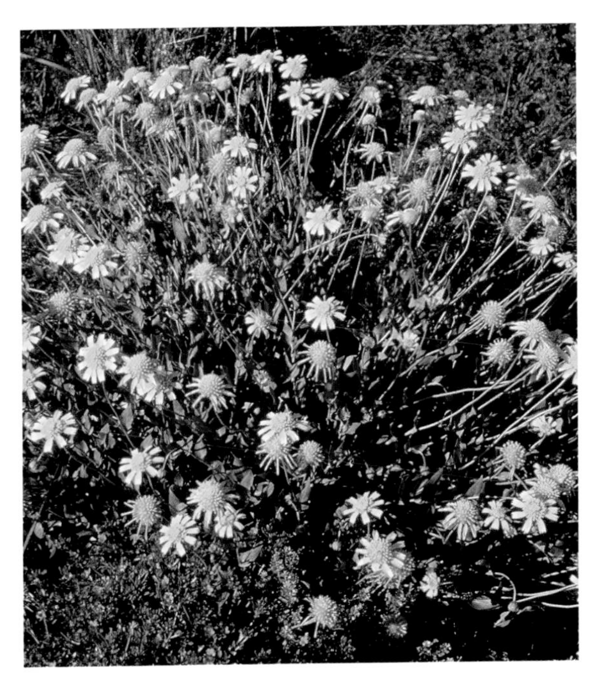

HUISACHE DAISY

until late summer, when they begin to sprawl. Keep all seed heads clipped or pinched off until late fall for the plants to keep producing flowers. These plants reseed very well, and a plentiful supply of plants for the garden should be available if the September flowers are allowed to make seeds.

Use these plants in groupings scattered throughout the beds or as an edging. They are especially attractive among rocks or pieces of weathered wood. They are also a good choice for hillsides, slopes, or natural areas, alone or in combination with other natives.

NOTE: In the wild, Huisache Daisy grows in colonies; when the plants are in flower, the entire area becomes almost intoxicating with the delightful fragrance. Many species of butterflies are attracted to the plants by the scent and stay around to sample the nectar.

Joe-Pye Weed

(Eutrochium fistulosum)

Family: Aster (Asteraceae)
Class: Native
Height: 3–10 feet
Bloom period: July–August
Range: 3, 4 (5, 6)

Stiffly upright, robust perennial, often forming large colonies. Leaves thick, conspicuously veined, arranged in a whorl around the purplish or reddish stem, which is usually hollow. Flowers in large masses, bright lilac-pink to purple or brownish-lavender, held in broadly rounded or dome-shaped terminal clusters.

CULTIVATION: Bedding plants of Joe-Pye Weed may be easily obtained from seeds. Best results are from seeds gathered in late August or as soon as mature and sown immediately or no later than mid-September for germination the following spring. Plants from seeds sown in flats indoors in late winter will be ready to transplant into the garden after the last frost. This plant is now frequently offered in the nursery trade.

Usually growing in well-drained marshy or boggy areas, Joe-Pye Weed grows quite well in rich, loamy garden soil but needs a bit of extra moisture. If allowed to dry out, especially during the hot, dry summer months, plants quickly die. Place near a dripping faucet, or, better yet, combine with other moisture lovers, such as Blue Waterleaf (*Hydrolea ovata*), Cardinal Flower (*Lobelia cardinalis*), Salt-marsh Mallow (*Kosteletzkya virginica*), Swamp Milkweed (*Asclepias incarnata*), Swamp Sunflower (*Helianthus angustifolius*), the vine Climbing Hempweed (*Mikania scandens*), and various mistflowers (*Conoclinium* spp.).

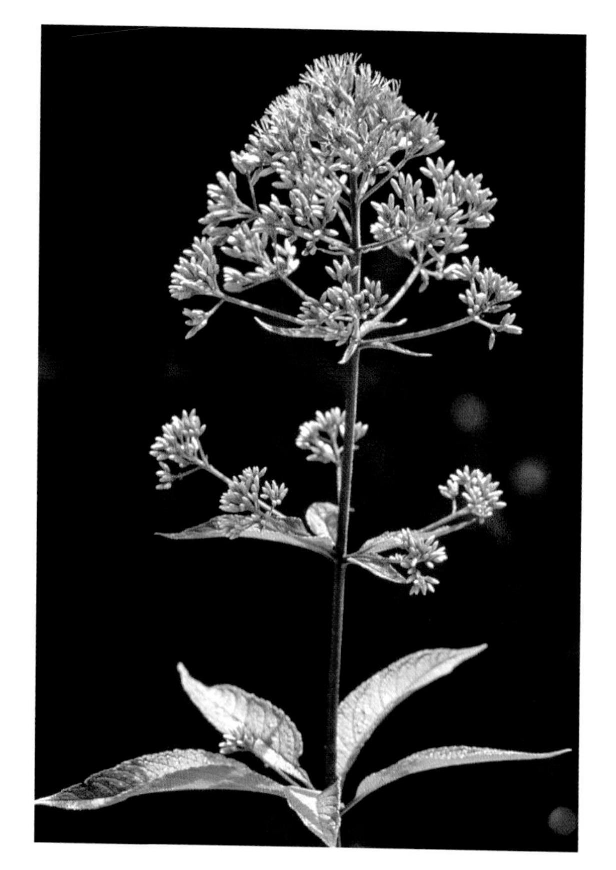

JOE-PYE WEED

Use two or three species in a bed of their own, work in lots of well-rotted humus, mulch deeply, and give them a separate soaker hose that can be run more often than the other ones in the garden.

Joe-Pye Weed eventually forms a nice clump of several stems. Once the plant has become well established, the clump can be lifted and divided either in fall or early spring just as new growth appears. Dig the clump, and cut or separate with a large, sharp knife, leaving one stem and a portion of the fibrous roots in each section. Replant immediately and water well.

Use these majestic plants at the back of the border, allowing at least three feet between plants. If the bed is in the open, plant Joe-Pye Weed in the middle, using lower-growing plants on either side. If in an excessively windy or exposed site, staking may be necessary. Naturally adapted to filtered sunlight, if given plenty of moisture and good drainage, Joe-Pye Weed grows even better in full sun and does not become quite as tall.

NOTE: Almost all of the large butterflies such as Swallowtails (family Papilionidae), Sulphurs (family Pieridae), Common Buckeye (*Junonia coenia*), Gulf Fritillary (*Agraulis vanillae incarnata*), Monarch (*Danaus plexippus*), Painted Lady (*Vanessa cardui*), and Red Admiral (*V. atalanta rubria*) commonly nectar on Joe-Pye Weed.

Mexican Sunflower (Tithonia)
(Tithonia rotundifolia)

Family: Aster (Asteraceae)
Class: Cultivated
Height: To 6 feet
Bloom period: June–frost
Range: Throughout

Upright, widely branched, robust annual. Leaves and stems velvety. Flowers large, vivid, orange-scarlet, terminal on long, hollow branches.

CULTIVATION: Mexican Sunflower is very easy to grow from seeds. Seeds will need to be ordered from catalogs, as they are not generally offered by nurseries. Although fast growing, this plant takes three or four months to begin flowering, so it is best to start seeds as early as possible. Plant seeds in flats or individually sectioned cartons indoors in late February for transplanting to the garden after the last frost or freeze date.

In the garden, place Mexican Sunflower

Mexican Sunflower (Tithonia)

in full sun. Water well with a root stimulator when setting into the beds, and fertilize occasionally during the entire growing season. Do not let plants wilt from lack of water, but be careful not to give too much moisture or fertilizer, which will cause lush, weak growth, and the plant will fall or break. Staking is sometimes necessary because branches are hollow; strong winds can easily bend or break the plants.

Use Mexican Sunflower at the back of the border, allowing plenty of room to expand. Crowding will cause it to drop the lower leaves and be rather unsightly. It works equally well in a solid border, as a screen, or as a hedge.

Once it is growing in the garden, leave a few choice flowers to mature seeds for the following year. Otherwise, keep seed heads clipped for more abundant flowering. In some areas this plant readily reseeds, but do not depend on this.

Despite its few drawbacks, Mexican Sunflower is an easily grown plant, produces over a long period, is very floriferous, and makes an absolutely spectacular splash in the garden.

NOTE: As far as butterflies are concerned, you cannot have too many Mexican Sunflowers. They are an excellent nectar source, and butterflies literally swarm around them continuously. All of the Swallowtails (family Papilionidae), Sulphurs (family Pieridae), Gulf Fritillary (*Agraulis vanillae incarnata*), Red Admiral (*Vanessa atalanta rubria*), Monarch (*Danaus plexippus*), and Queen (*D. gilippus thersippus*) seem especially attracted. Its foliage is readily eaten by larvae of the Bordered Patch (*Chlosyne lacinia adjutrix*). The original, orange-flowered cultivar "Torch" is shown here.

RELATED SPECIES: The cultivars 'Sundance,' a three-foot form; 'Goldfinger,' a twenty-four- to thirty-inch form; and 'Yellow Torch,' a three- to four-foot yellow-flowered version of the original 'Torch,' have recently been offered by the trade. Butterflies seem to prefer the orange-colored ones to the yellow.

Pentas

(Pentas lanceolata)

Family: Madder (Rubiaceae)
Class: Cultivated
Height: 1–2½ feet
Bloom period: April–frost (all year)
Range: Throughout

Upright to somewhat sprawling, many-branched annual or perennial herb, becoming woody near base. Leaves opposite, conspicuously veined, dark green. Flowers five-lobed,

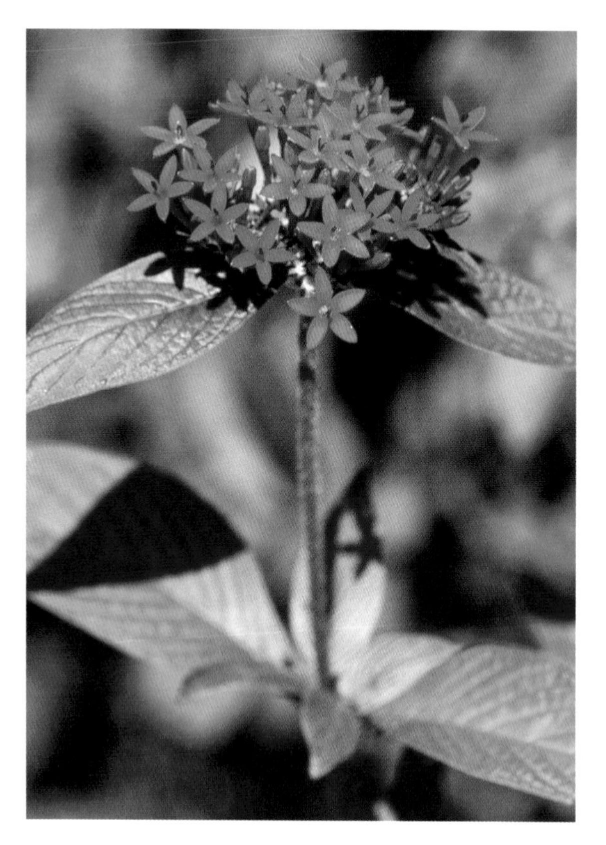
PENTAS

trumpet-shaped, numerous, and borne in rounded terminal clusters from upper portion of stem.

CULTIVATION: This short-lived native of tropical Africa has numerous clusters of starlike flowers, adding wonderful drifts of bright color to gardens all year long in some areas and from early spring until hard freezes in others. It is usually obtained from nurseries.

Pentas may be planted in full sun if given plenty of moisture, but ideally it should receive morning sun and open, dappled shade during the hottest part of the day. Its flowers are such favorites that butterflies continuously nectar from even the lightly shaded plants, feeding from them until almost dark every day.

Although tolerant of a fairly wide range of growing conditions, Pentas like a deep, rich garden loam with excellent drainage and like moisture when the ground dries out. Well-rotted barnyard manure is an excellent fertilizer when applied in moderation.

In the extreme southernmost regions, Pentas may be evergreen, eventually becoming almost shrublike. In other areas, new plants can be purchased each spring and treated as annuals. Often Pentas is listed and sold as a greenhouse or indoor plant, and it is a wonderful subject when used this way. If you want to carry your garden plants over the winter instead of purchasing anew each spring, lift them from the garden in early fall, trim back if needed, pot them, and bring inside to be placed in a sunny window. Fertilize lightly during the winter months.

New plants for potting can also be obtained by taking tip cuttings in early fall and rooting in either water or a peat and soil mix. Repot in a mixture of loam, sand, and peat after they are well rooted. Set winter plants in the garden after the soil becomes thoroughly warm in spring, or gradually expose them to outside conditions and leave in the pots for patio or porch use. Whether indoors or out, after the plants have begun to show good growth, pinch out all growing tips for low, bushier plants that will produce even more flowering clusters.

In the garden, massed plantings of Pentas produce the most striking effects and of course attract the most butterflies. Cluster them in groupings of six to ten plants intermittently throughout the border, edge a walk or driveway, or use extravagantly along the edge of a wooded area or in a solid bed of their own. Flower color is usually a vibrant

pink, rose, magenta, or red, so use Pentas with the more subtle colors of pink-flowered Sweet Alyssum (*Lobularia maritima*), Summer Phlox (*Phlox paniculata*), White-flowered Plumbago (*Plumbago scandens*), Gregg's Mistflower (*Conoclinium dissectum*), or yellow-flowered columbines (*Aquilegia* spp.).
NOTE: Try to find the old-fashioned, red-flowered form of Pentas if possible—they are shrublike, attain a height of four feet, live up to seven years, are hardier, and bear more nectar. These can easily be started from seeds.

PHACELIA

Phacelia
(Phacelia congesta)

Family: Waterleaf (Hydrophyllaceae)
Class: Native
Height: To 3 feet
Bloom period: March–June
Range: Throughout

Upright, leafy, several-stemmed annual or biennial. Leaves soft, sticky. Flowers numerous, small, blue to purplish, borne in conspicuously coiled terminal clusters that uncurl as flowers open.
CULTIVATION: As Phacelia is not commonly available either from nurseries or seed dealers, seeds will almost surely have to be gathered from a wild source. Since Phacelia begins opening flowers at the base of the curled racemes and continues for some time, seeds will be mature in the basal capsules of the spike while the tip is still in tightly coiled buds, so flowering plants must be watched closely. Collect seeds when the first basal capsules have begun to split open.

To collect seeds, clip only the most mature racemes, and only one from each plant, allowing plenty to remain in the wild for reseeding. Keep racemes in an open paper bag a few days for seeds to continue maturing and drying. Most viable capsules open within a week. Shake the plant against the bag to loosen all the seeds, and then store them in the refrigerator until time to plant. After all danger of frost is past, plant seeds in prepared beds where plants are to remain. Thin as soon as plants are three or four inches high, but do not wait long to transplant, as older plants do not like being moved.

Phacelia grows well in many soil types but does best in a loose, rich soil. Place it where it will receive morning or late afternoon sun, but do not expose it to hot midday sun. Keep it moist and fertilize occasionally. A good mulch helps keep roots cool and retains moisture.

Phacelia readily reseeds itself each year, so once plants are established, yearly seed collecting and sowing should not be necessary. Simply transplant new seedlings to the desired sites each spring. Germination is much better in disturbed ground, so each fall scratch or rake the ground thoroughly around the plants.

Phacelia is most effective when displayed in groups of several plants to form large masses. It is lovely combined with Prairie Verbena (*Glandularia bipinnatifida*), Purple Coneflower (*Echinacea purpurea*), Scarlet Muskflower (*Nyctaginia capitata*), Western Peppergrass (*Lepidium alyssoides*), or various columbines (*Aquilegia* spp.). Clumps of Phacelia in front of shrubs such as Cherry Sage (*Salvia greggii*), Beebrush (*Aloysia gratissima*), Texas Kidneywood (*Eysenhardtia texana*), and Agarita (*Mahonia trifoliolata*) make an impressive showing.

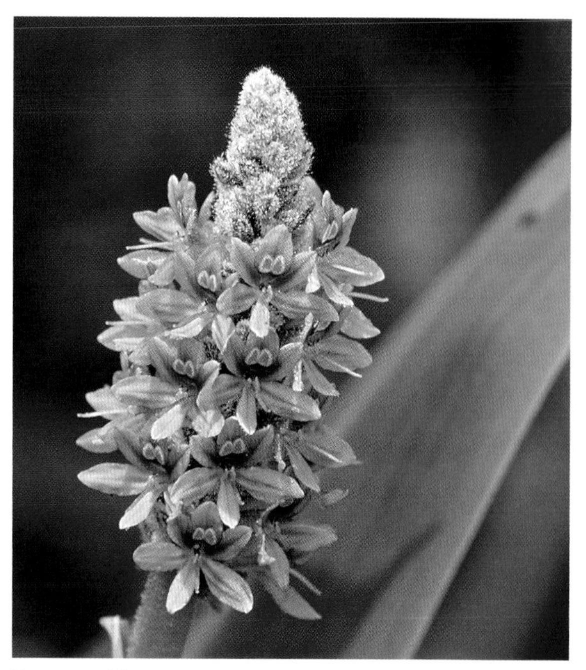

PICKEREL WEED

Pickerel Weed

(Pontederia cordata)

Family: Pickerel Weed (Pontederiaceae)
Class: Native
Height: 2–4 feet above water
Bloom period: April–October
Range: 2, 3, 4, 6

Strictly upright, stout, smooth aquatic perennial from creeping rootlike rhizome, rooted in mud, usually forming colonies. Leaves and flowering spike extending above water. Leaves long, thick, glossy, deeply notched at base and forming long lobes. Flowers violet-blue, small, numerous, congested in slender, elongating spikes at tip of long stalk.

CULTIVATION: These plants can possibly be started from fully ripened seeds but not from cuttings. First plants of Pickerel Weed should be purchased at a nursery specializing in native or aquatic plants. After becoming fully established and putting forth new shoots, the plants can be taken up and divided, either to obtain new plants or to keep the original plant in bounds. They can easily be divided even when in full bloom without damage if done carefully and replanted immediately. Under good growing conditions, Pickerel Weed may possibly overcrowd other plants in a small pond but would never be considered aggressive.

When placing these in a home pond, prepare soil to a depth of at least four to six inches. Soils can be acidic, lime, sandy, or clay. A soil comparable to that of a rich floodplain would be ideal, or use a good garden soil with a small addition of clay. Plants can be left in their own separate pots if desired.

Pickerel Weed can be used alone or in company with the taller-growing Powdery Thalia (*Thalia dealbata*), Cardinal Flower (*Lobelia cardinalis*), Lizard's-tail (*Saururus cernuus*), Spider Lily (*Hymenocallis liriosme*), or sagittarias (*Sagittaria* spp.), as well as floating plants of Floating-heart (*Nymphoides aquatica*), Bogmoss (*Mayaca aubletii*), or various Water-lilies (*Nymphaea* spp.).

Prairie Liatris
(Liatris pycnostachya)

Family: Aster (Asteraceae)
Class: Native
Height: 3–5 feet
Bloom period: June–October
Range: 2, 3, 4

Stiffly upright, unbranched perennial somewhat rough to the touch, growing from a woody corm or rootstock. Leaves narrow, in dense grasslike basal clump and along stem, becoming smaller toward flowering spike. Flowers lavender to dark purple, in small heads, the heads numerous and densely crowded in a long, slender, wandlike spike; flowers open from tip downward, the flowering period lasting for a month or more.
CULTIVATION: Prairie Liatris may be obtained either from seeds or corms or as potted nursery plants. After flowering, spikes usually remain green until after the first frost or two; then the purplish pappus hairs on the seed begin changing to a grayish-white as they become dry and fluffy. After the entire stalk has taken on this fluffy appearance, clip the entire seed portion and place in an open paper bag. Place it in a warm, dry place for a few days to become completely dry. Seeds should then be easy to remove by shaking or hitting the spike against the side of the sack. Removing fluff or other debris is not necessary unless desired. Place seeds in a dry container, and store in the refrigerator. Seeds may be planted in flats in the house or greenhouse in late winter and then transplanted to pots after three or four leaves have developed. Transplant to the garden in May or early June. Better germination may come from seeds a year or two old.

A simpler planting method is to clip the dried seed spike, lay the entire stalk on the ground, cover with soil, and mark the spot well. Seedlings (which resemble grass or onions) should come up the following spring. After seedlings have developed true leaves, transplant to permanent places in the garden, spacing at least two feet apart. Plant in groups of five to seven or farther apart for the best show of flowers. Give the plants good to moderately rich, well-drained soils in full sun. They like a bit of moisture now and then, but soggy winter soils cause the corms to rot. Depending on growing conditions, plants may

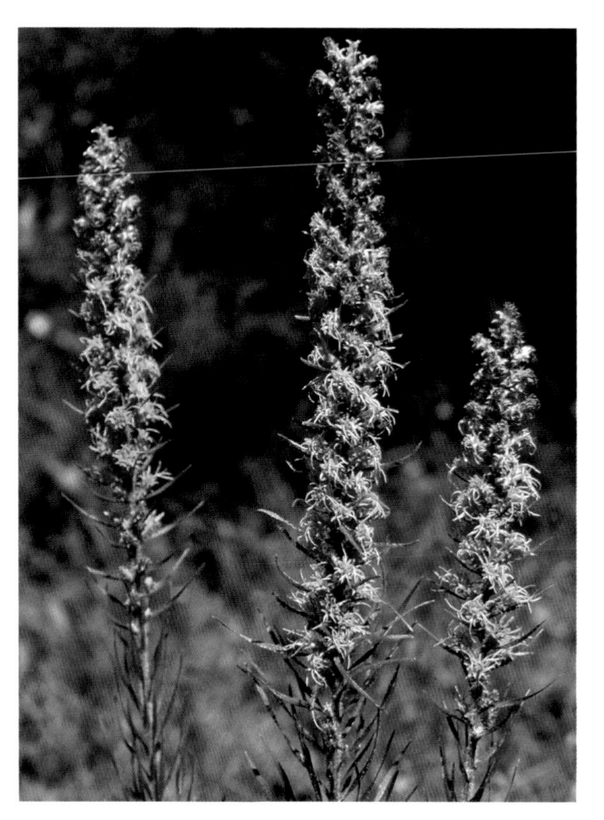

PRAIRIE LIATRIS

become so tall and top-heavy that staking may be needed. Do this early because the spikes may bend or twist and, once twisted, will not straighten even if staked.

The purple flowers of Prairie Liatris are absolutely spectacular in late summer when combined with other nectar-rich plants such as pink-flowered Butterfly Bush (*Buddleja davidii*), Purple Coneflower (*Echinacea purpurea*), Brown-eyed Susan (*Rudbeckia hirta*), Bush Sunflower (*Simsia calva*), Summer Phlox (*Phlox paniculata*), Tooth-leaved Goldeneye (*Viguiera dentata*), Golden Crownbeard (*Verbesina encelioides*), Hispid Wedelia (*Wedelia acapulcensis* var. *hispida*), Pentas (*Pentas lanceolata*), White-flowered Plumbago (*Plumbago scandens*), Two-leaved Senna (*Senna roemeriana*), or goldenrods (*Solidago* spp.).

NOTE: There are many native species of *Liatris* in the state. Choose the showiest ones in your area, and use them lavishly, for they take only a small amount of space in the garden and butterflies absolutely flock to them during the entire time of blooming. The flowering period is somewhat variable, depending on species and climatic and garden conditions.

Prairie Verbena
(Glandularia bipinnatifida)

Family: Verbena (Verbenaceae)
Class: Native
Height: 6–12 inches
Bloom period: February–December (all year)
Range: Throughout

Low-growing, sprawling or trailing, evergreen, short-lived perennial, blooming almost all year; sometimes dies out in winter from extreme cold. Leaves ferny, deeply incised. Flower clusters of numerous, short-tubed, five-petaled, lavender to pale purple blossoms rising above foliage; almost cover the ground-hugging plants; flower clusters flat to roundish when first opening, with the cluster continually elongating as new flowers open; nectar-producing clusters of small purple flowers a virtual feast for butterflies.

CULTIVATION: Prairie Verbena readily establishes itself in the garden from wind-blown seeds, and these new little plants are easily transplanted where wanted anytime throughout the growing season. They will bloom the first season and can become a three-foot or larger mass by season's end.

These plants can also be propagated by layering. Scratch or loosen the soil beneath the nodes on an elongated branch that is already lying on the ground. Press each node into the loosened soil, cover lightly with soil, and lay a small rock on top of the branch to hold the branch in firm contact with the soil. When nodes show healthy roots, clip apart between

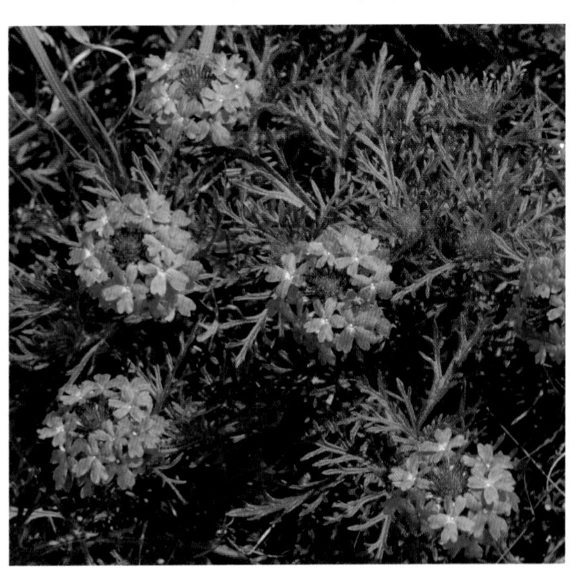

PRAIRIE VERBENA

each node, replanting each rooted node. Plant in full sun, in almost any soil, but plants will be especially luxuriant in moist but well-drained limestone.

For best and most continuous flowering, keep spent, elongating flower clusters clipped off so the plant does not make seeds. Do not pamper these plants with extra fertilizer, excessive watering, or extra-rich soils.

The lavender coloring of these flowers blends with almost any color combination in the garden or in the wild. In the garden they are especially lovely backed by white, soft yellow, or pink-colored flowers. The plant can be used almost anywhere—as a border edging, in hanging pots, or trailing over a wall. Wherever used, the butterflies will find it and visit frequently.

Purple Coneflower
(Echinacea purpurea)

Family: Aster (Asteraceae)
Class: Native/Cultivated
Height: 1–3 feet
Bloom period: June–October
Range: Throughout

A hardy, stocky, long-lived perennial. Leaves long, dark green, rough, mostly in a basal clump. Flowers large, four inches, sunflower-like, pink to dark rose, borne in abundance atop slender stems; central disk or cone consists of numerous small, bright golden-bronze flowers surrounded by numerous ray flowers of purplish-pink; flower heads particularly long lasting, often remaining in good condition a month or more.

CULTIVATION: The Purple Coneflower offered by nurseries and through catalogs is simply an "improved" form of one of the native species. The most common cultivar, 'Bright Star,' has dark pink flowers, but white-, red-, and purple-flowered forms are also available.

If cultivars are wanted, purchase them as small potted seedlings for spring planting—they will flower the first year. If plants are already established in the garden, even more plants can be obtained by digging a two- or three-year-old clump in the fall and dividing. After digging, shake the dirt loose and gently separate the multiple crowns into separate sections, making sure each section includes several well-developed roots. Replant divisions with tips about one inch below the soil surface. Water thoroughly. For best growth and flowering effect, space plants eighteen to twenty-four inches apart, and divide every three or four years. These plants die back to ground level each winter and are late to leaf out in the spring.

It is always best to start the native species from seeds. Tag a flowering plant, and return four or five weeks later. Give the cones plenty

Purple Coneflower

of time to mature and to loosen the seeds, for cones are prickly when dry and not pleasant to work with. Bend ripe seed heads over an open paper bag and knock against the sides to loosen seeds, or clip the cone from the plant and shake vigorously in the bag. Some seeds will remain in the seed head, so scatter heads around the mother plant for natural propagation. Immediately after harvesting, sow gathered seeds in a prepared seedbed and mark well. Seedlings should appear the following spring. Move to permanent places in the garden when four true leaves appear. Seeds can also be held over for planting in spring. If started after the last frost, plants should bloom by fall.

Although flowering stalks of Purple Coneflower are tall and slender, wind does not seem to bother them. This plant can survive in dry, droughty soil but thrives best in well-drained, fertile, somewhat limy soil, so work a handful of limestone amendments around each plant in the spring. Plant in full sun. Purple Coneflower tolerates some light shade but will become somewhat taller and scraggly, and the color will be paler. If not given good drainage, plants will simply rot away; if there is any doubt, add a shovelful of sand to each planting hole.

Scatter Purple Coneflowers in small groups throughout the middle of a bed or border. For a truly stunning effect, try planting taller native ones at the back of a group of cultivars and border with Sweet Alyssum (*Lobularia maritima*) or verbenas (*Verbena* spp.). The natives are also very drought tolerant and wonderful used in naturalized plantings.

Purple Marsh-fleabane (Canela)
(Pluchea odorata)

Family: Aster (Asteraceae)
Class: Native
Height: To 6 feet
Bloom period: May–frost
Range: Throughout

Upright, stout, lush, deeply rooted, solitary-stemmed annual, branching in upper portion. Leaves alternate, gray-green, downy, toothed along margin, exuding sticky resin. Flower heads, small, pinkish-purple, forming numerous flat-topped clusters. Seeds numerous, narrow, black, ridged. All parts of this plant emit a strong, pungent, or unpleasant scent when disturbed.

CULTIVATION: Purple Marsh-fleabane will

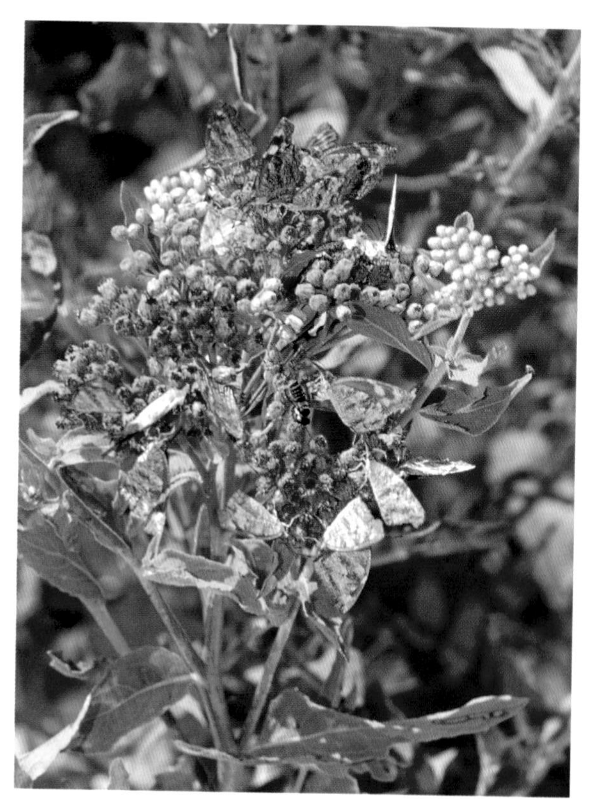

PURPLE MARSH-FLEABANE (CANELA)

naturally be found inhabiting low moist sites such as swales and depressions and along borders of resacas, lakes, and marshes. In a cultivated site it will tolerate drier situations but will need to be placed where moisture can be given when needed.

The best and easiest method for propagation is to gather old flowering clusters after they become tannish and fluffy. Clip the entire cluster, hold it over a container, and shake the seeds loose. Keep the cluster in an open container for few days for more seeds to mature. Sow some seeds immediately where plants are wanted; plant other seeds in fall if needed. They can also be started indoors in pots or flats, using moist potting soil. Seeds usually germinate readily, and plants grow quickly. Once started, plants will self-reproduce by rhizomes and form colonies.

NOTE: Purple Marsh-fleabane may not be a plant for the border but is a strong butterfly attractant if growing in a more "wilderness-like" area, such as along the edge of a large pond or around outbuildings if moisture can be provided. If any place can be found for this plant, it should be used. When in flower, it is usually covered by several species of butterflies and is continually used throughout the day.

Salt-marsh Mallow
(Kosteletzkya virginica)

Family: Mallow (Malvaceae)
Class: Native
Height: 3–6 feet
Bloom period: June–October
Range: 4

Robust, upright, shrublike, herbaceous perennial from tough roots. Stems thick, much-branched, usually several and fast growing, quickly forming large clump. Leaves large, gray-green, somewhat rough to the touch. Flowers large, cupped, shell pink to soft rose, borne in terminal portion of leafy stems and branches; last only a day, with new flowers opening each morning.

CULTIVATION: Blooming from late spring to early fall, the softly colored and attractive blossoms of this little-used native perennial are a welcome contrast to the usual reds and yellows of the garden. Naturally occurring in saline soils along brackish coastal marshes, Salt-marsh Mallow performs admirably under ordinary garden conditions throughout the coastal portion of the state. The tough roots grow deep, so plants usually require no additional watering except during really dry periods. Plant in good but sandy soil—if the soil is not already sandy, add a couple of shovelfuls to each planting hole. Yearly additions of compost (along with natural deterioration of the mulch) provide all the added nutrients needed. Mulching is very important, so apply fresh material as often as needed to help keep the roots cool. If fertilizing, do so sparingly,

SALT-MARSH MALLOW

as plants can easily be burned and can also be caused to produce foliage instead of flowers. An exposure of full sun is ideal; plants in the shade will be weak, leggy, and with only a few pale flowers. Place plants at least four feet apart, allowing plenty of room to spread.

Potted plants of Salt-marsh Mallow are occasionally sold by nurseries offering native plants. New plants are easy to grow from seeds. Each section of the five-part, brown capsule contains a dark brown seed. Collect capsules after they begin to split open. These seeds are frequently attacked by weevils, so put seeds in a glass or plastic container along with a portion of insect strip, seal, and store in a dark closet for a couple of weeks. At end of the two weeks, remove the strip and store seeds in the refrigerator until spring. Plant seeds in a marked location in the garden after soil has become thoroughly warm. Seedlings should be well rooted and large enough to transplant to permanent locations in the garden six to eight weeks later. Weekly applications of a very weak, all-purpose liquid fertilizer hasten seedling growth. New plants can also be started by taking tip cuttings before flowering begins and rooting in a sand and perlite medium. Even first-year plants, whether from seeds or cuttings, can be expected to flower modestly. Once the root system is well developed, vegetative growth is vigorous and flowering profuse.

Use Salt-marsh Mallow in liberal masses in the border, where butterflies can easily find its lovely pink blossoms. A single large grouping against a fence draped in Autumn Clematis (*Clematis paniculata*), Climbing Hempweed (*Mikania scandens*), or Queen's Wreath (*Antigon leptopus*) can be spectacular. It is late to break dormancy so can be interplanted with shallow-rooted spring- and early-summer-flowering plants.

For complementary fall companions, try Mexican Orchid Tree (*Bauhinia mexicana*), Sweet Almond Verbena (*Aloysia virgata*), Wright's Snakeroot (*Ageratina wrightii*), Summer Phlox (*Phlox paniculata*), or goldenrods (*Solidago* spp.), mistflowers (*Conoclinium* spp.), or some of the blue-colored salvias (*Salvia* spp.).

Sedum
(Sedum spectabile)

Family: Stonecrop (Crassulaceae)
Class: Cultivated
Height: 12–24 inches
Bloom period: August–November
Range: Throughout

Strong, stocky, succulent perennial. Leaves thick, leathery, silvery-green. Flowers numerous, small, glowing pink, tightly compacted into large, flattish, terminal clusters; deepen into darker hues with age.

CULTIVATION: Sedum is not readily propagated from seeds, but plants are available from almost all local nurseries and through catalogs. Many different growth forms and colors are available, but for butterflies the taller ones are the best. Also, butterflies do not seem to be very fond of the spring-flowering yellow one or the red-flowered form 'Autumn Joy.' For best results, stick to the tall-growing, mass-flowering, pink-flowered sedums.

Sedum is exceptionally easy to grow, re-

SEDUM

quiring little or no attention, and plants may be left undisturbed for years. They do best when planted in infertile, gravelly soils with excellent drainage. Under ideal conditions flower clusters reach dinner-plate size. Be careful not to overfertilize or overwater these plants, or growth will become too lush and succulent, plants will sprawl and break, and only a few flowers will be produced, if any at all. Under good growing conditions, Sedum is usually hardy, practically insect- and disease-free, and, once well established, actually quite drought tolerant.

Flower color is the brightest when Sedum grows in full sun. Sedum tolerates some light shade, but stems are generally weaker, and flowers are fewer and paler in color.

To increase the number of plants, lift and divide clumps in the spring. Place new divisions in small groups, spacing the divisions eighteen to twenty inches apart. Stem cuttings can also be taken in summer. As soon as they are rooted, place cuttings in the garden in permanent locations. They will bloom the following year. Once established, Sedum does not like to be moved, so unless new plants are needed, leave clumps undisturbed.

In the border, Sedum blends beautifully with Hispid Wedelia (*Wedelia acapulcensis* var. *hispida*), Tooth-leaved Goldeneye (*Viguiera dentata*), Lindheimer's Senna (*Senna lindheimeri*), New England Aster (*Symphyotrichum novae-angliae*), Texas Aster (*S. drummondii* var. *texana*), Tahoka Daisy (*Machaeranthera tanacetifolia*), Plains Zinnia (*Zinnia grandiflora*), or verbenas (*Verbena* spp.), liatris (*Liatris* spp.), goldenrods (*Solidago* spp.), and mistflowers (*Conoclinium* spp.). It is especially lovely planted in great masses in front of shrubs such as Beebrush (*Aloysia gratissima*), Blue Mistshrub (*Caryopteris incana*), Butterfly Bush (*Buddleja davidii*), Texas Kidneywood (*Eysenhardtia texana*), and Wright's Snakeroot (*Ageratina wrightii*).

NOTE: This sedum is one of the best cultivated late fall-flowering plants for attracting butterflies. It draws in just about every species still flying at the time of its flowering. Insects frequently use the large, flat flower clusters for basking as well as nectaring. The cultivar 'Meteor' is shown here.

Showy Bergamot
(Monarda didyma)

Family: Mint (Lamiaceae)
Class: Native/Cultivated
Height: To 4 feet
Bloom period: June–August
Range: Throughout

Lush, robust, square-stemmed perennial, forming colonies. Leaves large, opposite, emitting a strong minty fragrance when crushed. Flowers numerous, lavender to rose-pink, tubular, forming flat, terminal clusters, each cluster surrounded by several green, leaflike bracts.
CULTIVATION: If given proper growing conditions, Showy Bergamot can be a spectacular plant for the garden. And considering its attractiveness to butterflies, it is well worth whatever extra effort is necessary. Well-established plants or dormant roots are readily obtained from nurseries or from catalogs, and this is the best method for obtaining a start. Many different color forms are available in the cultivars, ranging from pure white to darkest purple to the scarlet of the wild, native form.

Showy Bergamot

Propagation of more plants can be by seeds, cuttings, or root division. Seeds are usually mature three or four weeks after flowers fall. Sow seeds indoors in January, using a good potting mixture and barely covering the seeds with soil. Seedlings are slow growing and benefit from occasional applications of a starter solution. Transfer plants to the garden as soon as all danger of frost has passed and three or four true leaves are showing. Seeds may be sown directly in the garden in spring or early summer for flowering plants the following year.

When well-established clumps have obtained a growth of six to eight inches in the spring, pinch or clip each new shoot back to three or four inches. This will cause the plants to form several new branches and will ensure a lower-growing and bushier plant. This also provides cuttings for new plants. Strip the lower leaves from the pinched tips, dip stems in rooting powder, and place in clay pots filled with a good rooting mixture. Sink pots in the ground where they can be watered frequently and will receive humidity. Cuttings should be rooted in four or five weeks and can then be transferred to the garden.

Showy Bergamot spreads by shallow underground runners, and new plants can be obtained by simply lifting one of the young plants along with a good ball of soil. Mature clumps can be divided in early spring before new shoots appear. If Showy Bergamot spreads more than desired, when replanting, place three or four plants in a large fifteen- or twenty-gallon plastic pot with the bottom removed and sink to ground level.

These hardy plants like continually moist soils during the growing season, and their lushness of foliage and amount of flowering depend on how much moisture they get. Try using them in a raised bed with rich, loamy, well-drained soil full of organic matter such as peat moss or leaf mold. Add sand for drainage if necessary. Mulch heavily and use a soaker hose. With extra watering, Showy Bergamot becomes spectacular even in full sun instead of the usually recommended semishade. Keep faded flowers clipped to prolong flowering. NOTE: The cultivar 'Crofway Pink' is shown here.

Stokes Aster

(Stokesia laevis)

Family: Aster (Asteraceae)
Class: Native/Cultivated
Height: To 2 feet
Bloom period: May–November
Range: 2, 3, 4, 6

Strong, clump-forming perennial. Leaves dark green, leathery, in dense rosettes. Flowers in tight four- to five-inch clusters, large, lavender or blue-violet, atop long, leafy stems, with one flower head opening at a time; flower heads surrounded by numerous stiff, leaflike bracts. CULTIVATION: Stokes Aster is a very hardy, easily grown plant that often remains evergreen in the southeastern portion of the state. It is another of the wildings that has been "improved" for lower growth, longer flowering, and a generally darker flower color. It is a common native in the southeastern states, ranging naturally as far west as central Louisiana.

Whether seeds are gathered from the wild or cultivars are purchased from catalogs or nurseries, the growing methods are the same. Do not be in a hurry to collect the seeds, for it will probably be at least two months after the petals fall from the head before seeds are mature. Once the bracts curl away from the head, clip the fruiting spikes and hang them head down in a paper bag. Let them air-dry for several days. Shake vigorously and crush the heads if necessary. The seeds may be sown directly in the garden in late spring.

Once a few plants are established, new plants are easily obtained by divisions, since the clumps should be divided every three or four years to maintain vigor. In the spring or fall, take up the clump and with a sharp knife divide the clump into sections, leaving plenty of roots with each section. Replant all sections immediately, watering in well.

Stokes Aster generally grows best and produces the most flowers with open, dappled afternoon shade and morning sun. It is not

Stokes Aster

particular about soil but must have excellent drainage. If soil is extra heavy, add a shovelful of coarse sand when planting. This plant is tolerant of heat, drought, and abuse, but it responds with lusher foliage and more and larger flowers if provided a little tender, loving care.

If seeds are not wanted, keep all spent flowers clipped for longer flowering. Once flowering has slowed down, usually in early summer, trim the plant back, and it will bloom again in the fall.

This is a wonderful, low-maintenance plant that makes a beautiful addition to the perennial border, providing a wealth of remarkable long-lasting blooms. Place it near the middle of the border, interplanted with other sun-loving species such as Butterfly Weed (*Asclepias tuberosa*), Purple Coneflower (*Echinacea purpurea*), Brown-eyed Susan (*Rudbeckia hirta*), and Pentas (*Pentas lanceolata*). Or use this versatile plant in large containers for patio or deck gardening, edging with Sweet Alyssum (*Lobularia maritima*), Garden Pansy (*Viola × wittrockiana*), or the purple-flowered Prairie Verbena (*Glandularia bipinnatifida*).

NOTE: Colors of the cultivars range from white to pink and various blues and lavenders. Butterflies prefer the darker lavenders and pinks, but some of the lighter blues might be tried. If any of the cultivars are advertised as being fragrant, try those.

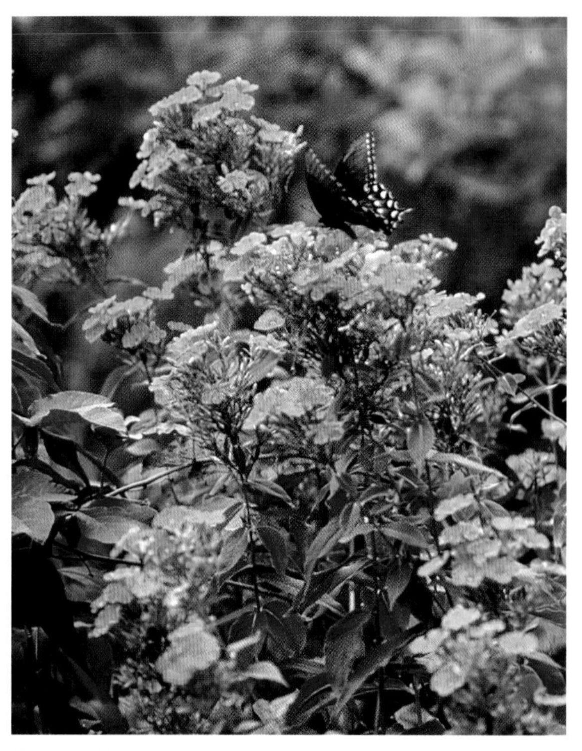

Summer Phlox

Summer Phlox
(Phlox paniculata)

Family: Phlox (Polemoniaceae)
Class: Cultivated
Height: 3–6 feet
Bloom period: June–October
Range: Throughout

A robust, upright, clump-forming, long-lived perennial. Leaves pale to dark green, opposite. Flowers trumpet-shaped, numerous, borne in large, somewhat elongated or dome-shaped terminal clusters.

CULTIVATION: Summer Phlox is a common native wildflower of the southeastern states that has been "improved." It is now used frequently in cultivation, and the spectacular clusters of rich color add greatly to the summer border. The flowers are strongly and sweetly scented and lure butterflies like magnets. At any one time, there will usually be a multitude of the insects, of all colors and sizes, for this phlox is well liked by just about every nectar-using butterfly.

Give Summer Phlox an extra-sandy soil, good drainage, full sun, and adequate moisture, and the plants will be absolutely spectacular. If the soil is more compacted, for example, contains a lot of clay, flowering and foliage are still excellent, but plants are much shorter. There seems to be a direct relation between sandy soil producing the tallest growth and compacted soil producing shorter plants.

New plants can be started from seeds, but colors rarely come true and are often muddy or generally undesirable. For new plants it is much better to take cuttings or divide plants. Start four- to six-inch cuttings in May or June, placing them in the garden in fall for flowering plants the following year. As there are large clumps of Summer Phlox wherever it grows, the gardener possessing these plants is usually happy to share a sprout or two. Young offshoots are easily separated from the mother plant and, with good care, will bloom the first year. Space new plants at least two feet apart, and leave them uncrowded. Keep old flowering heads trimmed to prevent seed set and ensure continued bloom. Lift and divide clumps every two or three years.

These plants are susceptible to mildew if not given good air circulation. Use a soaker hose for watering, or lay a regular hose on the ground, keeping foliage as dry as possible at all times. A deep, loose mulching conserves moisture and prevents dirt from splashing on the leaves, which can also cause mildew. Although it is imperative to keep water off the foliage of these plants, a deep, regular watering is absolutely essential during excessively dry periods.

Summer Phlox is the perfect border plant, with its strong, upright growth form and long blooming period. Place the plants in groups toward the back of the border if the soil is sandy or in the middle if the soil is heavier with clay. Combine with other summer- and fall-flowering species such as Butterfly Bush (*Buddleja davidii*), Sweet Almond Verbena (*Aloysia virgata*), White-flowered Plumbago (*Plumbago scandens*), Globe Thistle (*Echinops banaticus*), New England Aster (*Symphyotrichum novae-angliae*), Salt-marsh Mallow (*Kosteletzkya virginica*), Mexican Sunflower (*Tithonia rotundifolia*), the single-flowered hollyhocks (*Alcea* spp.), mistflowers (*Conoclinium* spp.), and the vines Climbing Hempweed (*Mikania scandens*) and Queen's Wreath (*Antigon leptopus*).

NOTE: *Phlox paniculata* can occasionally be found in local nurseries, but a wider choice will be from catalogs. There are many varieties available, but always choose ones especially noted for fragrance and butterfly usage. Some of them no longer produce abundant nectar.

RELATED SPECIES: Two other native and similar phlox, *P. subulata* and *P. maculata*, are also much used by butterflies. They, too, have been "improved," so selection must be made carefully.

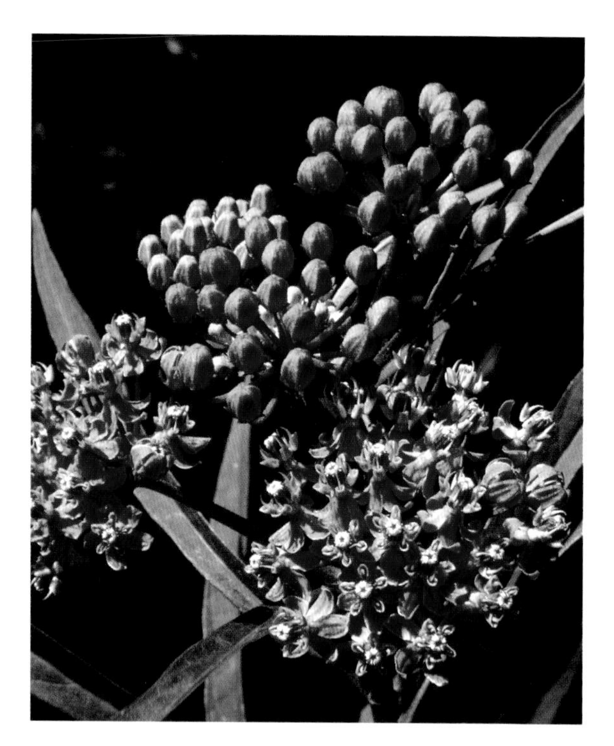

Swamp Milkweed

Swamp Milkweed

(Asclepias incarnata)

Family: Milkweed (Asclepiaceae)
Class: Native
Height: 1½–6 feet
Bloom period: July–October
Range: 1, 2, 6

An erect, stout, usually several-stemmed herbaceous perennial, in the wild usually found along the moist edges of creeks and rivers. Leaves opposite, usually upright, mostly smooth, lance-shaped, cover stalks and upper branches. Flowers in short-stalked clusters, pink to purplish (rarely white), forming large masses at top of round, hollow stems. Almost all parts contain milky sap typical of milkweeds, although not as abundantly as with most species.

CULTIVATION: For best growth and most flowers, select a site in full sun in an organically rich, very moist but well-drained soil for this one. Placing it in an especially prepared bed along with Powdery Thalia (*Thalia dealbata*), Pickerel Weed (*Pontederia cordata*), Joe-Pye Weed (*Eutrochium fistulosum*), and Cardinal Flower (*Lobelia cardinalis*) would be ideal.

This plant is now often offered in the trade, which is the best method of obtaining plants. New plants can be started from seeds if collected when the fruiting pods are fully ripe. Seeds need a period of cold stratification for at least two to three weeks. Plant in either fall or early spring. If the soil is warm, the seeds should germinate in one to two weeks. Ordinarily, it takes three years to get flowering plants from seeds.

Root division of healthy, well-established plans in either fall or early spring is a much quicker method for obtaining more flowering plants.

In some situations, Swamp Milkweed may actually spread more than wanted. If this should happen, simply pull or break off any unwanted sprouts. The clumps can be prevented from spreading by planting the original plant in a very large plastic planter with the bottom removed.

NOTE: Swamp Milkweed is a wonderful plant for areas such as a septic field or around water tanks or leaky hoses. It is more tolerant of salt buildup from low-lying or frequently moist areas than most plants. Nectar from these plants appears to be an even better attraction than Butterfly Weed (*A. tuberosa*) or Tropical Milkweed (*A. curassavica*), which are hard to beat.

Tahoka Daisy (Tansy Aster)
(Machaeranthera tanacetifolia)
Family: Aster (Asteraceae)
Class: Native
Height: To 16 inches
Bloom period: May–frost
Range: 1, 7 (6)

TAHOKA DAISY

Low annuals, somewhat sprawling or more often forming well-rounded mounds. Leaves pale green, deeply divided, fernlike. Flowers large, purple and yellow, in masses that overtop leaves.

CULTIVATION: Tahoka Daisy is one of the showiest of the western annuals and is easily cultivated. It is easy to grow from seeds and is also now being commonly offered as a container plant in the nursery trade. In the wild this plant generally grows in large colonies or masses and should be used in the garden in this way to attract butterflies. Plant as many as you can possibly allow space for, placing no fewer than five or six in a group—a dozen or more is even better. Plant fourteen to sixteen inches apart for a wonderful solid mass of color. They are excellent in several large containers for patio, deck, or porch plantings. Plant in full sun in any good garden soil, hold the fertilizer and water, keeping them on the light and dry side, and they should perform beautifully. Lighten extra-heavy clay garden soils by adding generous amounts of sand and coarse compost when planting. If an excessive amount of water is given these plants, they eventually "grow" themselves to death, producing a lot of foliage and few flowers.

First peak bloom is usually in May, but if spent flower heads are regularly clipped, a good showing will continue through summer with another peak period in the fall. If spent flowers are not cut, plants will continue flowering but not as abundantly. For a natural effect, scatter seeds in early fall in undisturbed beds, mark well, watch for the ferny rosettes in early spring, and thin as needed. Extra plants can be transplanted to other areas. If seeds are to be sown in the spring, place between paper towels, put towels between two layers of very wet peat moss in a plastic bag, and chill in the refrigerator at least two weeks before planting. As these plants vary in flower color, ranging from a dark royal purple all the way through pinkish to pure white, choose plants whose seeds you want while the plants are in flower.

NOTE: Although used by many species of butterflies, Tahoka Daisy seems to be especially favored by Skippers (family Papilionidae) and Hairstreaks (family Lycaenidae).

Tall Goldenrod
(Solidago altissima)

Family: Aster (Asteraceae)
Class: Native
Height: To 7 feet
Bloom period: September–November
Range: Throughout

A tall, stiff, robust, upright, many-stemmed deciduous perennial with stems branching in upper portion; spreads by underground runners, forming large colonies. Leaves alternate, numerous, rough, pale green, becoming smaller toward flowering portion of stem. Flowers numerous, small, yellow, fragrant, on upperside of branch in flattened racemes forming large terminal cluster.

CULTIVATION: Various species of goldenrod are commonly offered by nurseries and through catalogs, many of them an "improved" variety or cultivar. Be very careful of these—often they have been bred for showiness, and there is no nectar to be had.

Propagation of Tall Goldenrod can be by seeds or stem cuttings. A few weeks after flowering, maturing flower clusters become brown and fluffy. Clip the entire cluster, and place in a paper bag to let seeds become fully mature and thoroughly dry. At this time seeds can be separated and planted individually or seed heads can be divided and sections can be planted outside where plants are wanted. Due to the unpredictability of seed germination, sow generously.

To take stem cuttings, remove the lower leaves from six to eight inches of tip growth, dip in rooting hormone, and place in either a sand/perlite mixture or moist potting soil.

After plants are established, young root sprouts can be separated from the mother plant and new colonies started. Choose the permanent placement of this plant with care. With good soil and full sun it will grow very tall and quickly spread by underground runners to form colonies of several feet across. It continues to spread, the colony becoming larger each year. This is definitely not a plant for a garden border but makes a stunning show in a somewhat isolated place.

NOTE: Individual blossoms of Tall Goldenrod contain little nectar, but the large mass of flowers compensates for this. These plants are a major source of fall nectar for numerous species of butterflies, especially the migrating Monarchs (*Danaus plexippus*).

More than thirty native species of goldenrod are found in the state, so finding a seed source is not difficult. Butterflies will readily use the one they are most familiar with, so try to use a local species. Tall Goldenrod is the largest, most robust, and aggressive of the natives—there are many smaller ones quite adaptable for use in the garden.

TALL GOLDENROD

Thread-leaf Groundsel
(Senecio flaccidus)

Family: Aster (Asteraceae)
Class: Native
Height: To 2½ feet
Bloom period: April–September
Range: 1, 6, 7

An erect to somewhat spreading or bushy perennial, usually with several woody-based stems; covered in silvery hairs. Leaves numerous, crowded, finely lobed near base. Flower heads yellow, numerous, in branched terminal clusters.

CULTIVATION: Thread-leaf Groundsel may be used in the poorest, hottest, and driest portion of the garden and can be started either from seeds or root cuttings. The easiest method is to gather seeds as soon as mature and scatter in late fall in prepared beds where plants are to remain. Thin seedlings to at least three feet apart, transplanting the extras to other locations. Do not crowd this plant, for it likes breathing room. It is not particular about soil but definitely requires good drainage. Rocky or gravelly soil where there is no problem with water retention is ideal. Do not overwater, but occasional extra waterings between rains in the middle of summer keep plants in vigorous flower. The plant ordinarily performs admirably without fertilizing.

THREAD-LEAF GROUNDSEL

Use Thread-leaf Groundsel in groups toward the back of the border or as specimen plantings, maybe in a grouping with rocks or silvery, weathered wood. Plants are extremely hardy under the most trying conditions and, when well grown, form beautiful mounds of true yellow throughout the year. Peak bloom is about mid-April or early May, then again in September, but if spent flower clusters are regularly clipped, there will be continuous color all summer. In protected areas plants may be evergreen, with scattered flowers occurring even during the winter months.

Tooth-leaved Goldeneye
(Viguiera dentata)

Family: Aster (Asteraceae)
Class: Native
Height: To 6 feet
Bloom period: September–frost
Range: 2, 6, 7

A tall, robust perennial forming colonies. Leaves sunflower-like, occurring opposite on the lower portion of the stalk, becoming alternate in upper portion. Flower heads numerous, golden-yellow, forming loose clusters at tips of long, slender branches.

CULTIVATION: The best method for obtaining Thread-leaved Goldeneye for the garden is by gathering the dry, stiff, seedlike achenes when dark brown and fully mature. Place "seeds" in an open paper bag for a few days until they are completely dry. Plant in prepared beds in late September or when good fall rains begin. Seeds can be held over and planted in the spring, but more robust plants producing the most flowers come from fall-planted seeds.

Goldeneye grows beautifully in its native limestone soil of Central and West Texas but will grow admirably in good garden soil past its normal range. In more northern areas, give it a protective, loose straw mulch for the winter; to the south, give some semishade; to the east, provide sandy soil with extra drainage, and add a bit of lime from time to time.

Because of its height, use Goldeneye toward the back of the border, against a fence, or as accent clumps. The plants are wonderful for hiding compost heaps, trash receptacles, brush piles, small storage buildings, or any other unsightly object. As these plants form large colonies, they are also useful on slopes to help control erosion. If Goldeneye is being used in a naturalizing planting, sow seeds toward the outer fringes of the planting, as their height and natural tendency to multiply may shade or crowd out lower-growing and less robust wildings.

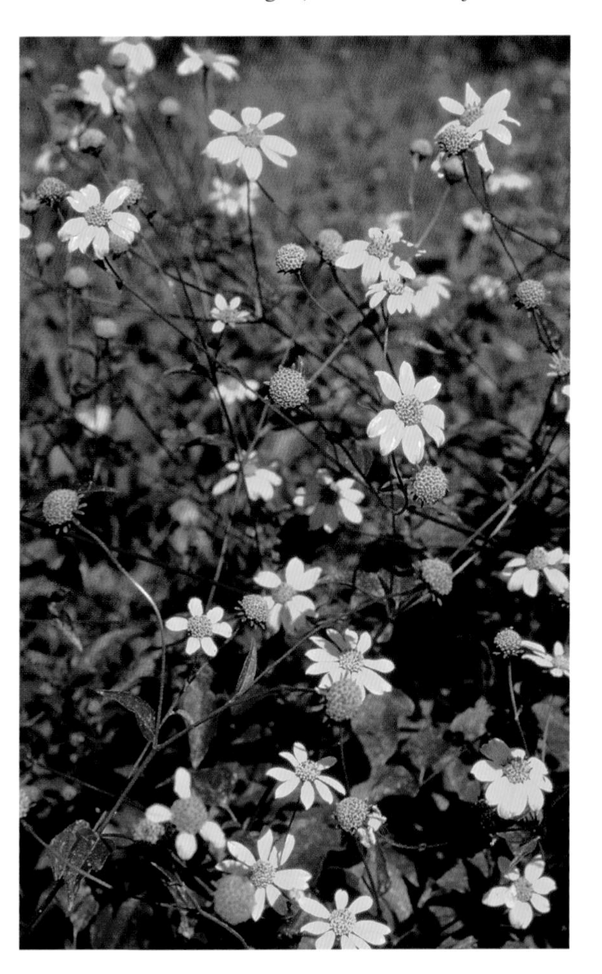

TOOTH-LEAVED GOLDENEYE

NOTE: This is one of the best western native fall wildflowers for attracting butterflies. Some of the most frequent visitors are Bordered Patch (*Chlosyne lacinia adjutrix*), Gray Hairstreak (*Strymon melinus franki*), Pearl Crescent (*Phyciodes tharos tharos*), Texan Crescent (*Anthanassa texana*), Sachem (*Atalopedes campestris huron*), and Fiery Skipper (*Hylephila phyleus*). The Gulf Fritillary (*Agraulis vanillae incarnata*), Great Purple Hairstreak (*Atlides halesus halesus*), Julia Longwing (*Dryas iulia*), Monarch (*Danaus plexippus*), Orange Sulphur (*Colias eurytheme*), and Sleepy Orange (*Abaeis nicippe nicippe*) are also commonly seen nectaring on the flowers. It is one of the larval food plants of the Bordered Patch butterfly.

RELATED SPECIES: Skeleton-leaf Goldeneye (*V. stenoloba*) is a three-foot-tall, many-branched shrub found in Regions 1, 5, 6, and 7 and is excellent when used in landscaping. It is covered in nectar-rich yellow flowers throughout the flowering season, especially after rainfall, and is visited by numerous species of butterflies.

Tropical Sage
(Salvia coccinea)

Family: Mint (Lamiaceae)
Class: Native
Height: 1–3 feet
Bloom period: February–December (all year)
Range: 1, 2, 3, 4, 5, 6

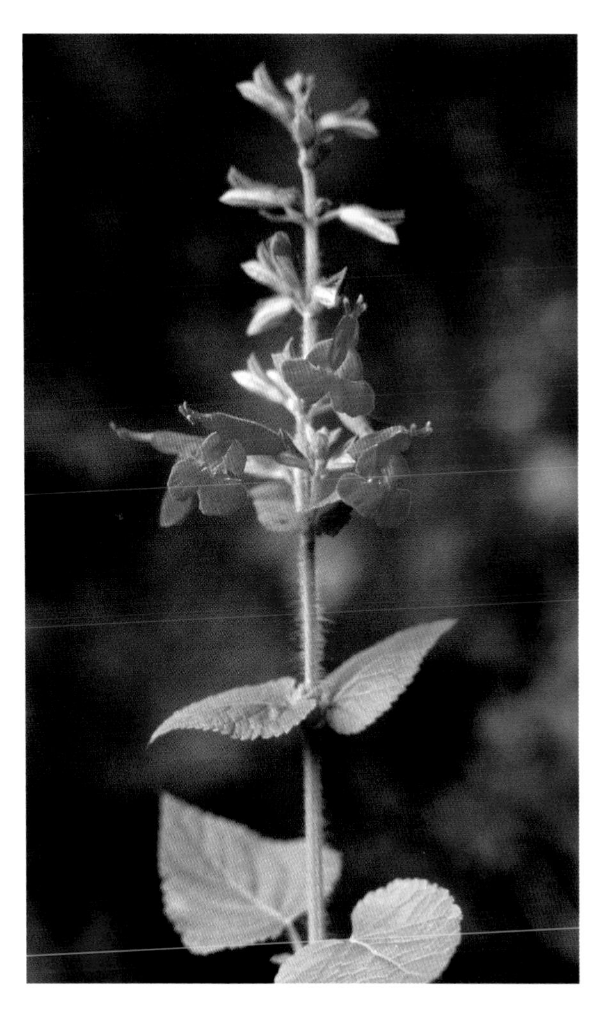

Tropical Sage

In the wild, usually forms large, erect or somewhat reclining stands displaying brilliant red patches rising above lush green foliage; an annual in the northernmost region of its range (Region 1); acts as a perennial in the middle of the state, usually freezing to the ground each winter; may remain evergreen and flowering the entire year in the Valley area. Leaves opposite on square stems. Flowers on a terminal spike, opposite, brilliant red, two-lobed, tubular, in separated whorls or clusters.
CULTIVATION: Tropical Sage readily reseeds and also root-sprouts, so obtaining new plants should be no problem. Transplant new seedlings to desired areas, or healthy clumps can be lifted and divided for larger plants. Water well to get roots started. Ideally, soil for Tropical Sage should be very well drained. If the soil is too rich, the plants tend to become larger and coarser than generally wanted in flower beds. In typical garden soils and with normal watering, it makes a spectacular showing; if it becomes too tall, simply trim back. The plants will grow back bushier, lusher, and full of blooms.

This is one plant that does well even in the poorest, driest soils so can be used in problem spots. After the first frost and when flowering has finished, trim the dead stalks and cover the plants with a layer of loose, dried leaves as a protective winter mulch.
NOTE: This is a favorite of butterflies having a long proboscis such as the Swallowtails (family Papilionidae), the larger Yellows and Sulphurs (family Pieridae), and some of the Skippers (family Hesperiidae). It is also avidly used by hummingbirds. In the trade, white- and pink-flowered forms are generally offered.
RELATED SPECIES: Cherry Sage (*S. greggii*), a shrubbier form, is a western native common in the trade and is offered in colors from white to deep magenta to red.

Virginia Frostweed

Virginia Frostweed

(Verbesina virginica)

Family: Aster (Asteraceae)
Class: Native
Height: 3–6 feet
Bloom period: August–December
Range: 2, 3, 4, 6

Abundant, late-summer and fall-blooming perennial herb attracting many species of butterflies, both "regulars" and the rare "strays" from the Valley; growing in large colonies beneath trees or edges and openings of woodlands; in the wild often common growing to fullest height along shaded creek banks. Leaves dull green, alternate, extending down the stem as narrow wings, upper leaves much smaller and stalkless. Stiffly upright stems branching only when flowering into short stems bearing clusters of white flowers, the whole forming a huge flat to rounded terminal mass.

CULTIVATION: Only occasionally offered at native plant sales, Virginia Frostweed will most probably need to be obtained from the wild. Because it is easily started by seeds, find a stand of plants along a roadside or in a friend's pasture. After the plants have been flowering for a while but before a frost, wrap fine netting over the entire head and tie beneath. After all flowering is finished, clip the entire head and hang upside down until the stalk is completely dry. Remove netting, and shake seeds loose. Plant some seeds immediately where plants are wanted in the garden, and save some seeds for planting in the spring. Once seedlings are up, they can be thinned to one to two feet apart. Plant at the back of the border along with Mexican Sunflower (*Tithonia rotundifolia*) and Joe-Pye Weed (*Eutrochium fistulosum*). Generally the plants are strong and sturdy and do not require staking. Their height is usually based on the amount of water they receive. Never overwater these plants, for they usually do best if somewhat on the dry side.

Virginia Frostweed blooms only during the latter part of the flowering season, but the massive flower heads are outstanding, and the beautiful dark green foliage makes a perfect background for lower-growing species.

NOTE: The common name "frostweed" comes from the phenomenon of freezing sap oozing from bursting stems after a hard freeze, forming beautiful and unusual formations. Frostweed is a larval food plant for the Silvery Checkerspot (*Chlosyne nycteis*), Bordered Patch (*C. lacinia adjutrix*), and Spring/Summer Azure (*Celastrina ladon*).

RELATED SPECIES: A very similar frostweed (*V. microptera*) with larger, whiter, and showier flowers occurs in the Rio Grande Valley area (Region 5) and is used by the Bordered Patch and Silvery Checkerspot.

Woolly Paper-flower (Paper-Daisy)

(Psilostrophe tagetina)

Family: Aster (Asteraceae)
Class: Native
Height: 4–24 inches
Bloom period: March–October
Range: 1, 5, 6, 7

WOOLLY PAPER-FLOWER (PAPER-DAISY)

A low, upright or sprawling, many-stemmed biennial or perennial from woody taproot. Stems usually much-branched, with the plant forming clumps or mounds. Leaves in basal portions of plant covered in soft, woolly hairs, stalked, sometimes lobed; upper leaves stalkless, smaller, greener. Flower heads yellow, several, and closely congested in terminal clusters.

CULTIVATION: Woolly Paper-flower is commonly found throughout the northern and western portions of the state, usually in gypsum or sandy soil, and is very drought resistant. In the wild it often forms extensive and showy masses along roadsides or in fields and meadows. When introduced into the garden, the plant's appearance becomes even more dramatic. After flowering, the numerous dried flowers become papery and can be left on the plant for months, the gold gradually fading to beautiful browns and tans.

The best method for propagating this plant is by seeds, collecting the seeds in late summer or early fall after they have become fully mature. Cut stems below the entire head, and hang upside down for a few days to air-dry. Plant in late fall for flowering plants the following season. In early spring young plants can be moved to desired locations in the garden.

Woolly Paper-flower is ideal as an edging for a sunny perennial border or along a path. It is also good in masses as filler plants when other plants finish flowering and are removed from the beds. Try using Woolly Paper-flower in large containers and in natural plantings. It is especially lovely placed close to rocks and in the company of Cenizo (*Leucophyllum frutescens*), Snapdragon Vine (*Maurandya antirrhiniflora*), Downy Paintbrush (*Castilleja sessiliflora*), Fine-leaf Woolly-white (*Hymenopappus flavescens*), New England Aster (*Symphyotrichum novae-angliae*), Phacelia (*Phacelia congesta*), and any of the verbenas (*Verbena* spp.).

A dense, woolly coating of soft hairs on the foliage of Woolly Paper-flower enables it to stand long periods of drought by reducing moisture lost by the leaves through evaporation. This ability to retain moisture could be fatal if plants were kept continually moist. Let the ground become completely dry between waterings, and add sand to the soil for good drainage when planting. Use only very light applications of fertilizer, if any at all. After a plant is growing well, it can be trimmed back to encourage even lusher growth, but this is usually not necessary.

Additional Information and More Nectar Plants

Bloom periods given here are only approximate.

Native Trees

Tree	***Height***	***Color***	***Bloom period***
Anacahuita *(Cordia boissieri)*	to 24 feet	white	throughout year
Basswood, Carolina *(Tilia americana* var. *caroliniana)*	to 120 feet	white or yellowish	May–July
Blackhaw, Rusty *(Viburnum rufidulum)*	to 20 feet	whitish	March–June
Bumelia *(Sideroxylon celastrina)*	to 25 feet	whitish, yellowish, or greenish	May–November
Bumelia, Gum *(Sideroxylon lanuginosum)*	to 45 feet	greenish-yellow	May–July
Catalpa, Northern *(Catalpa speciosa)*	to 60 feet	white to cream to pinkish	May and June
Cherry, Wild Black *(Prunus serotina)*	to 90 feet	white	March and April
Chinquapin, Allegheny *(Castanea pumila)*	to 30 feet	cream to yellowish-green	March–June
Dogwood, Flowering *(Cornus florida)*	to 36 feet	greenish-yellow and white	March and April
Ebony, Texas *(Ebenopsis ebano)*	to 30 feet	white	April–July
Lead Tree, Golden-ball *(Leucaena retusa)*	to 25 feet	bright yellow	April–August
Mesquite, Honey *(Prosopis glandulosa)*	to 30 feet	yellowish-green	May–September
Palo Verde *(Parkinsonia texana)*	to 25 feet	bright yellow	spring and summer
Plum, Mexican *(Prunus mexicana)*	to 25 feet	white	March
Redbud *(Cercis canadensis)*	to 35 feet	pink to dark rose or purplish	March and April
Retama *(Parkinsonia aculeata)*	to 30 feet	bright yellow	spring–fall
Serviceberry, Downy *(Amelanchier arborea)*	to 20 feet	white	March–June

Tree	Height	Color	Bloom period
Silverbells, Carolina *(Halesia carolina)*	to 36 feet	snowy-white	April and May
Silverbells, Two-winged *(Halesia diptera)*	to 30 feet	snowy-white	March–June
Tenaza *(Havardia pallens)*	to 18 feet	white to cream	May–September
Willow, Desert *(Chilopsis linearis)*	to 30 feet	white to dark rosy-pink or violet-purple	April–November

Nonnative/Cultivated Trees

Tree	Height	Color	Bloom period
Apple *(Malus pumila)*	to 30 feet	white-pink	April and May
Bird-of-Paradise, Mexican *(Caesalpinia mexicana)*	to 15 feet	red and yellow	throughout year
Chaste Tree *(Vitex agnus-castus)*	to 20 feet	lilac to lavender	June–October
Chinaberry *(Melia azedarach)*	to 45 feet	purplish	March–June
Grapefruit *(Citrus maxima × paradisi)*	to 20 feet	white	throughout year
Lemon *(Citrus limon)*	to 15 feet	white and purplish	throughout year
Lime *(Citrus aurantifolia)*	to 15 feet	white	throughout year
Mimosa, Flowering *(Albizia julibrissin)*	to 25 feet	pale pink to dark rose	April–September
Orange, Sweet *(Citrus sinensis)*	to 20 feet	white	throughout year
Peach *(Prunus persica)*	to 25 feet	pink	March–June
Plum, American *(Prunus americana)*	to 25 feet	white	March–June

Native Shrubs

Shrub	*Height*	*Color*	*Bloom period*
Acacia, Cat-claw *(Acacia greggii)*	to 6 feet	creamy-white to yellow	April–November
Acacia, Twisted *(Acacia schaffneri)*	to 5 feet	bright yellow	spring
Acanthus, Flame *(Anisacanthus quadrifidus* var. *wrightii)*	to 5 feet	deep orange to orange-red	June–frost
Agarita *(Mahonia trifoliolata)*	to 6 feet	yellow	February–May
Anisacanth, Chinati *(Anisacanthus puberulus)*	to 6 feet	white, pink to purple	April–July (after rains)
Anisacanth, Dwarf *(Anisacanthus linearis)*	to 5 feet	golden-yellow to orange-red	summer (after rains)
Azalea, Piedmont *(Rhododendron canescens)*	to 10 feet	pale pink to dark rose	March–May
Baccharis, Willow *(Baccharis salicina)*	to 6 feet	greenish-yellow	May–July
Beebrush, Common *(Aloysia gratissima)*	to 9 feet	white	March–December
Beebrush, Rio Grande *(Aloysia macrostachya)*	to 3 feet	pink to red or lavender	January–November
Beebrush, Wright's *(Aloysia wrightii)*	to 6 feet	white	June–November
Brasil *(Condalia hookeri)*	to 15 feet	yellowish to greenish	April–October
Buckeye, Mexican *(Ungnadia speciosa)*	to 15 feet	pink to lavender	March–June
Butterfly Bush, Woolly *(Buddleja marrubiifolia)*	to 3 feet	golden-yellow to orange-red	March–September
Buttonbush, Common *(Cephalanthus occidentalis)*	to 15 feet	creamy-white to pinkish	June–October
Ceanothus, Desert *(Ceanothus greggii)*	to 6 feet	white	spring
Ceanothus, Fendler's *(Ceanothus fendleri)*	to 3 feet	white	April–September
Cherry, Barbados *(Malpighia glabra)*	to 8 feet	pink to reddish or purplish	March–November
Chokeberry, Red *(Aronia arbutifolia)*	to 20 feet	white to pale pink	March–May
Colubrina, Texas *(Colubrina texensis)*	to 6 feet	greenish-yellow	April–July
Crucita *(Chromolaena odorata)*	to 6 feet	lilac to purplish-blue	August–frost

Shrub	*Height*	*Color*	*Bloom period*
Dewberry, Southern (*Rubus trivialis*)	to 3 feet	white	February–May
Elbow-bush, Fall (*Forestiera ligustrina*)	to 12 feet	greenish-yellow	August–frost
Elbow-bush, Spring (*Forestiera pubescens*)	to 8 feet	greenish-yellow	January–April
Fendler-bush, Cliff (*Fendlera rupicola*)	to 6 feet	white to pinkish	March–July
Goldeneye, Skeleton-leaf (*Viguiera stenoloba*)	to 4 feet	yellow	May–September
Guajillo (*Acacia berlandieri*)	to 6 feet	creamy-white	November–April
Hawthorn, Parsley-leaved (*Crataegus marshallii*)	to 25 feet	white	March and April
Honeysuckle, Bush (*Lonicera albiflora*)	to 6 feet	white, creamy to yellow	March–May
Indigo, False (*Amorpha fruticosa*)	to 15 feet	dark purple	April–July
Kidneywood, Texas (*Eysenhardtia texana*)	to 10 feet	white to creamy	April–November
Lantana, Desert (*Lantana achyranthifolia*)	to 5 feet	pink or lavender to purple and yellow	February–frost
Lantana, Texas (*Lantana urticoides*)	to 6 feet	yellow to orange to red	May–frost
Lippia, Scented (*Lippia graveolens*)	to 9 feet	yellowish or white	March–frost
Lippia, White-flowered (*Lippia alba*)	to 6 feet	white, pink, violet to purple	March–November
Mimosa, Cat's-claw (*Mimosa aculeaticarpa* var. *biuncifera*)	to 3 feet	whitish to pink	April–October
Mimosa, Fragrant (*Mimosa borealis*)	to 3 feet	white to pink	April–August
Mimosa, Velvet-pod (*Mimosa dysocarpa*)	to 6 feet	purplish-pink	June and July
Orange, Mexican (*Choisya dumosa*)	to 6 feet	white	June–frost
Pepper-bush, Sweet (*Clethra alnifolia*)	to 9 feet	white	July–September
Plum, Chickasaw (*Prunus angustifolia*)	to 12 feet	creamy-white	March and April
Plum, Creek (*Prunus rivularis*)	to 6 feet	white	March

Shrub	*Height*	*Color*	*Bloom period*
Plum, Oklahoma (*Prunus gracilis*)	to 5 feet	creamy-white	March and April
Plumbago, White-flowered (*Plumbago scandens*)	to 4 feet	white	almost all year
Rabbit-brush, Southwest (*Chrysothamnus pulchellus*)	to 3½ feet	golden-yellow	August–November
Redroot (*Ceanothus herbaceus*)	to 4 feet	white	May–August
Roosevelt-weed (*Baccharis neglecta*)	to 9 feet	yellowish-white to creamy	September and October
Rose, Carolina (*Rosa carolina*)	to 3 feet	pink	May–August
Rose, Prairie (*Rosa setigera*)	to 15 feet	white to rose-pink	April–July
Rose-mallow, Desert (*Hibiscus coulteri*)	to 3 feet	pale to dark yellow	April–August
Rosemary-mint, Woolly (*Poliomintha incana*)	to 3 feet	white, bluish to pale purple	April–July
Sage, Cherry (*Salvia greggii*)	to 2 feet	white to red	April–frost
Sage, Shrubby Blue (*Salvia ballotiflora*)	to 6 feet	bluish or purple	January–frost
Seep-willow (*Baccharis salicifolia*)	to 10 feet	yellowish-white or cream	summer through fall
Senna, Wislizenus (*Senna wislizeni*)	to 10 feet	yellow	May–August
Silverbells, American (*Styrax americana*)	to 15 feet	snowy-white	April and May
Snakeroot, Havana (*Ageratina havanensis*)	to 10 feet	white	September–frost
Snakeroot, Wright's (*Ageratina wrightii*)	to 4 feet	white	September–frost
Snakewood, Texas (*Colubrina texensis*)	to 6 feet	greenish-yellow	March–June
Sweetspire, Virginia (*Itea virginica*)	to 9 feet	white	April–June
Tea, New Jersey (*Ceanothus americanus* var. *pitcherii*)	to 4 feet	white	May and July
Wax-mallow, Drummond's (*Malvaviscus drummondii*)	to 10 feet	red	throughout year
Yellow Bells (*Tecoma stans*)	to 6 feet	yellow	April–frost

Nonnative/Cultivated Shrubs

Shrub	***Height***	***Color***	***Bloom period***
Abelia, Glossy *(Abelia × grandiflora)*	to 8 feet	white to pale pink	June–November
Azalea, Garden *(Rhododendron* spp.)	to 8 feet	white, pink to magenta, salmon to yellow or orange	April–July
Bird-of-Paradise, Barbados *(Caesalpinia pulcherrima)*	to 15 feet	red	May–October (all year)
Bird-of-Paradise, Mexican *(Caesalpinia mexicana)*	to 12 feet	red and yellow	most of year
Bougainvillea *(Bougainvillea glabra)*	to 12 feet	white, yellow, red to purplish	spring–frost (all year)
Butterfly Bush *(Buddleja davidii)*	to 12 feet	white, pink, blue, lavender to purple	May–frost
Butterfly Bush, Lindley's *(Buddleja lindleyana)*	to 6 feet	reddish to purplish or violet	summer through fall
Duranta, Golden Dew-drop *(Duranta erecta)*	to 18 feet	white, blue to lavender	almost all year
Lantana, Trailing *(Lantana montevidensis)*	to 10 feet	pale lilac to purple	February–October
Lantana, West Indian *(Lantana camara)*	to 6 feet	yellow to orange or red	May–frost
Lilac, Fragrant *(Syringa vulgaris)*	to 12 feet	white, rose to purple	early spring
Mistshrub, Blue *(Caryopteris incana)*	to 2 feet	white or blue	late summer through fall
Mock-orange *(Philadelphus* spp.)	to 6 feet	white to pink	April–June
Privet, Chinese *(Ligustrum sinense)*	to 20 feet	white	March–May
Spiraea *(Spiraea japonica)*	to 3 feet	white, rose, and pink	summer through fall
Verbena, Sweet Almond (*Aloysia virgata*)	to 15 feet	white	May–frost

Native Vines

Vine	*Length*	*Color*	*Bloom period*	*Life cycle*
Alamo Vine *(merremia dissecta*	to 12 feet	white and red	May–frost	perennial
Bindweed, Field *(Convolvulus arvensis)*	to 4 feet	white or pink	April–frost	perennial
Bindweed, Texas *(Convolvulus equitans)*	to 6 feet	white and reddish-maroon	April–frost	perennial
Climbing Milkweed *(Funastrum cynanchoides)*	to 4 feet	white to pink or purplish	April–October	perennial
Grape, Fox *(Vitis vulpine)*	to 20 feet	creamy to greenish	May–June	perennial
Hempweed, Climbing *(Mikania scandens)*	to 15 feet	whitish to pinkish	June–Frost	perennial
Honeysuckle, Coral *(Lonicera sempervirens)*	to 10 feet	red or yellow	March–June	perennial
Morning Glory, Lindheimer's *(Ipomoea lindheimeri)*	to 6 feet	pale blue to lavender	April–frost	perennial
Morning Glory, Salt-marsh *(Ipomoea sagittata)*	to 6 feet	dark pink to red-purple	April–frost	perennial
Morning Glory, Sharp-pod *(Ipomoea cordatotriloba)*	to 6 feet	lavender to rose-purple	April–frost	perennial
Pepper-vine *(Ampelopsis arborea)*	to 15 feet	greenish	June and July	perennial
Wild Potato Vine *(Ipomoea pandurata)*	to 10 feet	white and wine red	June–October	perennial

Nonnative/Cultivated Vines

Vine	*Length*	*Color*	*Bloom period*	*Life cycle*
Bean, Scarlet Runner *(Phaseolus coccineus)*	to 8 feet	red	summer–frost	annual
Clematis, Autumn *(Clematis paniculata)*	to 30 feet	white	August and September	perennial
Honeysuckle, Yellow *(Lonicera flava)*	to 10 feet	yellow	April and May	perennial
Jasmine, Star *(Trachelospermum jasminoides)*	to 20 feet	white to creamy	May and June	evergreen
Mexican Flame Vine *(Pseudogynoxys chenopodioides)*	to 10 feet	orange	June–October	perennial
Pea, Annual Sweet *(Lathyrus odoratus)*	to 6 feet	white, pink to red or purple	spring or fall	annual
Pea, Perennial Sweet *(Lathyrus latifolius)*	to 7 feet	white, pink to red or purple	June–October	perennial
Queen's Wreath (Coral Vine) *(Antigonon leptopus)*	to 40 feet	white to rose-pink	almost all year	perennial
Wisteria, Chinese *(Wisteria sinensis)*	to 40 feet	white, blue to purplish	April and May	perennial

NATIVE HERBS

Herb	Height	Color	Bloom period	Life cycle
Anemone, Carolina *(Anemone caroliniana)*	to 1 foot	white, pink to dark violet	February–May	perennial
Anemone, Ten-petal *(Anemone berlandieri)*	to 15 inches	white, pink, lavender, pale to dark blue	February–May	perennial
Aster, Tall Blue *(Symphyotrichum praealtum)*	to 4 feet	purple and yellow	October–frost	perennial
Aster, Texas *(Symphyotrichum drummondii* var. *texana)*	to 3 feet	lavender to purple and yellow	September–frost	perennial
Beebalm, Lemon *(Monarda citriodora)*	to 2 ½ feet	pink to purple	May–July	annual or biennial
Beebalm, Spotted *(Monarda punctata)*	to 3 feet	white to pale pink	April–July	annual to perennial
Bergamot, Lindheimer's *(Monarda lindheimeri)*	to 2 feet	creamy-white	April–August	perennial
Bergamot, Wild *(Monarda fistulosa)*	to 5 feet	dark pink to lavender	May–July	perennial
Blue-eyed Grass, Bray *(Sisyrinchium dimorphum)*	to 1 foot	white to dark blue	April–July	perennial
Blue-eyed Grass, Dotted *(Sisyrinchium pruinosum)*	to 1 foot	blue, blue-purple to purple-violet	April and May	perennial
Blue-eyed Grass, Langlois's *(Sisyrinchium langloisii)*	to 14 inches	blue	March–June	perennial
Blue-eyed Grass, Swordleaf *(Sisyrinchium ensigerum)*	to 1 ½ feet	blue	March and April	perennial
Bluets, Needle-leaf *(Houstonia acerosa)*	to 1 foot	white, pink to violet	May–October	perennial
Bluets, Nodding *(Houstonia subviscosa)*	to 6 inches	white to pale pink	March–June	annual
Bluets, Small *(Houstonia pusilla)*	to 4 inches	lilac to blue-violet	January–May	annual
Boneset, Trailing *(Fleischmannia incarnata)*	to 3 feet	white	October–January	perennial
Brown-eyed Susan *(Rudbeckia hirta)*	to 3 feet	yellow and brown	May–frost	biennial
Bush-sunflower, Awnless *(Simsia calva)*	to 3 feet	yellow	April–November	perennial
Cardinal Flower *(Lobelia cardinalis)*	to 4 feet	scarlet-red	August–frost	perennial
Coneflower, Purple *(Echinacea purpurea)*	to 3 feet	pink to dark rose	May–frost	perennial
Crownbeard, Dwarf *(Verbesina nana)*	to 8 inches	orangish	June–November	perennial
Crownbeard, Golden *(Verbesina encelioides)*	to 4 feet	yellow	almost all year	annual or perennial

Herb	Height	Color	Bloom period	Life cycle
Daisy, Chocolate *(Berlandiera lyrata)*	to 1 foot	yellow	spring–frost	perennial
Daisy, Tahoka *(Machaeranthera tanacetifolia)*	to 16 inches	white to purple	May–November	annual
Dalea, Purple *(Dalea purpurea)*	to 3 feet	rose-purple	June and July	perennial
Dalea, White *(Dalea candida)*	to 3 ½ feet	white	May–October	perennial
Dogweed, Prickle-leaf *(Thymophylla acerosa)*	to 6 inches	lemon-yellow	March–frost	perennial
Frogfruit *(Phyla* spp.)	4–24 inches	white	May–November (all year)	perennial
Frostweed, Virginia *(Verbesina virginica)*	to 6 feet	white	August–frost	perennial
Gaillardia, Fragrant *(Gaillardia suavis)*	to 2 feet	yellow to red and brown	March–June	perennial
Garlic, False *(Nothoscordum bivalve)*	to 22 inches	white to creamy-yellow	throughout year	perennial
Goldeneye, Tooth-leaved *(Viguiera dentata)*	to 6 feet	yellow	September–frost	perennial
Goldenrod *(Solidago* spp.)	to 7 feet	yellow	summer and fall	perennial
Groundsel, Broom *(Senecio riddellii)*	to 3 ½ feet	yellow	April–October	perennial
Groundsel, Texas *(Senecio ampullaceus)*	to 3 feet	yellow	February–June	annual
Groundsel, Thread-leaf *(Senecio flaccidus)*	to 2 ½ feet	yellow	April–October	perennial
Heliotrope, Seaside *(Heliotropium curassavicum)*	to 16 inches	white	March–frost	perennial
Hyacinth, Wild *(Camassia scilloides)*	to 2 feet	pale blue to violet	March–June	perennial
Ironweed *(Vernonia* spp.)	to 7 feet	dark rose to dark purple	July–frost	perennial
Liatris, Narrow-leaf *(Liatris punctata* var. *mucronata)*	to 3 feet	purple	August–frost	perennial
Liatris, Prairie *(Liatris pycnostachya)*	to 5 feet	lavender to dark purple	June–November	perennial
Liatris, Rough *(Liatris aspera)*	to 4 feet	lavender	July–November	perennial
Mallow, Salt-marsh *(Kosteletzkya virginica)*	to 6 feet	pale pink to rose-pink	June–October	perennial

Herb	***Height***	***Color***	***Bloom period***	***Life cycle***
Marigold, Desert *(Baileya multiradiata)*	to 1 foot	bright yellow	throughout year	perennial
Marsh-fleabane, Purple *(Pluchea odorata)*	to 6 feet	pinkish-purple	May–August	annual
Milkweed, Antelope-horns *(Asclepias asperula)*	to 1 feet	greenish-yellow	March–frost	perennial
Milkweed, Purple *(Asclepias purpurascens)*	to 3 feet	dark rose	April–August	perennial
Milkweed, Swamp *(Asclepias incarnata)*	to 5 feet	pale to dark pink	June–November	perennial
Mint, Wild *(Mentha arvensis)*	to 2 feet	white to lilac	May–frost	perennial
Mistflower, Betony-leaf *(Conoclinium betonicifolium)*	to 3 feet	blue to purplish	throughout year	perennial
Mistflower, Blue *(Conoclinium coelestinum)*	to 6 feet	blue to purplish-blue	April–frost	perennial
Mistflower, Gregg's *(Conoclinium dissectum)*	to 3 feet	blue to purplish	April–frost	perennial
Mountain-mint, Slender-leaf *(Pycnanthemum tenuifolium)*	to 6 feet	white to lavender, spotted with purple	June–frost	perennial
Onion, Drummond's Wild *(Allium canadense)*	to 1 foot	white, pink to reddish	March–June	perennial
Onion, Nodding Wild *(Allium ceruum)*	to 20 inches	white to pink	July–October	perennial
Onion, Yellow-flowered Wild *(Allium coryi)*	to 8 inches	yellow	April and May	perennial
Paper-flower, Woolly *(Psilostrophe tagetina)*	to 2 feet	yellow	March–November	perennial
Parsley, Nuttall's Prairie *(Polytaenia nuttallii)*	to 3 feet	yellow	April–July	perennial
Parsley, Texas Prairie *(Polytaenia texana)*	to 3 feet	yellow	April–August	perennial
Peppergrass, Western *(Lepidium alyssoides)*	to 28 inches	white	February–September	perennial
Phacelia *(Phacelia congesta)*	to 3 feet	blue to purplish	March–July	annual or biennial
Phlox, Downy *(Phlox pilosa)*	to 2 feet	pale pink to purple	April and May	perennial
Sage, Cedar *(Salvia roemeriana)*	to 2 feet	brilliant red	March–September	perennial
Sage, Mealy *(Salvia farinacea)*	to 3 feet	violet-blue	March–frost	perennial
Sage, Tropical *(Salvia coccinea)*	to 3 feet	red to dark scarlet; rarely, white to pink	almost all year	perennial

Herb	*Height*	*Color*	*Bloom period*	*Life cycle*
Sand-verbena, Amelia's *(Abronia ameliae)*	to 2 feet	lavender to violet-purple	March–July	perennial
Sand-verbena, Sweet *(Abronia fragrans)*	to 2 feet	white tinged with pink or lavender	December–September	perennial
Selfheal *(Prunella vulgaris)*	to 2 feet	purple or violet and lavender	April–July	perennial
Senna, Lindheimer's *(Senna lindheimeriana)*	to 6 feet	yellow	August–November	perennial
Sida, Spreading *(Sida abutifolia)*	to 6 inches	yellow to orange-yellow	March–October	perennial
Spring Beauty *(Claytonia virginica)*	to 1 foot	white to rose	January–May	perennial
Star-grass, Yellow *(Hypoxis hirsuta)*	to 8 inches	yellow	March–June	perennial
Star-violet *(Stenaria nigricans)*	to 20 inches	white, pink to purplish	April–frost	perennial
Stenandrium, Shaggy *(Stenandrium barbatum)*	to 4 inches	lavender to purple	March–frost	perennial
Sunflower, Annual *(Helianthus annuus)*	to 6 feet	yellow and red or purple	May–frost	annual
Sunflower, Maximilian *(Helianthus maximiliani)*	to 10 feet	bright yellow	August–frost	perennial
Sunflower, Swamp *(Helianthus angustifolius)*	to 6 feet	yellow and purplish-red	August–frost	perennial
Thistle, Texas *(Cirsium texanum)*	to 6 feet	pink to rose-purple	April–September	biennial or perennial
Verbena, Prairie *(Glandularia bipinnatifida)*	to 1 foot	lavender to purple	almost all year	perennial
Verbena, Rose *(Glandularia canadensis)*	to 16 inches	pink to rose or purple	March and April	perennial
Vervain, Gulf *(Verbena xutha)*	to 6 feet	blue to purple	March–November	perennial
Violet, Wild *(Viola* spp.*)*	to 8 inches	white, blue to purple or yellow	February–June	annual or perennial
Wallflower, Plains *(Erysimum asperum)*	to 2 feet	yellow to orange-red	April–August	biennial or perennial
Waterleaf, Blue *(Hydrolea ovata)*	to 2 ½ feet	blue	September–November	perennial
Wedelia, Hispid *(Wedelia acapulcensis* var. *hispida)*	to 3 feet	yellow-orange	March–frost	perennial
Weed, Butterfly *(Asclepias tuberosa)*	to 3 feet	yellow to orange or red	June–September	perennial

Herb	Height	Color	Bloom period	Life cycle
Weed, Joe-Pye *(Eutrochium fistulosum)*	to 10 feet	lilac-pink to purple or brownish-lavender	July and August	perennial
Weed, Pickerel *(Pontederia cordata)*	to 4 feet	violet-blue	June–October	perennial
Zinnia, Desert *(Zinnia acerosa)*	to 10 inches	white and yellow	June–November	perennial
Zinnia, Plains *(Zinnia grandiflora)*	to 9 inches	yellow	June–November	perennial

Nonnative/Cultivated Herbs

Herb	Height	Color	Bloom period	Life cycle
Allium *(Allium spp.)*	to 4 feet	white through reds and purples	spring and summer	perennial
Allium, Giant *(Allium giganteum)*	to 5 feet	purple	June	perennial
Alyssum *(Lobularia saxatile)*	to 1 foot	yellow	spring	perennial
Alyssum, Sweet *(Lobularia maritima)*	3–10 inches	white, pink, purple	spring–frost	annual
Amaranth, Globe *(Gomphrena globosa)*	to 3 feet	white, pink to apricot, wine or maroon to purple	April–frost	annual
Asperula *(Asperula azurea)*	to 1 foot	lavender-blue	midsummer	annual
Aster, Fall Purple *(Symphyotrichum oblongifolia)*	to 2 ½ feet	bluish-lavender to purple	early fall–frost	perennial
Bergamot, Showy *(Monarda didyma)*	to 4 feet	red	June–September	perennial
Borage *(Borage officinalis)*	to 3 feet	blue	spring–late fall	perennial
Camassia *(Camassia leichtlinii)*	to 3 feet	lavender-violet	March–June	perennial
Catmint *(Nepeta mussinii)*	to 1 foot	white to lavender-blue	May–October	perennial
Catnip *(Nepeta cataria)*	to 3 feet	white	spring and summer	perennial
Chives, Garden *(Allium schoenoprasum)*	to 1 foot	lavender	May and June	perennial
Cigar Plant *(Cuphea micropetala)*	to 2 feet	orange-yellow	fall	perennial

Herb	*Height*	*Color*	*Bloom period*	*Life cycle*
Clover, White *(Trifolium repens)*	to 10 inches	white to pinkish	April–October	perennial
Clover, White Sweet *(Melilotus albus)*	to 8 feet	white	May–frost	annual or biennial
Clover, Yellow Sweet *(Melilotus officinalis)*	to 5 feet	yellow	May–November	biennial
Coneflower, Sweet *(Rudbeckia subtomentosa)*	to 5 feet	yellow	summer–frost	perennial
Cosmos *(Cosmos bipinnatus)*	to 4 feet	white to pink or rose, yellow to red	spring–frost	annual
Cuphea, Bat-faced *(Cuphea llavea)*	to 3 feet	red and purple	June–frost	perennial
Cupid's Dart *(Catanache caerulea)*	to 2 feet	sky-blue	June–frost	perennial
Garlic, Society *(Tulbaghia violacea)*	to 2 feet	lilac	summer	perennial
Heliotrope, Common *(Heliotropium arborescens)*	to 16 inches	lavender to purple	summer and fall	perennial
Horehound, Common *(Marrubium vulgare)*	to 3 feet	white	May–frost	perennial
Hyssop, Anise *(Agastache foeniculum)*	to 3 feet	pale lavender to bluish	spring–frost	annual
Iris, Yellow *(Iris pseudacorus)*	to 3 feet	yellow	April and May	perennial
Jupiter's Beard *(Centranthus rubra)*	to 3 feet	white, pink, red	June–frost	perennial
Lavender *(Lavendula latifolia)*	to 2 feet	lavender to purple	summer–frost	perennial
Liatris *(Liatris scariosa)*	to 5 feet	pink to lavender	August–October	perennial
Liatris, Dense *(Liatris spicata)*	to 6 feet	rose-purple	July–October	perennial
Mallow, Rose *(Lavatera trimestris)*	to 5 feet	pink to rose	June–October	annual
Maltese Cross *(Lychnis chalcedonica)*	to 3 feet	vivid scarlet	June and July	perennial
Marigold, Single *(Tagetes patula)*	to 1 ½ feet	yellow to maroon	spring–frost	annual
Marjoram *(Origanum vulgare)*	to 15 inches	white	summer	annual
Milkweed *(Asclepias physocarpa)*	to 2½ feet	creamy-white	summer–fall	perennial

Herb	***Height***	***Color***	***Bloom period***	***Life cycle***
Milkweed, Showy *(Asclepias speciosa)*	to 4 feet	pink to rose	May–September	perennial
Milkweed, Tropical *(Asclepias curassavica)*	to 3 feet	yellow and orange	spring–frost	annual or perennial
Moss Rose *(Portulaca grandiflora)*	to 10 inches	white, pink to red, yellow to orange	spring–frost	annual
Oregano, Mexican *(Poliomintha longiflora)*	to 3 feet	pinkish-purple	July–frost	perennial
Pentas *(Pentas lanceolata)*	to 2½ feet	pink to rose, magenta to red	April–frost (all year)	annual or perennial
Petunia, Old-fashioned *(Petunia axillaris)*	to 2 feet	white to purple	April–frost	annual
Phlox, Smooth *(Phlox glaberrima)*	to 5 feet	white to pink, lavender to purple	April–September	perennial
Phlox, Summer *(Phlox paniculata)*	to 3 feet	white to purple or magenta	June–November	perennial
Pink, Old-fashioned *(Dianthus barbatus)*	to 2 feet	pink to deep magenta	January–July	perennial
Rocket, Sweet *(Hesperis matronalis)*	to 2 feet	white, lilac to purple	April–July	perennial
Sedum *(Sedum spectabile)*	to 2 feet	pink	August–frost	perennial
Shepherd's-needle *(Bidens alba)*	to 4 feet	white	almost all year	annual
Shrimp Plant *(Justicia brandegeana)*	to 3 feet	white, pink to salmon and reddish-brown	April–frost	perennial
Thistle, Globe *(Echinops banaticus)*	to 3 feet	powdery-blue	July–September	perennial
Tithonia *(Tithonia rotundifolia)*	to 6 feet	yellow to orange	May–frost	annual
Verbena, Argentina *(Verbena bonariensis)*	to 4 feet	purple	May–frost	annual or perennial
Weed, Butterfly *(Asclepias tuberosa cultivar* 'Gay Butterflies')	to 2 feet	pink to scarlet, yellow to gold	midsummer–frost	perennial
Zinnia, Classic *(Zinnia linearis)*	to 15 inches	orange	spring–frost	annual
Zinnia, Mexican *(Zinnia angustifolia)*	to 10 inches	yellow, gold, and red	spring–frost	annual

Janais Patch
(Chlosyne janais)

Appendix

How to Photograph Butterflies

Having enticed butterflies into your garden, you may now be interested in photographing them. After attempting once or twice, however, you may feel like giving up butterfly photography as being too frustrating. Butterflies just do not ordinarily stay still long enough to have their pictures taken.

Patience and persistence are the first and most basic requirements in good butterfly photography. Following are some tips that may help—even with first efforts. The information given here is for photographing in the field or garden, for it is here in the butterflies' natural habitat that the most exciting opportunities for recording their natural behavioral activities and beauty are presented.

Cameras

The 35 mm camera is still the choice of a few photographers, but digital cameras are the overwhelming choice today. Many models are on the market, and the best method of choosing is to give serious thought to your aim or reason for photographing and what the photos will be used for. In the choosing, the camera body may not be as important as the lenses you purchase. For really spectacular shots of butterflies you will want close-ups, which will require a macro lens of some length. And for the serious photographer, a flash will be mandatory.

Practice

After the camera and lens are purchased, one of the best things you can do to ensure good pictures of butterflies is to practice before ever going into the field or garden. Collect a dead butterfly from the grill of your car or from along the roadside. If a real butterfly cannot be found, cut a butterfly shape from stiff, colored paper or a picture from an old magazine. Either glue or wire it to a stick or dried flower, and start working with the camera, using your equipment in various situations. Do this with different sizes of butterflies, different colors, and with wings both open and closed. Working outside, place the butterfly against different backgrounds, such as tree bark, grasses, flowers, earth, and rocks. Constantly view your shots on the screen. Do this until you know instantly which lenses and settings give you the results you like under various conditions.

In the Field

Photographing active butterflies in the field requires some physical stamina and expertise. It also demands patience and the ability and agility to move slowly and to remain in odd, uncomfortable positions for several minutes at a time. Butterflies are endowed with a well-developed sense of self-preservation, and any sudden movement will send them fluttering out of range, so sneak up on the wary insects with all the stealth of a predator. Make all your movements with fluidity and slow deliberation in order not to betray your presence. This requires rigid self-discipline and concentration. Especially watch where you step (or crawl) when approaching a butterfly, for you may inadvertently move a branch of the plant the butterfly is resting on. Any movement of the resting or feeding place—or even of a branch of a nearby plant—that is faster than the present wind movement will usually cause the butterfly to take flight. Also, take care that your shadow does not fall across a basking or feeding butterfly, because this will either send it off in fright or at least cause it to close its wings.

Once you have slowly worked your way near the subject, and if it is close to the ground, ever so slowly sink down until one knee is on the ground and the other knee is bent toward your chin. The bent knee offers an excellent support or "flexible tripod." You can either rest your arms on top of the knee or clasp the knee between the elbows, whichever puts the film plane parallel with the wings of the butterfly and helps in steadying the camera.

Now that you are at least in partial position, you must quickly decide on the portion of the butterfly's body you want in focus along with its eyes. As the human eye is always drawn to a creature's eyes in a photograph, concentrate first on getting the eyes of your butterfly in sharp focus. Then you can begin to move the camera a few millimeters one way or another to get the wings and body in focus. But do not take too long in moving the camera about. The more time you spend squinting through the viewfinder, the more your eyes will tire and become unable to focus properly, and your back and arm muscles will begin to tremble so that you can no longer hold the camera steady. It is far better to focus on the eyes, move the camera a time or two for maximum wing coverage, hold your breath, and gently press the shutter. If the butterfly does not fly away, then try for even better shots with more of the butterfly in focus.

Spending time watching the butterfly's actions is often beneficial. Some species, such as the small Yellows, Blues, Hairstreaks, and some Skippers, take off when disturbed, fly back and forth or round and round, then return and alight in exactly the same spot as before. In some instances, do not move in pursuit but remain perfectly still, and chances are you will get the shot after all. There are some species that allow a fairly close approach, while others, once disturbed, take off never to be seen again. Ever. It is good to know something of the habits of each species in order to save time waiting for one not likely to return.

Within each species there are "personalities" that show exceptions to the general species behavior. With persistence (or luck), there will occasionally be a butterfly belonging to a group generally noted for their flightiness that will sit for a long period of time, letting you get shot after shot after shot. At such times you almost wish it *would* leave.

Using digital cameras, most people shoot in automatic focus, but when photographing butterflies, this is often too slow. One easy method is to quickly pick an object close by about the same size as the butterfly being stalked. Then focus the camera on the object to include approximately the amount of butterfly and background wanted in the photograph, add extension tubes if needed, then slowly approach the insect with everything preset. Some literature suggests that you start shooting when still far away from the insect, then continue to move forward, refocusing and shooting. In my opinion this is a terrible waste of time. Spend the time on slow movements getting closer to the subject. You will avoid moving your hands in constant refocusing, for one of the things you do not want to do when photographing butterflies is to move any more than absolutely necessary. This applies especially to your camera and hands, which are the closest things to the subject.

Also, before approaching a butterfly, note its general outline, always keeping in mind what you intend to do with your photo, and choose the desired format, whether vertical or horizontal. Since it is usually easier to hold a camera in the horizontal position, do not forget that some things lend themselves more satisfactorily to a vertical format—butterflies newly emerged with downward-hanging wings or at rest with wings closed, for example.

Background often helps decide the format. If the background is cluttered or distracting, then having more of the butterfly in the frame eliminates more of the background. In this case, use vertical format if the insect's wings are closed, horizontal if the wings are spread.

Often the background can enhance the coloring of the butterfly, the shape of its wings, or some other detail. If the background happens to be grass or leaves, throw it out of focus for a wonderful textured and mottled effect. If the butterfly is on a flower and there are other flowers in the background, these can be thrown out of focus by using a smaller f-stop (opening the lens) and will become beautiful spots of blurred coloring with no distracting detail.

Since butterflies can see colors exceptionally well, the clothing you wear while photographing is extremely important. This is a time to see and not be seen, so bright clothing is best left for other occasions. Drab greens, browns, and khaki work best, thereby allowing you to blend in with the surroundings.

Your clothing should be loose fitting enough to allow for extra maneuverability, but nothing should remain loose, such as shirttails or unbuttoned sleeves, to flap about in the wind. Keep bandannas or photo equipment in pockets or bags and not loose to move or rattle. Hats are sometimes useful in helping shade and disrupt the face outline. They also help cut the glare from glasses and prevent sunburn and sunstroke. Avoid floppy brims, however, whose movement may scare the butterfly away. Knee pads, the type worn by athletes and that can be purchased at a sporting goods store, are most welcome when kneeling or crawling about on rocks or among stickers.

A small camera bag belted about the waist is very handy for carrying extension tubes, extra lenses, and CompactFlash (CF) cards. Keep a couple of small cotton cloths or bandannas in the bag for cleaning glasses and cameras—*never* the lens—and wiping a sweaty brow. It is often handy to have a small case that can be opened and closed quickly and easily and fitted with an assortment of small tools. Include a pair of small scissors for clipping twigs and grass blades, tweezers for

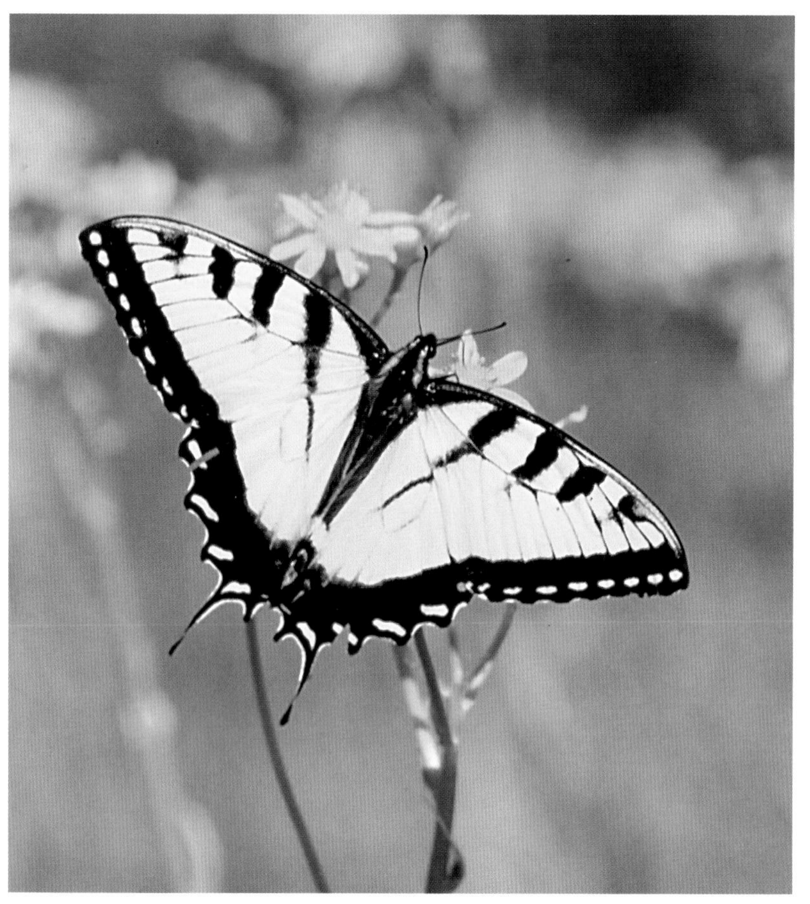

Eastern Tiger Swallowtail *(Papilio glaucus)*

removing debris, and an assortment of small artist's brushes for removing spider webs, dust, and so on. These items are really appreciated when photographing eggs, larvae, and pupae.

Butterflies are easily spooked by anything bright or flashing. Leave at home all shiny jewelry such as watches, rings, or dangling earrings. It is worth the extra time to cover the silver-colored metal parts of your camera and flash with electrician's tape. This can be removed easily after a photo session, if desired. If purchasing a new camera, consider a black model, though it will be slightly more expensive. If you have an off-flash and the bracket is of shiny metal, paint it matte black. Everything you can possibly do to eliminate brightness, flashing, and movement is that much more to your advantage.

Do not leave the lens cap dangling from the lens. If you usually use this type of cap, remove it completely before venturing forth in pursuit of butterflies, or pursue is all you are likely to do. Use a regular, unattached cap for this work, easing it into your pocket after removal from the lens.

Whatever you do, do not spray yourself with insect repellent before heading out to photograph butterflies. Butterflies *are* insects, so they are very sensitive to the smell of repellent. One whiff of Deepwoods Off, and the butterflies will head for unpolluted areas, not to be seen again while you are around.

One of the best times of day for photographing butterflies is early in the morning. At this time they are very hungry and allow a closer approach while nectaring. It is also the time when they are more likely to be basking. In the spring and fall months, especially, they spend much time in the cool mornings with wings spread wide, bringing their body temperature up to optimal operating levels. Late afternoons are also excellent during the summer, for then they have a tendency to "cool down," their feeding motions becoming much slower. It is at such times you can get the easiest and most spectacular shots.

For the best color rendition, choose early morning, late afternoon, or brightly overcast days for your butterfly photography. During these periods the light is much softer and the shadows are less harsh or nonexistent.

Gardening Tips for Easier Butterfly Photography

If photos are one of the objectives when planting a garden to attract butterflies, then remember when laying out the flower beds to leave a walk space around some of the insects' favorite plants. This walk space could be bare ground, native grasses, or leaf litter. These give plenty of room for moving with the butterfly and a background that appears natural if included in the photo. A walkway of brick, gravel, or hay offers a convenient kneeling place on damp or dewy mornings without getting your clothes wet or dirty. On the other hand, bricks and gravel can be unappealing in a photo if you are trying for the pure, naturalistic approach for your photos.

In border plantings there are advantages to three different types of arrangements:

1. Plants spaced close together in a narrow border, giving a lot of flowers and good opportunities for photos by walking in front of the border. Long portions of the border should be composed of all the same species and color of flowers so that the background in the photo is not a clash of colors or shapes. This, of course, is how the butterflies also like the plants to be.
2. Narrow borders in an open lawn space, preferably running north and south, which provide space to walk on both sides and with the east side receiving morning sun and the west side the afternoon sun. This eliminates some of the problems with deep shadows.
3. Plants spaced widely apart in a border or bed, leaving stepping space between in order to maneuver between the plants without demolishing the adjacent plants and the chance for future photos.

Keep the area around the butterflies' best-liked plants free of garden litter or unsightly debris. There is nothing more frustrating than to see a perfect butterfly specimen with wings calmly stretched out in the sun, only to find when you look through the viewfinder that a water hose, plastic plant container, or plant markers are causing ugly and distracting blobs in the background.

If photographing is done mostly in early morning, make it a habit to patrol the beds the evening before, removing spent flowers, seedpods, dead leaves, unsightly grass clumps, and so on. At this time, view every flower as a potential shot with a butterfly perched on it and clean up the area accordingly.

Before starting your photo session, place a few branches of greenery to hide such things as piles of fermenting fruit. Also, use the f-stop, controlling the depth of field to throw the background out of focus if an object cannot be moved, covered up, or otherwise eliminated.

If you have a new garden with small perennials, then plant tall annuals, such as Mexican Sunflower, or hollyhocks, sunflowers, cosmos, dill, or fennel, behind them for an attractive background.

Consider wooden fences in your garden. They are more attractive in general than wire and provide a more natural background. If you have wire fences, try covering them by planting with perennial or annual vines that grow well in your area. Or again, plant in the staggered-height method, tall plants at the back and lower-growing plants in front. Inexpensive reed fencing can be attached to an existing wire fence, making an excellent natural-colored background. Any vines allowed to grow on the reed fence should be of the annual sort, however, as the reed does not usually last but two to five years.

If there is no way to keep a wire fence out of an otherwise perfectly wonderful shot, then choose an angle where the sun does not reflect off the wire, select a small f-stop for a shallow depth of field, and throw the wire out of focus. Check the depth-of-field button often to be sure when the wire becomes acceptably blurred.

If there is a particular plant or group of plants in the border, such as Flame Acanthus, Mexican Sunflower, or various lantanas, mistflowers, and verbenas, it is possible to set up a blind and wait for butterflies to come to you. While this lets you rest more comfortably, it greatly restricts the butterfly poses obtainable. The ability to move around, get the wings of the butterfly in a flatter plane, or move for better lighting is greatly reduced from a blind. However, you can achieve more shots from the blind because you are concealed and will not frighten the insects.

If larval food plants are included in the garden plantings, photos of the entire life cycle, from egg to adult can be taken. Chrysalides can be closely watched, and photos of the emerging adult butterfly can be taken. After emerging, the butterfly remains clinging to the chrysalis case or a nearby leaf or branch for quite some time, resting and letting its wings become completely dry. During this period, excellent photos may be obtained of the perfect specimen if the surface it rests on is not moved, causing the butterfly to take flight. Close and frequent observations of the garden can sometimes net some unusual or especially interesting shots not usually obtainable on ordinary, casual walks or hikes. To capture permanently such fleeting beauty will probably become quite addictive.

Sources for Special Photo Equipment

Blacklock Photo Equipment, Inc.
P.O. Box 560
Moose Lake, MI 55767

Leonard Rue Enterprises, Inc.
138 Millbrook Road
Hardwick, NJ 07825
www.rue.com

Lepp and Associates
P.O. Box 6240
Los Osos, CA 93412

Really Right Stuff
www.reallyrightstuff.com

Sources for Information

www.blackrabbit.com
www.moosepeterson.com
www.outdoorphotographer.com
zinio.com/outdoorphotographer

Nature Photography Publications

Nature Photographer
P.O. Box 220
Lubec, ME 04652
www.naturephotographermag.com
(This is the best magazine for nature photography and has helpful articles on becoming a better photographer.)

Outdoor Photographer
Werner Publishing Company
12121 Wilshire Blvd., Suite 1200
Los Angeles, CA 90025-1176
www.outdoorphotographer.com/
(This is an excellent magazine on nature photography and frequently has articles on photographing insects, including butterflies.)

Seed and Plant Sources

For sources in the state, start by visiting all local nurseries, local plant sales, and seed exchanges. If more information or sources are needed, and for a complete listing of nurseries throughout the state, refer to the excellent guide by Nan Booth Simpson and Patricia Scott McHargue, *The Texas Garden Resource Book—a Guide to Garden Resources across the State* (Houston: Bright Sky Press, 2009). A good source of information is the Lady Bird Johnson Wildflower Center (4801 La Crosse Avenue, Austin, TX 78739; telephone 512-232-0100; www.wildflower.org). Seeds of Texas exchanges seeds only and charges a $20 annual membership fee, which includes publications (P.O. Box 9882, College Station, TX 77842).

ZEBRA SWALLOWTAIL
(Eurytides marcellus)
WITH A BROKEN TAIL

Some of the following out-of-state nurseries offer mail-order sales, but others do not, so contact them for information first. Companies who advertise that they collect plants from the wild are not listed here.

Dry Country Plants
Las Cruces, NM 88001

Goodness Grows, Inc.
332 Elberton Road
P.O. Box 311
Lexington, GA 30648–0311

Mail Order Natives
P.O. Box 9366
Lee, FL 32059
www.mailordernatives.com

Native Seeds/Search
3061 N. Campbell Avenue
Tucson, AZ 85719
www.nativeseeds.org

Natural Gardens
607 Barbrow Land
Knoxville, TN 33932

Niche Gardens
1111 Dawson Road
Chapel Hill, NC 27516
www.nichegardens.com

Oak Hill Farm
204 Pressley Street
Clover, SC 29710
www.lgyp.com/brochure.asp?c = 411762

Park Seed Company
1 Parkdon Avenue
Greenwood, SC 29647
www.parkseed.com

Passiflora Wildflower Co.
Route 1, Box 190-A
Germantown, NC 27019
919-591-5816

Plant Delights Nursery, Inc.
9241 Sauls Rd.
Raleigh, NC 27603-9326
www.plantdelights.com/

Plants of the Southwest
3095 Agua Fria Street
Santa Fe, NM 87507
www.plantsofthesouthwest.com

Prairie Basse Native Plants
217 St. Fidelis Street
Carencro, LA 70520
337-896-9187

Seeds of Change
621 Old Santa Fe Trail #10
Santa Fe, NM 87501
www.seedsofchange.com

Southwestern Native Seeds
Box 50503
Tucson, AZ 85703
www.nativeseeds.org/

Sunlight Gardens
174 Golden Lane
Andersonville, TN 37705
www.sunlightgardens.com

Thompson & Morgan
P. O. Box 1308
Jackson, NJ 08527
www.thompson-morgan.com

W. Atlee Burpee & Co.
300 Park Avenue
Warminster, PA 18974
www.burpee.com

Wayside Gardens
1 Garden Lane
Hodges, SC 29695
www.waysidegardens.com

Woodlanders, Inc.
1128 Colleton Avenue
Aiken, SC 29801-4728
www.woodlanders.net
(offers yellow-flowered *Asclepias*)

Butterfly Organizations, Societies, and Publications

American Entomological Society
1900 Benjamin Franklin Parkway
Philadelphia, PA 19103-1101
http://rex.ansp.org/hosted/aes/

Butterfly Lovers International
268 Bush Street
San Francisco, CA 94104
415-864-1169

Entomological Society of America
10001 Derekwood Lane #100
Lanham, MD 20706-4876
www.entsoc.org/
(publishes several periodicals, including *Annals of the Entomological Society of America, Journal of Economic Entomology, Environmental Entomology, American Entomologist,* and *Common Names of Insects and Related Organisms*)

The Entomological Society of Canada
393 Winston Avenue
Ottawa, ON K2A 1Y8, Canada
www.esc-sec.ca
(publishes *The Canadian Entomologist* and *Memoirs of the Canadian Entomological Society of Canada*)

The Lepidoptera Research Foundation, Inc.
c/o Santa Barbara Museum of Natural History
2559 Puesta del Sol Road
Santa Barbara, CA 93105

www.lepidopteraresearchfoundation.org/
(publishes *Journal of Research on the Lepidoptera*)

The Lepidopterists' Society
c/o Los Angeles County Museum of Natural History
900 Exposition Blvd.
Los Angeles, CA 90007-4057
www.lepsoc.org
(publishes *The Journal of the Lepidopterists' Society* and *The News of the Lepidopterists' Society*)

National Garden Clubs, Inc.
Preservation of Butterflies
4401 Magnolia Avenue
St. Louis, MO 63110-3492
www.gardenclub.org/
(a special committee on the conservation, preservation, and attracting of butterflies)

North American Butterfly Association
4 Delaware Road
Morristown, NJ 07960
www.naba.org/
(publishes *American Butterflies* and *Butterfly Gardener* and coordinates information from yearly July butterfly counts)

Southern Lepidopterists' Society
c/o Marc Minno
600 NW 34 Terrace
Gainesville, FL 32607
www.southernlepsoc.org/
(publishes *Southern Lepidopterists' News*)

Texas Organization for Endangered Species
P.O. Box 12773
Austin, TX 78711-2773
www.tpwd.state.tx.us/

The Western Monarch Migration Project
Oregon Department of Agriculture
635 Capital Street NE
Salem, OR 97310-0110
503-986-4663

The Xerces Society
628 NE Broadway, Suite 200
Portland, OR 97232
www.xerces.org
(publishes the newsletter *Wings*)

Young Entomologists' Society
c/o Department of Entomology
Michigan State University
East Lansing, MI 48824-1115
(publishes *Y. E. S. Quarterly*)

Plant Organizations, Societies, and Publications

Texas

Fort Worth Nature Center & Refuge
9601 Fossil Ridge Road
Fort Worth, TX 76135
www.fwnaturecenter.org

Native Plant Project
P.O. Box 2742
San Juan, TX 78589
www.nativeplantproject.org
(publishes *Sabal,* a monthly newsletter)

Native Plant Society of New Mexico, El Paso Chapter
P.O. Box 221036
El Paso, TX 79913
http://npsnm.unm.edu

Native Plant Society of Texas
320 West San Antonio Street
Fredericksburg, TX 78624-3727
www.npsot.org/
(publishes *NPSOT News* quarterly)

Native Prairies Association of Texas
2002-A Guadalupe Street PMB 290
Austin, TX 78705-5609
http://texasprairie.org

Outdoor Nature Club
P.O. Box 270894
Houston, TX 77277-0894
www.outdoornatureclub.org
(publishes the newsletter *Nature Notes*; owns Little Thicket Nature Sanctuary; sponsors many field trips throughout Texas)

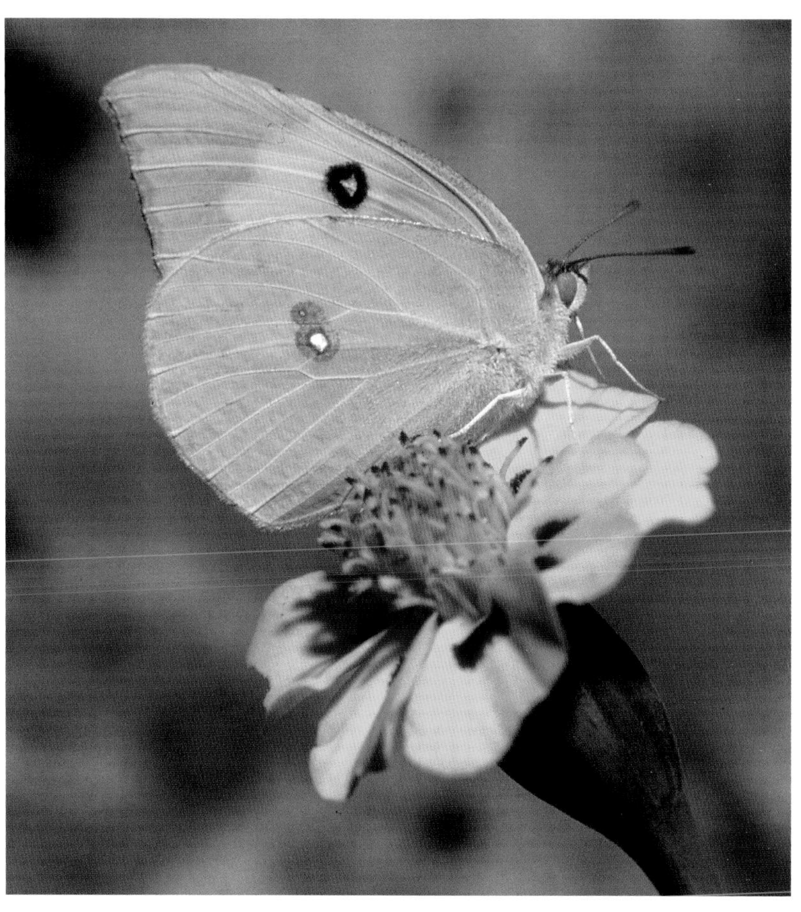

Southern Dogface
(Zerene cesonia)

South Texas Botanical Gardens & Nature Center
8545 S. Staples
Corpus Christi, TX 78413
www.stxbot.org

Texas Garden Clubs, Inc.
3111 Botanic Garden Road
Fort Worth, TX 76107
www.texasgardenclubs.org

Texas Master Gardener
http://mastergardener.tamu.edu/

Texas Master Gardeners Association
http://txmg.org/

Outside Texas

Arkansas Native Plant Society
c/o Department of Forest Resources
University of Arkansas at Monticello
P.O. Box 3468
Monticello, AR 71655
http://anps.org/

Kansas Native Plant Society
McGregor Herbarium
2045 Constant Avenue
Lawrence, KS 66047-3729
www.kansasnativeplantsociety.org

Louisiana Native Plant Society
216 Caroline Dormon Road
Saline, LA71070
www.lnps.org/
(publishes *Newsletter*)

National Garden Clubs
4401 Magnolia Avenue
St. Louis, MO 63110
www.gardenclub.org/

Oklahoma Native Plant Society
2435 South Peoria Avenue
Tulsa, OK 74114
http://oknativeplants.org/

White Peacock
(Anartia jatrophae luteipicta)

Places of Interest

Antique Rose Emporium
9300 Luechemeyer Road
Brenham, TX 77833-6453
www.antiqueroseemporium.com

Antique Rose Emporium
7561 East Evans Road
San Antonio, TX 78266
www.antiqueroseemporium.com/san-antonio-display-gardens

Armand Bayou Nature Center
8500 Bay Area Boulevard
Pasadena, TX 77507
www.abnc.org

Brierwood Caroline Dormon Nature Preserve
216 Caroline Dormon Road
Saline, LA 71070
318-576-3379

Butterfly World
Tradewinds Park South
3600 West Sample Road
Coconut Creek, FL 33066
www.butterflyworld.com

Chihuahuan Desert Nature Center & Botanical Gardens
43869 SH 118
Fort Davis, TX 79734
http://cdri.org/

Cibolo Nature Center
Hwy 46
Boerne, TX 78006
www.cibolo.org/

Day Butterfly Center
Callaway Gardens
Pine Mountain, GA 31822
www.callawaygardens.com

Houston Arboretum and Nature Center
4501 Woodway Drive
Houston, TX 77024
www.houstonarboretum.org

Houston Museum of Natural Science
One Hermann Circle Drive
Houston, TX 77030
www. hmns.org/

Lady Bird Johnson Wildflower Center
4801 La Crosse Avenue
Austin, TX 78739-1702
www.wildflower.org

Louisiana Nature and Science Center
Joe W. Brown Memorial Park
11000 Lake Forest Boulevard
New Orleans, LA 70127

LSU Hilltop Arboretum
11855 Highland Road
Baton Rouge, LA 70810
www.lsu.edu/hilltop

Mercer Arboretum & Botanic Gardens
22306 Aldine Westfield Road
Humble, TX 77338-1071
www.hcp4.net/mercer

Mizell Farms, Inc.
83211 Highway 25
P.O. Box 484
Folsom, LA 70437
www.mizellfarm.com
(Annual Butterfly Extravaganza)

Moody Gardens
One Hope Boulevard
Galveston, TX 77554
www.moodygardens.com

National Butterfly Center
3333 Butterfly Park Drive
Mission, TX 78572
http://nationalbutterflycenter.org/

North American Butterfly Park
3333 Butterfly Park Dr.
Mission, TX 78572

San Antonio Botanical Garden
555 Funston Place
San Antonio, TX 78209-6635
www.sabot.org

Texas Discovery Gardens
3601 Martin Luther King Boulevard
Dallas, TX 75210
www.texasdiscoverygardens.org

Wild Basin Wilderness Preserve
805 N. Capitol of Texas Hwy
Austin, TX 78746
www.wildbasin.org

Zilker Botanical Gardens
2100 Barton Springs Road
Austin, TX 78746
www.austintexas.gov/department/zilker-metropolitan-park
(includes Doug Blachly Butterfly Trail)

Gardening Magazines and Newsletters

Birds & Blooms
5925 Country Lane
Greendale, WI 53129
www.birdsandblooms.com

Fine Gardening
The Taunton Press
Box 355
63 South Main Street
Newton, CT 06470
www.finegardening.com

Flower and Garden Magazine
4251 Pennsylvania Avenue
Kansas City, MO 64111

Garden Design
2 Park Avenue, 10th Floor
New York, NY 10016
www.gardendesign.com

Gardens & More
P.O. Box 864
McKinney, TX 75069
972-238-6474

Growing from Seed
Thompson & Morgan
P.O. Box 1308
Jackson, NJ 08527

Horticulture
P.O. Box 53880
Boulder, CO 80321-3880

Neil Sperry's Gardens Magazine
P.O. Box 864
McKinney, TX 75070
www.neilsperry.com/gardens/

Rodale's Organic Gardening
Emmaus, PA 18099-0003
www.organicgardening.com

Southern Living
P.O. Box 523
Birmingham, AL 35201
www.southernliving.com/magazine/

Texas Gardener
P.O. Box 9005
Waco, TX 76714
www.texas.gardener.com

Wildflower
4801 La Crosse Avenue
Austin, TX 78739
(publication of the Lady Bird Johnson Wildflower Center)
www.wildflower.org/magazine/

Gardening Supplies

CobraHead LLC
P.O. Box 519
Cambridge, WI 53523
www.cobraheadllc.com

Gardener's Marketplace
Story Communications, Inc.
Schoolhouse Road
Pownal, VT 95261

Gardener's Supply Company
128 Intervale Road
Burlington, VT 05401
www.gardeners.com

Garden-Ville
4001 Ranch Road 620 South
Austin, TX 78738
www.garden-ville.com
(organic gardening supplies)

Natural Gardening Research Center
Hwy 48
P.O. Box 149
Sunman, IN 47041
(information and supplies for biological control of insects)

Park Seed
P.O. Box 46
Cokesbury Road
Greenwood, SC 29648-0046
www.parkseed.com/

Smith & Hawken
1330 10th Street
Berkeley, CA 94710
www.smithandhawken.com/

Wayside Gardens
Hodges, SC 29695-0001
www.waysidegardens.com/

Habitat Preservation

Heard Natural Science Museum & Wildlife Sanctuary
One Nature Place
McKinney, TX 75069-8840
www.heardmuseum.org/

Native Prairies Association of Texas
The Program Committee
Texas Woman's University
P.O. Box 22675
Denton, TX 76204

Texas Conservation Alliance
Suite 3B
5518 Dyer
Dallas, TX 75206

Texas Land Conservancy
P.O. Box 162481
Austin, TX 78716
www.texaslandconservancy.org

Texas Natural Heritage Program
General Land Office
Stephen F. Austin Building
1700 Congress Avenue
Austin, TX 78767

The Nature Conservancy of Texs
503 B East Sixth Street
Austin, TX 78701

Further Information

Kika de la Garza Plant Materials Center, NRCS
3409 N FM 1355
Kingsville, TX 78363
www.tx.nrcs.usda.gov/technical/pmc/kingsville.html

Lady Bird Johnson Wildflower Center
4801 La Crosse Avenue
Austin, TX 78739-1702
www.wildflower.org

Louisiana Project Wildflower
Lafayette Natural History Museum
637 Girard Park Drive
Lafayette, LA 70504

The National Xeriscape Council, Inc.
940 East Fifty-first Street
Austin, TX 78751-2241
512-454-8626

Native Plant Project
P.O. Box 2742
San Juan, TX 78589
www.nativeplantproject.org

Natural Resources Conservation Service, USDA
http://www.tx.nrcs.usda.gov/
(click on Find a Service Center for the county offices)

The Nature Conservancy
1815 North Lynn Street
Arlington, VA 22209
www.tnc.org

NatureServe
www.natureserve.org

Rodale's Organic Gardening
Resources for Organic Pest Control
Rodale Press, Inc.
Emmaus, PA 18049
www.organicgardening.com
(an important publication containing sources for natural pest control supplies, pest control guidelines, and so on)

Soil and Water Conservation Society
945 SW Ankeny Road
Ankeny, IA 50023
http://www.swcs.org

Texas A&M AgriLife Extension Service
Texas A&M University System
http://agrilifeextension.tamu.edu/
(AgriLife Extension county offices directory at http://counties.agrilife.org/)

Texas Association of Nurserymen, Inc.
512 East Riverside Drive, Suite 207
Austin, TX 78704

Texas Botanical Garden Society
P.O. Box 5642
Austin, TX 78763
http://texasbot.tripod.com

Texas Department of Agriculture
P.O. Box 12847
Austin, TX 78711
http://texasagriculture.gov
(Texas Native Plant Directory, an important source for nursery plants)

Texas Department of Transportation
Landscape Division
Eleventh and Brazos
Austin, TX 78701
www.txdot.gov/

Texas Parks & Wildlife Department
4200 Smith School Road
Austin, TX 78744
www.tpwd.state.tx.us

Texas Wildflower Hotline
Texas Department of Highways
800-452-929
(information for best wildflower viewing)

US Government Printing Office
Superintendent of Documents
710 North Capitol Street
Washington, DC 20402
Ask for these free government pamphlets:
PL National Parks
PL 41 Insects
PL 43 Forestry
PL 44 Plants
PL 46 Soils and Fertilizers
PL 88 Ecology
Home Garden Brochure

Bibliography

Butterflies and Insects

Allen, Thomas J., Jim P. Brock, and Jeffery Glassburg. *Caterpillars in the Field and Garden.* New York: Oxford University Press, 2005.

Barth, Friedrich G. *Insects and Flowers.* Princeton, NJ: Princeton University Press, 1985.

Bordelon, Charles, and Ed Knudson. *Lepidoptera of the Big Thicket National Preserve and Adjacent Regions of Southeast Texas.* Houston: Self-published, 1999.

Brewer, Jo. *Wings in the Meadow.* New York: Houghton Mifflin, 1967.

Brewer, Jo, and Kjell Bloch Sandved. *Butterflies.* New York: Harry N. Abrams, 1976.

Brewer, Jo, and Dave Winter. *Butterflies and Moths.* New York: Prentice Hall, 1986.

Brock, Jim P., and Kenn Kaufman. *Butterflies of North America.* New York: Houghton Mifflin, 2003.

DeVries, Philip J. *Butterflies of Costa Rica.* Princeton, NJ: Princeton University Press, 1987.

Dole, M., W. B. Gerard, and J. M. Nelson. *Butterflies of Oklahoma, Kansas and North Texas.* Norman: University of Oklahoma Press, 2004.

Douglas, Mathew M. *The Lives of Butterflies.* Ann Arbor: University of Michigan Press, 1987.

Ehrlich, P., and A. Ehrlich. *How to Know the Butterflies.* Dubuque, IA: William C. Brown, 1961.

Emmel, Thomas C. *Butterflies—Their World, Their Life Cycle, Their Behavior.* New York: Alfred A. Knopf, 1975.

Faegri, K., and L. van der Pijl. *The Principles of Pollination Ecology.* New York: Pergamon Press, 1971.

Feltwell, John. *The Natural History of Butterflies.* New York: Facts on File, 1986.

Field, William D. *A Manual of the Butterflies and Skippers of Kansas.* Bulletin of the University of Kansas 39/10 (1975): 3–328. Originally published 1938.

Forey, Pamela, and Cecilia Fitzsimons. *An Instant Guide to Butterflies.* New York: Bonanza Books, 1987.

Freeman, H. A. "The Distribution and Flower Preferences of the Theclinae of Texas." *Field and Lab* 18 (1950): 65–72.

———. "Ecological and Systematic Study of the Hesperioidea of Texas." *Southern Methodist University Studies* 6 (1951): 1–67.

Friedlander, Timothy Paul. "A Taxonomic Revision of *Asterocampa* Rober 1916 (Insecta, Lepidoptera, Nymphalidae)." PhD diss., Texas A&M University, 1985.

Gerberg, Eugene J., and Ross H. Arnette Jr. *Florida Butterflies.* Baltimore: Natural Science Publications, 1989.

Glassberg, Jeffery. *Butterflies through Binoculars—the East.* Oxford: Oxford University Press, 1999.

———. *Butterflies through Binoculars—the West.* Oxford: Oxford University Press, 2001.

Harris, Lucien, Jr. *Butterflies of Georgia.* Norman: University of Oklahoma Press, 1972.

Holland, W. J. *The Butterfly Book.* Garden City, NY: Doubleday, Doran, 1931.

Howe, William H. *The Butterflies of North America.* New York: Doubleday, 1975.

Kevan, P. G. "Floral Coloration, Its Colorimetric Analysis and Significance in Anthecology." *In The Pollination of Flowers by Insects,* edited by A. J. Richards, 51–78. New York: Academic Press, 1978.

Kimball, Charles P. *The Lepidoptera of Florida.* Gainesville: State of Florida Department of Agriculture, Division of Plant Industry, 1965.

Klots, Alexander B. *A Field Guide to the Butterflies of North America, East of the Great Plains.* Boston: Houghton Mifflin, 1951.

Knudson, Ed, and Charles Bordelon. *Checklist of Lepidoptera of the Texas Hill Country: Illustrated Checklist of Lepidoptera from the Texas Hill Country and Adjacent Areas.* Publication 8. Texas Lepidoptera Survey. Houston: Self-published, 2001.

Knuth, Paul. *Handbook of Flower Pollination.* Vols. 1, 11, 111. Oxford: Clarendon Press, 1906.

Kulman, H. M. "Butterfly Production Management." In *Insect Ecology.* University of Minnesota Agricultural Experiment Station Technical

Bulletin. St. Paul: University of Minnesota Agricultural Experiment Station, 1977.
Lovell, John H. *The Flower and the Bee: Plant Life and Pollination.* New York: Charles Scribner's Sons, 1918.
Martin, E. C., E. Oertel, N. P. Nye et al. *Beekeeping in the United States.* USDA Agriculture Handbook No. 35, rev. Washington, DC: US Department of Agriculture, 1980.
Miller, Lee D., and F. Martin Brown. *A Catalogue/Checklist of the Butterflies of America North of Mexico.* The Lepidopterists' Society, Memoir No. 2. Sarasota, FL: Serbin Printing, 1981.
Minno, Marc C., Jerry F. Butler, and Donald W. Hall. *Florida Butterfly Caterpillars and Their Host Plants.* Gainesville: University Press of Florida, 2005.
Opler, Paul A., and George O. Krizek. *Butterflies East of the Great Plains.* Baltimore: Johns Hopkins University Press, 1984.
Orsak, L. J. *Butterfly Production Management in California with Emphasis on Native Plants: One Form of Urban Wildlife Enhancement.* Sacramento: California Native Plant Society, 1982.
Owen, Denis. *Camouflage and Mimicry.* Oxford: Oxford University Press, 1980.
Pelham, Jonathan P. *A Catalogue of the Butterflies of the United States and Canada.* Beverly Hills, CA: Lepidoptera Research Foundation, 2008.
Philbrick, Helen, and John Philbrick. *The Bug Book.* Pownal, VT: Story Communications, 1974.
Pyle, Robert M. *The Audubon Society Field Guide to North American Butterflies.* New York: Alfred A. Knopf, 1981.
———. *The Audubon Society Handbook of Butterfly Watchers.* New York: Charles Scribner's Sons, 1984.
Quinn, Mike, and Mark Klym. *An Introduction to Butterfly Watching.* Austin: Texas Parks and Wildlife Department, 2009.
Richard, J., and Joan E. Heitzman. *Butterflies and Moths of Missouri.* Jefferson City: Missouri Department of Conservation, 1987.
Saunders, Aretas A. *Butterflies of the Allegany State Park.* Albany: University of the State of New York, 1932.
Schappert, Phil. *A World for Butterflies.* Buffalo, NY: Firefly Books, 2005.
Scott, James A. *The Butterflies of North America.* Stanford, CA: Stanford University Press, 1986.
Scudder, Samuel Hubbard. *The Butterflies of the Eastern United States and Canada.* Cambridge, MA: Self-published, 1889.
Shields, Oakley. "Flower Visitation Records for Butterflies." *Pan-Pacific Entomologist* 48 (1972): 189–203.
Tilden, J. W., and Arthur C. Smith. *A Field Guide to Western Butterflies.* Boston: Houghton Mifflin, 1986.
Tveten, John, and Gloria Tveten. *Butterflies of Houston and Southeast Texas.* Austin: University of Texas Press, 1996.
Tyler, Hamilton A. *The Swallowtail Butterflies of North America.* Healdsburg, CA: Naturegraph Publishers, 1975.
Van-Wright, Richard I., and Philip R. Ackery, eds. *The Biology of Butterflies.* Symposium of the Royal Entomological Society Series. London: Academic Press, 1984.
Wagner, David L. *Caterpillars of Eastern North America.* Princeton, NJ: Princeton University Press, 2005.
Wauer, Roland H. *Butterflies of the Lower Rio Grande Valley.* Boulder, CO: Johnson Books, 2004.
———. *Butterflies of West Texas Parks and Preserves.* Lubbock: Texas Tech University Press, 2002.
———. *Finding Butterflies in Texas.* Boulder, CO: Johnson Books, 2006.

Also consulted were regional butterfly and food plant lists, as well as many articles from *Journal of Research on the Lepidoptera, News of the Lepidopterists' Society, Bulletin of Southern California Academy of Science, and Psyche,* especially the works of Roy O. Kendall and Raymond Neck.

Gardening

Abbott, Carroll. *How to Know and Grow Texas Wildflowers.* Kerrville, TX: Green Horizons Press, 1988.
Adams, W. D. "Propagation of Annual Plant Ornamentals." *In Proceedings of the Texas Turfgrass Conference* 28, 85–89. College Station: Texas A&M University, Texas Turfgrass Association, 1974.
Beckett, Kenneth A., David Carr, and David Stevens. *The Contained Garden.* New York: Viking Press, 1982.
Brooklyn Botanic Garden. *Gardening with Wildflowers.* Handbook No. 38. Brooklyn, NY: Brooklyn Botanic Garden, 1979.

Bruce, Hal. *How to Grow Wildflowers and Wild Shrubs and Trees in Your Own Garden.* New York: Alfred A. Knopf, 1976.

Campbell, Stu. *The Mulch Book.* Charlotte, VT: Garden Way, 1974.

Crockett, James Underwood, E. Oliver Allen, and the Editors of Time-Life Books. *Wildflower Gardening.* Alexandria, VA: Time-Life Books, 1977.

Curtis, Will C. *Propagation of Wild Flowers.* Farmingham, ME: New England Wild Flower Society, 1978.

Daniels, J. C. *Your Florida Guide to Butterfly Gardening: A Guide for the Deep South.* Gainesville: University Press of Florida, 2000.

Dormon, Caroline. *Natives Preferred.* Baton Rouge, LA: Claitor's Publishing Division, 1965.

Easey, Ben. *Practical Organic Gardening.* London: Faber and Faber, 1976.

Editors of Sunset Books and Sunset Magazine. *Garden Color, Annuals and Perennials.* Menlo Park, CA: Lane Publishing, 1981.

———. *Gardening in Containers.* Menlo Park, CA: Lane Publishing, 1970.

———. *New Western Garden Book.* Menlo Park, CA: Lane Publishing, 1986.

Everett, Thomas H., and Roy Hay, eds. *Illustrated Guide to Gardening.* Pleasantville, NY: Reader's Digest, 1983.

Fontenot, William R. *Native Gardening in the South.* Carencro, LA: Prairie Basse, 1992.

Foster, Catherine Osgood. *Organic Flower Gardening.* Emmaus, PA: Rodale Press, 1975.

Free, Montague. *Plant Propagation in Pictures.* Garden City, NY: Doubleday, 1957.

Garrett, Howard. *Texas Gardening the Natural Way: The Complete Handbook.* Austin: University of Texas Press, 2004.

Gilkeson, Linda, Pam Peirce, and Miranda Smith. *Rodale's Pest & Disease Problem Solver.* Emmaus, PA: Rodale Press, 1996.

Golueke, Clarence G. *Composting: A Study of the Process and Its Principles.* Emmaus, PA: Rodale Press, 1972.

Gregg, Evelyn. *Bio-Dynamic Sprays.* Stroudsburg, PA: Bio-Dynamic Farming and Gardening Association.

Gregg, Richard. *Companion Plants and How to Use Them.* Old Greenwich, CT: Devon-Adair, 1966.

Groom, Dale. *Lowe's Texas Gardening Handbook.* Brentwood, TN: Cool Springs Press, 1999.

Haring, Elda. *The Complete Book of Growing Plants from Seed.* Grandview, MO: University Books, 1967.

Hartmann, Hudson T., and Dale E. Kester. *Plant Propagation, Principles and Practices.* Englewood Cliffs, NJ: Prentice Hall, 1959.

Hazeltine, Cheryl. *Cheryl Hazeltine's Central Texas Gardener.* College Station: Texas A&M University Press, 2010.

Heger, M. "Propagating Perennial Plants." *Minnesota Horticulture* 105, no. 7 (1977): 194–195.

Hill, Madelene, Gwen Barclay, and Jean Hardy. *Southern Herb Growing.* Fredericksburg, TX: Shearer Publishing, 1987.

Howes, F. N. *Plants and Beekeeping.* London: Faber and Faber, 1979.

Kramer, Jack. *Hanging Gardens.* Piscataway, NJ: New Century Publishers, 1982.

———. *The Natural Way to Pest-Free Gardening.* New York: Charles Scribner's Sons, 1972.

Martin, Laura C. *The Wildflower Meadow Book, a Gardener's Guide.* Charlotte, NC: East Woods Press, 1986.

Minnich, Jerry. *The Earthworm Book.* Emmaus, PA: Rodale Press, 1977.

Nancarrow, Loren, and Janet Hogan Taylor. *The Worm Book.* Berkeley, CA: Ten Speed Press, 1998.

Newman, L. H. *Create a Butterfly Garden.* London: Billing and Sons, 1967.

Nokes, Jill. *How to Grow Native Plants of Texas and the Southwest.* Austin: Texas Monthly Press, 1986.

———. *How to Grow Native Plants of Texas and the Southwest.* Rev. and updated ed. Austin: University of Texas Press, 2001.

Odenwald, Neil, and James Turner. *Identification, Selection, and Use of Southern Plants for Landscape Design.* Baton Rouge, LA: Claitor's Publishing Division, 1987.

Organic Gardening and Farming Magazine. *The Complete Book of Composting.* Emmaus, PA: Rodale Press, 1971.

———. *The Organic Way to Plant Protection.* Emmaus, PA: Rodale Press, 1966.

Ortho Books. *Landscaping with Wildflowers and Native Plants.* San Francisco: Ortho Books, 1984.

Phillips, Harry R. *Growing and Propagating Wild Flowers.* Chapel Hill: University of North Carolina Press, 1985.

Phillips, Judith. *New Mexico Gardener's Guide.* Franklin, TN: Cool Springs Press, 1998.

———. *Southwestern Landscaping with Native Plants.*

Santa Fe: Museum of New Mexico Press, 1987.
Proctor, John, and Susan Proctor. *Color in Plants and Flowers*. New York: Everest House, 1978.
River Oaks Garden Club. *A Garden Book for Houston and the Gulf Coast*. Houston: Pacesetter Press, 1975.
Rodale, J. I., ed. *Getting the Bugs out of Organic Gardening*. Emmaus, PA: Rodale Press, 1973.
———. *The Organic Way to Mulching*. Emmaus, PA: Rodale Press, 1972.
Ross, Gary Noel. *Gardening for Butterflies in Louisiana*. Louisiana Department of Wildlife and Fisheries. Baker, LA: Baker Printing, 1994.
Sedenko, Jerry. *The Butterfly Garden: Creating Beautiful Gardens to Attract Butterflies*. New York: Villard Books (Random House), 1991.
Smyser, Carol A. *Nature's Design*. Emmaus, PA: Rodale Press, 1982.
Sperry, Neil. *Neil Sperry's Complete Guide to Texas Gardening*. 2nd ed., rev. Dallas: Taylor Publishing, 1991.
Stewart, Amy. *The Earth Moved*. Chapel Hill, NC: Algonquin Books of Chapel Hill, 2004.
Sullivan, Gene A., and Richard H. Daley. *Directory to Resources on Wildflower Propagation*. National Council of State Garden Clubs. St. Louis: Missouri Botanical Garden, 1981.
Taylor, Kathryn S., and Stephen F. Hamblin. *Handbook of Wild Flower Cultivation*. New York: Macmillan, 1963.
Tekulsky, Mathew. *The Butterfly Garden*. Boston: Harvard Common Press, 1985.
Tenebaum, Frances. *Gardening with Wild Flowers*. New York: Charles Scribner's Sons, 1973.
Wasowski, Sally, and Julie E. Ryan. *Landscaping with Native Texas Plants*. Austin: Texas Monthly Press, 1985.
Wasowski, Sally, and Andy Wasowski. *Gardening with Native Plants of the South*. Dallas: Taylor Publishing, 1994.
———. *Texas Native Gardens*. Houston: Gulf Publishing, 1997.
Wasowski, Sally, with Andy Wasowski. *Native Gardens for Dry Climates*. New York: Clarkson Potter Publishers, 1995.
———. *Native Texas Plants: Landscaping Region by Region*. Austin: Texas Monthly Press, 1988.
Webster, Bob. *The South Texas Garden Book*. San Antonio: Corona Publishing, 1980.
Welch, William C. *Perennial Garden Color*. Dallas: Taylor Publishing, 1989.
Wildseed. *A Grower's Guide to Wildflowers*. Houston: Wildseed, 1988.
Wilson, Jim. *Landscaping with Wildflowers*. New York: Houghton Mifflin, 1992.
Wilson, William H. W. *Landscaping with Wildflowers and Native Plants*. San Francisco: Ortho Books, 1984.
Xerces Society in association with the Smithsonian Institution. *Butterfly Gardening—Creating Summer Magic in Your Garden*. San Francisco: Sierra Club Books in association with the National Wildlife Federation, 1990.

Plant and Wildflower Guides

Ajilvsgi, Geyata. *Wild Flowers of the Big Thicket, East Texas, and Western Louisiana*. College Station: Texas A&M University Press, 1979.
———. *Wildflowers of Texas*. Fredericksburg, TX: Shearer Publishing, 1984.
———. *Wildflowers of Texas*. Rev. ed. Fredericksburg, TX: Shearer Publishing, 2002.
Arzeni, Charles B., and Terri M. Simon. *Plants of Mexico*. Charleston: Eastern Illinois University, 1974.
Brown, C. A. *Wildflowers of Louisiana and Adjoining States*. Baton Rouge: Louisiana State University Press, 1973.
Cheatham, Scooter, Marshall C. Johnson, and Lynn Marshall. *Useful Wild Plants of Texas, the Southeastern and Southwestern United States, the Southern Plains, and Northern Mexico*. Vols. 1–2. Austin: Useful Wild Plants, 2000.
Correll, D. S., and M. C. Johnston. *Manual of the Vascular Plants of Texas*. Renner: Texas Research Foundation, 1970.
Cox, Paul, and Patty Leslie. *Texas Trees: A Friendly Guide*. San Antonio: Corona Publishing, 1988.
Dean, Blanche, Amy Mason, and Joab Thomas. *Wildflowers of Alabama and Adjoining States*. Tuscaloosa: University of Alabama Press, 1973.
Diggs, George M., Jr., Barney L. Lipscomb, and Robert J. O'Kennon. *Shinners and Mahler's Illustrated Flora of North Central Texas*. Dallas: Austin College Center for Environmental Studies and the Botanical Research Institute of Texas, 1999.
Enquist, Marshall. *Wildflowers of the Texas Hill Country*. Austin: Lone Star Botanical, 1987.
Everitt, James H., and D. Lynn Drawe. *Trees, Shrubs and Cacti of South Texas*. Lubbock: Texas Tech University Press, 1993.

Fleetwood, Raymond J. *Plants of Santa Ana National Wildlife Refuge, Hidalgo County, Texas*. Alamo, TX: US Department of the Interior, Fish and Wildlife Service, 1973.

Gould, Frank W. *The Grasses of Texas*. College Station: Texas A&M University Press, 1975.

Hatch, Stephen L., Kancheepuram N. Gandhi, and Larry E. Brown. *Checklist of the Vascular Plants of Texas*. College Station: Texas A&M University, Texas Agricultural Experiment Station, 1990.

Jones, Fred B. *Flora of the Texas Coastal Bend*. Sinton, TX: Welder Wildlife Foundation, 1975.

Jones, Stanley D., Joseph K. Wipff, and Paul Montgomery. *Vascular Plants of Texas: A Comprehensive Checklist including Synonymy, Bibliography, and Index*. Austin: University of Texas Press, 1997.

Kartesz, John T., and Rosemarie Kartesz. *A Synonymized Checklist of the Vascular Flora of the United States, Canada, and Greenland*. Chapel Hill: University of North Carolina Press, 1980.

Kirkpatrick, Zoe Merriman. *Wildflowers of the Western Plains*. Austin: University of Texas Press, 1992.

Loflin, Brian, and Shirley Loflin. *Grasses of the Texas Hill Country*. College Station: Texas A&M University Press, 2006.

Lonard, Robert L. *Guide to Grasses of the Lower Rio Grande Valley, Texas*. Edinburg: University of Texas–Pan American Press, 1993.

Loughmiller, Campbell, and Lynn Loughmiller. *Texas Wildflowers*. Austin: University of Texas Press, 1984.

Lynch, Brother Daniel. *Native and Naturalized Woody Plants of Austin & the Hill Country*. Austin: Acorn Press, 1981.

———. *Plants of Austin, Texas*. Austin: St. Edwards University, 1974.

Mahler, William F. *Flora of Taylor County, Texas*. Dallas: Southern Methodist University, 1973.

———. *Keys to the Vascular Plants of the Black Gap Wildlife Management Area, Brewster County, Texas*. Dallas: Privately published by William F. Mahler, 1973.

———. *Shinners' Manual of the North Central Texas Flora*. Fort Worth: Botanical Research Institute of Dallas, 1988.

Martin, William C., and Charles R. Hutchins. *Spring Wild Flowers of New Mexico*. Albuquerque: University of New Mexico Press, 1984.

Mason, Charles T., Jr., and Patricia B. Mason. *A Handbook of Mexican Roadside Flora*. Tucson: University of Arizona Press, 1987.

McCoy, Doyle. *Roadside Trees and Shrubs of Oklahoma*. Norman: University of Oklahoma Press, 1981.

McDougall, W. B., and Omer E. Sperry. *Plants of Big Bend National Park*. Washington, DC: US Government Printing Office, 1951.

Meier, Leo, and Jan Reid. *Texas Wildflowers*. New York: News America Publishing; Sydney, Australia: Weldon Owen Publishing, 1989.

Midgley, Jan W. *Louisiana Wildflowers*. Raleigh, NC: Sweet Water Press, 1999.

National Wildflower Research Center. *Wildflower Handbook*. Austin: Texas Monthly Press, 1989.

Nieland, Lashara J., and Willa F. Finley. *Lone Star Wildflowers*. Lubbock: Texas Tech University Press, 2009.

Niering, William A., and Nancy C. Olmstead. *The Audubon Society Field Guide to North American Wildflowers—Eastern Region*. New York: Alfred A. Knopf, 1979.

Nixon, Elray. *Trees, Shrubs and Woody Vines of East Texas*. Nacogdoches, TX: Bruce Lyndon Cunningham Productions, 1985.

Parks, H. B. *Valuable Plants Native to Texas*. Bulletin No. 551. College Station: Texas Agricultural Experiment Station, 1937.

Pellett, Frank C. *American Honey Plants*. Hamilton, IL: American Bee Journal, 1930.

Pesmen, Walter M. *Flora Mexicana*. Globe, AZ: Dale S. King, 1962.

Peterson, Charles D., and Larry E. Brown. *Vascular Flora of the Little Thicket Nature Sanctuary, San Jacinto County, Texas*. Houston: Brunswick Press, 1983.

Powell, A. Michael. *Trees and Shrubs of the Trans-Pecos and Adjacent Areas*. Austin: University of Texas Press, 1998.

Rechenthin, C. A. *Native Flowers of Texas*. Temple, TX: US Department of Agriculture, Soil Conservation Service, 1972.

Rickett, H. W. *Wild Flowers of the United States*. Vol. 3, *Texas*. New York: McGraw-Hill, 1969.

Sanborn, C. E. *Texas Honey Plants*. Bulletin No. 102. College Station: Texas Agricultural Experiment Station, 1908.

Simpson, Benny J. *A Field Guide to Texas Trees*. Austin: Texas Monthly Press, 1988.

Spellenberg, Richard. *The Audubon Society Field Guide to North American Wildflowers—Western Region*.

New York: Alfred A. Knopf, 1979.
Stanley, Paul C. *Trees and Shrubs of Mexico*. Washington, DC: Contributions, US National Herbarium, US Government Printing Office, 1920–1926.
Stevens, William Chase. *Kansas Wild Flowers*. Lawrence: University Press of Kansas, 1948.
Taylor, R. John, and Constance E. S. Taylor. *An Annotated List of the Ferns, Fern Allies, Gymnosperms, and Flowering Plants of Oklahoma*. Durant: Southeastern Oklahoma State University, 1991.
Texas Department of Agriculture. *Texas Native Tree and Plant Directory*. Austin: Texas Department of Agriculture, 1986.
Thomas, R. Dale, and Charles M. Allen. "A Preliminary Checklist of the Dicotyledons of Louisiana." In *Contributions of the Herbarium of Northeast Louisiana University*, no. 3. Monroe: Northeast Louisiana University, Department of Biology, 1982.
Turner, B. L., Holly Nichols, Geoffrey Denny, and Oded Doron. *Atlas of the Vascular Plants of Texas*. Vols. 1–2. Fort Worth: BRIT Press, 2003.
Tveten, J., and G. Tveten. *Wildflowers of Houston*. Houston: Rice University Press, 1993.
United States Department of Agriculture. *Common Weeds of the United States*. New York: Dover, 1971.
———. *Seeds of Woody Plants in the United States*. Forest Service, US Department of Agriculture, Agriculture Handbook No. 450. Washington, DC: US Government Printing Office, 1974.
Vines, Robert A. *Trees, Shrubs and Woody Vines of the Southwest*. Austin: University of Texas Press, 1974.
Warnock, Barton H. *Wildflowers of the Big Bend Country, Texas*. Alpine, TX: Sul Ross State University, 1970.
———. *Wildflowers of the Davis Mountains and Marathon Basin, Texas*. Alpine, TX: Sul Ross State University, 1977.
———. *Wildflowers of the Guadalupe Mountains and the Sand Dune Country, Texas*. Alpine, TX: Sul Ross State University, 1974.
Whitehouse, Eula. *Common Fall Flowers of the Coastal Bend of Texas*. Sinton, TX: Rob and Bessie Welder Wildlife Foundation, 1962.

Photography

Angel, Heather. *Photographing Nature: Insects*. Hertfordshire, UK: Fountain Press, 1975.
Bauer, Erwin, and Peggy Bauer. *Photographing Wild Texas*. Austin: University of Texas Press, 1985.
Blacklock, Craig, and Nadine Blacklock. *Photographing Wildflowers*. Minneapolis, MN: Voyageur Press, 1987.
Gerlach, John, and Barbara Gerlach. *Digital Nature Photography: The Art and the Science*. Waltham, MA: Focal Press, 2007.
Miller, George Oxford. *Texas Photo Safaris*. Austin: Texas Monthly Press, 1986.
Natural Image [quarterly newsletter]. Published by George Lepp & Assoc., P.O. Box 6240, Los Osos, CA 99652.
O'Toole, Christopher, and Ken Preston-Mafham. *Insects in Camera*. Oxford: Oxford University Press, 1985.
Rotenberg, Nancy, and Michael Lustbader. *How to Photograph Close-ups in Nature*. Mechanicsburg, PA: Stackpole Books, 1999.
Shaw, John. *Closeups in Nature*. New York: American Photographic Book Publishing, 1987.
———. *The Nature Photographer's Complete Guide to Professional Field Techniques*. New York: American Photographic Book Publishing, 1984.
West, Larry, with Julie Ridl. *How to Photograph Insects and Spiders*. Mechanicsburg, PA: Stackpole Books, 1994.

Index

Note: Page references in *italics* refer to photographs and figures. Page references in **boldface** refer to tables.